Fundamentals of Finite Mathematics

James Shaedel
Lynn Jay Wolfmeyer

Nelson-Hall
Chicago

Many people deserve special recognition for the help they have given during the preparation of this book. Thanks to all those who reviewed the manuscript and made numerous helpful suggestions. Special thanks to Dr. Charles Oprian, Dr. A. Jerry Shryock, and Dr. Arnold Wendt who test-taught a portion of the material. Their comments have been greatly appreciated. Special thanks also go to Dr. Edwin Sims for his suggestions regarding some of the exercises, and to Linda McDonald and Mary Wolfmeyer for typing the manuscript.

Finally, we wish to express our heartfelt gratitude to our families. Without their patience and support this project would not have been possible.

LIBRARY OF CONGRESS CATALOGING IN PUBLICATION DATA

Schaedel, James.
Fundamentals of finite mathematics.

Includes index.
1. Mathematics—1961- I. Wolfmeyer, Lynn.
II. Title.
QA39.2.S33 1985 510 84-25394
ISBN 0-8304-1105-4

Manufactured in the United States of America

10 9 8 7 6 5 4 3 2 1

The paper in this book is pH neutral (acid-free).

Cover painting and chapter opening details: *New York City* by Ralph Fasanella

Contents

6 Probability 326

7 Descriptive Statistics 396

8 Probability Distributions 448

9 Markov Chains 482

Preface

Finite mathematics, unlike algebra, calculus, geometry, or statistics, is not well defined. In fact, it is a variable set of topics which has been evolving, at an introductory level, into a practical applications course. Such topics as operations research, information processing, quantitative information sciences, and optimization theory are within the grasp of students who do not have extensive mathematical backgrounds. By exposing students to the type of finite mathematics used to solve many real-world applied problems, we hope that students will recognize problem types that arise later in their careers and, recognizing the problem as one which has a solution, will seek that solution.

In this text we have assumed a background of two years of high school algebra or the equivalent. We have assumed no prior exposure to the book's main concepts or their application to real-world problems. Our chief aim has been to meet the wide range of needs reflected by different syllabi found across the nation.

Finite mathematics has been heavily influenced by computers because real-world, business-based problem-solving techniques have been influenced by computers. Within the next decade data processing systems will be in a large percentage of American homes and in nearly every American business. Our schools are already beginning to reflect these changes.

This book is not the cutting edge of a new wave of untried mathematical tricks. The material is easy to teach and easy to learn. We have paid much attention to the arithmetic of solving the problems, and the success obtained from successfully solving the problems is a tremendous motivator for the student.

The book itself is divided into five main topics, not mutually separate: matrix algebra, linear programming, applied topics, probability and statistics, and mathematics of finance. A short description of each of these topics follows.

Matrix algebra includes chapter 1 (where matrices are introduced) and chapters 2, 3, 9, and 10 (where matrix ideas are used in problem-solving techniques).

Linear programming: chapter 3 introduces the concept of linear programming. Ideas related to linear programming can be found in chapters 4 and 10.

Applied topics are a theme throughout this text. Special applications are included as chapters 4, 7, 8, 9, 10, and 11.

Probability and statistics as a broad field are introduced in chapters 6, 7, and 8. Chapter 5 helps pave the way for the study of probability.

Mathematics of finance is somewhat unrelated to the rest of the topics: it is, however, a part of the syllabus of some finite math courses and is therefore included here.

To cover all of this material would require about one full year. Several sequencing plans are possible for a one term (quarter or semester) course. The following diagram illustrates the chapter dependence. For example, material in chapters 1 and 6 is prerequisite to understanding the material in chapter 9.

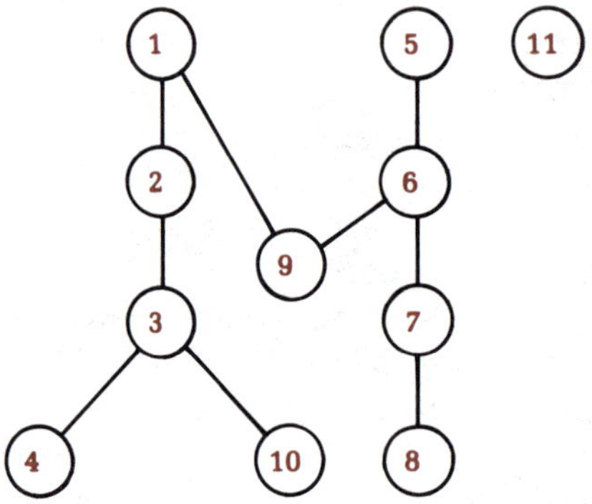

The following are some suggestions for a one term course:

Chapters 1–6 (time = 1 semester)
Chapters 5, 6, 1, 2, 3, and 4 (time = 1 semester)
Chapters 5, 6, 7, 8, and 11 (time = 1 semester)
Chapters 1, 2, 3, 4, 9, and 10 (time = 1 semester)
Chapters 1–4 (time = 1 quarter)
Chapters 1, 2, 5, 6 (time = 1 quarter)
Chapters 1, 2, 3, and 11 (time = 1 quarter)

Other plans are possible. Usually five or six chapters will fill a semester while four will make up a quarter. Chapters 6 and 11 may take longer to work through than the others.

Special Note to the Student

In order to aid you in your study of this text, we have provided a substantial number of completed examples. In addition, many of the exercises, including all of the review exercises, have answers in the back of the book.

In the Algebra Review, chapter 0, we have also included several self-reviews to provide an instant check on your understanding of the concepts. When you come to a self-review, cover the answer in the left margin and work the problem. Then uncover the answer and check your result. If your answer is incorrect, review the preceding examples to check your understanding of the concepts previously covered.

A student solution manual is available for this book. This manual contains answers to exercises that are also found in the back of the book, with complete solutions for some of the more complex problems.

Throughout the text, special notes, remarks, and cautions have been marked with special symbols (⇩) (✔). These are intended to point out special problems or information. They should be read carefully.

0 Algebra Review

Introduction

The algebraic concepts necessary for working with linear equations are central to the type of mathematical analysis discussed in this book. Many of you have probably covered some of these ideas in previous algebra courses. If so, this chapter will be a quick review of key ideas. If you have not studied this material before, this chapter will provide you with necessary background for studying the topics in this book.

0.1 Linear Equations and Inequalities in One Variable

Two of the key topics studied in this book are systems of linear equations and linear programming. These topics require a knowledge of linear equations and inequalities in one or more variables.

In this section, we will consider linear equations and inequalities in just one variable. In the remaining sections of the chapter we will discuss linear equations and inequalities in two variables.

Linear Equations in One Variable

A **linear equation in one variable** is an equation which has the form $ax + b = c$, where a, b, and c are real numbers, $a \neq 0$, and x is a variable representing a real number.

NOTE: The choice of x to represent the variable is customary. We will follow this convention. In some situations where the variable represents time, t will be used instead of x.

In this text, a, b, and c, the constants in the equation, will be **rational numbers.** Rational numbers are numbers that can be expressed as m/n where m and n are whole numbers (positive or negative) and n cannot be zero.

When we solve a linear equation, we are trying to find all numbers that make the equality true. That is, we want to find all values of x such that $ax + b = c$ is a true statement.

■ Example 1. Solve $x - 2 = 1$ for x.

Solution: Here $a = 1$, $b = -2$, and $c = 1$

Since $3 - 2 = 1$, $x = 3$ is a solution to the linear equation. This is the only solution. ■

Our objective here is to develop a systematic procedure for solving linear equations in one variable. Before doing this, we shall consider some basic properties of equality of real numbers. (Recall that x, a, b, and c all represent real numbers; hence the expression $ax + b$ also represents a real number.) The properties of equality of real numbers can be used to develop a systematic procedure for solving linear equations in one variable.

Property 0.1 Let a, b, c, d be real numbers.

1. If $a = b$, then $a + c = b + c$.
2. If $a = b$, then $ac = bc$.
3. If $a = b$ and $b = c$, then $a = c$.
4. If $a = b$ and $c = d$, then $a + c = b + d$.

Review property 0.1 states that we can add the same (or equal) quantities to both sides of an equality, or multiply both sides of an equality by the same quantity, and still maintain the equality. Part 3 of the review property is called the **transitive property** and allows us to string together several statements to obtain a single equality.

Although no restrictions are placed on the real numbers a, b, c, or d in review property 0.1, we would normally not add or multiply using zero since this would not aid in finding a solution to an equation. Adding zero to both sides of an equation or inequality changes nothing, and multiplying by zero would give $0 = 0$ which is certainly true, but not enlightening.

We now consider several examples where property 0.1 is used to solve a linear equation.

■ **Example 2.** Solve $2x + 3 = 5$ for x.

Solution:

$2x + 3 = 5$	Given.
$2x + 3 + (-3) = 5 + (-3)$	Add -3 to both sides.
$2x = 2$	$3 + (-3) = 0$ and $5 + (-3) = 2$.
$\frac{1}{2} \cdot 2x = \frac{1}{2} \cdot 2$	Multiply both sides by $\frac{1}{2}$.
$x = 1$	$\frac{1}{2} \cdot 2x = x$ and $\frac{1}{2} \cdot 2 = 1$.

The solution is $x = 1$. This is the only solution. ■

■ **Example 3.** Solve $3x + 2 = 7$ for x.

Solution:

$3x + 2 = 7$	Given.
$3x = 5$	Add -2 to both sides.
$x = \frac{5}{3}$	Multiply both sides by $\frac{1}{3}$.

The solution is $x = \frac{5}{3}$. ■

Self-Review 1

(Cover the answers in the left margin until you have completed the solution.)

a. Solve $3x + 7 = 3$ for x.
b. Solve $8t - 5 = 7$ for t.

Answers:
a. $x = -\frac{4}{3}$
b. $x = \frac{3}{2}$

Sometimes the equation we encounter is not written in the exact form given earlier; the equation may look like $\frac{1}{x} + 5 = 6$ or $2x - 3 = 5x + 4$. Many such equations can still be solved using the techniques which apply to the standard form.

■ Example 4. Solve $2x - 3 = 5x + 4$ for x.

Solution:

$2x - 3 = 5x + 4$	Given.
$-3x - 3 = 4$	Add $-5x$ to both sides.
$-3x = 7$	Add 3 to both sides.
$x = -\frac{7}{3}$	Multiply both sides by $-\frac{1}{3}$.

The solution is $x = -\frac{7}{3}$. ■

■ Example 5. Solve $2x - 3 = 2x + 4$ for x.

Solution:

$2x - 3 = 2x + 4$	Given.
$-3 = 4$	Add $-2x$ to both sides.

The statement $(-3 = 4)$ is never true, therefore the original equation has no solution. ■

Equations such as $\frac{1}{x} + 5 = 6$ must be handled with care. We want to be very careful not to multiply by zero since we would get $0 = 0$, which is true regardless of what x is. We will comment further on this after looking at the next example.

■ Example 6. Solve $\frac{1}{x} + 5 = 6$ for x.

Solution:

$\frac{1}{x} + 5 = 6$	Given.
$\frac{1}{x} = 1$	Add -5 to both sides.
$1 = x$	Multiply both sides by x.

The solution is $x = 1$. ■

Note that in example 6 we multiplied by x, the variable. If the value of x had turned out to be zero, we would have had a problem. That is, if

$x = 0$, $\frac{1}{x}$ is not defined. Then there would have been no real solution, even though the solution *process* would have given one.

It is a good idea to check your solution to be sure it satisfies the original equation and that all terms in the original equation are defined. The check is done by using the value obtained in the solution process for x in the original equation.

■ Example 7. Solve $\frac{3}{x} - 5 = 8$ for x.

Solution:

$\frac{3}{x} - 5 = 8$	Given.
$\frac{3}{x} = 13$	Add 5 to both sides.
$3 = 13x$	Multiply both sides by x.
$\frac{3}{13} = x$	Multiply both sides by $\frac{1}{13}$.

Therefore, the solution is $x = \frac{3}{13}$.

To check, we substitute $\frac{3}{13}$ for x in the original equation.

$$\begin{aligned}\frac{3}{\frac{3}{13}} - 5 &= 3 \cdot \frac{13}{3} - 5 \\ &= 13 - 5 \\ &= 8.\end{aligned}$$

The solution checks. ■

Self-Review 2

Solve $\frac{3}{t - 2} + 5 = 2$ for t.

$t = 1$

■ Example 8. Solve $\frac{1}{y - 2} + \frac{2}{y - 3} = \frac{1}{y - 2}$ for y.

Solution:

$$\frac{1}{y-2} + \frac{2}{y-3} = \frac{1}{y-2} \qquad \text{Given.}$$

$$1 + \frac{2(y-2)}{y-3} = 1 \qquad \text{Multiply both sides by } (y-2).$$

$$y - 3 + 2(y-2) = y - 3 \qquad \text{Multiply both sides by } (y-3).$$

$$y - 3 + 2y - 4 = y - 3 \qquad \text{Simplify } 2(y-2).$$

$$3y - 7 = y - 3 \qquad \text{Combine terms.}$$

$$2y - 7 = -3 \qquad \text{Add } -y \text{ to both sides.}$$

$$2y = 4 \qquad \text{Add 7 to both sides.}$$

$$y = 2 \qquad \text{Multiply both sides by } \frac{1}{2}.$$

We obtain the solution $y = 2$. However, if we substitute 2 for y in the original equation, the two terms $\frac{1}{y-2}$ are not defined $(2 - 2 = 0)$. Hence there is no real solution. ■

■ Example 9. The length of a rectangle is 6 meters more than its width. If the perimeter of the rectangle (distance around) is 52 meters, what are the length and width of the rectangle?

Solution: The perimeter of a rectangle is given by $P = 2l + 2w$ where l is the length and w is the width. We let w be the width of the given rectangle. Since the length is 6 meters more than the width, we have $l = w + 6$.

$$\begin{aligned} \text{Then } P &= 2l + 2w \\ &= 2(w + 6) + 2w = 52 \\ \text{Then } 2w + 12 + 2w &= 52 \\ 4w + 12 &= 52 \\ 4w &= 40 \\ w &= 10 \end{aligned}$$

Therefore $l = w + 6 = 10 + 6 = 16$ and the dimensions of the rectangle are 16m and 10m. ■

Linear Inequalities in One Variable

We turn our attention now to **linear inequalities** in one variable. Here we are concerned with statements of the form $ax + b < c$, $ax + b > c$, $ax + b \leq c$, or $ax + b \geq c$.

Suppose we consider a real number line and some number k on the line as in the figure below. The number k separates the line into three distinct sets of values. These sets can be described as the values less than $k (x < k)$, the value k itself $(x = k)$, and the values greater than $k (x > k)$.

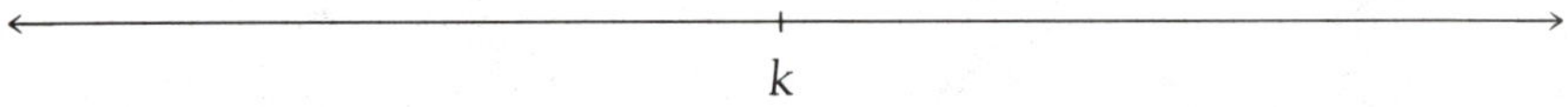

We can also indicate the values that are either greater than or equal to k by writing $x \geq k$.

In the above illustration, the number k is called the **boundary point** of the sets of values for which $x > k$ and $x < k$. The number k is not included when we write either $x > k$ or $x < k$. However, when we write $x \leq k$ or $x \geq k$, k is included.

When working with linear inequalities, we are always concerned about whether or not the boundary value is included. We can represent inequalities in one variable on a line by using some type of notation to indicate whether or not the boundary value is included. The following figures illustrate one way this can be done.

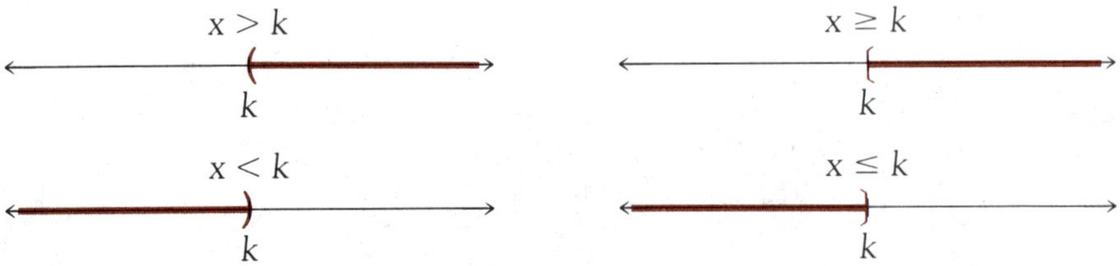

We use a bracket symbol when the boundary point is included and a parenthesis symbol when the boundary point is not included.

■ **Example 10.** Show each of the following on a number line.

a. $x \geq 5$ **c.** $x < 7$
b. $x \leq -3$ **d.** $x > -1$

Solutions: First draw the line for each and then locate a reference point to be 0.

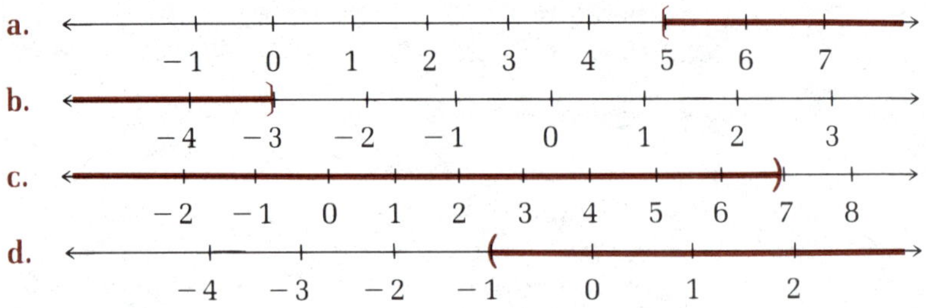

■

NOTE: We are including all real numbers satisfying the inequality in each of these, not just the whole numbers. For part *a*, 7.5, 6.834, $\sqrt{51}$, etc. are included. We identify only whole numbers for reference but other numbers are also included.

We are now ready to consider solving linear inequalities in one variable. The solution technique is based on the following properties. These properties are stated only for "less than" ($<$) but they are also true for the other order relations ($>$, $\leq$, $\geq$).

Property 0.2 Let a, b, c, d be real numbers.

1. If $a < b$ and $b < c$, then $a < c$.
2. If $a < b$, then $a + c < b + c$.
3. If $a < b$ and $c < d$, then $a + c < b + d$.
4. If $a < b$ and $c > 0$, then $a \cdot c < b \cdot c$.
5. If $a < b$ and $c < 0$, then $a \cdot c > b \cdot c$.

It is especially important to look carefully at parts 4 and 5. The other parts are the same as for equality, but there is a difference when multiplying on both sides of an inequality.

If both sides of an inequality are multiplied by a positive number ($c > 0$), the inequality remains the same (part 4). When both sides of an inequality are multiplied by a negative number ($c < 0$) however, the inequality sign must be reversed. The following illustration will help show why this is true.

We know that $3 < 7$. Suppose we multiply both sides of this inequality by 2. Then $2 \cdot 3 = 6$ and $2 \cdot 7 = 14$ and $6 < 14$. If, however, we multiply both sides by -2, we have $(-2)3 = -6$ and $(-2)7 = -14$ and $-14 < -6$ or $-6 > -14$. The order relation sign is reversed when we multiply by a negative number on both sides.

The following examples will illustrate the techniques involved in solving a linear inequality in one variable. We have also included a graph of the solution.

■ Example 11. Solve $2x - 3 > 5$ for x.

Solution:

$2x - 3 > 5$	Given.
$2x > 8$	Add 3 to both sides.
$x > 4$	Multiply both sides by $\frac{1}{2}$.

The solution consists of all values greater than 4 as shown on the graph.

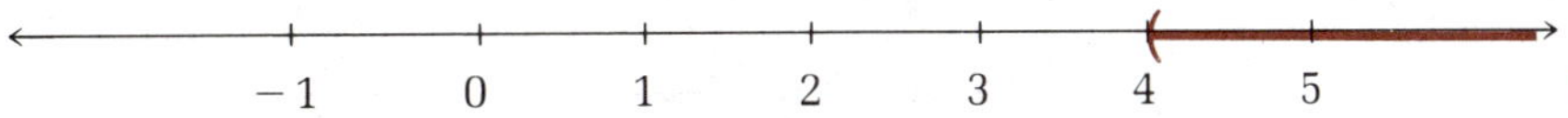

■

Self-Review 3

Solve each of the following and sketch the graph of the solution on a number line.

a. $3x - 2 < 7$.
b. $3t + 4 > 2$.

a. $x < 3$

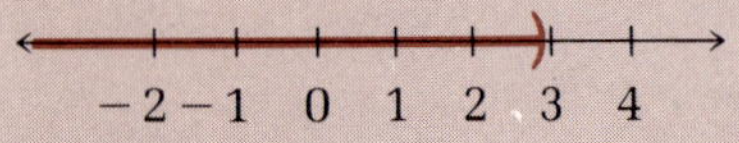

b. $x > -\frac{2}{3}$

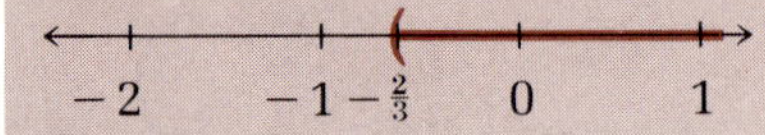

■ **Example 12.** Solve $3y + 7 \le 5$ for y.

Solution:

$3y + 7 \le 5$	Given.
$3y \le -2$	Add -7 to both sides.
$y \le -\frac{2}{3}$	Multiply both sides by $\frac{1}{3}$.

The solution consists of all values less than or equal to $-\frac{2}{3}$.

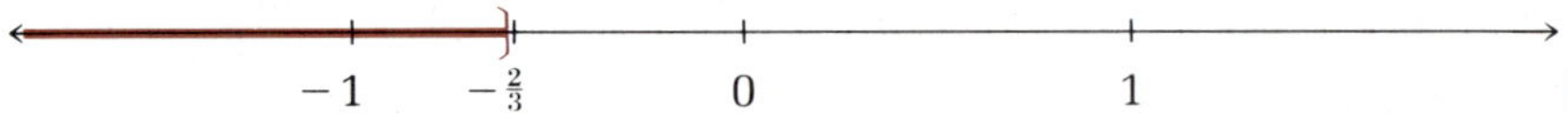

■

Self-Review 4

Solve $4u + 3 \ge 1$ for u and sketch the graph of the solution.

$u \ge -\frac{1}{2}$

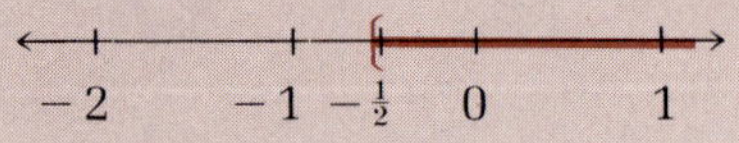

■ **Example 13.** Solve $-2x + 3 > 1$ for x.

Solution:

$-2x + 3 > 1$	Given.
$-2x > -2$	Add -3 to both sides.
$x < 1$	Multiply both sides by $-\frac{1}{2}$. (Inequality is reversed.)

The solution consists of all real numbers less than 1.

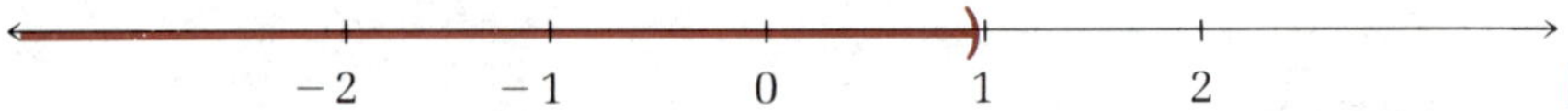

■

Self-Review 5

Solve $-3x + 4 \le 15$ for x and sketch the graph of the solution.

$x \ge -\frac{11}{3}$

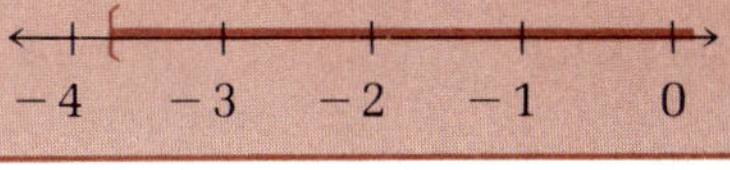

In some of the examples involving linear equations, we multiplied by an expression involving the variable. This can also happen when we work with inequalities. However, when inequalities are involved, we must consider whether the quantity we are multiplying by is positive or negative.

The situation is handled by considering two cases, one with the quantity positive and one with the quantity negative. The following examples will help illustrate.

■ Example 14. Solve $\frac{1}{x} + 1 > 2$ for x.

Solution:

$\frac{1}{x} + 1 > 2$	Given.
$\frac{1}{x} > 1$	Add −1 to both sides.

We now wish to multiply both sides by the variable x. To do so we consider two cases, x positive $(x > 0)$ and x negative $(x < 0)$.

Case 1

$x > 0$	
$\frac{1}{x} > 1$	from first part of solution.
$1 > x$	Multiply both sides by x ($x > 0$, so inequality doesn't change).

Hence, we obtain a solution of $x < 1$, but we also must have $x > 0$ (we assumed this for case 1). The solution is, then, all real numbers greater than 0 and less than 1 or the numbers between 0 and 1.

Case 2

$x < 0$

$\frac{1}{x} > 1$ — from first part of solution.

$1 < x$ — Multiply both sides by x ($x < 0$, so inequality is reversed).

We obtain as a solution all real numbers greater than 1 but less than 0 (we assumed $x < 0$ for case 2). However, there are no real numbers that satisfy both of these conditions. Hence, there is no solution from case 2. The solution is all real numbers between 0 and 1, as obtained from case 1. ■

■ **Example 15.** Solve $\frac{2}{x-3} + 5 < 3$ for x.

Solution:

$\frac{2}{x-3} + 5 < 3$	Given.
$\frac{2}{x-3} < -2$	Add -5 to both sides.

Case 1

$x - 3 > 0$ or $x > 3$

$\frac{2}{x-3} < -2$	from first part.
$2 < -2(x-3)$	Multiply both sides by $x - 3$. ($x - 3 > 0$)
$2 < -2x + 6$	$-2(x - 3) = -2x + 6$.
$-4 < -2x$	Add -6 to both sides.
$2 > x$	Multiply both sides by $-\frac{1}{2}$.

We obtain $x < 2$ and $x > 3$, which is not possible, so there is no solution for case 1.

Case 2

$x - 3 < 0$ or $x < 3$

$\frac{2}{x-3} < -2$	from first part.
$2 > -2(x-3)$	Multiply both sides by $x - 3$. ($x - 3 < 0$)
$2 > -2x + 6$	
$-4 > -2x$	Add -6 to both sides.
$2 < x$	Multiply both sides by $-\frac{1}{2}$.

Here we obtain $x > 2$ and $x < 3$, so the solution for case 2 (and the original problem) is all real numbers between 2 and 3.

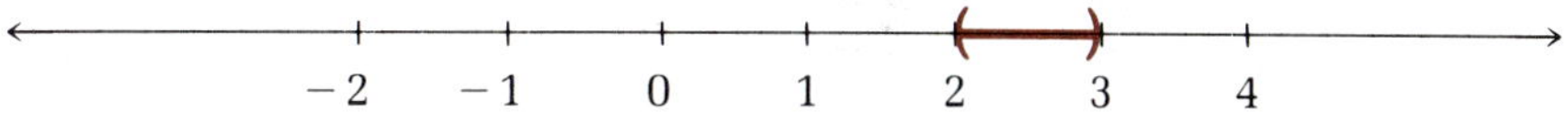

Example 16. Solve $\dfrac{2t + 1}{t + 3} > 1$ for t.

Solution: **Case 1**

$t + 3 > 0$ or $t > -3$

$\dfrac{2t + 1}{t + 3} > 1$	Given.
$2t + 1 > t + 3$	Multiply both sides by $t + 3$. $(t + 3 > 0)$
$t + 1 > 3$	Add $-t$ to both sides.
$t > 2$	Add -1 to both sides.

This gives $t > 2$ and $t > -3$ which are both true for all $t > 2$.

Case 2

$t + 3 < 0$ or $t < -3$

$\dfrac{2t + 1}{t + 3} > 1$	Given.
$2t + 1 < t + 3$	Multiply both sides by $t + 3$. $(t + 3 < 0)$
$t + 1 < 3$	Add $-t$ to both sides.
$t < 2$	Add -1 to both sides.

We obtain $t < 2$ and $t < -3$ which are both true when $t < -3$.

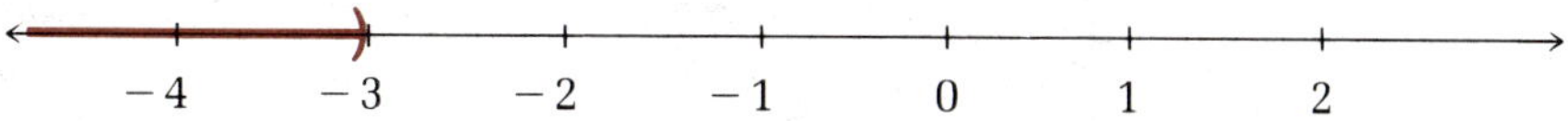

The solution for the original inequality is obtained by using the values that satisfy at least one of the cases. Hence the solution to the original inequality consists of all real numbers that are less than -3 or greater than 2.

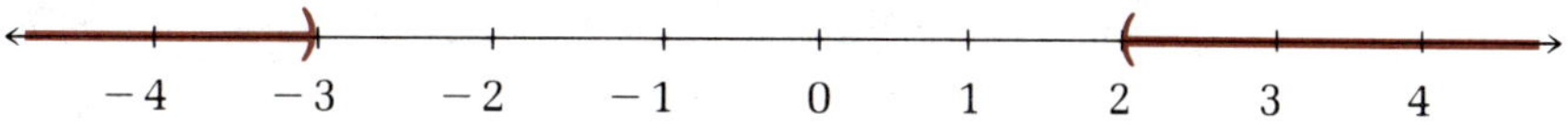

A second method for solving single variable inequalities is to determine the *boundary numbers* for the inequality and test a single number in the interior of each of the regions defined by these boundaries. The rationale is that an inequality does not change, either from greater than to less than, or from less than to greater than, except at these boundary numbers. It follows that when the boundary numbers have divided the number line into regions, every number in any specific region will produce the same kind of inequality. Put another way, if an inequality statement is true for any one number in a region, it will be true for all numbers in that region, and conversely, if any number in a region fails to make the inequality true then every number in that region will fail.

What are boundary numbers? There are two kinds of **boundary numbers:** **1.** numbers which are the solutions to the equation formed by replacing the inequality with an equal sign, and **2.** numbers which produce division by zero. We will re-examine some of the previous examples using this method.

Example 17. Solve $3x - 2 < 7$ for x.

Solution: First find the boundary numbers.

$$3x - 2 = 7$$
$$3x = 9$$
$$x = 3$$

Since there is no division by zero, $x = 3$ is the only boundary number. We locate 3 on a number line and examine, with test numbers, the two regions formed.

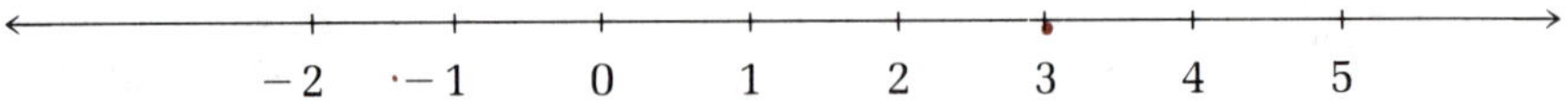

Take any number in the region to the right of 3 and test in the original statement $3x - 2 < 7$. We arbitrarily select 4 as our test number. Since $3(4) - 2 < 7$ is a false statement, we conclude that every number to the right of 3 will also fail.

Next we try a test number to the left of 3, we let $x = 2$ and since $3(2) - 2 < 7$ is a true statement we conclude that every number in this region will work. Our solution then consists of all numbers less than 3.

CAUTION: Test each boundary number itself. Since $3(3) - 2 \not< 7$, 3 is not in the solution.

Example 18. Solve $\dfrac{2}{x - 3} + 5 < 3$ using the boundary number method.

Solution: To find the boundary numbers solve $\frac{2}{x-3} + 5 = 3$, and notice $x = 3$ produces division by zero.

$$\begin{aligned} \frac{2}{x-3} &= -2 \\ 2 &= -2x + 6 \\ 2x &= 4 \\ x &= 2 \end{aligned}$$

These numbers, $x = 2$ and $x = 3$, are the boundary numbers. The boundary numbers divide the line into three regions which we test one at a time.

Take a number larger than 3 and test that region.

Substituting $x = 4$ into $\frac{2}{x-3} + 5 < 3$ we see that

$$\begin{aligned} \frac{2}{4-3} + 5 &< 3 \\ 2 + 5 &< 3 \\ 7 &< 3 \text{ which is a false statement.} \end{aligned}$$

No solution exists in this region. Next select a number between 2 and 3, say $x = 2\frac{1}{2}$ and test that region by again substituting $2\frac{1}{2}$ for x in the example.

$$\begin{aligned} \frac{2}{x-3} + 5 &< 3. \\ \frac{2}{2\frac{1}{2} - 3} + 5 &< 3 \\ \frac{2}{-\frac{1}{2}} + 5 &< 3 \\ -4 + 5 &< 3 \\ 1 &< 3. \text{ This is a true statement.} \end{aligned}$$

We conclude that every number between 2 and 3 will be a solution.

Finally, we test the region less than 2. We select as a test number $x = 0$.

$$\begin{aligned} \frac{2}{0-3} + 5 &< 3 \\ -\frac{2}{3} + 5 &< 3 \\ 4\frac{1}{3} &< 3. \text{ This is a false statement.} \end{aligned}$$

Therefore our solution consists of the numbers between 2 and 3.

In summary when solving inequalities, first locate all boundary numbers by solving the corresponding equation and noting any number that produces division by zero. Test each of the regions defined by these numbers separately. Test the boundary numbers themselves.

Self-Review 6

Consider the inequality

$$\frac{2x - 7}{x - 5} \geq 1.$$

a. What are the boundary numbers?
b. Which boundary numbers, if any, are in the solution?
c. What is the solution?

a. 2,5
b. 2
c. $x \leq 2$ or $x > 5$

Exercise Set 0.1

In exercises 1–30 solve the equation for x.

1. $3x + 5 = 8$
2. $4x - 3 = 5$
3. $2x - 3 = 7$
4. $6x + 1 = 7$
5. $-3x + 5 = 8$
6. $-2x + 3 = 9$
7. $-2x + 3 = 1$
8. $-3x - 2 = 7$
9. $2x + 3 = 8$
10. $3x + 4 = 11$
11. $5x - 7 = 3$
12. $7x - 3 = 15$
13. $-3x + 2 = -4$
14. $-2x + 3 = -7$
15. $-11x + 2 = 4$
16. $-5x + 1 = 8$
17. $3x + 2 = 2x - 3$
18. $4x - 3 = 3x + 2$
19. $2x - 5 = 3x + 5$
20. $4x + 2 = 5x - 3$
21. $-2x + 4 = 3x - 5$
22. $-8x + 7 = 2x - 13$
23. $7x - 3 = 2x - 15$
24. $8x - 3 = 3x - 5$
25. $\dfrac{2}{x + 4} + 4 = 2$
26. $\dfrac{3}{x - 3} - 2 = -5$
27. $\dfrac{3x + 1}{x - 2} + 3 = 8$
28. $\dfrac{2x + 3}{x - 1} + 2 = 6$
29. $\dfrac{3}{x - 1} + \dfrac{2}{x - 2} = \dfrac{3}{x - 1}$
30. $\dfrac{2}{x - 4} + \dfrac{3}{x + 2} = \dfrac{3}{x + 2}$

In exercises 31–56 solve the inequality for x and graph the solution.

31. $2x + 5 < 7$
32. $3x - 2 < 7$
33. $4x - 3 > 5$
34. $5x + 3 > 13$
35. $3x + 4 \leq 11$
36. $2x - 5 \leq 12$
37. $3x - 4 \geq 5$
38. $5x + 4 \geq 6$
39. $2x + 3 > 1$
40. $3x + 7 > 2$
41. $3x + 11 > 4$
42. $5x + 7 \geq 1$
43. $-2x + 3 < 6$
44. $-3x + 5 > 8$
45. $-3x + 12 \geq 5$
46. $-2x + 11 < 4$
47. $-7x + 8 \leq 6$
48. $-3x + 2 \geq 5$
49. $-8x + 3 < 15$
50. $-4x + 15 \leq 8$
51. $\dfrac{3}{x - 2} + 5 < 7$
52. $\dfrac{2}{x + 1} + 4 < 10$
53. $\dfrac{5}{x + 2} - 7 > 5$
54. $\dfrac{4}{x - 1} - 3 > 6$
55. $\dfrac{3x + 5}{x - 2} + 2 > 8$
56. $\dfrac{2x + 3}{x - 1} + 2 > 5$
57. The length of a rectangle is 5 meters more than the width. If the perimeter is 110 meters, what are the length and width?
58. The width of a rectangle is 8 meters less than the length. If the perimeter is 56 meters, what are the length and width?
59. The length of a rectangle is 3 times the width. If the perimeter is 800 yards, what are the length and width?
60. The length of a rectangle is 5 yards more than twice the width. If the perimeter is 154 yards, what are the length and width?

0.2 Linear Equations in Two Variables

In this section, we want to consider some properties of **linear equations in two variables.** We are particularly interested in techniques used to write a linear equation that satisfies certain specified conditions. In chapter 3 this technique will be used to find equations of lines.

It can be shown that the graph of an equation of the form $ax + by + c = 0$, where a and b are not both zero, is a straight line. We can find points on the line by substituting a value for x (or y) and solving for the corresponding value of y (or x). For each value we substitute, we get a point of the form (x,y) on the graph of the line.

■ **Example 19.** Consider the equation $3x + 2y - 6 = 0$.

If we let $x = 0$, we have

$$\begin{aligned} 3 \cdot 0 + 2y - 6 &= 0, \\ \text{or } 2y &= 6, \\ \text{or } y &= 3. \end{aligned}$$

Hence, the point (0,3) is on the line. Letting $x = 2$, we obtain

$$\begin{aligned} 3 \cdot 2 + 2y - 6 &= 0, \\ \text{or } 6 + 2y - 6 &= 0, \\ \text{or } 2y &= 0, \\ \text{and } y &= 0. \end{aligned}$$

The point (2,0) is on the line. We can find other points on the line by choosing additional values for x (or y). The points $(-2,6)$, $\left(-1,\frac{9}{2}\right)$, $(4,-3)$, and $\left(\frac{2}{3},2\right)$ can be obtained by letting x be -2, -1, 4, and $\frac{2}{3}$ respectively, and solving for y. ■

Two distinct points determine a line. To graph a particular line, it is only necessary to find two of its points. Once these are located on the graph, we can draw the line.

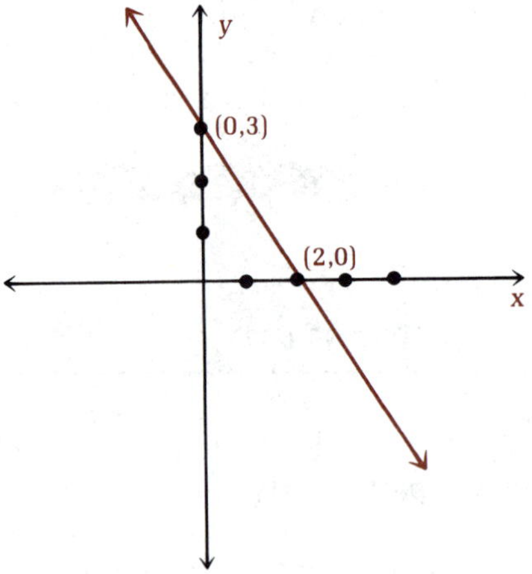

We found previously that the points (0,3) and (2,0) are on the line with equation $3x + 2y - 6 = 0$. To graph this line, we could locate these two points on the graph and draw the straight line through these points.

Consider the equation $ax + by + c = 0$. This is a linear equation in two variables. Assuming b is not 0, we can solve this equation for y in terms of x.

$$ax + by + c = 0$$
$$by = -ax - c$$
$$y = \frac{-a}{b}x + \frac{-c}{b}, \; b \neq 0$$

The two fractions $-\frac{a}{b}$ and $-\frac{c}{b}$ are special values for the line. The fraction $-\frac{a}{b}$ is called the **slope** of the line. Slope gives an indication of how rapidly the line "rises" or "falls." That is, for each one unit increase in x, y changes by $-\frac{a}{b}$ units. The letter m is frequently used to represent slope $\left(m = -\frac{a}{b}\right)$. If m is positive, y increases by m units as x is increased by one unit. If m is negative, y decreases by m units as x is increased by one unit.

For example, consider the equation $2x - y + 5 = 0$. Solving this equation for y in terms of x, we obtain the equation $y = 2x + 5$. The following table gives some points on this line. The points were obtained by picking values of x and finding the corresponding y value.

x	−3	−2	−1	$-\frac{1}{2}$	0	$\frac{1}{2}$	1	2	3	5
y	−1	1	3	4	5	6	7	9	11	15

A definite pattern appears as we read across the table. Each time we increase x by 1 unit, y increases by 2 units (the coefficient of x in the equation $y = 2x + 5$). If x increases by $\frac{1}{2}$ unit $\left(\text{e.g., from 0 to } \frac{1}{2}\right)$, y increases by 1 unit and when x increases by 2 units (e.g., from 3 to 5) y increases by 4 units. However, the ratio of the change in y to the change in x is still 2 in both cases. That is, $\frac{1}{\frac{1}{2}} = 2$ and $\frac{4}{2} = 2$.

Regardless of what two points we choose on the line, this ratio of the change in y (rise) to the change in x (run) is the same value. This ratio of the "rise" to the "run" is called the slope of the line. If the equation is solved for y in terms of x, the ratio (slope) will be the coefficient of x.

■ Example 20. Consider the equation $3x + y - 6 = 0$. Solving for y in terms of x gives $y = -3x + 6$. Some of the points on this line are given in the following table.

x	−3	−2	−1	0	$\frac{1}{3}$	1	2	3	5
y	15	12	9	6	5	3	0	−3	−9

For this line, each time x is increased by 1 unit, y is *decreased* by 3 units. Again, picking any two points on the line, the ratio of the change in y (rise) to the change in x (run) is the constant value -3 (the minus sign indicates a decrease in y for an increase in x). The slope of this line is -3 where the 3 indicates how much change and the minus sign indicates that y decreases as x increases. Notice again that -3 is the coefficient of x in the equation $y = -3x + 6$. ■

The slope of a line is the ratio of the "rise" to the "run" for any non-vertical line. A vertical line has an equation of the form $x = c$ where c is a constant. The slope of a vertical line is *undefined*. A horizontal line has an equation of the form $y = c$, c is a constant, and has slope zero.

If the equation of a line is written in the form $y = mx + b$, then m is the slope of the line. That is,

$$m = \frac{\text{rise}}{\text{run}} = \frac{\text{change in } y}{\text{change in } x}.$$

NOTE: The value of b in the equation $y = mx + b$ is not the same as the value of b in the equation $ax + by + c = 0$. The value of b in the equation $y = mx + b$ is the same as the value of $\frac{-c}{b}$ in the equation $ax + by + c = 0$. Since all of the coefficients will normally be known values in either equation, the confusion should be minimal.

We return now to the equation $ax + by + c = 0$. When we solved for y in terms of x to obtain $y = \frac{-a}{b}x + \frac{-c}{b}$, $b \neq 0$, we indicated that $\frac{-c}{b}$ was also a special value.

If $x = 0$, then $y = -\frac{c}{b}$. That is, the point $\left(0, -\frac{c}{b}\right)$ is on the graph of the line. The value $-\frac{c}{b}$ is called the **y-intercept** of the line.

When the equation is in the form $y = mx + b$, the y-intercept is b. That is, the line crosses the y-axis at the point (0,b). For the line with equation $2x - y + 5 = 0$ ($y = 2x + 5$) discussed earlier, the y-intercept is 5. The y-intercept of the line with equation $3x + y - 6 = 0$ ($y = -3x + 6$) is 6.

■ Example 21. Find the slope and y-intercept of each line and sketch the graph of the line.

a. $2x + y - 3 = 0$.

b. $x - y + 2 = 0$.

Solutions: **a.**

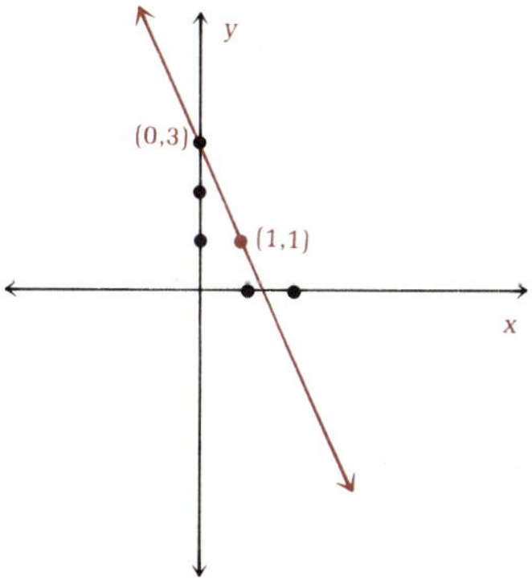

$2x + y - 3 = 0$. Solve for y in terms of x. Then $y = -2x + 3$. Hence the slope is -2, and the y-intercept is 3. If $x = 0$, then $y = 3$, and if $x = 1$, then $y = 1$. So (0,3) and (1,1) are on the graph. We see that the line "falls" as x is increased. This is indicated by the negative slope ($m = -2$).

b.

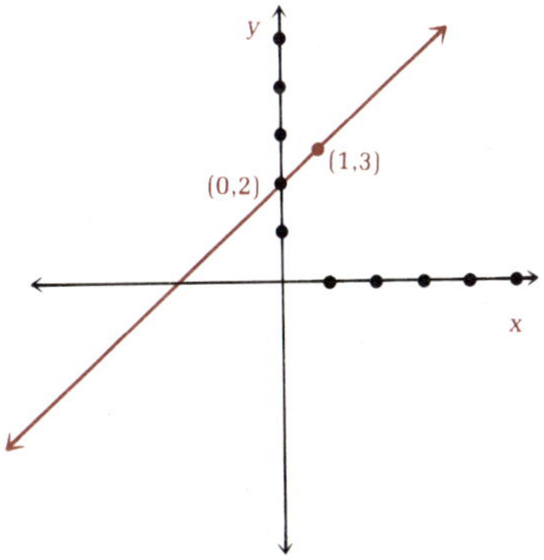

$x - y + 2 = 0$. Solving for y in terms of x, $-y = -x - 2$ and, hence, $y = x + 2$. The slope is 1, and the y-intercept is 2. The points (0,2) and (1,3) are on the graph. This line "rises" to the right, as indicated by its positive slope ($m = 1$). ■

Self-Review 7

a. $y = \frac{2}{3}x - \frac{5}{3}$

b. $m = \frac{2}{3}$

c. $b = -\frac{5}{3}$

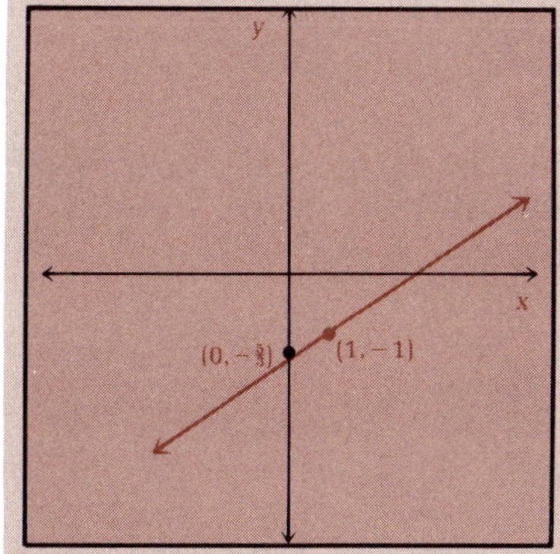

Consider the line with equation $2x - 3y - 5 = 0$.

a. Solve for y in terms of x.
b. What is the slope of the line?
c. What is the y-intercept of the line?
d. Sketch a graph of the line.

The slope of a line is the ratio of the change in the y value to the change in the x value. For any given line, this value (denoted m) is constant. If (x_1,y_1) and (x_2,y_2) are any two distinct points on a line, then the slope is given by $\boxed{m = \dfrac{y_2 - y_1}{x_2 - x_1}, \text{ providing } x_2 \neq x_1.}$

If $x_2 = x_1$, then the line is a vertical line and the slope is undefined. The equation of a vertical line is of the form $x = c$, where c is some constant. For example, $x = 3$ is a vertical line. Two points on this line are (3,0) and (3,1). A horizontal line has an equation of the form $y = k$, with k a constant, and has slope equal to zero. The equation $y = 2$ is a horizontal line. Two points on this line are (0,2) and (1,2).

Writing Equations of Lines

When we write an equation such as $ax + by + c = 0$, $b \neq 0$, in the form $y = \left(-\dfrac{a}{b}\right)x - \dfrac{c}{b}$ we are using the **slope-intercept form.** We frequently will write $\boxed{y = mx + b}$ where m is the slope and b is the y-intercept. The "b" in $y = mx + b$ is not the same as the "b" in $ax + by + c = 0$.

Given an equation of a line, we can find the slope and y-intercept by putting the equation in slope-intercept form. By substituting values for one of the variables and solving for the remaining variable, we can find two points on the line and draw the graph.

Conversely, given two points or the slope and y-intercept, we can write an equation of the line. We will consider three methods of writing an equation of a line. The particular method used is normally dependent on the available information. The three forms for writing an equation of a line will be given as properties, without proof.

Property 0.3 ***Two-Point Form.*** If A and B are two points with coordinates (x_1,y_1) and (x_2,y_2) respectively, then an equation of the line through A and B is given by

$$\frac{y - y_1}{x - x_1} = \frac{y_2 - y_1}{x_2 - x_1} \text{ or } y - y_1 = \frac{y_2 - y_1}{x_2 - x_1}(x - x_1) \text{ if } x_2 \neq x_1.$$

In this equation, x_1, x_2, y_1, y_2 are the given constant values that are the coordinates of the points A and B and x and y are variables.

■ Example 22. Write an equation of the line through each of the following pairs of points.

a. $A(2,3)$, $B(-2,2)$

b. $A(1,3)$, $B(1,-3)$

Solutions: **a.** Let $(x_1,y_1) = (2,3)$ and $(x_2,y_2) = (-2,2)$. Then $\frac{y-3}{x-2} = \frac{2-3}{-2-2}$ is an equation of the line. This can be simplified to $\frac{y-3}{x-2} = \frac{-1}{-4}$

$$\text{or } y - 3 = \frac{1}{4}(x-2)$$
$$y - 3 = \frac{1}{4}x - \frac{1}{2}$$
$$y = \frac{1}{4}x + \frac{5}{2}.$$

There are many other forms that could be used for the final form of the equation.

It should be noted that the choice of which point to use for (x_1,y_1) is arbitrary. If we choose $(x_1,y_1) = (-2,2)$ and $(x_2,y_2) = (2,3)$ we have

$$\frac{y-2}{x-(-2)} = \frac{3-2}{2-(-2)}$$
$$\frac{y-2}{x+2} = \frac{1}{4}$$
$$y - 2 = \frac{1}{4}(x+2)$$
$$y - 2 = \frac{x}{4} + \frac{1}{2}$$
$$y = \frac{1}{4}x + \frac{5}{2},$$

which is the same equation as before. One important point to remember is that in the ratio $\frac{y_2 - y_1}{x_2 - x_1}$, the values chosen for y_1 and x_1, as well as those chosen for y_2 and x_2, must come from the same points. In addition, in the ratio $\frac{y - y_1}{x - x_1}$, the values x_1 and y_1 must come from the same point.

b. Let $(x_1,y_1) = (1,3)$ and $(x_2,y_2) = (1,-3)$. Since $x_1 = x_2 = 1$, the form does not apply. The line containing these two points is a vertical line that has an equation of the form $x = c$, where c is a constant. Since $x_1 = 1$, an equation of the line is $x = 1$. ■

Property 0.4 ***Point-Slope Form.*** If A is a point with coordinates (x_1,y_1), then an equation of the line through A with slope m is given by the equation $y - y_1 = m(x - x_1)$.

This form can be obtained from the two-point form given in Property 0.3 since the slope is given by $m = \frac{y_2 - y_1}{x_2 - x_1}$.

■ Example 23. Find an equation of the line through $A(-1,3)$ having slope $m = \frac{-1}{3}$.

Solution: Using the point-slope form, we have

$$y - 3 = \frac{-1}{3}(x - (-1))$$

$$y - 3 = \frac{-1}{3}x - \frac{1}{3}$$

$$y = \frac{-1}{3}x + \frac{8}{3}.$$

■

Property 0.5 ***Slope-Intercept Form.*** An equation of the line with slope m and y-intercept b is given by $y = mx + b$.

Since the y-intercept is b, the point $(0,b)$ is on the line. Replacing (x_1,y_1) by $(0,b)$ in the point-slope form gives $y - b = m(x - 0)$ or $y = mx + b$.

■ Example 24. Find an equation of the line with slope $\frac{1}{2}$ and y-intercept -2.

Solution: Using the slope-intercept form with $m = \frac{1}{2}$ and $b = -2$, we have $y = \frac{1}{2}x + (-2)$ or $y = \frac{x}{2} - 2$. ■

The lines that satisfy the conditions of the previous properties are unique. That is, there is exactly one line satisfying the specified conditions. However, the equations can have many forms. The equations $y = \frac{x}{2} - 2$, $2y = x - 4$, $x = 2y + 4$, and $2x - 4y - 8 = 0$ all represent the same line even though they look somewhat different.

Self-Review 8

Write an equation of the line satisfying the given conditions.

a. Through the points $A(-3,-1)$ and $B(3,-2)$.

b. Through the point $A(3,5)$ having slope $m = \frac{2}{3}$.

c. Slope $m = \frac{3}{2}$ and y-intercept $b = -3$.

a. $y = -\frac{1}{6}x - \frac{3}{2}$

b. $y = \frac{2}{3}x + 3$

c. $y = \frac{3}{2}x - 3$

One additional property is sometimes helpful when dealing with lines.

Property 0.6 If l_1 and l_2 are two distinct lines in a plane with nonzero slopes m_1 and m_2 respectively, then

a. l_1 is parallel to l_2 if and only if $m_1 = m_2$.*

b. l_1 is perpendicular to l_2 if and only if $(m_1)(m_2) = -1$ $\left(m_1 = \frac{-1}{m_2} \text{ or } m_2 = \frac{-1}{m_1}\right)$.‡

Note here that if m_1 or m_2 is equal to zero, the corresponding line is horizontal. Any perpendicular to this line is vertical and has undefined slope.

Example 25. Consider lines with equations $2x + y - 5 = 0$ and $x - 2y + 3 = 0$. Rewriting these equations in slope-intercept form we have $y = -2x + 5$ and $y = \frac{1}{2}x + \frac{3}{2}$. The slopes are -2 and $\frac{1}{2}$ and since $(-2)\left(\frac{1}{2}\right) = -1$, the two lines are perpendicular. ■

*Of course, two vertical lines (undefined slopes) are parallel as are two lines both having slope zero (horizontal lines).

‡A vertical line (undefined slope) and a horizontal line (slope zero) are also perpendicular.

■ Example 26. Write equations of lines satisfying the given conditions:

a. through the point $(1, -3)$, parallel to the line with equation $x - 2y + 3 = 0$.

b. through the point $(-1,3)$ perpendicular to the line with equation $3x + y - 2 = 0$.

Solutions:

a. Since the line is to be parallel to $x - 2y + 3 = 0$, we first find the slope of this line:

$x - 2y + 3 = 0, \quad -2y = -x - 3, \quad y = \frac{x}{2} + \frac{3}{2}$, so the slope is $\frac{1}{2}$. The slope of any line parallel to this line will also be $\frac{1}{2}$.

Using $m = \frac{1}{2}$ and $A(1, -3)$ in the point-slope form gives $y - (-3) = \frac{1}{2}(x - 1)$ or $y + 3 = \frac{x}{2} - \frac{1}{2}, \quad y = \frac{x}{2} - \frac{7}{2}$ which is the line we want.

b. The slope of $3x + y - 2 = 0$ can be found by putting the equation in slope-intercept form:

$3x + y - 2 = 0, \quad y = -3x + 2$ so the slope is -3. The slope of any line perpendicular to this line is $m = -\left(\frac{1}{-3}\right) = \frac{1}{3}$. Using $m = \frac{1}{3}$ and $A(-1,3)$ we have $y - 3 = \frac{1}{3}(x - (-1)), \quad y - 3 = \frac{1}{3}(x + 1), \quad y - 3 = \frac{x}{3} + \frac{1}{3}, \quad y = \frac{x}{3} + \frac{10}{3}$. ■

Self-Review 9

Write equations satisfying the given conditions:

a. Through the point $A(2,3)$ parallel to the line with equation $2x + y - 3 = 0$.

b. Through the point $A(3,1)$ perpendicular to the line with equation $2x - 3y + 3 = 0$.

a. $y = -2x + 7$

b. $y = -\frac{3}{2}x + \frac{11}{2}$

Linear Equations in Two Variables: Applications

Linear equations in two variables have many applications. In chapter 2 we will look at some situations involving more than one equation. For now, we look at some simple situations involving one equation.

The concept of a *function* is useful for describing this situation. In simplified terms, a **function** is a rule that assigns to each number in some set of numbers exactly one number in another set, possibly the same set as the first one.

If we consider a linear equation in two variables, written in slope-intercept form as $y = mx + b$, then it is easy to see that for each value we choose for x, we obtain exactly one value for y. That is, this equation represents a function. We sometimes say that y is a function of x and this equation represents a *linear function*.

Several examples of functions can be given.

1. Suppose you work part time, earning \$10 per hour. Your week's pay is a function of the number of hours worked.
2. If you put \$500 in a savings account at 6% interest, the interest you earn is a function of the length of time you leave the money in the account.
3. Your water bill is a function of the number of cubic feet of water you used.
4. The cost of filling your car with gasoline is a function of the amount you put in the car.

In the remaining examples of this section, we will be concerned with writing linear functions (equations in two variables) based on given information.

■ **Example 27.** Gasoline costs \$1.25 per gallon. Write a linear equation expressing cost in terms of the quantity and use this equation to find the cost of 6 gallons and 10 gallons.

Solution: The equation will be of the form $C = cx$ where C is total cost, c is cost per gallon, and x is the number of gallons. The equation is $C = 1.25\ x$.

The cost of 6 gallons is

$$C = 1.25\ (6) = \$7.50.$$

The cost of 10 gallons is

$$C = 1.25\ (10) = \$12.50.$$

■

■ Example 28. A local water company has a minimum charge of \$2.00 per month plus \$.15 per cubic foot for water used. Write an equation expressing the total cost in terms of the amount of water used and find the total cost if the consumption is 50 cubic feet. What is the cost for a consumption of 80 cubic feet? 100 cubic feet?

Solution: Since the minimum charge is \$2.00, when no water is used the cost will be \$2.00. That is (0,2.00) is a point on the graph so the y-intercept is 2.00. The equation will be of the form $y = mx + b$ where y is the total cost, m is the cost per cubic foot, x is the consumption, and b is the minimum cost. The equation is

$$y = 0.15x + 2.00.$$

The cost of 50 cubic feet is

$$\begin{aligned} y &= 0.15(50) + 2.00 \\ &= 7.50 + 2.00 \\ &= \$9.50 \end{aligned}$$

The cost of 80 cubic feet is

$$\begin{aligned} y &= 0.15(80) + 2.00 \\ &= 12.00 + 2.00 \\ &= \$14.00 \end{aligned}$$

The cost of 100 cubic feet is

$$\begin{aligned} y &= 0.15(100) + 2.00 \\ &= 15.00 + 2.00 \\ &= \$17.00 \end{aligned}$$

■

■ Example 29. A toy manufacturer has costs of \$5.00 per unit in the production of one of its dolls. It also has fixed costs of \$500. Fixed costs are costs incurred whether any dolls are produced or not. Write an equation expressing total cost in terms of the number of dolls produced and find the cost of producing 200 dolls.

Solution: Let x be the number of dolls produced and let y be the total cost. The equation will be of the form $y = mx + b$. Since the cost is \$500 when no dolls are produced, the point (0,500) is on the graph so 500 is the y-intercept (b). The equation is

$$y = 5x + 500.$$

The cost of producing 200 dolls is

$$\begin{aligned} y &= 5(200) + 500 \\ &= 1000 + 500 \\ &= \$1500 \end{aligned}$$

■

Self-Review 10

A plumber charges \$20 per hour for work plus a \$15 service call fee. Write an equation that expresses the total cost of service (C) in terms of the time worked (t). What is the total cost for service that required $2\frac{1}{2}$ hours to complete?

$C = 20t + 15$
$C = \$65$

Linear equations are sometimes used for prediction purposes. That is, we assume a relationship is linear, or at least approximately linear. If we have enough information to write a linear equation, then we can use this equation to predict one of the variables based on known values of the other variable.

What we consider here is an oversimplification of the situation. The actual problem is more complex and is treated in courses in statistics under the title linear regression. We will assume here that two pairs of corresponding information are sufficient to determine the linear equation, even though the actual regression model normally involves more than two pairs.

■ Example 30. Assume a linear relationship exists between a student's grade point average and his score on a 100 point math exam. Write an equation expressing the exam score in terms of the grade point average if a student with a G.P.A. of 2.5 scored 75 and a student with a G.P.A. of 3.5 scored 85. Predict the exam score for students with G.P.A.'s of 2.0 and 3.8 respectively.

Solution: We first want to write the equation. We let x represent G.P.A. and y represent exam score. Based on the information given for the two students, the points (2.5,75) and (3.5,85) are on the line. Using the two-point form for the equation of a line (property 3), we obtain

$$y - 75 = \frac{85 - 75}{3.5 - 2.5}(x - 2.5)$$

$$y - 75 = \frac{10}{1}(x - 2.5)$$

$$y - 75 = 10x - 25$$

$$y = 10x + 50$$

For a student with a G.P.A. of 2.0, the predicted exam score is $y = 10(2.0) + 50 = 70$.

For a student with a G.P.A. of 3.8, the predicted exam score is $y = 10(3.8) + 50 = 88$. ■

Self-Review 11

It is thought that a linear relationship exists between the amount of time spent exercising per week following a prescribed exercise program and weight loss. One person averaged 20 hours per week exercising and lost an average of 2 pounds per week. Another person exercised an average of 10 hours a week and lost an average of $\frac{3}{4}$ pounds per week.

a. Write an equation that expresses weight loss (l) in terms of time exercised (t).

b. How much weight would you expect a person averaging 16 hours per week to lose, on the average?

a. $l = \frac{1}{8}t - \frac{1}{2}$

b. $1\frac{1}{2}$ pounds

Exercise Set 0.2

1. For each of the following, write an equation of the line satisfying the given conditions.

a. Through points $(2, -4)$ and $(-3, -3)$.

b. Through points $(1, -2)$ and $(-3, -1)$.

c. Through points $\left(\frac{1}{2}, -\frac{1}{4}\right)$ and $\left(\frac{3}{4}, \frac{1}{4}\right)$.

d. Through points $\left(2, -\frac{1}{2}\right)$ and $\left(\frac{1}{4}, -\frac{1}{2}\right)$.

e. Slope $\frac{1}{3}$ through point $(3,2)$.

f. Slope -2 through point $(-2, -3)$.

g. Slope $-\frac{1}{2}$, y-intercept -2.
h. Through $(2, -3)$ parallel to $2x - 3y - 5 = 0$.
i. Through $(3, -3)$ parallel to $x + 3y - 3 = 0$.
j. Through $(1, -4)$ perpendicular to $2x - y + 3 = 0$.
k. Through $(-3, 2)$ perpendicular to $x + 2y - 2 = 0$.

2. For each of the following, determine if the lines are parallel, perpendicular, or neither.
   a. $2x - 3y + 5 = 0$ and $3x + 2y - 3 = 0$
   b. $x - 2y + 3 = 0$ and $-2x + 4y + 3 = 0$
   c. $x + 2y - 5 = 0$ and $x - 2y + 5 = 0$
   d. $-2x + 3y = 0$ and $4x - 6y - 13 = 0$
   e. $3x + 3y - 2 = 0$ and $2x - 2y + 2 = 0$
3. Find the slope and y-intercept of each line and sketch the graph of the line.
   a. $2x + 3y = 6$
   b. $5x - 4y + 7 = 0$
   c. $x + y - \frac{1}{2} = 0$
   d. $3x - 2y + 8 = 0$
   e. $3y = 12$
   f. $4x = 8$
   g. $x + y = 0$
   h. $3x + 2y = 0$
4. Given p = price per unit and q = quantity of units sold, then an equation where p is solved in terms of q is called a demand equation.
   a. Find the price per unit if $q = 10$ and the demand equation is $p = 3q + 2$.
   b. Given $p = aq + 4$, find a when it is known that $p = 24$ when $q = 5$.
5. Write the equation of each line in the form $y = mx + b$ given
   a. $ax + y = 8$ and $y = 2$ when $x = 3$.
   b. the line $3x + dy = 14$, where the point $(4,2)$ is on the line.
   c. $ax + by = 12$ where $a = 2b$ and the point $(1,1)$ is on the line.
6. Which of the following are *not* equations of straight lines?
   a. $2x + 3y = \frac{1}{2}$
   b. $\frac{1}{2}x + 5y = 5$
   c. $x^2 + 2y = 6$
   d. $x^2 = 9$
   e. $(y^2 - 2y + 1) = 4$
   f. $x^3 - 2x + 1 = 0$
   g. $y = x^2$
   h. $\frac{1}{x} + y = 7$
7. Write an equation of the line containing the point $(2, -5)$ and parallel to the vertical line through the point $(3,4)$.
8. Write an equation of the line containing the point $(4, -3)$ that is perpendicular to the vertical line through the point $(2, -1)$.
9. Miguel works part-time in a grocery store making \$6.50 per hour. Write a linear equation expressing his week's earnings in terms of the number of hours he works and find his earnings for a week in which he worked 25 hours.

10. Rita bakes special occasion cakes and sells them for $12.50 each. Write an equation expressing her receipts in terms of the number of cakes sold and find her receipts for a week in which she sold 7 cakes.
11. Mrs. Sanchez works as a salesperson at an appliance store. She receives a salary of $100 per week plus a commission of 5% on goods sold. Write an equation expressing her weekly earnings in terms of the dollar value of goods sold. What were her earnings for a week in which her sales were $1000? $5500?
12. The local electric company has a monthly charge of $3.50 plus $0.06 per kilowatt hour used. Write an equation that expresses the total charge in terms of usage. What is the total bill if the monthly consumption is 782 kilowatt hours? 583 kilowatt hours?
13. Assume that height in inches and weight in pounds for men follows a linear relationship. Suppose, also, that a man 70 inches tall weighs 166 pounds and a man 65 inches tall weighs 157 pounds.
    a. Write an equation that expresses weight in terms of height.
    b. What would be the predicted weight of a man 60 inches tall? 68 inches tall? 75 inches tall?
14. Suppose that the relationship between a student's score on a placement examination in mathematics and the score on the final examination is assumed to be linear. It is known that a student with a score of 27 on the placement examination got an 88 on the final and a student with a score of 21 on the placement examination got a 70 on the final.
    a. Write an equation expressing final examination score in terms of placement examination score.
    b. What is the predicted final examination score for a student with a placement examination score of 10? 15? 24? 30?

0.3 Linear Inequalities in Two Variables

One of the major concepts discussed in finite mathematics is linear programming (chapter 3). Linear programming involves the use of linear inequalities in two or more variables. In this section, **linear inequalities in two variables** will be considered with special emphasis on the graph of the solution.

Our interest here is on inequalities of the form $ax + by < c, ax + by > c$, $ax + by \geq c$, or $ax + by \leq c$. In each case, a, b, c are real numbers (actually rational numbers in this text) with not both a and b equal to zero. We will consider all four of the inequality types here, even though in chapter 3 only those that include equality will be considered.

Recall from section 1 that a point separates, or divides, a line into three distinct sets of points. In the same way, a line separates a plane into three distinct sets of points. These three sets of points are the line itself, the set of points on one side of the line, and the set of points on the other side of the line.

The line is called the **boundary line** and the sets of points on each side of the line are called **half-planes**. The line is not included in either of the half-planes. The following figure illustrates a line and the two associated half-planes.

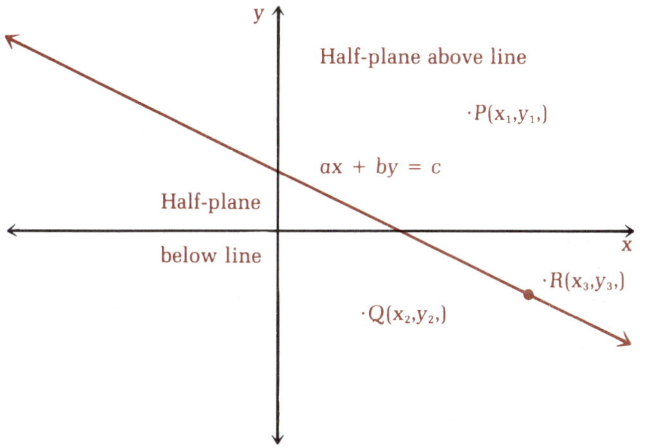

Three points (P, Q, R) have also been identified in the figure. In particular, R is on the line with equation $ax + by = c$, P is in one of the half-planes, and Q is in the other half-plane.

Since R is on the line, we know that the coordinates of R, (x_3,y_3), satisfy the equation of the line. That is, $ax_3 + by_3 = c$. However, since P is not on the line, the coordinates of P do not satisfy the equation of the line. That is, $ax_1 + by_1 \neq c$. Hence, either $ax_1 + by_1 > c$ or $ax_1 + by_1 < c$.

In the same way, since Q is not on the line, either $ax_2 + by_2 > c$ or $ax_2 + by_2 < c$. Suppose we know that $ax_1 + by_1 > c$. Is it possible that $ax_2 + by_2 > c$? The answer is no. That is, if $ax_1 + by_1 > c$, then it can be shown that $ax_2 + by_2 < c$. Similarly, if $ax_1 + by_1 < c$, it can be shown that $ax_2 + by_2 > c$.

In fact, if one point in a half-plane determined by the line with equation $ax + by = c$ satisfies the inequality $ax + by > c$, then every point in that half-plane satisfies this inequality. Furthermore, no point in the other half-plane can satisfy this inequality.

If, then, we start with a linear equation $ax + by = c$, the corresponding line separates the plane into the line and the two half-planes. These three sets of points can be described as the set of points for which $ax + by = c$ (the line), the set of points for which $ax + by > c$ (one half-plane), and the set of points for which $ax + by < c$ (the other half-plane).

To graph a linear inequality in two variables we proceed as follows:

1. we first graph the line associated with the inequality. (This line is the

boundary line that determines the two half-planes), **2.** we then choose any point not on this line (any point that does not satisfy the equation), and **3.** we test this point to see if it satisfies the given inequality. If it does, the solution includes the half-plane containing this point. If not, the solution includes the half-plane not containing this point.

The point we choose for testing the inequality is called the *test point.* While any point not on the line will work, the point (0,0) is an easy point to use when it is not on the line. We will normally use this point when possible.

We will adopt the procedure, here, of shading the portion of the graph that is *not included.* That is, the unshaded portion of the graph is the solution. The reason for this technique will be clearer in chapter 3 when we consider more than one inequality. For inequalities of the form $ax + by \leq c$ or $ax + by \geq c$, where the line is a part of the solution, we will graph the line as a solid line. For strict inequalities, $ax + by < c$ or $ax + by > c$, we will use a dotted line since the line is not a part of the solution.

We will now consider some examples to illustrate the ideas we have been discussing. In some of the examples we will graph solutions to inequalities while in others we will write inequalities for given graphs.

■ **Example 31.** Sketch the graph of the solution of the inequality $2x + 3y < 6$.

Solution:

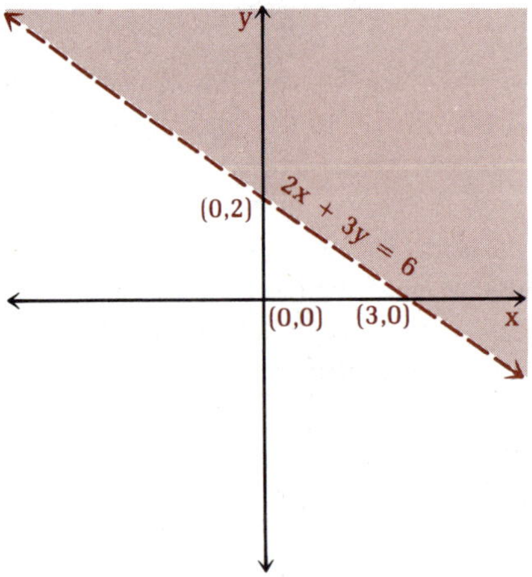

The line associated with the inequality has equation $2x+3y=6$. Letting $x = 0$ we have $2 \cdot 0 + 3y = 6$ or $3y = 6$ or $y = 2$. Hence the point (0,2) is on the line. If $y = 0$, we obtain $2x + 3 \cdot 0 = 6$ or $2x = 6$ or $x = 3$, so the point (3,0) is on the line. Since we are working with strict inequality ($<$), the line is sketched as a dotted line (not included in the solution) through the points (0,2) and (3,0). The point (0,0) is not on the line ($2 \cdot 0 + 3 \cdot 0 \neq 6$) so we will use (0,0) as a test point. Since $2 \cdot 0 + 3 \cdot 0 = 0 < 6$, (0,0) satisfies the inequality and this point is in the solution. We shade the half-plane not containing (0,0), leaving the solution unshaded. ■

■ **Example 32.** Sketch the graph of the solution to the inequality $6x + 5y \geq 24$.

Solution:

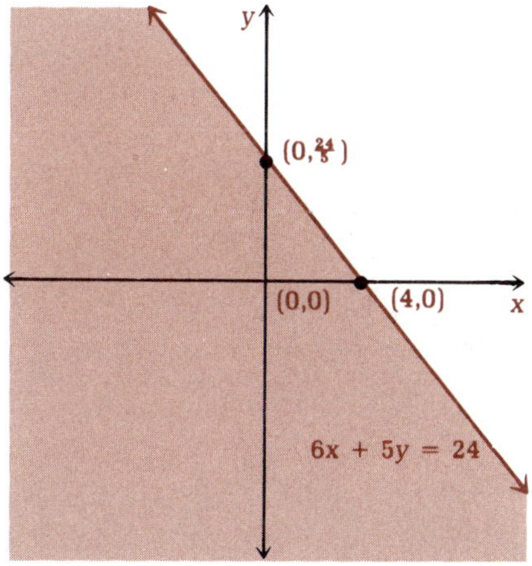

The boundary line has equation $6x + 5y = 24$. Points $\left(0,\frac{24}{5}\right)$ and $(4,0)$ are on the line (check for yourself). The point $(0,0)$ is not on the line (why?) so we use $(0,0)$ as a test point. This point does not satisfy the original inequality (why?). Since the original inequality is of the form $\geq$, the line is included in the solution so we sketch $6x + 5y = 24$ as a solid line and shade the half-plane containing $(0,0)$. ■

Self-Review 12

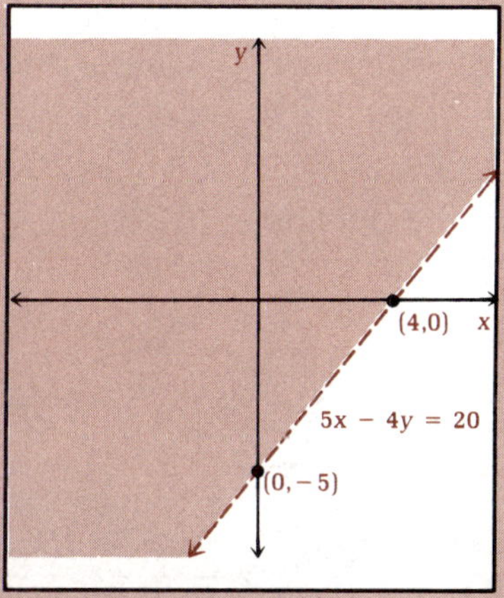

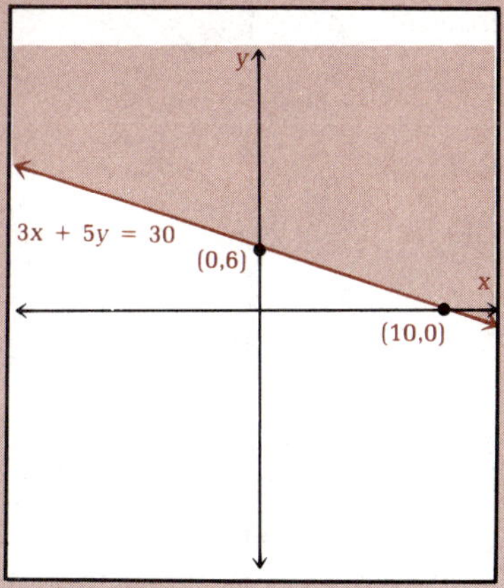

Sketch the graph of the solution of each of the following inequalities.

a. $5x - 4y > 20$
b. $3x + 5y \leq 30$

■ Example 33. Write the inequality whose graph is shown in the following figure.

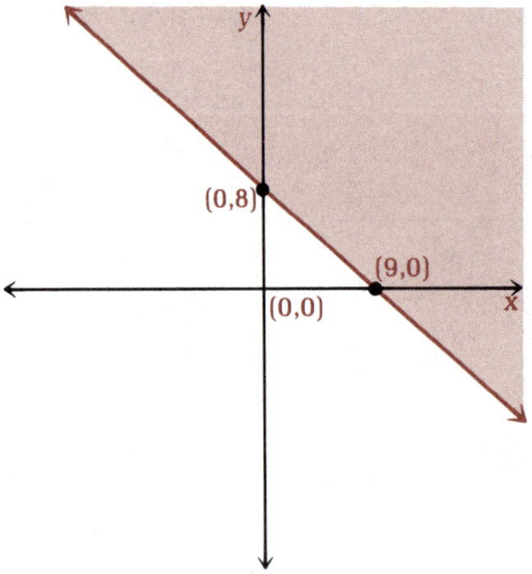

Solution:

$$y - 8 = \frac{0 - 8}{9 - 0}(x - 0)$$

$$y - 8 = \frac{-8}{9}x$$

$$9y - 72 = -8x$$

$$8x + 9y = 72$$

We first find an equation of the boundary line, which is included since it is sketched as a solid line. We use the two point form of section 0.2, with the points (0,8) and (9,0).

The half-plane we want contains (0,0). Using $x = 0$, $y = 0$, we have $8 \cdot 0 + 9 \cdot 0 = 0 < 72$. Therefore the inequality is $8x + 9y \leq 72$. ■

Self-Review 13

$8x - 11y > 44$

Write an inequality for the graph given in the following figure.

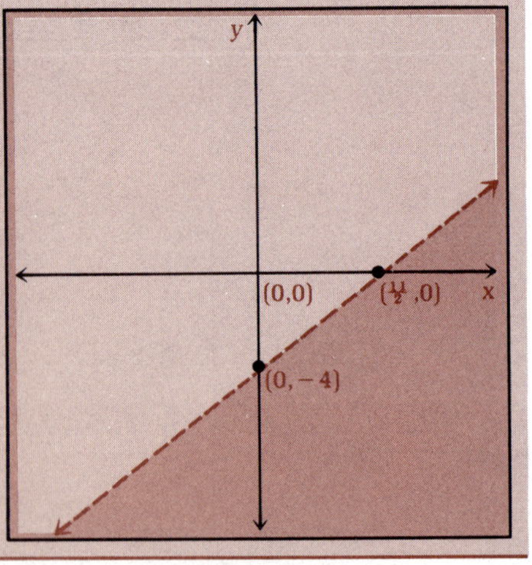

To close this section, we will consider some elementary examples of writing inequalities based on given information. We will encounter more problems of this type in chapter 3.

■ Example 34. Angela and Juan have \$2.00 to spend for candy. They decide they want to buy some large gumballs which cost 10¢ each, and some lollipops which cost 30¢ each. Write an inequality based on this information and then sketch the graph of its solution.

Solution: Since they only have \$2.00, the total amount they spend must be less than or equal to 2.00. Letting x be the number of gumballs and y the number of lollipops purchased, we see that $.10x$ is the total cost of gumballs and $.30y$ is the total cost of lollipops. Hence, the total cost (we will ignore tax) is given by $.10x + .30y$. The inequality is, then, $.10x + .30y \leq 2.00$ or, after multiplying through by 10 to remove decimals, $x + 3y \leq 20$. The graph is given below.

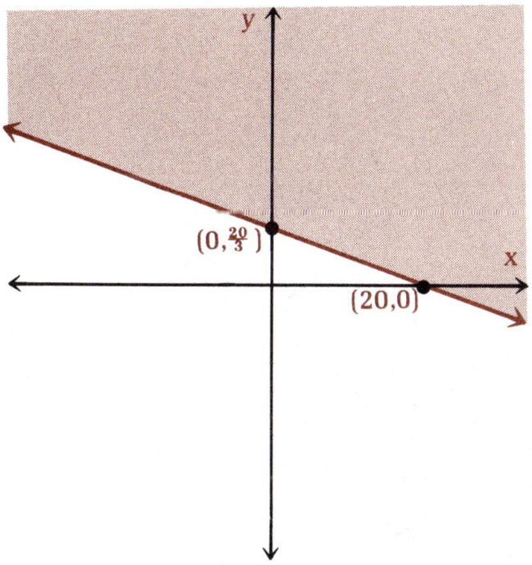

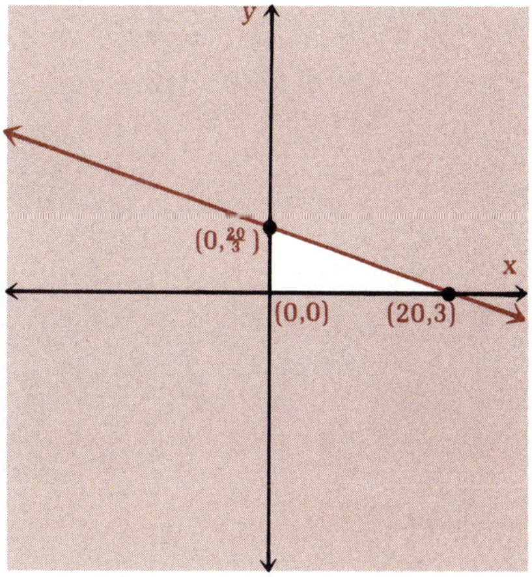

However, if we think about the problem, we can be a little more specific. Since x represents the number of gumballs, x can never be negative. Therefore, no points to the left of the y-axis need to be considered. Similarly, no points below the x-axis should be included since y also cannot be negative. Using this additional information, based on practicality, we can use the graph above right. ■

Self-Review 14

$15x + 20y \geq 40$

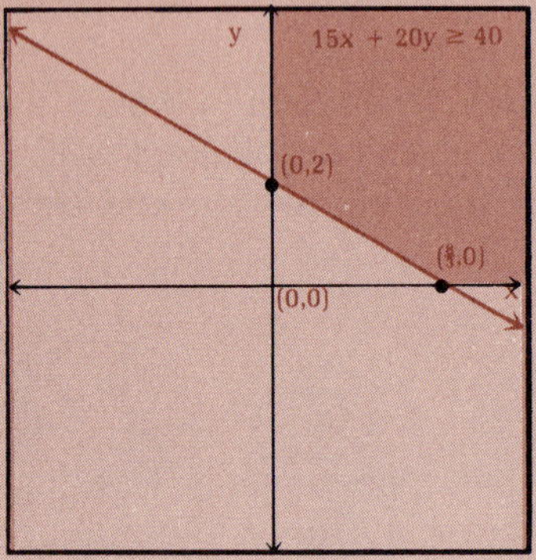

A man has been instructed by his doctor to take at least 40 units of iron per day. The man has two brands of vitamins with iron available, Multivi containing 15 units of iron per tablet and V-Tab containing 20 units of iron per tablet. Write an inequality based on this information and sketch a graph showing the practical solution to the inequality.

Exercise Set 0.3

For each of exercises 1–16, sketch the graph of the solution for the given inequality.

1. $3x + 5y < 15$
2. $2x + 6y < 18$
3. $5x - 2y < 20$
4. $4x - 3y < 24$
5. $x + 5y > 12$
6. $4x + 3y > 18$
7. $6x - 5y > 30$
8. $7x - 3y > 21$
9. $-3x + 7y \leq 21$
10. $-4x + 3y \leq 12$
11. $4x + 5y \leq 20$
12. $6x + 7y \leq 21$
13. $10x + 3y \geq 15$
14. $5x + 9y \geq 27$
15. $-3x + 2y \geq 12$
16. $-5x + 3y \geq 15$

For each of exercises 17–24, write an inequality whose solution is represented by the given graph.

17.

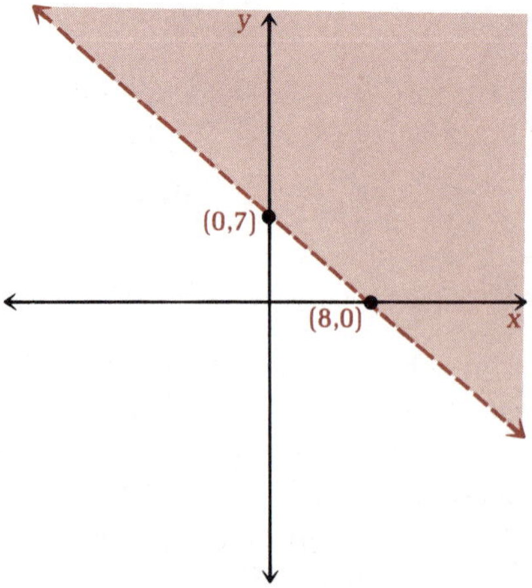

18.

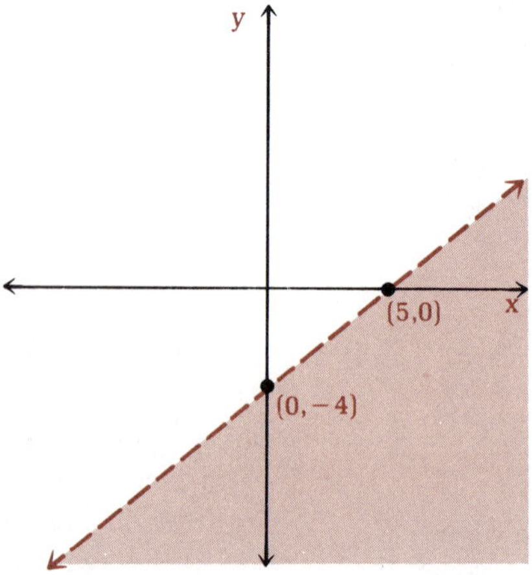

19.

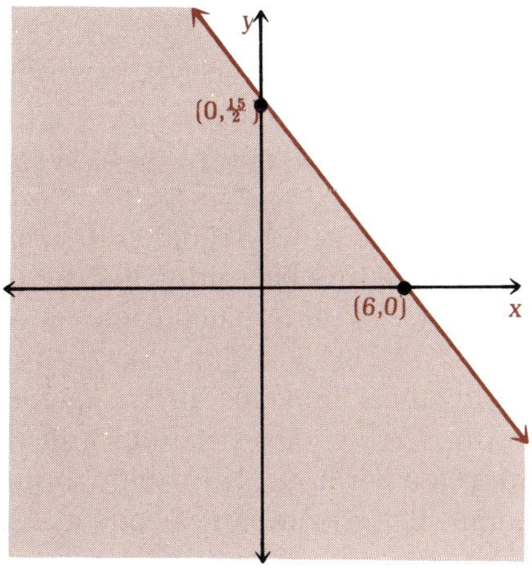

20.

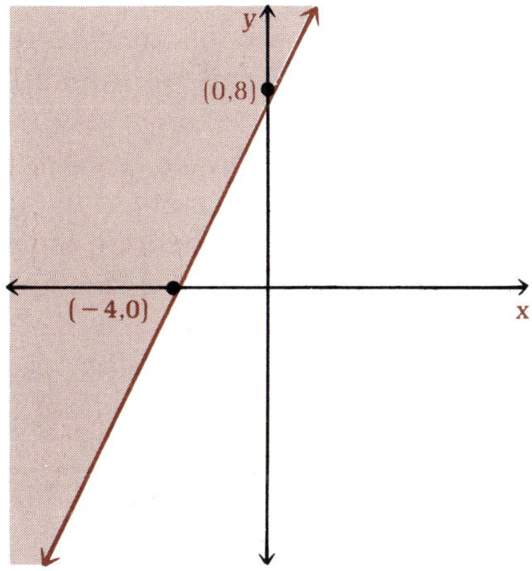

21.

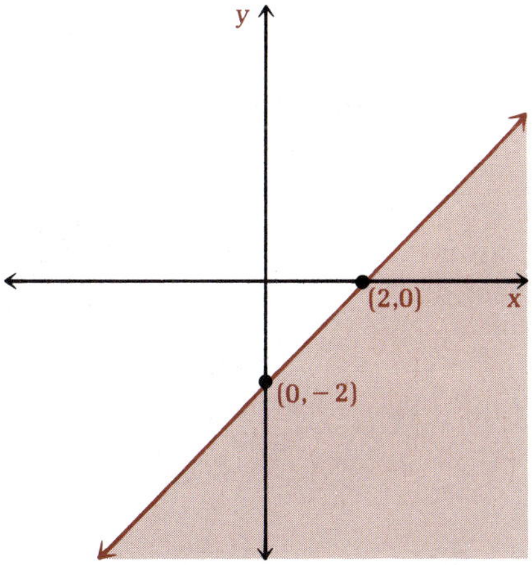

22.

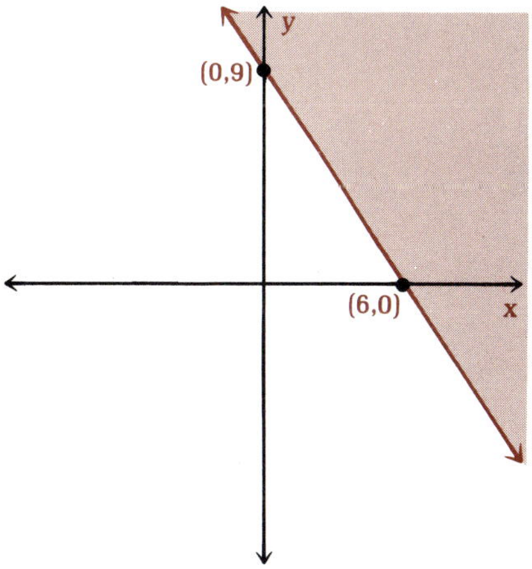

23.

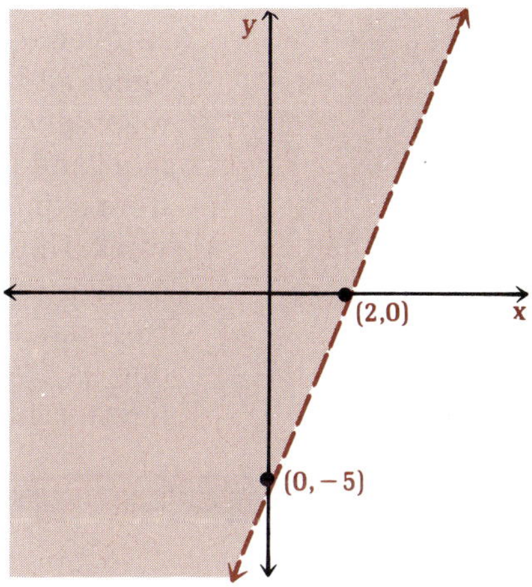

24.

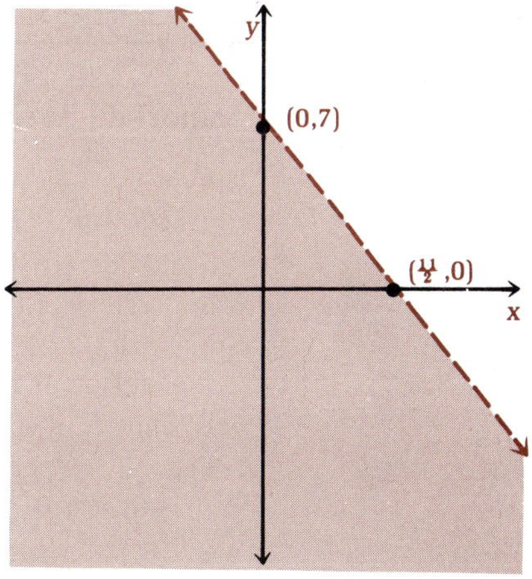

For each of exercises 25–30, write an inequality based on the given information and sketch the graph of the inequality.

25. Denise and Danny have been given \$10.00 to buy party favors for their birthday party. They have agreed to buy noisemakers costing \$.50 each and party hats costing \$.40 each. How many of each can they buy?

26. A college dean has been allotted \$150,000 to hire new faculty members for next year. He can hire beginning instructors for \$15,000 each or experienced instructors for \$18,000 each. How many of each can he hire?

27. A company manufactures bicycles at two factories. The factory at Houston can produce 2 bicycles per hour and the factory at Tulsa can produce 3 bicycles per hour. How many hours per week should each factory operate if the company needs at least 75 bicycles per week?

28. Isaac has been instructed by his doctor to supplement his diet with at least 100 units of vitamin C per day. He has two multiple vitamin supplements available. Brand *X* contains 40 units of vitamin C per tablet and brand *Y* contains 30 units of vitamin C per tablet. How many of each tablet should Isaac take per day?

29. Janet and Jean bake cookies to sell at the farmer's market. They know from experience that they can't sell more than 30 dozen cookies per week. How long should each woman work if Janet can make 5 dozen cookies per hour and Jean can make 4 dozen cookies per hour?

30. How much money should an investor place in each of two accounts, one paying 6% annual interest and the other paying 8% annual interest, if he wants the combined interest to be at least \$480 per year?

0.4 Chapter Review

Important Terms and Formulas in Chapter 0.

linear equation in one variable (section 0.1)

rational numbers (section 0.1)

transitive property (section 0.1)

linear inequality in one variable (section 0.1)

boundary point (section 0.1)

linear equations in two variables (section 0.2)

two-point form (section 0.2)

$\dfrac{y - y_1}{x - x_1} = \dfrac{y_2 - y_1}{x_2 - x_1}$, if $x_2 \neq x$ (section 0.2)

$y - y_1 = \dfrac{y_2 - y_1}{x_2 - x_1}(x - x_1)$, if $(x_2 \neq x_1)$ (section 0.2)

point-slope form (section 0.2)

$y - y_1 = m(x - x_1)$ (section 0.2)

slope (section 0.2)
y-intercept (section 0.2)
$m = \frac{y_2 - y_1}{x_2 - x_1}$, providing $x_2 \neq x_1$ (section 0.2)
slope-intercept form (section 0.2)
$y = mx + b$ (section 0.2)
slope-intercept form (section 0.2)
linear inequalities in two variables (section 0.3)
boundary line (section 0.3)
half-plane (section 0.3)
test point (section 0.3)

Review Exercises

For exercises 1–24 solve the equation.

1. $3x + 5 = 9$
2. $6x - 3 = 5$
3. $2t + 1 = 4$
4. $3t - 1 = 5$
5. $-2y + 3 = 9$
6. $-2y - 3 = -7$
7. $4t + 3 = 11$
8. $-3t + 8 = 10$
9. $5y + 4 = 14$
10. $-3t + 5 = 9$
11. $3y - 3 = 2y + 4$
12. $2x - 4 = 4x + 6$
13. $-2y + 3 = y + 4$
14. $3t - 5 = t - 3$
15. $2x + 6 = 2(x + 3)$
16. $2x - 5 = -2x + 3$
17. $2x + 3 = 2(x + 2)$
18. $-3x + 4 = -2x + 5$
19. $\frac{2}{x + 3} = 2$
20. $\frac{-t}{t + 1} = 5$
21. $\frac{3t - 1}{t + 3} = 4$
22. $\frac{3}{t + 1} + \frac{2}{t - 1} = \frac{1}{t + 1}$
23. $\frac{3}{x - 3} = \frac{4}{x - 2}$
24. $\frac{3}{y + 1} + \frac{2}{y + 2} = \frac{3}{y + 1}$

For exercises 25–50 solve the inequality.

25. $3y + 5 < 8$
26. $2y - 3 \leq 7$
27. $2x + 3 \leq 1$
28. $5y - 7 < 3$
29. $4x + 3 > 6$
30. $3y - 2 \geq 5$
31. $2t - 3 < 5$
32. $3y + 4 > 1$
33. $-3x + 2 < 7$
34. $-9y + 5 \geq 8$
35. $-7x + 3 > 4$
36. $-2x + 3 \leq 6$
37. $3x + 2 \leq 4x - 3$
38. $2x + 4 \geq 3x + 8$
39. $5x - 3 > 2x + 4$
40. $3x + 2 < 2x - 3$
41. $-3x + 4 < 2x - 5$
42. $-3t + 5 > 2t - 4$

43. $3x - 3 \leq x - 4$

44. $2x + 4 \geq 3x + 2$

45. $-7x + 4 \geq 2x - 5$

46. $3x + 2 \leq -4x + 5$

47. $\dfrac{3}{2x - 4} \geq 6$

48. $\dfrac{4}{-3x + 6} < 5$

49. $\dfrac{2x + 5}{x - 3} \leq 6$

50. $\dfrac{3x - 2}{2x + 3} > 3$

In exercises 51–80 write an equation of the line satisfying the given conditions.

51. Through the points (5,3) and (−2,−1).
52. Through the points (−4,3) and (3,−4).
53. Through the points (3,−5) and (3,7).
54. Through the points (2,3) and (6,1).
55. Through the points (3,−4) and (−2,−4).
56. Through the points (0,7) and (4,0).
57. With slope −3 through the point (4,3).
58. With slope $\frac{1}{2}$ through the point (−1,1).
59. With slope $\frac{-2}{3}$ through the point (1,4).
60. With slope $\frac{4}{5}$ through the point (−3,−4).
61. With slope $\frac{-3}{2}$ through the point (3,−2).
62. With slope $\frac{5}{2}$ through the point (−1,−2).
63. With slope −2 and *y*-intercept 3.
64. With slope $\frac{1}{2}$ and *y*-intercept −2.
65. With slope $\frac{-3}{2}$ and *y*-intercept −1.
66. With slope $\frac{2}{3}$ and *y*-intercept 6.
67. Through (3,−1) parallel to $2x - y + 3 = 0$.
68. Through (−2,3) parallel to $x - 2y + 2 = 0$.
69. Through (3,2) parallel to $3x + 2y + 5 = 0$.
70. Through (−3,−1) parallel to $4x + 3y - 5 = 0$.
71. Through (3,−2) parallel to the *y*-axis.
72. Through (−3,−4) parallel to the *x*-axis.
73. Through (1,3) perpendicular to $2x + y + 3 = 0$.
74. Through (−3,−1) perpendicular to $4x - 2y - 5 = 0$.
75. Through (4,4) perpendicular to $3x + 4y - 3 = 0$.
76. Through (3,−2) perpendicular to $5x + 3y - 8 = 0$.

77. Through $\left(\frac{1}{2},\frac{3}{2}\right)$ perpendicular to $2x + 3y - 12 = 0$.
78. Through $(0,7)$ perpendicular to $3x - 2y + 8 = 0$.
79. Through $(3,-5)$ perpendicular to the y-axis.
80. Through $(-4,-3)$ perpendicular to the x-axis.

In exercises 81–90 sketch the graph of the solution for the given inequality.

81. $3x - 2y \leq 18$
82. $4x + 3y > 12$
83. $3x + 5y \geq 15$
84. $7x + 2y \leq 21$
85. $3x - 4y < 12$
86. $6x - 5y \geq 30$
87. $-3x + 7y > 14$
88. $-2x + 15y < 15$
89. $3x - 2y \geq 17$
90. $7x + 4y \leq 28$

In exercises 91–100 write an inequality whose solution has the given graph.

91.

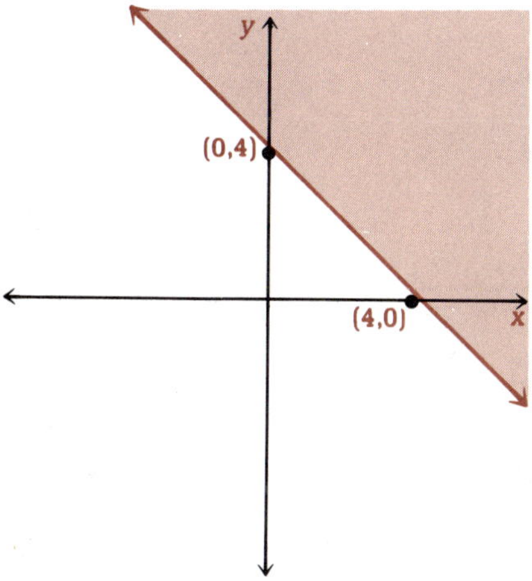

92.

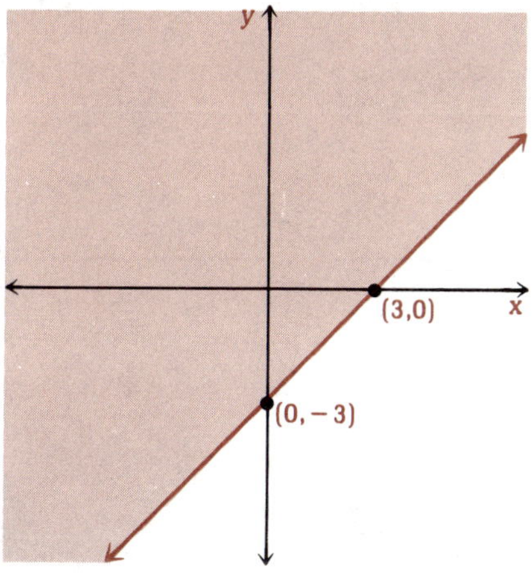

93.

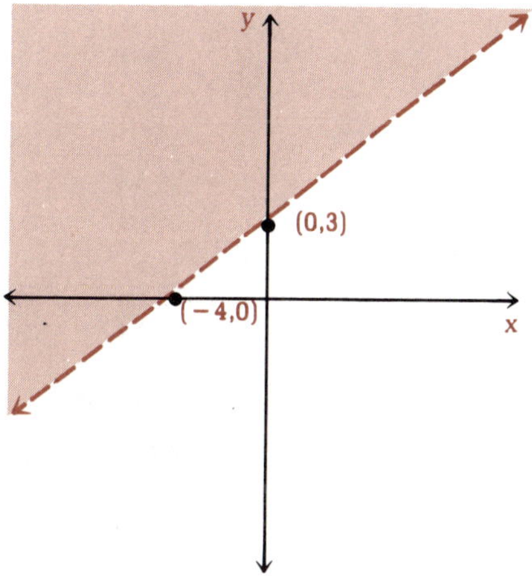

94.

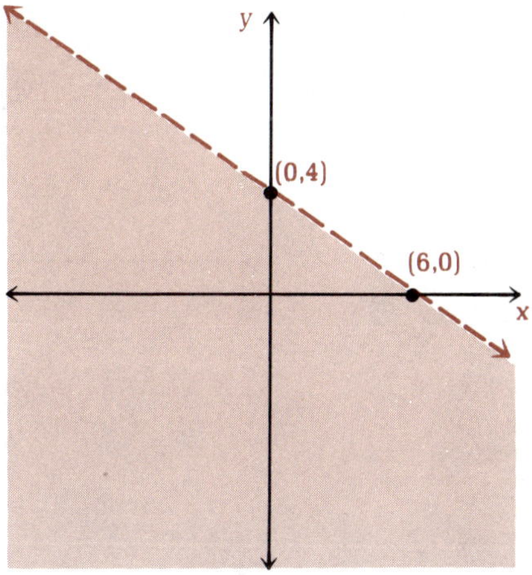

95.

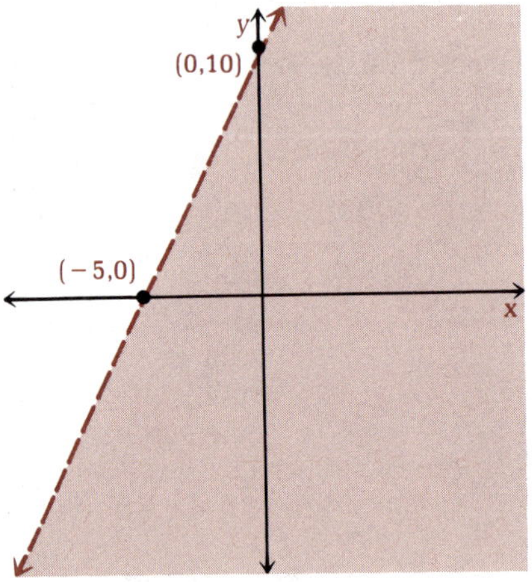

96.

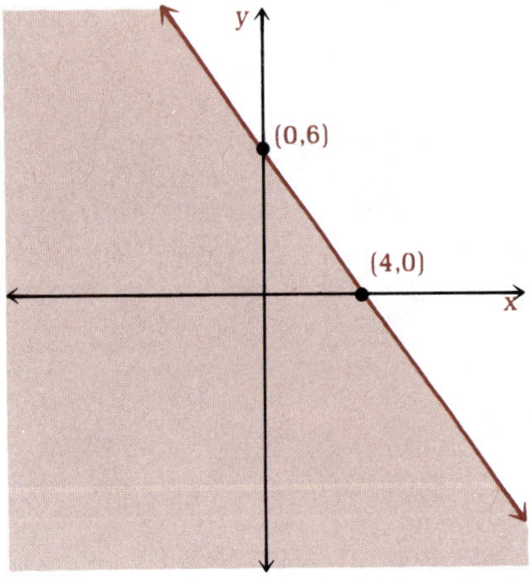

97.

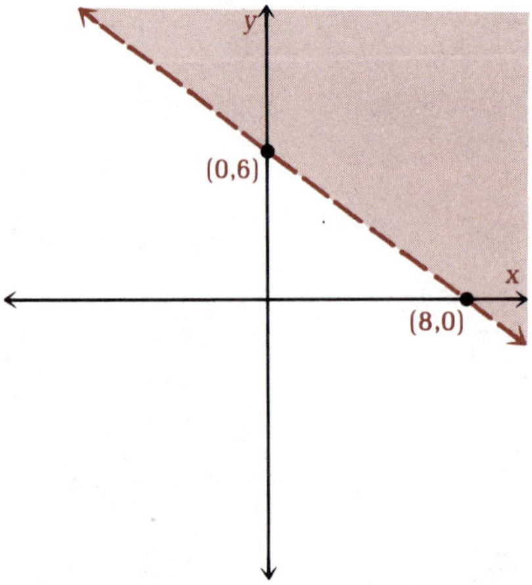

98.

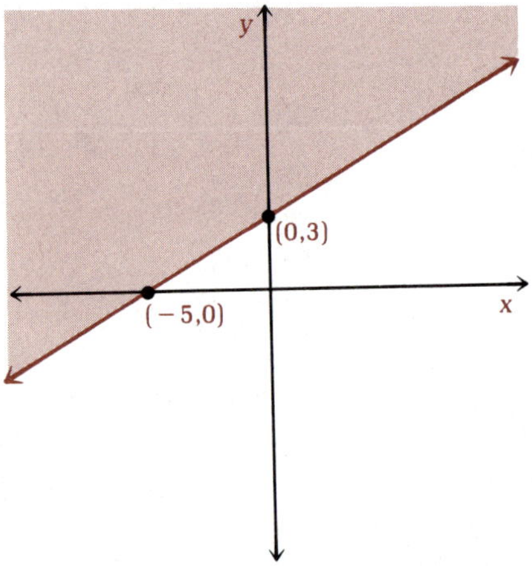

99.

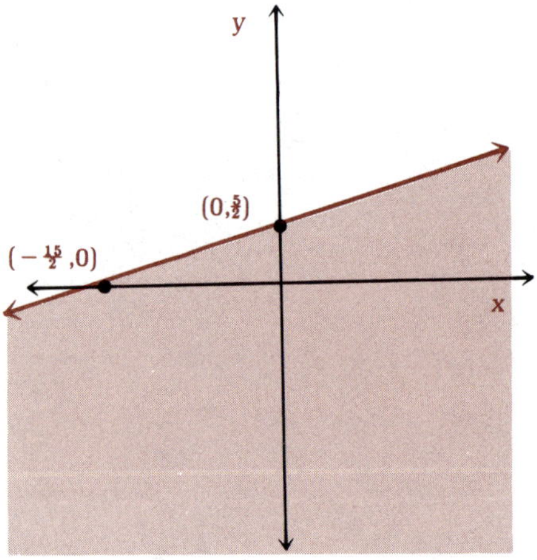

100.

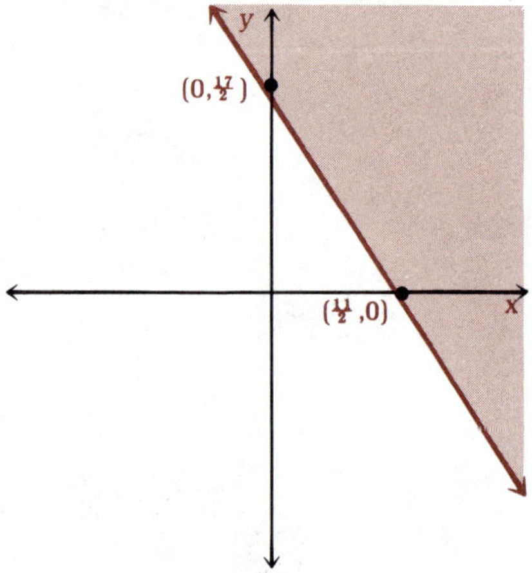

101. The perimeter of a rectangle is 120 meters and the length is twice the width. What are the length and width of the rectangle?

102. The length of a rectangle exceeds 3 times its width by 5 meters. What are the length and width if the perimeter of the rectangle is 170 meters?

103. Mr. Smith has 600 meters of fence. He wants to fence off a pasture in the shape of a square. What will be the length of one side of the square if he uses all of the fence?

104. The length of a rectangle is 10 meters less than 5 times the width. If the perimeter of the rectangle is 580 meters, what are the length and width?

105. Sally works part-time while she is in school. She is paid $4.20 per hour.

a. Write an equation expressing her earnings in terms of the number of hours worked.

b. What were her earnings last week if she worked 20 hours?

c. How much will Sally make in a week in which she works 25 hours?

106. Jim charges $6.00 per hour to mow lawns.

a. Write an equation expressing Jim's charge in terms of hours worked.

b. How much did he charge Mrs. Smith if he spent 2 hours mowing?

c. How much did Mr. Jones pay if Jim worked $1\frac{1}{2}$ hours on his lawn?

d. How much should Jim charge Mr. Rich if it takes $4\frac{1}{2}$ hours to mow his lawn?

107. Mrs. Sanchez works as a salesperson in an appliance store. She receives a salary of $200 a week plus a 5% commission on sales.

a. Write an equation expressing her weekly earnings in terms of sales.

b. How much did Mrs. Sanchez earn last week if her sales totaled $5,000?

c. The week after Thanksgiving Mrs. Sanchez had sales totaling $13,500. What were her earnings for the week?

108. A local plumber has a $15 trip charge. In addition, he charges $25 per hour for labor. What is your trip plus labor charge if the plumber worked

a. $\frac{1}{2}$ hour?

b. 2 hours?

c. $3\frac{1}{4}$ hours?

109. Temperature is commonly measured in degrees Celsius (°C) or degrees Fahrenheit (°F). The relationship between the two scales is linear with 0°C equivalent to 32°F and 100°C equivalent to 212°F.

a. Write an equation that expresses degrees Fahrenheit (F) in terms of degrees Celsius (C).

b. What Fahrenheit temperature is equivalent to 20°C?
c. What Fahrenheit temperature is equivalent to 30°C?

110. Refer back to exercise 109.
a. Write an equation that expresses degrees Celsius (C) in terms of degrees Fahrenheit (F).
b. What Celsius temperature is equivalent to 77°F?
c. What Celsius temperature is equivalent to 98.6°F?

111. There is one temperature that is the same on both temperature scales (see exercises 109, 110). Using your result from part (a) of either exercise 109 or 110, find this temperature.

112. A company has two factories manufacturing a certain product. Factory A produces 15 units per hour and factory B produces 12 units per hour. The company needs at least 1000 units per week.
a. Write an inequality based on the given information.
b. Sketch a graph of this inequality.

113. A dog needs at least 40 units of meat protein per day. It regularly eats dry dog food which contains 2 units of meat protein per ounce and canned dog food which contains 3 units of meat protein per ounce.
a. Write an inequality based on the given information.
b. Sketch a graph of this inequality.

114. To keep active in his retirement, Mr. Green makes and sells birdhouses and bird feeders. Each house requires 2 square feet of plywood and each feeder requires $2\frac{1}{2}$ square feet of plywood. Mr. Green has available 64 square feet of plywood.
a. Write an inequality based on the given information.
b. Sketch a graph of this inequality.

115. The advertising department of a company has been allotted $250,000 for radio and television ads to promote a new product. Each 30 second spot on radio costs $10,000 and each 30 second spot on television costs $30,000.
a. Write an inequality based on this information.
b. Sketch a graph of this inequality.

116. A company believes a linear relationship exists between consumer demand for its product and the price charged. When the price was $3 per unit the demand was 500 units per week. However, when the price increased to $4, the demand dropped to 300 units.
a. Write a linear equation expressing demand in terms of price.
b. What is the estimated demand per week if the price is increased to $5?
c. What is the estimated demand if the price is decreased to $3.75?
d. If the company can comfortably produce 400 units per week, what should the price be to sell all 400 units?

117. A college dean believes a linear relationship exists between students' scores on a standardized test and their college grade point averages. A student with a test score of 70 had a grade point average of 2.00, while a student with a test score of 80 had a grade point average of 2.75.

a. Write an equation expressing grade point average in terms of test score.

b. Predict a grade point average for students with test scores of 75, 85, 90, and 95.

118. A linear relationship exists between a company's production of a certain product and the selling price of the product. When the selling price was $2.50 per unit, they were willing to produce 800 units per week. However, when the price increased to $2.75 per unit, they were willing to produce 900 units per week.

a. Write an equation expressing production in terms of price.

b. How many units would the company be willing to produce per week if the price is $2.40? $2.90? $3.25?

1 Matrices

Introduction

The focus of this text is on application of mathematics to problem situations. The idea is to develop mathematical models and techniques that can be used to solve "real world" problems. Many of these "real world" problems can be solved longhand or using a hand calculator. However, when the size of the problem becomes prohibitively large, computers may become a necessary tool. The ability to organize data, recognize problem types, and identify specific computer software is all that is necessary to solve many complicated practical problems.

The techniques developed here are, in many cases, the techniques used by the computer to solve problems. In situations where it is appropriate, we have tried to relax some restrictions in an attempt to simplify computations.

As you work problems using the various techniques, you are working through some of the steps that a computer would perform to solve the same problem. You should become proficient with these techniques. In addition, you should learn to recognize problem types so you know which technique to use to solve a problem.

Several of the problem-solving techniques developed in this text involve the use of a matrix. (The plural of matrix is matrices.) A **matrix** is a rectangular array of numbers. Many tables can be, and in some cases are, written in matrix form. Table 1.1 is an example of information that could be written as a matrix.

Table 1.1

Light Bulb Inventory by Cases

	Brand X	*Brand Y*
60-Watt	3	22
75-Watt	4	9
100-Watt	8	15
Three-Way	7	11

■ Example 1. Using matrix notation, this information would be written as

$$\begin{bmatrix} 3 & 22 \\ 4 & 9 \\ 8 & 15 \\ 7 & 11 \end{bmatrix}$$

Note that the column and row headings are missing, but the other information is included. This is permissible with the understanding that you remember what these headings are. The data contained in a matrix may be combined in different ways with other matrices to provide insight and answers to problems. The rules governing this process are referred to as *matrix arithmetic*. ■

In the remainder of this chapter, we will develop several ideas relating to matrix arithmetic and reduction techniques. In subsequent chapters some of the problems that use matrix methods for their solution will be explored.

1.1 Basic Notation and Terminology

In this book when a matrix is written out completely, brackets ([ ]) will be used as the matrix symbol. The numbers in a matrix are called **entries of a matrix.** In the preceding example the numbers 3, 4, 22, 9, etc., are called entries.

The **size of a matrix** is determined by the number of rows and columns it has. The matrix in Example 1 is 4 by 2 (written 4×2) because it has 4 rows and 2 columns. In general, a matrix is $m \times n$ if it has m rows and n columns. The number of rows is given first. If a matrix is $n \times n$ then it is called a **square matrix.**

Capital letters such as $A,B,C,\ldots$ or subscripted capitals such as $A_1,A_2,\ldots$ will be used to represent general matrices when the entries are not needed. Two matrices, M and N, are **equal** ($M = N$) if they are identical. That is, they are the same size and all of their corresponding entries are equal.

A **row matrix** is a matrix whose size is $1 \times n$. Referring to Table 1.1, the number of cases of 60-watt light bulbs could be represented by the row matrix $[3 \;\; 22]$. A matrix with size $m \times 1$ is called a **column matrix.** The number of cases of Brand X light bulbs in various sizes could be represented by the column matrix

$$\begin{bmatrix} 3 \\ 4 \\ 8 \\ 7 \end{bmatrix}$$

The **address of an entry** is a way of indicating the entry's position in the matrix. An ordered pair (i,j) will be used where 'i' indicates what row the entry is in and 'j' indicates the column. The address of the number 9 in Example 1 is given by (2,2) while the address of 7 is (4,1). We write a_{ij} to represent the entry at the (i,j) address of a matrix.

For square matrices, the entries at addresses of the form (i,i) are said to be on the **main diagonal.** These entries have the same row and column designation.

■ Example 2.

$$M = \begin{bmatrix} \textcircled{2} & -1 & 3 & 4 \\ 3 & \textcircled{3} & -1 & 5 \\ -2 & 3 & \textcircled{4} & 1 \\ 5 & 1 & 3 & \textcircled{4} \end{bmatrix}$$

The main diagonal entries, circled for identification, are 2, 3, 4, and 4 at the addresses (1,1), (2,2), (3,3), and (4,4) respectively. ■

Of special interest are square matrices with all main diagonal entries equal to 1 and all other entries equal to 0.

$$I_2 = \begin{bmatrix} 1 & 0 \\ 0 & 1 \end{bmatrix} \quad I_3 = \begin{bmatrix} 1 & 0 & 0 \\ 0 & 1 & 0 \\ 0 & 0 & 1 \end{bmatrix} \quad I_4 = \begin{bmatrix} 1 & 0 & 0 & 0 \\ 0 & 1 & 0 & 0 \\ 0 & 0 & 1 & 0 \\ 0 & 0 & 0 & 1 \end{bmatrix}$$

are examples. These matrices are called **identity matrices.** The symbol I_n will be used to emphasize the size of the particular identity matrix. Remember, identity matrices are always square, so I_n means the $n \times n$ identity matrix.

A matrix such as $\begin{bmatrix} 0 & 0 & 0 \\ 0 & 0 & 0 \end{bmatrix}$ with all of its entries equal to zero is called a **zero matrix.**

The **transpose of a matrix** A is the matrix A^T with the rows and columns of A interchanged. That is, the rows of A are the columns of A^T, and the columns of A are the rows of A^T. Hence if A is $m \times n$, then A^T is $n \times m$.

■ Example 3. Find the transpose of each matrix.

a. $A = \begin{bmatrix} 1 & 3 & -1 \\ 2 & -1 & 3 \end{bmatrix}$ **b.** $B = \begin{bmatrix} 2 & 1 & 4 \\ 1 & -1 & 2 \\ -3 & 1 & 2 \end{bmatrix}$

c. $C = \begin{bmatrix} 2 & 1 \\ -1 & -1 \\ 1 & 3 \end{bmatrix}$ **d.** $D = \begin{bmatrix} 2 \\ 7 \\ 4 \end{bmatrix}$

Solutions:

a. $A^T = \begin{bmatrix} 1 & 2 \\ 3 & -1 \\ -1 & 3 \end{bmatrix}$ A is 2×3, A^T is 3×2.

b. $B^T = \begin{bmatrix} 2 & 1 & -3 \\ 1 & -1 & 1 \\ 4 & 2 & 2 \end{bmatrix}$ B is 3×3, B^T is 3×3.

c. $C^T = \begin{bmatrix} 2 & -1 & 1 \\ 1 & -1 & 3 \end{bmatrix}$ C is 3×2, C^T is 2×3.

d. $D^T = [2 \quad 7 \quad 4]$ D is 3×1, D^T is 1×3. ■

Exercise Set 1.1

1. Give the size of each of the following matrices.

a. $\begin{bmatrix} 1 & -1 & 3 & 1 \\ 2 & -1 & 4 & 5 \\ -6 & 2 & -2 & -3 \end{bmatrix}$ **b.** $[1 \quad 6 \quad 1 \quad 2]$ **c.** $\begin{bmatrix} a & b \\ c & d \\ e & f \end{bmatrix}$

d. $\begin{bmatrix} 2 & 1 & 3 & 4 \\ 1 & -1 & 2 & 3 \\ 4 & 2 & 3 & 1 \\ 6 & 1 & 4 & 1 \\ -1 & 3 & 1 & -3 \end{bmatrix}$ **e.** $\begin{bmatrix} 1 \\ 2 \\ 3 \\ 4 \end{bmatrix}$ **f.** $\begin{bmatrix} 1 & 2 \\ 3 & 4 \\ 5 & 6 \\ 7 & 8 \end{bmatrix}$

2. For the given matrix A, find the entry at each indicated address.

$$A = \begin{bmatrix} 2 & 1 & 3 & 4 \\ 1 & -1 & -2 & 3 \\ 4 & 2 & 3 & 1 \\ 6 & -1 & 4 & -1 \\ -1 & 3 & 1 & -3 \end{bmatrix}$$

a. (2,3) **b.** (5,4) **c.** (3,2) **d.** (3,3)
e. (1,3) **f.** (4,1) **g.** (4,4) **h.** (4,5)

3. Write I_5.

4. For each matrix, identify the main diagonal entries.

a. $\begin{bmatrix} 1 & 2 \\ 2 & 1 \end{bmatrix}$ **b.** $\begin{bmatrix} -3 & 1 & 2 \\ 1 & 2 & 3 \\ -3 & -1 & -2 \end{bmatrix}$

c. $\begin{bmatrix} 1 & -1 & 3 & -1 \\ -2 & -1 & -2 & 2 \\ 1 & -1 & 3 & 2 \\ -2 & 1 & 4 & 4 \end{bmatrix}$ **d.** $\begin{bmatrix} 1 & 3 & -1 & 4 \\ 2 & 1 & 3 & 2 \\ 1 & -1 & -1 & 3 \end{bmatrix}$

5. Find the transpose of each matrix.

a. $\begin{bmatrix} 3 & -1 & 4 & 2 \\ 5 & 2 & 3 & -5 \\ 3 & 1 & 4 & 1 \end{bmatrix}$ **b.** $\begin{bmatrix} 6 & -1 \\ 1 & 5 \\ 2 & 3 \end{bmatrix}$ **c.** $\begin{bmatrix} 3 & 1 & -1 \\ 2 & -2 & 2 \end{bmatrix}$

d. $\begin{bmatrix} 3 & -1 & -2 & 4 \\ 4 & -2 & 1 & 1 \\ 6 & 3 & -2 & -1 \\ 1 & 4 & 2 & 1 \end{bmatrix}$ **e.** I_3 **f.** $\begin{bmatrix} 2 & 3 & 4 \\ 1 & 0 & 2 \end{bmatrix}^T$

1.2 Matrix Arithmetic: Addition, Subtraction, Scalar Multiplication

Suppose a company has three factories producing four products. Table 1.2 gives the January production and Table 1.3 gives the February production. What is the total production for January and February?

Table 1.2

January Production

Factory	Product 1	2	3	4
A	27	31	15	10
B	20	22	31	15
C	8	32	21	16

Table 1.3

February Production

Factory	Product 1	2	3	4
A	22	25	10	5
B	16	20	27	12
C	6	26	15	12

To find the total production of factory A for product 1, we would add the corresponding table entries $(27 + 22 = 49)$ and determine that factory A produced a total of 49 units of product 1 in the two months. In a like manner we could determine the total production of each factory for each product. This information is given in Table 4.

Table 1.4

Total Production for January and February

Factory	Product 1	2	3	4
A	49	56	25	15
B	36	42	58	27
C	14	58	36	28

■ Example 4. This same result could be obtained by using matrix notation and an operation called **matrix addition.** Let J be the matrix representing January production and F the matrix representing February production. Hence

$$J = \begin{bmatrix} 27 & 31 & 15 & 10 \\ 20 & 22 & 31 & 15 \\ 8 & 32 & 21 & 16 \end{bmatrix}, \qquad F = \begin{bmatrix} 22 & 25 & 10 & 5 \\ 16 & 20 & 27 & 12 \\ 6 & 26 & 15 & 12 \end{bmatrix}.$$

$$\text{Then } J + F = \begin{bmatrix} 49 & 56 & 25 & 15 \\ 36 & 42 & 58 & 27 \\ 14 & 58 & 36 & 28 \end{bmatrix}.$$

The following definition will be used for matrix addition. ■

Definition 1.1 If A and B are both $m \times n$ matrices, then $A + B$ is the $m \times n$ matrix whose entries are the sums of the corresponding entries of A and B.

That is, the (i,j) entry of the matrix $A + B$ is the sum $a_{ij} + b_{ij}$ where a_{ij} is the (i,j) entry of A and b_{ij} is the (i,j) entry of B.

CAUTION: For $A + B$ to be defined, A and B must be exactly the same size.

While it is true that any two matrices of the same size may be added, in practical situations the sum may not have any meaning.

■ Example 5. Consider Table 1.5 and Table 1.6.

Table 1.5

High Temperature on the 5th of the Month in Degrees Fahrenheit

City	Jan.	Apr.	July	Oct.
New York	32	52	77	58
Kansas City	28	55	79	59
Denver	30	48	73	52
Seattle	38	49	65	52

Table 1.6

Precipitation in Inches on the 5th of the Month

City	Jan.	Apr.	July	Oct.
New York	.9	.05	.2	0
Kansas City	.3	0	1.2	0
Denver	.6	0	.1	.1
Seattle	.05	2.5	.7	1.1

If the information in these tables is written in matrix form, both matrices are 4×4 and they could be added. However, there would be no meaning to the sum. In practical situations, for the sum of two matrices to have meaning, the corresponding column headings and row headings must represent the same items. (You can't compare apples and oranges.) ■

■ Example 6. Find $A + B$ for each.

a. $A = \begin{bmatrix} 2 & -1 & 3 \\ 3 & 3 & -2 \\ -2 & 1 & 4 \end{bmatrix} \qquad B = \begin{bmatrix} 1 & 3 & 1 \\ 2 & -1 & -1 \\ 1 & -1 & 3 \end{bmatrix}$

b. $A = \begin{bmatrix} 1 & 3 & 2 & 4 \\ -1 & 4 & 1 & 3 \end{bmatrix}$ $B = \begin{bmatrix} -2 & -2 & 3 & 1 \\ -1 & -2 & -1 & 1 \end{bmatrix}$

c. $A = \begin{bmatrix} -1 & 2 \\ 3 & 1 \\ 1 & -2 \end{bmatrix}$ $B = \begin{bmatrix} 4 & 1 \\ -1 & -1 \\ 2 & 2 \end{bmatrix}$

d. $A = \begin{bmatrix} 1 & 3 & -1 & -2 \\ -2 & 1 & 3 & 4 \\ 1 & -1 & 1 & 1 \end{bmatrix}$ $B = \begin{bmatrix} -1 & 3 & 2 \\ 1 & 2 & 3 \\ 3 & 2 & 1 \end{bmatrix}$

Solutions:

a.
$$\begin{bmatrix} 2 & -1 & 3 \\ 3 & 3 & -2 \\ -2 & 1 & 4 \end{bmatrix} + \begin{bmatrix} 1 & 3 & 1 \\ 2 & -1 & -1 \\ 1 & -1 & 3 \end{bmatrix}$$
$$= \begin{bmatrix} 2+1 & -1+3 & 3+1 \\ 3+2 & 3+(-1) & -2+(-1) \\ -2+1 & 1+(-1) & 4+3 \end{bmatrix}$$
$$= \begin{bmatrix} 3 & 2 & 4 \\ 5 & 2 & -3 \\ -1 & 0 & 7 \end{bmatrix}$$

b.
$$\begin{bmatrix} 1 & 3 & 2 & 4 \\ -1 & 4 & 1 & 3 \end{bmatrix} + \begin{bmatrix} -2 & -2 & 3 & 1 \\ -1 & -2 & -1 & 1 \end{bmatrix}$$
$$= \begin{bmatrix} 1+(-2) & 3+(-2) & 2+3 & 4+1 \\ -1+(-1) & 4+(-2) & 1+(-1) & 3+1 \end{bmatrix}$$
$$= \begin{bmatrix} -1 & 1 & 5 & 5 \\ -2 & 2 & 0 & 4 \end{bmatrix}$$

c.
$$\begin{bmatrix} -1 & 2 \\ 3 & 1 \\ 1 & -2 \end{bmatrix} + \begin{bmatrix} 4 & 1 \\ -1 & -1 \\ 2 & 2 \end{bmatrix}$$
$$= \begin{bmatrix} -1+4 & 2+1 \\ 3+(-1) & 1+(-1) \\ 1+2 & -2+2 \end{bmatrix} = \begin{bmatrix} 3 & 3 \\ 2 & 0 \\ 3 & 0 \end{bmatrix}$$

d. Since the size of A is 3×4 and the size of B is 3×3, $A + B$ is not defined. ■

Subtraction of matrices can be defined in a manner similar to addition.

Definition 1.2 If A and B are both $m \times n$ matrices, then $A - B$ is the $m \times n$ matrix whose (i,j) entry is found by subtracting the (i,j) entry of B from the (i,j) entry of A.

That is, the (i,j) entry of $A - B$ is $a_{ij} - b_{ij}$ where a_{ij} is the (i,j) entry of A and b_{ij} is the (i,j) entry of B.

■ Example 7. A company has three plants producing 4 products. Their production goal for the month of January is given in Table 1.7. The production for the first three weeks is given in Table 1.8. What should their production be the last week of January to meet their goal?

Table 1.7

Production Goal for January

	Product			
Factory	1	2	3	4
Red	62	49	81	33
Blue	35	29	50	73
Green	45	57	39	60

Table 1.8

Production for First Three Weeks of January

	Product			
Factory	1	2	3	4
Red	41	39	50	31
Blue	21	29	21	40
Green	40	21	27	40

Let A be the production goal matrix and B be the three week production matrix.

$$A = \begin{bmatrix} 62 & 49 & 81 & 33 \\ 35 & 29 & 50 & 73 \\ 45 & 57 & 39 & 60 \end{bmatrix}, B = \begin{bmatrix} 41 & 39 & 50 & 31 \\ 21 & 29 & 21 & 40 \\ 40 & 21 & 27 & 40 \end{bmatrix}.$$

$$A - B = \begin{bmatrix} 62 & 49 & 81 & 33 \\ 35 & 29 & 50 & 73 \\ 45 & 57 & 39 & 60 \end{bmatrix} - \begin{bmatrix} 41 & 39 & 50 & 31 \\ 21 & 29 & 21 & 40 \\ 40 & 21 & 27 & 40 \end{bmatrix}$$

$$= \begin{bmatrix} 62 - 41 & 49 - 39 & 81 - 50 & 33 - 31 \\ 35 - 21 & 29 - 29 & 50 - 21 & 73 - 40 \\ 45 - 40 & 57 - 21 & 39 - 27 & 60 - 40 \end{bmatrix}$$

$$= \begin{bmatrix} 21 & 10 & 31 & 2 \\ 14 & 0 & 29 & 33 \\ 5 & 36 & 12 & 20 \end{bmatrix}$$

The matrix $\begin{bmatrix} 21 & 10 & 31 & 2 \\ 14 & 0 & 29 & 33 \\ 5 & 36 & 12 & 20 \end{bmatrix}$ gives the needed production for the last week so that the production goal is met. We see from Table 1.9 that factory *Red* needs to produce 10 units of product 2 whereas factory *Blue* needs to produce none. This matrix, $(A - B)$, leads directly to the production schedule for the last week.

Table 1.9

Production Schedule for Last Week of January

	Product			
Factory	*1*	*2*	*3*	*4*
Red	21	10	31	2
Blue	14	0	29	33
Green	5	36	12	20

■

■ Example 8. Find $A - B$ for each of the following.

a. $A = \begin{bmatrix} 2 & 1 & 3 & 4 \\ 2 & 1 & 1 & 2 \\ -3 & 1 & 4 & 1 \end{bmatrix}$, $B = \begin{bmatrix} -1 & 1 & 3 & 2 \\ 1 & 2 & 3 & 4 \\ -1 & 3 & 1 & 2 \end{bmatrix}$

b. $A = \begin{bmatrix} 1 & 3 \\ -2 & -1 \\ 2 & 4 \end{bmatrix}$, $B = \begin{bmatrix} 2 & 1 \\ -1 & -3 \\ 3 & 2 \end{bmatrix}$

Solutions:

a.
$$\begin{bmatrix} 2 & -1 & 3 & 4 \\ 2 & 1 & 1 & 2 \\ -3 & 1 & 4 & 1 \end{bmatrix} - \begin{bmatrix} -1 & 1 & 3 & 2 \\ 1 & 2 & 3 & 4 \\ -1 & 3 & 1 & 2 \end{bmatrix}$$
$$= \begin{bmatrix} 2-(-1) & -1-1 & 3-3 & 4-2 \\ 2-1 & 1-2 & 1-3 & 2-4 \\ -3-(-1) & 1-3 & 4-1 & 1-2 \end{bmatrix}$$
$$= \begin{bmatrix} 3 & -2 & 0 & 2 \\ 1 & -1 & -2 & -2 \\ -2 & -2 & 3 & -1 \end{bmatrix}$$

b. $\begin{bmatrix} 1 & 3 \\ -2 & -1 \\ 2 & 4 \end{bmatrix} - \begin{bmatrix} 2 & 1 \\ -1 & -3 \\ 3 & 2 \end{bmatrix}$

$$= \begin{bmatrix} 1-2 & 3-1 \\ -2-(-1) & -1-(-3) \\ 2-3 & 4-2 \end{bmatrix} = \begin{bmatrix} -1 & 2 \\ -1 & 2 \\ -1 & 2 \end{bmatrix}$$ ■

Consider again the production in Table 1.4 and suppose the company wants to double this production for the months of March and April. How many items of each type would each factory have to produce to accomplish this production goal? Since factory *Red* produced 49 units of product 1, they should produce 2 × 49 or 98 units during March and April. In a similar manner, we can determine the production goal given in Table 1.10.

Table 1.10

Production Goal for March and April

	Product			
Factory	1	2	3	4
A	98	112	50	30
B	72	84	116	54
C	28	116	72	56

This is an example of the use of another matrix arithmetic operation called **scalar multiplication.**

Definition 1.3 If A is an $m \times n$ matrix and k is a real number, then the scalar product kA is the $m \times n$ matrix whose entries are k times the corresponding entries of A.

That is, if a_{ij} is the (i,j) entry of A, the (i,j) entry of kA is ka_{ij}.

■ Example 9.

a. $$-3\begin{bmatrix} 3 & 4 & -1 \\ 0 & -2 & 1 \\ -1 & 3 & -2 \end{bmatrix} = \begin{bmatrix} -9 & -12 & 3 \\ 0 & 6 & -3 \\ 3 & -9 & 6 \end{bmatrix}$$

b. $$\begin{bmatrix} \frac{3}{2} & 4 \\ \frac{1}{4} & 2 \end{bmatrix} = \begin{bmatrix} \frac{6}{4} & \frac{16}{4} \\ \frac{1}{4} & \frac{8}{4} \end{bmatrix} = \frac{1}{4}\begin{bmatrix} 6 & 16 \\ 1 & 8 \end{bmatrix}$$

c. $$3\begin{bmatrix} 5 & 7 \\ 2 & -1 \end{bmatrix} - \frac{1}{2}\begin{bmatrix} 8 & 6 \\ 4 & -4 \end{bmatrix}$$

$$= \begin{bmatrix} 15 & 21 \\ 6 & -3 \end{bmatrix} - \begin{bmatrix} 4 & 3 \\ 2 & -2 \end{bmatrix} = \begin{bmatrix} 11 & 18 \\ 4 & -1 \end{bmatrix}$$ ■

Property 1.1: ***Properties of addition and scalar multiplication:***

1. If A and B are both $m \times n$ matrices, then $A + B = B + A$. (Commutative Property of Addition)
2. If A, B, and C are all $m \times n$ matrices, then $(A + B) + C = A + (B + C)$. (Associative Property of Addition)
3. If A and B are $m \times n$ matrices and s is a real number, $s(A + B) = sA + sB$. It is also correct to think of this property as $sA + sB = s(A + B)$. (Distributive Property)
4. If A is a matrix and s and t are real numbers, then $(s + t)A = sA + tA$. (Distributive Property)
5. If A is a matrix and s and t are real numbers, then $s(tA) = (st)A$.

CAUTION: In general, $A - B \neq B - A$.

Exercise Set 1.2

1. For each pair of matrices, find $A + B$, $A - B$, $B - A$, $2A$, and $-3B$.

a. $A = \begin{bmatrix} 2 & 1 & 3 & 4 \\ -1 & 3 & -2 & 1 \\ -4 & 3 & 4 & 1 \end{bmatrix}$, $B = \begin{bmatrix} 3 & 1 & -1 & -2 \\ -1 & 3 & 1 & 2 \\ -2 & -1 & 3 & 1 \end{bmatrix}$

b. $A = \begin{bmatrix} 2 & 1 \\ -1 & 4 \\ 3 & -4 \end{bmatrix}$, $B = \begin{bmatrix} -1 & -2 \\ 3 & -1 \\ -2 & 2 \end{bmatrix}$

c. $A = \begin{bmatrix} \frac{1}{2} & -\frac{1}{4} & 1 \\ -\frac{1}{2} & \frac{1}{2} & -\frac{1}{4} \\ \frac{1}{4} & \frac{1}{2} & 1 \end{bmatrix}$, $B = \begin{bmatrix} -\frac{1}{4} & -\frac{1}{4} & -\frac{1}{2} \\ \frac{1}{2} & -1 & -\frac{1}{4} \\ \frac{3}{4} & \frac{3}{2} & -\frac{1}{2} \end{bmatrix}$

d. $A = \begin{bmatrix} -2 & 1 & 3 \\ 3 & -1 & -1 \\ 1 & 1 & -2 \\ -2 & -1 & 1 \end{bmatrix}$, $B = \begin{bmatrix} -1 & 2 & -1 \\ 3 & 1 & -1 \\ 2 & 2 & -3 \\ -1 & -1 & 3 \end{bmatrix}$

e. $A = \begin{bmatrix} 3 & -1 \\ 1 & 1 \\ -2 & -1 \end{bmatrix}$, $B = \begin{bmatrix} 1 & -3 & -1 \\ -2 & 4 & -3 \end{bmatrix}$

2. For each pair of matrices in problem 1, find $A - 3B$, $2A - 3B$, and $A + 2B$.

3. Using

$$A = \begin{bmatrix} -1 & 3 & 2 \\ 1 & -1 & -2 \\ 2 & 1 & -3 \end{bmatrix}, B = \begin{bmatrix} 1 & -1 & 1 \\ 3 & 2 & -1 \\ 1 & 4 & 2 \end{bmatrix}, \text{ and}$$

$$C = \begin{bmatrix} -3 & 1 & 4 \\ 1 & -1 & 2 \\ -2 & 2 & 3 \end{bmatrix}, \text{ verify part 2 of property 1.}$$

4. Verify part 3 of property 1 using

$$A = \begin{bmatrix} 2 & -1 & 3 & 4 \\ 1 & -2 & -1 & 2 \end{bmatrix},$$

$$B = \begin{bmatrix} -1 & -1 & 2 & 1 \\ 5 & 1 & -2 & 3 \end{bmatrix}, \text{ and } s = 3.$$

5. A new factory has opened and the number of workers, by shift, in each of four departments is given in the table.

	Shift		
Dept.	*1*	*2*	*3*
A	15	10	5
B	23	20	12
C	19	19	2
D	27	31	5

The company plans to double the number of employees in each department on each shift by the end of six months, and to double this number by the end of one year. What will the employment table be at the end of one year?

6. The factory in problem 5 manufactures four products. The total production for the first quarter is given by the table.

	Shift		
Product	*1*	*2*	*3*
A	60	50	20
B	35	30	10
C	55	50	15
D	80	60	20

During the second quarter, because of the increase in employees, production increases. The average monthly production for the second quarter is

	Shift		
Product	*1*	*2*	*3*
	35	25	10
	25	20	10
	25	20	10
	35	30	15

What is the total production for the first six months?

7. The company in problems 5 and 6 has 5 job classifications plus shift differentials. The hourly salary schedule is

	Shift		
Job Class	*1*	*2*	*3*
I	5.40	5.50	5.60
II	5.80	5.90	6.00
III	6.00	6.10	6.20
IV	6.20	6.30	6.40
V	6.40	6.50	6.60

The new contract calls for a 10% across-the-board wage increase. What is the new salary schedule?

8. Given $A = \begin{bmatrix} 1 & 5 & 3 \\ 2 & 1 & 0 \\ 1 & 0 & 4 \end{bmatrix}$, compute $3A - 2A^T$.

9. Given $A = \begin{bmatrix} 5 & 3 & 1 \\ -4 & 3 & 2 \end{bmatrix}$ and $B = \begin{bmatrix} 1 & 2 & 1 \\ 0 & 1 & 1 \end{bmatrix}$, show that $(A + B)^T = A^T + B^T$

10. Given $A = \begin{bmatrix} .02 & .30 & .5 \\ .40 & .20 & .02 \\ .35 & .60 & .40 \end{bmatrix}$, find $A - I_3$ and $I_3 - A$.

1.3 Matrix Multiplication

One other important matrix arithmetic operation needs to be considered here. This operation is **matrix multiplication.** Whereas the other matrix arithmetic operations are defined in a very natural way, such is not necessarily the case with matrix multiplication. The following example will serve to introduce the technique used in matrix multiplication.

Example 10. Suppose a factory has three workers producing an item. Worker D can produce 5 units per hour; worker E, 3 units per hour; and worker F, 3 units per hour. The time available for D is 2 hours; for E, 3 hours; and for F, 2 hours. What is the total number of units that can be produced?

Solution: We see that D produces 5 units per hour for 2 hours, so she can produce $5 \times 2 = 10$ units. Similarly, E can produce $3 \times 3 = 9$ units and F can produce $3 \times 2 = 6$ units. Hence the total number that can be produced is $10 + 9 + 6 = 25$ units.

Suppose we let the number each worker can produce per hour be the row matrix $[5 \quad 3 \quad 3]$ and the available time per worker be the column matrix $\begin{bmatrix} 2 \\ 3 \\ 2 \end{bmatrix}$. We then write $[5 \quad 3 \quad 3]\begin{bmatrix} 2 \\ 3 \\ 2 \end{bmatrix}$. When we perform operations on two objects of the same kind we want the result to be an object of the same kind. That is, the sum or product of two numbers is a number. Then the product of two matrices should be a matrix. But we also want the answer in this case to be a number. What we get, then, is a matrix with one entry, namely the number we are seeking.

$$[5 \quad 3 \quad 3]\begin{bmatrix} 2 \\ 3 \\ 2 \end{bmatrix} = [5 \times 2 + 3 \times 3 + 3 \times 2]$$
$$= [10 + 9 + 6] = [25]$$

■

When we consider multiplying two matrices A and B to obtain the product AB, the number of columns of A (the matrix on the left) must be equal to the number of rows of B (the matrix on the right). That is, if A is $m \times n$, B must be $n \times p$. The product matrix will be $m \times p$. If the number of columns in the left matrix does not equal the number of rows in the right matrix, the product is not defined.

Before proceeding to larger matrices, we consider the general case of a row matrix times a column matrix. Consider the matrix $R = [r_1 r_2 r_3 \ldots r_n]$ and the matrix $C = \begin{bmatrix} c_1 \\ c_2 \\ \vdots \\ c_n \end{bmatrix}$. Note that R and C each have n entries. That is, R has n columns and C has n rows.

The product RC (R is $1 \times n$ and C is $n \times 1$) is to be a 1×1 matrix whose entry is given by $r_1c_1 + r_2c_2 + r_3c_3 + \ldots + r_nc_n$. That is,

$$[r_1 r_2 r_3 \ldots r_n] \begin{bmatrix} c_1 \\ c_2 \\ c_3 \\ \vdots \\ c_n \end{bmatrix} = [r_1c_1 + r_2c_2 + r_3c_3 + \ldots + r_nc_n]$$

■ Example 11. Find the following row by column products.

a. $\begin{bmatrix} 3 & 1 & 2 \end{bmatrix} \begin{bmatrix} -2 \\ 1 \\ 2 \end{bmatrix}$

b. $\begin{bmatrix} 1 & -1 & 3 & 2 & 1 \end{bmatrix} \begin{bmatrix} 4 \\ 1 \\ -2 \\ 3 \\ 1 \end{bmatrix}$

c. $\begin{bmatrix} 1 & 2 & -1 \end{bmatrix} \begin{bmatrix} 3 \\ -2 \end{bmatrix}$

Solutions:

a. $\begin{bmatrix} 3 & 1 & 2 \end{bmatrix} \begin{bmatrix} 2 \\ 1 \\ 2 \end{bmatrix} = [3(-2) + 1 \times 1 + 2 \times 2]$

$= [-6 + 1 + 4] = [-1]$

b. $\begin{bmatrix} 1 & -1 & 3 & 2 & 1 \end{bmatrix} \begin{bmatrix} 4 \\ 1 \\ -2 \\ 3 \\ 1 \end{bmatrix}$

$= [1 \times 4 + (-1) \times 1 + 3(-2) + 2 \times 3 + 1 \times 1]$

$= [4 - 1 - 6 + 6 + 1] = [4]$

c. Since the left matrix has 3 columns but the right matrix has only 2 rows, the product is not defined. ■

The product of a row matrix times a column matrix has been written in the proceding example as a 1×1 matrix, Using this idea, we now define matrix multiplication.

Definition 1.4. If A is an $m \times n$ matrix and B is an $n \times p$ matrix, then the product AB is an $m \times p$ matrix whose (i, j) entry is the sum of the products of the entries in the ith row of A and the corresponding entries in the jth column of B.

In other words, the (1,1) entry in AB is the entry obtained by taking the product of the first row of A and the first column of B, the (3,2) entry of AB is the entry obtained from the product of the third row of A and the second column of B, and so on.

■ Example 12. Find AB for each pair of matrices, if it exists.

a. $A = \begin{bmatrix} 2 & -1 \\ 3 & 4 \end{bmatrix}, B = \begin{bmatrix} -1 & 3 \\ 2 & 1 \end{bmatrix}$

b. $A = \begin{bmatrix} 3 & 1 \\ -2 & 2 \\ 1 & -1 \end{bmatrix}, B = \begin{bmatrix} 4 & -1 & 3 \\ 2 & 5 & 2 \end{bmatrix}$

c. $A = \begin{bmatrix} 2 & -1 & 3 \\ 3 & 3 & 2 \end{bmatrix}, B = \begin{bmatrix} 3 & 1 \\ -2 & -1 \\ 1 & 3 \end{bmatrix}$

d. $A = \begin{bmatrix} 1 & 3 & -2 \\ 2 & 1 & 3 \\ 3 & 2 & 4 \end{bmatrix}, B = \begin{bmatrix} 3 & 1 & 2 & 2 \\ 1 & 3 & 1 & 4 \\ -2 & -2 & 3 & 1 \end{bmatrix}$

e. $A = \begin{bmatrix} 3 & 1 & -1 \\ 2 & 1 & 3 \\ -3 & 2 & 4 \end{bmatrix}, B = \begin{bmatrix} 2 & -2 & 1 \\ 1 & 4 & 2 \\ 5 & -1 & 3 \end{bmatrix}$

f. $A = \begin{bmatrix} \frac{1}{2} & \frac{1}{3} & -\frac{1}{2} \\ \frac{2}{3} & -\frac{3}{4} & \frac{1}{4} \\ \frac{1}{3} & -\frac{2}{3} & \frac{1}{2} \end{bmatrix}, B = \begin{bmatrix} \frac{2}{3} & -\frac{1}{4} & \frac{1}{2} \\ -\frac{3}{4} & \frac{2}{3} & \frac{1}{3} \\ \frac{1}{4} & \frac{2}{3} & \frac{1}{3} \end{bmatrix}$

g. $A = \begin{bmatrix} -2 & 3 \\ 5 & 1 \\ 3 & 5 \end{bmatrix}, B = \begin{bmatrix} 3 & 1 \\ -2 & 4 \\ 1 & 4 \end{bmatrix}$

h. $A = \begin{bmatrix} 2 \\ 1 \\ -1 \end{bmatrix}, B = [3 \quad 1 \quad -1 \quad 2]$

Solutions:

a.
$$\underset{2 \times 2}{\begin{bmatrix} 2 & -1 \\ 3 & 4 \end{bmatrix}} \underset{2 \times 2}{\begin{bmatrix} -1 & 3 \\ 2 & 1 \end{bmatrix}} = \begin{bmatrix} 2(-1) + (-1)2 & 2 \cdot 3 + (-1)1 \\ 3(-1) + 4 \cdot 2 & 3 \cdot 3 + 4 \cdot 1 \end{bmatrix}$$
$$= \underset{2 \times 2}{\begin{bmatrix} -4 & 5 \\ 5 & 13 \end{bmatrix}}$$

b.
$$\underset{3 \times 2}{\begin{bmatrix} 3 & 1 \\ -2 & 2 \\ 1 & -1 \end{bmatrix}} \underset{2 \times 3}{\begin{bmatrix} 4 & -1 & 3 \\ 2 & 5 & 2 \end{bmatrix}}$$
$$= \begin{bmatrix} 3 \cdot 4 + 1 \cdot 2 & 3(-1) + 1 \cdot 5 & 3 \cdot 3 + 1 \cdot 2 \\ -2 \cdot 4 + 2 \cdot 2 & (-2)(-1) + 2 \cdot 5 & -2 \cdot 3 + 2 \cdot 2 \\ 1 \cdot 4 + (-1)2 & 1 \cdot (-1) + (-1)5 & 1 \cdot 3 + (-1)2 \end{bmatrix}$$
$$= \underset{3 \times 3}{\begin{bmatrix} 14 & 2 & 11 \\ -4 & 12 & -2 \\ 2 & -6 & 1 \end{bmatrix}}$$

c.
$$\underset{2 \times 3}{\begin{bmatrix} 2 & -1 & 3 \\ 3 & 3 & 2 \end{bmatrix}} \underset{3 \times 2}{\begin{bmatrix} 3 & 1 \\ -2 & -1 \\ 1 & 3 \end{bmatrix}}$$
$$= \begin{bmatrix} 2 \cdot 3 + (-1)(-2) + 3 \cdot 1 & 2 \cdot 1 + (-1)(-1) + 3 \cdot 3 \\ 3 \cdot 3 + 3(-2) + 2 \cdot 1 & 3 \cdot 1 + 3(-1) + 2 \cdot 3 \end{bmatrix}$$
$$= \underset{2 \times 2}{\begin{bmatrix} 11 & 12 \\ 5 & 6 \end{bmatrix}}$$

d.

$$\overset{3\times 3}{\begin{bmatrix} 1 & 3 & -2 \\ 2 & 1 & 3 \\ 3 & 2 & 4 \end{bmatrix}} \overset{3\times 4}{\begin{bmatrix} 3 & 1 & 2 & 2 \\ 1 & 3 & 1 & 4 \\ -2 & -2 & 3 & 1 \end{bmatrix}}$$

$$= \begin{bmatrix} 1\cdot 3 + 3\cdot 1 + (-2)(-2) & 1\cdot 1 + 3\cdot 3 + (-2)(-2) & 1\cdot 2 + 3\cdot 1 + (-2)3 & 1\cdot 2 + 3\cdot 4 + (-2)1 \\ 2\cdot 3 + 1\cdot 1 + 3(-2) & 2\cdot 1 + 1\cdot 3 + 3(-2) & 2\cdot 2 + 1\cdot 1 + 3\cdot 3 & 2\cdot 2 + 1\cdot 4 + 3\cdot 1 \\ 3\cdot 3 + 2\cdot 1 + 4(-2) & 3\cdot 1 + 2\cdot 3 + 4(-2) & 3\cdot 2 + 2\cdot 1 + 4\cdot 3 & 3\cdot 2 + 2\cdot 4 + 4\cdot 1 \end{bmatrix}$$

$$= \overset{3\times 4}{\begin{bmatrix} 10 & 14 & -1 & 12 \\ 1 & -1 & 14 & 11 \\ 3 & 1 & 20 & 18 \end{bmatrix}}$$

e.

$$\overset{3\times 4}{\begin{bmatrix} 3 & 1 & -1 \\ 2 & 1 & 3 \\ -3 & 2 & 4 \end{bmatrix}} \overset{3\times 3}{\begin{bmatrix} 2 & -2 & 1 \\ 1 & 4 & 2 \\ 5 & -1 & 3 \end{bmatrix}}$$

$$= \begin{bmatrix} 3\cdot 2 + 1\cdot 1 + (-1)5 & 3(-2) + 1\cdot 4 + (-1)(-1) & 3\cdot 1 + 1\cdot 2 + (-1)\,3 \\ 2\cdot 2 + 1\cdot 1 + 3\cdot 5 & 2(-2) + 1\cdot 4 + 3(-1) & 2\cdot 1 + 1\cdot 2 + 3\cdot 3 \\ -3\cdot 2 + 2\cdot 1 + 4\cdot 5 & (-3)(-2) + 2\cdot 4 + 4(-1) & -3\cdot 1 + 2\cdot 2 + 4\cdot 3 \end{bmatrix}$$

$$= \overset{3\times 3}{\begin{bmatrix} 2 & -1 & 2 \\ 20 & -3 & 13 \\ 16 & 10 & 13 \end{bmatrix}}$$

f.

$$\overset{3\times 3}{\begin{bmatrix} \frac{1}{2} & \frac{1}{3} & -\frac{1}{2} \\ \frac{2}{3} & -\frac{3}{4} & \frac{1}{4} \\ \frac{1}{3} & -\frac{2}{3} & \frac{1}{2} \end{bmatrix}} \overset{3\times 3}{\begin{bmatrix} \frac{2}{3} & -\frac{1}{4} & \frac{1}{2} \\ -\frac{3}{4} & \frac{2}{3} & \frac{1}{3} \\ \frac{1}{4} & \frac{2}{3} & \frac{1}{3} \end{bmatrix}}$$

$$= \begin{bmatrix} \frac{1}{2}\cdot\frac{2}{3} + \frac{1}{3}(-\frac{3}{4}) + (-\frac{1}{2})\frac{1}{4} & \frac{1}{2}(-\frac{1}{4}) + \frac{1}{3}\cdot\frac{2}{3} + (-\frac{1}{2})(\frac{2}{3}) & \frac{1}{2}\cdot\frac{1}{2} + \frac{1}{3}\cdot\frac{1}{3} + (-\frac{1}{2})\frac{1}{3} \\ \frac{2}{3}\cdot\frac{2}{3} + (-\frac{3}{4})(-\frac{3}{4}) + \frac{1}{4}\cdot\frac{1}{4} & \frac{2}{3}(-\frac{1}{4}) + (-\frac{3}{4})\frac{2}{3} + \frac{1}{4}\cdot\frac{2}{3} & \frac{2}{3}\cdot\frac{1}{2} + (-\frac{3}{4})\frac{1}{3} + \frac{1}{4}\cdot\frac{1}{3} \\ \frac{1}{3}\cdot\frac{2}{3} + (-\frac{2}{3})(-\frac{3}{4}) + \frac{1}{2}\cdot\frac{1}{4} & \frac{1}{3}(-\frac{1}{4}) + (-\frac{2}{3})\frac{2}{3} + \frac{1}{2}\cdot\frac{2}{3} & \frac{1}{3}\cdot\frac{1}{2} + (-\frac{2}{3})\frac{1}{3} + \frac{1}{2}\cdot\frac{1}{3} \end{bmatrix}$$

$$= \begin{bmatrix} \frac{2}{6} - \frac{3}{12} - \frac{1}{8} & -\frac{1}{8} + \frac{2}{9} - \frac{2}{6} & \frac{1}{4} + \frac{1}{9} - \frac{1}{6} \\ \frac{4}{9} + \frac{9}{16} + \frac{1}{16} & -\frac{2}{12} - \frac{6}{12} + \frac{2}{12} & \frac{2}{6} - \frac{3}{12} + \frac{1}{12} \\ \frac{2}{9} + \frac{6}{12} + \frac{1}{8} & -\frac{1}{12} - \frac{4}{9} + \frac{2}{6} & \frac{1}{6} - \frac{2}{9} + \frac{1}{6} \end{bmatrix}$$

$$= \begin{bmatrix} \frac{8 - 6 - 3}{24} & \frac{-9 + 16 - 24}{72} & \frac{9 + 4 - 6}{36} \\ \frac{64 + 81 + 9}{144} & \frac{-2 - 6 + 2}{12} & \frac{4 - 3 + 1}{12} \\ \frac{16 + 36 + 9}{72} & \frac{-3 - 16 + 12}{36} & \frac{3 - 4 + 3}{18} \end{bmatrix}$$

3×3

$$= \begin{bmatrix} \frac{-1}{24} & \frac{-17}{72} & \frac{7}{36} \\ \frac{77}{72} & \frac{-1}{2} & \frac{1}{6} \\ \frac{61}{72} & \frac{-7}{36} & \frac{1}{9} \end{bmatrix}$$

$$= \frac{1}{72} \begin{bmatrix} -3 & -17 & 14 \\ 77 & -36 & 12 \\ 61 & -14 & 8 \end{bmatrix}$$

g. A is 3×2 and B is 3×2, so AB is not defined.

h. A is 3×1 and B is 1×4 so AB is 3×4.

$$\begin{bmatrix} 2 \\ 1 \\ -1 \end{bmatrix} [3 \quad 1 \quad -1 \quad 2]$$

$$= \begin{bmatrix} 2 \cdot 3 & 2 \cdot 1 & 2 \cdot (-1) & 2 \cdot 2 \\ 1 \cdot 3 & 1 \cdot 1 & 1 \cdot (-1) & 1 \cdot 2 \\ (-1)3 & (-1)1 & (-1)(-1) & (-1)2 \end{bmatrix}$$

$$= \begin{bmatrix} 6 & 2 & -2 & 4 \\ 3 & 1 & -1 & 2 \\ -3 & -1 & 1 & -2 \end{bmatrix}$$

■

■ Example 13. Suppose in Example 10 worker D has 3 hours available on Monday, 2 hours on Wednesday, and 4 hours on Friday. Worker E has 2 hours on Monday, 3 hours on Wednesday, and 4 hours on Friday. On Monday, F has 4 hours available, on Wednesday 3 hours, and on Friday 2 hours. What is the total production for each day?

Solution: Let M be the matrix of available time where the columns represent days and the rows represent workers, and let A be the row matrix giving per hour production. Then $A = [5 \quad 3 \quad 3]$ and $M = \begin{bmatrix} 3 & 2 & 4 \\ 2 & 3 & 4 \\ 4 & 3 & 2 \end{bmatrix}$.

$$\begin{aligned} AM &= [5 \quad 3 \quad 3] \begin{bmatrix} 3 & 2 & 4 \\ 2 & 3 & 4 \\ 4 & 3 & 2 \end{bmatrix} \\ &= [(5\cdot3+3\cdot2+3\cdot4) \quad (5\cdot2+3\cdot3+3\cdot3) \quad (5\cdot4+3\cdot4+3\cdot2)] \\ &= [(15+6+12) \quad (10+9+9) \quad (20+12+6)] \\ &= [33 \quad 28 \quad 38]. \end{aligned}$$

The columns of the product matrix represent days, so we see that total production is 33 units on Monday, 28 units on Wednesday, and 38 units on Friday. ■

It is extremely important to remember that the number of columns in the matrix on the left must be equal to the number of rows in the matrix on the right. If this relationship is not satisfied, then the product is undefined.

CAUTION: In general, $AB \neq BA$. See Examples 14, 15, and 16.

■ **Example 14.** If A is 3×4 and B is 4×5, AB is a 3×5 matrix, but BA is not defined. Even if both products are defined, they may not be equal. ■

■ **Example 15.** If A is 3×4 and B is 4×3, AB is 3×3, but BA is 4×4. ■

If A and B are both square matrices of the same size, is AB equal to BA?

■ **Example 16.** Let $A = \begin{bmatrix} 2 & -1 \\ 3 & 2 \end{bmatrix}$ and $B = \begin{bmatrix} 1 & 3 \\ -1 & 2 \end{bmatrix}$.

Solution:

$$AB = \begin{bmatrix} 2\cdot1 + (-1)(-1) & 2\cdot3 + (-1)(2) \\ 3\cdot1 + 2(-1) & 3\cdot3 + 2\cdot2 \end{bmatrix} = \begin{bmatrix} 3 & 4 \\ 1 & 13 \end{bmatrix}.$$

$$BA = \begin{bmatrix} 1\cdot2 + 3\cdot3 & 1(-1) + 3\cdot2 \\ -1\cdot2 + 2\cdot3 & (-1)(-1) + 2\cdot2 \end{bmatrix} = \begin{bmatrix} 11 & 5 \\ 4 & 5 \end{bmatrix}.$$

Hence $AB \neq BA$ since $\begin{bmatrix} 3 & 4 \\ 1 & 13 \end{bmatrix} \neq \begin{bmatrix} 11 & 5 \\ 4 & 5 \end{bmatrix}$. ■

Of special interest is multiplication by one of the special identity matrices I_n that we considered in Section 1.1. If A is an $n \times n$ (square) matrix, then $AI_n = I_nA = A$. Even when A is not a square matrix, we have a special product. If A is $m \times n$, then $I_mA = A$ and $AI_n = A$.

■ Example 17. Consider $I_3 = \begin{bmatrix} 1 & 0 & 0 \\ 0 & 1 & 0 \\ 0 & 0 & 1 \end{bmatrix}, \begin{bmatrix} 2 & -1 & 3 & 4 \\ 3 & 1 & 2 & -1 \\ -1 & 5 & 1 & -1 \end{bmatrix}$

and $I_4 = \begin{bmatrix} 1 & 0 & 0 & 0 \\ 0 & 1 & 0 & 0 \\ 0 & 0 & 1 & 0 \\ 0 & 0 & 0 & 1 \end{bmatrix}$.

Solution:

Then $I_3A = \begin{bmatrix} 2 & -1 & 3 & 4 \\ 3 & 1 & 2 & -1 \\ -1 & 5 & 1 & -1 \end{bmatrix} = A$ and

$AI_4 = \begin{bmatrix} 2 & -1 & 3 & 4 \\ 3 & 1 & 2 & -1 \\ -1 & 5 & 1 & -1 \end{bmatrix} = A$.

This is the reason these special matrices are called *identity matrices*. ■

Property 1.2 ***Properties of Matrix Multiplication***

1. If A is $m \times n$, B is $n \times p$, and C is $p \times q$, then $(AB)C = A(BC)$ and the product is $m \times q$. (Associative property of multiplication)
2. If A is $m \times n$ and B and C are both $n \times p$, then $A(B + C) = AB + AC$ and is $m \times p$. (Left distributive property)
3. If A and B are $m \times n$ and C is $n \times p$, then $(A + B)C = AC + BC$ and is $m \times p$. (Right distributive property)
4. If A is an $m \times n$ matrix, B is an $n \times p$ matrix and k is a constant, then $A(kB) = k(AB)$ and is $m \times p$.

■ Example 18. Let A and B be the matrices of Example 12 (f) of this section. Find AB.

Solution:

$$A = \begin{bmatrix} \frac{1}{2} & \frac{1}{3} & -\frac{1}{2} \\ \frac{2}{3} & -\frac{3}{4} & \frac{1}{4} \\ \frac{1}{3} & -\frac{2}{3} & \frac{1}{2} \end{bmatrix} = \frac{1}{12}\begin{bmatrix} 6 & 4 & -6 \\ 8 & -9 & 3 \\ 4 & -8 & 6 \end{bmatrix}$$

$$B = \begin{bmatrix} \frac{2}{3} & -\frac{1}{4} & \frac{1}{2} \\ -\frac{3}{4} & \frac{2}{3} & \frac{1}{3} \\ \frac{1}{4} & \frac{2}{3} & \frac{1}{3} \end{bmatrix} = \frac{1}{12}\begin{bmatrix} 8 & -3 & 6 \\ -9 & 8 & 4 \\ 3 & 8 & 4 \end{bmatrix}$$

$$AB = \left(\frac{1}{12}\begin{bmatrix} 6 & 4 & -6 \\ 8 & -9 & 3 \\ 4 & -8 & 6 \end{bmatrix}\right) \cdot \left(\frac{1}{12}\begin{bmatrix} 8 & -3 & 6 \\ -9 & 8 & 4 \\ 3 & 8 & 4 \end{bmatrix}\right)$$

by property 1.2 Part (4).

$$= \left(\frac{1}{12} \cdot \frac{1}{12}\right)\begin{bmatrix} 6 & 4 & -6 \\ 8 & -9 & 3 \\ 4 & -8 & 6 \end{bmatrix}\begin{bmatrix} 8 & -3 & 6 \\ -9 & 8 & 4 \\ 3 & 8 & 4 \end{bmatrix}$$

$$= \frac{1}{144}\begin{bmatrix} 48 - 36 - 18 & -18 + 32 - 48 & 36 + 16 - 24 \\ 64 + 81 + 9 & -24 - 72 + 24 & 48 - 36 + 12 \\ 32 + 72 + 18 & -12 - 64 + 48 & 24 - 32 + 24 \end{bmatrix}$$

scalar product

$$= \frac{1}{144}\begin{bmatrix} -6 & -34 & 28 \\ 154 & -72 & 24 \\ 122 & -28 & 16 \end{bmatrix} = \left(\frac{1}{144}\right) \cdot 2\begin{bmatrix} -3 & -17 & 14 \\ 77 & -36 & 12 \\ 61 & -14 & 8 \end{bmatrix}$$

$$= \frac{1}{72}\begin{bmatrix} -3 & -17 & 14 \\ 77 & -36 & 12 \\ 61 & -14 & 8 \end{bmatrix}$$

This result is the same as before, but the fractions were not a part of the multiplication. ■

Exercise Set 1.3

1. Find AB, if it exists, for each of the following.

a. $A = \begin{bmatrix} 1 & -1 \\ 3 & 2 \end{bmatrix}$, $B = \begin{bmatrix} 2 & 2 \\ -1 & 3 \end{bmatrix}$

b. $A = \begin{bmatrix} 2 & 3 \\ -1 & 4 \\ -5 & 2 \end{bmatrix}$, $B = \begin{bmatrix} 1 & 2 \\ -2 & 1 \end{bmatrix}$

c. $A = \begin{bmatrix} 3 & 1 & 4 \\ 1 & 2 & 1 \\ 2 & -2 & 3 \end{bmatrix}, B = \begin{bmatrix} -1 & 3 \\ 4 & 2 \\ -3 & 5 \end{bmatrix}$

d. $A = \begin{bmatrix} 1 & -2 & 1 \\ 3 & 1 & -2 \end{bmatrix}, B = \begin{bmatrix} 1 & 2 & -1 & -2 \\ 2 & -1 & 3 & 2 \\ 1 & 4 & 5 & -2 \end{bmatrix}$

e. $A = \begin{bmatrix} -2 & 1 & 3 \\ 3 & 2 & 2 \end{bmatrix}, B = \begin{bmatrix} -1 & 3 \\ 3 & -2 \\ 4 & 1 \end{bmatrix}$

f. $A = \begin{bmatrix} 1 & 3 & -1 \\ 2 & 4 & 2 \\ -2 & 1 & 3 \end{bmatrix}, B = \begin{bmatrix} 2 & 1 \\ 3 & -1 \end{bmatrix}$

g. $A = \begin{bmatrix} 1 & -2 & 3 & 1 \\ 4 & 2 & -2 & 3 \\ 1 & 2 & 4 & 1 \end{bmatrix}, B = \begin{bmatrix} -2 & 2 & 3 \\ 4 & 1 & 5 \\ 1 & 2 & -1 \\ -1 & 3 & -2 \end{bmatrix}$

h. $A = \begin{bmatrix} -\frac{1}{4} & \frac{1}{3} & \frac{3}{4} \\ 2 & \frac{2}{3} & -\frac{1}{4} \end{bmatrix}, B = \begin{bmatrix} \frac{1}{2} & -\frac{1}{4} \\ -\frac{3}{4} & \frac{2}{3} \\ 3 & 1 \end{bmatrix}$

2. Find BA, if it exists, for each part of Problem 1.

3. Let A be a 3×4 matrix and let $B = \begin{bmatrix} 3 & 0 & 0 \\ 0 & 3 & 0 \\ 0 & 0 & 3 \end{bmatrix}$. What is the effect of multiplying A on the left by B?

4. What is the effect of multiplying a 3×4 matrix on the left by $\begin{bmatrix} 1 & 0 & 0 \\ 0 & 2 & 0 \\ 0 & 0 & 1 \end{bmatrix}$?

5. Perform the indicated operations where possible.

a. $\begin{bmatrix} 2 & -1 & 3 \\ 1 & -1 & 1 \end{bmatrix} \left(\begin{bmatrix} 1 & 3 \\ -2 & 1 \\ 1 & 4 \end{bmatrix} - 2 \begin{bmatrix} 2 & -1 \\ 1 & -1 \\ -2 & 3 \end{bmatrix} \right)$

b. $\begin{bmatrix} 2 & -1 \\ 3 & 1 \\ -1 & 2 \end{bmatrix}\left(\begin{bmatrix} 1 & -1 & 2 \\ 3 & 1 & -1 \end{bmatrix} + 3\begin{bmatrix} 2 & 1 & -2 \\ 3 & -2 & 2 \end{bmatrix}\right)$

c. $\begin{bmatrix} 1 & 2 \\ 3 & -1 \end{bmatrix}\left(\begin{bmatrix} -1 & 2 & 1 \\ 3 & 5 & 2 \\ 1 & 2 & 1 \end{bmatrix} + 2\begin{bmatrix} -2 & 1 & 3 \\ 3 & 2 & 1 \\ -1 & 3 & 1 \end{bmatrix}\right)$

d. $\begin{bmatrix} 2 & 1 & -2 \\ -3 & 1 & 1 \end{bmatrix}\left(\begin{bmatrix} 1 & 2 & 1 & 1 \\ -2 & 1 & -1 & 3 \\ -1 & -2 & 1 & 3 \end{bmatrix} - 2\begin{bmatrix} 1 & 2 & 1 & -1 \\ -1 & 3 & 2 & -1 \\ 4 & -1 & 3 & 2 \end{bmatrix}\right)$

6. Let $A = \begin{bmatrix} 2 & 1 & -1 & 3 \\ 1 & 3 & 1 & 2 \\ -1 & 2 & 1 & 1 \end{bmatrix}$. Multiply A on the left by each of the following matrices.

a. $\begin{bmatrix} 1 & 0 & 0 \\ 0 & 0 & 1 \\ 0 & 1 & 0 \end{bmatrix}$

b. $\begin{bmatrix} 1 & -2 & 0 \\ 0 & 1 & 0 \\ 0 & 0 & 1 \end{bmatrix}$

c. $\begin{bmatrix} 1 & 0 & 0 \\ -3 & 1 & 0 \\ -2 & 0 & 1 \end{bmatrix}$

d. $\begin{bmatrix} 1 & 0 & -3 \\ 0 & 1 & -2 \\ 0 & 0 & 1 \end{bmatrix}$

7. Given $A = \begin{bmatrix} 2 & 1 & 3 \\ 0 & 1 & 4 \end{bmatrix}$ and $B = \begin{bmatrix} 3 & 1 \\ 1 & 1 \\ -1 & 0 \end{bmatrix}$, show that

a. $(AB)^T = B^T A^T$
b. $(BA)^T = A^T B^T$

1.4 Elementary Row Operations and Matrix Inverses

Consider the two 3×3 matrices

$$A = \begin{bmatrix} -1 & 1 & 1 \\ 1 & 4 & -5 \\ 1 & -2 & 0 \end{bmatrix} \text{ and } B = \begin{bmatrix} 10 & 2 & 9 \\ 5 & 1 & 4 \\ 6 & 1 & 5 \end{bmatrix}.$$

Since both are 3×3, we can find AB and BA.

$$AB = \begin{bmatrix} -1 & 1 & 1 \\ 1 & 4 & -5 \\ 1 & -2 & 0 \end{bmatrix} \begin{bmatrix} 10 & 2 & 9 \\ 5 & 1 & 4 \\ 6 & 1 & 5 \end{bmatrix}$$

$$= \begin{bmatrix} -10 + 5 + 6 & -2 + 1 + 1 & -9 + 4 + 5 \\ 10 + 20 - 30 & 2 + 4 - 5 & 9 + 16 - 25 \\ 10 - 10 + 0 & 2 - 2 + 0 & 9 - 8 + 0 \end{bmatrix}$$

$$= \begin{bmatrix} 1 & 0 & 0 \\ 0 & 1 & 0 \\ 0 & 0 & 1 \end{bmatrix} = I_3.$$

$$BA = \begin{bmatrix} 10 & 2 & 9 \\ 5 & 1 & 4 \\ 6 & 1 & 5 \end{bmatrix} \begin{bmatrix} -1 & 1 & 1 \\ 1 & 4 & -5 \\ 1 & -2 & 0 \end{bmatrix}$$

$$= \begin{bmatrix} -10 + 2 + 9 & 10 + 8 - 18 & 10 - 10 + 0 \\ -5 + 1 + 4 & 5 + 4 - 8 & 5 - 5 + 0 \\ -6 + 1 + 5 & 6 + 4 - 10 & 6 - 5 + 0 \end{bmatrix}$$

$$= \begin{bmatrix} 1 & 0 & 0 \\ 0 & 1 & 0 \\ 0 & 0 & 1 \end{bmatrix} = I_3.$$

Hence $AB = BA = I_3$. This situation is similar to multiplication of real numbers where $\frac{2}{3} \cdot \frac{3}{2} = 1$. We say that $\frac{2}{3}$ and $\frac{3}{2}$ are multiplicative inverses of each other.

In the same manner, the matrices A and B above are **inverses** of each other since $AB = BA = I_3$. Notationally we say $B = A^{-1}$ or $A = B^{-1}$.

Definition 1.5 If A is an $n \times n$ (square) matrix, then the *inverse of A* is a matrix A^{-1}, if it exists, such that $A \cdot A^{-1} = A^{-1} \cdot A = I_n$.

CAUTION: The inverse of a matrix can exist only if the original matrix is square. However, not all square matrices have inverses. If a square matrix has an inverse, the inverse is unique and we say the matrix is **invertible.**

To show uniqueness, suppose A, B, and C are $n \times n$ matrices such that $AB = BA = I_n$ and $AC = CA = I_n$. Then

$$\begin{aligned} AB &= AC \\ B(AB) &= B(AC) \\ (BA)B &= (BA)C && \text{by Property 1.2, part } a. \\ I_nB &= I_nC && \text{by assumption, } BA = I_n. \\ B &= C && I_n \text{ is the identity.} \end{aligned}$$

■ **Example 19.** Determine whether or not $B = A^{-1}$ for each of the following.

a. $A = \begin{bmatrix} 1 & 2 \\ 2 & 5 \end{bmatrix}$, $B = \begin{bmatrix} 5 & -2 \\ -2 & 1 \end{bmatrix}$

b. $A = \begin{bmatrix} 2 & -3 \\ 1 & 1 \end{bmatrix}$, $B = \begin{bmatrix} \frac{1}{5} & \frac{3}{5} \\ -\frac{1}{5} & \frac{2}{5} \end{bmatrix}$

c. $A = \begin{bmatrix} 2 & -1 \\ 3 & 2 \end{bmatrix}$, $B = \begin{bmatrix} 2 & -1 \\ -3 & 2 \end{bmatrix}$

d. $A = \begin{bmatrix} -\frac{7}{2} & -4 & -\frac{3}{2} \\ -\frac{3}{2} & -2 & -1 \\ -2 & -\frac{3}{2} & 0 \end{bmatrix}$, $B = \begin{bmatrix} 12 & -18 & -8 \\ -16 & 24 & 10 \\ 14 & -22 & -8 \end{bmatrix}$

Solutions:

a. $$AB = \begin{bmatrix} 1 & 2 \\ 2 & 5 \end{bmatrix}\begin{bmatrix} 5 & -2 \\ -2 & 1 \end{bmatrix} = \begin{bmatrix} 5-4 & -2+2 \\ 10-10 & -4+5 \end{bmatrix} = \begin{bmatrix} 1 & 0 \\ 0 & 1 \end{bmatrix}$$

$$BA = \begin{bmatrix} 5 & -2 \\ -2 & 1 \end{bmatrix}\begin{bmatrix} 1 & 2 \\ 2 & 5 \end{bmatrix} = \begin{bmatrix} 5-4 & 10-10 \\ -2+2 & -4+5 \end{bmatrix} = \begin{bmatrix} 1 & 0 \\ 0 & 1 \end{bmatrix}$$

so $B = A^{-1}$ or $A = B^{-1}$.

b. $$AB = \begin{bmatrix} 2 & -3 \\ 1 & 1 \end{bmatrix}\begin{bmatrix} \frac{1}{5} & \frac{3}{5} \\ -\frac{1}{5} & \frac{2}{5} \end{bmatrix} = \begin{bmatrix} \frac{2}{5}+\frac{3}{5} & \frac{6}{5}-\frac{6}{5} \\ \frac{1}{5}-\frac{1}{5} & \frac{3}{5}+\frac{2}{5} \end{bmatrix} = \begin{bmatrix} 1 & 0 \\ 0 & 1 \end{bmatrix}$$

and $$BA = \begin{bmatrix} \frac{1}{5} & \frac{3}{5} \\ -\frac{1}{5} & \frac{2}{5} \end{bmatrix}\begin{bmatrix} 2 & -3 \\ 1 & 1 \end{bmatrix} = \begin{bmatrix} \frac{2}{5}+\frac{3}{5} & -\frac{3}{5}+\frac{3}{5} \\ -\frac{2}{5}+\frac{2}{5} & \frac{3}{5}+\frac{2}{5} \end{bmatrix} = \begin{bmatrix} 1 & 0 \\ 0 & 1 \end{bmatrix},$$

so $B = A^{-1}$ or $A = B^{-1}$.

c. $AB = \begin{bmatrix} 2 & -1 \\ 3 & 2 \end{bmatrix}\begin{bmatrix} 2 & -1 \\ -3 & 2 \end{bmatrix}$

$= \begin{bmatrix} 4+3 & -2-2 \\ 6-6 & -3+4 \end{bmatrix} = \begin{bmatrix} 7 & -4 \\ 0 & 1 \end{bmatrix},$

so $B \neq A^{-1}$.

d. $AB = \begin{bmatrix} -\frac{7}{2} & -4 & -\frac{3}{2} \\ -\frac{3}{2} & -2 & -1 \\ -2 & -\frac{3}{2} & 0 \end{bmatrix}\begin{bmatrix} 12 & -18 & -8 \\ -16 & 24 & 10 \\ 14 & -22 & -8 \end{bmatrix}$

$= \begin{bmatrix} -42+64-21 & 63-96+33 & 28-40+12 \\ -18+32-14 & 27-48+22 & 12-20+8 \\ -24+24+0 & 36-36+0 & 16-15+0 \end{bmatrix}$

$= \begin{bmatrix} 1 & 0 & 0 \\ 0 & 1 & 0 \\ 0 & 0 & 1 \end{bmatrix}$

and $BA = \begin{bmatrix} 12 & -18 & -8 \\ -16 & 24 & 10 \\ 14 & -22 & -8 \end{bmatrix}\begin{bmatrix} -\frac{7}{2} & -4 & -\frac{3}{2} \\ -\frac{3}{2} & -2 & -1 \\ -2 & -\frac{3}{2} & 0 \end{bmatrix}$

$= \begin{bmatrix} -42+27+16 & -48+36+12 & -18+18+0 \\ 56-36-20 & 64-48-15 & 24-24+0 \\ -49+33+16 & -56+44+12 & -21+22+0 \end{bmatrix}$

$= \begin{bmatrix} 1 & 0 & 0 \\ 0 & 1 & 0 \\ 0 & 0 & 1 \end{bmatrix}.$

Hence $B = A^{-1}$ or $A = B^{-1}$. ■

Given a square matrix, is there any way to determine whether or not it is invertible? If we know a square matrix is invertible, is there any procedure for finding the inverse? The answer to both of the preceding questions is "yes," and our initial approach will be to find the inverse. If it doesn't exist, the procedure will indicate this to us at some stage of the computation.

The method of inverting a matrix used here involves three new concepts: **elementary row operations, equivalent matrices,** and **Gauss-Jordan reduction.** Each of these concepts should be mastered since they will be used repeatedly throughout much of this book.

There are three elementary row operations. They are as follows:

1. Replace a row by a nonzero constant times the row.
2. Interchange two rows.
3. Replace some specified row by adding a nonzero multiple of another row to the specified row. Or, more generally, replace a specified row by adding a nonzero multiple of another row to a nonzero multiple of the specified row.

CAUTION: Multiplying a row by a nonzero constant is not the same as scalar multiplication. In scalar multiplication, every row is multiplied by the constant.

When elementary row operations are used to change a matrix A into a new matrix B, the matrix B is said to be **row equivalent** to A. In general, a matrix B is row equivalent to a matrix A if B can be obtained from A by a sequence of one or more elementary row operations. We write $A \sim B$. Remember that $A \sim B$ is different in meaning than $A = B$.

■ Example 20. Perform the indicated elementary row operations on the matrix $A = \begin{bmatrix} 1 & -1 & 3 \\ 2 & 1 & 4 \\ -1 & 2 & 1 \end{bmatrix}$ to produce a matrix B that is row equivalent to A.

a. Multiply row 1 by 2. (Notation: 2R1)
b. Interchange rows 1 and 3. (Notation: R1 ↔ R3)
c. Replace row 3 by 2 times row 1 plus row 3. (Notation: 2R1 + R3)

When we write aRi + bRj, it is row j that is being replaced.

Solutions: a. Replacing row 1 by 2 times row 1 we get the matrix

$$B = \begin{bmatrix} 2 & -2 & 6 \\ 2 & 1 & 4 \\ -1 & 2 & 1 \end{bmatrix}.$$

b. Interchanging rows 1 and 3 of A gives the matrix

$$B = \begin{bmatrix} -1 & 2 & 1 \\ 2 & 1 & 4 \\ 1 & -1 & 3 \end{bmatrix}.$$

c. Replacing row 3 by 2 times row 1 plus row 3 we get

$$B = \begin{bmatrix} 1 & -1 & 3 \\ 2 & 1 & 4 \\ 2(1) + (-1) & 2(-1) + 2 & 2(3) + 1 \end{bmatrix}$$

$$= \begin{bmatrix} 1 & -1 & 3 \\ 2 & 1 & 4 \\ 1 & 0 & 7 \end{bmatrix}$$

Gauss-Jordan reduction is a procedure that uses a sequence of elementary row operations on a matrix to produce a row equivalent matrix in **completely reduced form.** A matrix is said to be in *completely reduced form* providing it satisfies each of the following four conditions.

1. The first nonzero entry in each row is a 1.
2. The first nonzero entry in each row is to the right of the first nonzero entry in each preceding row.
3. Each column containing the first nonzero row entry of any row is cleared. That is, it contains no other nonzero entry.
4. Any rows consisting entirely of zeros are at the bottom of the matrix.

Example 21. Determine whether or not each of the following matrices is in completely reduced form. If a matrix is not, indicate which conditions are not satisfied.

a. $\begin{bmatrix} 1 & 0 & 0 & 0 \\ 0 & 1 & 0 & 0 \\ 0 & 0 & 1 & 0 \\ 0 & 0 & 0 & 1 \end{bmatrix}$

b. $\begin{bmatrix} 1 & 0 & 1 & 2 \\ 0 & 1 & 0 & 0 \\ 0 & 0 & 0 & 0 \end{bmatrix}$

c. $\begin{bmatrix} 0 & 1 & 0 \\ 1 & 0 & 0 \\ 0 & 0 & 1 \end{bmatrix}$

d. $\begin{bmatrix} 1 & 0 & 1 \\ 0 & 2 & 0 \\ 0 & 0 & 1 \end{bmatrix}$

e. $\begin{bmatrix} 1 & 0 & 1 & 0 \\ 0 & 1 & 1 & 0 \\ 0 & 0 & 0 & 1 \end{bmatrix}$

f. $\begin{bmatrix} 0 & 1 & 1 & 2 \\ 0 & 0 & 0 & 0 \\ 1 & 0 & 0 & 1 \\ 0 & 0 & 1 & 1 \end{bmatrix}$

Solutions:

a. Yes, this matrix is in completely reduced form.
b. Yes, this matrix is in completely reduced form.
c. No. Condition 2 is not satisfied. Interchanging rows 1 and 2 would put the matrix in reduced form.
d. No. Row 2 violates condition 1 and column 3 violates condition 3. Adding -1 times row 3 to row 1 and multiplying row 2 by $\frac{1}{2}$ would put the matrix in reduced form.
e. Yes. The 1s in column 3 are acceptable, since column 3 does not contain the first nonzero entry of any row.
f. No. Row 2 violates condition 4, rows 3 and 1 violate condition 2, and column 3 violates condition 3. ■

The matrix in example 21(a) is the identity matrix I_4. We see that it is in completely reduced form. In fact, every identity matrix, I_n, is in completely reduced form.

As we try to find the inverse of a square matrix, we will try to reduce the given matrix to I_n using elementary row operations. If the given matrix is row equivalent to I_n, then it is invertible.

We now consider an example using Gauss-Jordan reduction to try to reduce a 3×3 matrix A to I_3. The successive matrices are on the left with the elementary row operation used to obtain these matrices on the right.

■ Example 22. Let $A = \begin{bmatrix} -1 & 1 & 1 \\ 1 & 4 & -5 \\ 1 & -2 & 0 \end{bmatrix}$. Use Gauss-Jordan reduction to reduce A.

Solution:

New Matrix	*Elementary Row Operations*
$A_1 = \begin{bmatrix} -1 & 1 & 1 \\ 0 & 5 & -4 \\ 1 & -2 & 0 \end{bmatrix}$	Replace row 2 by row 1 plus row 2
$A_2 = \begin{bmatrix} -1 & 1 & 1 \\ 0 & 5 & -4 \\ 0 & -1 & 1 \end{bmatrix}$	Replace row 3 by row 1 plus row 3
$A_3 = \begin{bmatrix} -1 & 0 & 2 \\ 0 & 5 & -4 \\ 0 & -1 & 1 \end{bmatrix}$	Replace row 1 by row 3 plus row 1
$A_4 = \begin{bmatrix} -1 & 0 & 2 \\ 0 & 0 & 1 \\ 0 & -1 & 1 \end{bmatrix}$	Replace row 2 by 5 times row 3 plus row 2

$$A_5 = \begin{bmatrix} -1 & 0 & 0 \\ 0 & 0 & 1 \\ 0 & -1 & 1 \end{bmatrix}$$ Replace row 1 by -2 times row 2 plus row 1

$$A_6 = \begin{bmatrix} -1 & 0 & 0 \\ 0 & 0 & 1 \\ 0 & -1 & 0 \end{bmatrix}$$ Replace row 3 by -1 times row 2 plus row 3

$$A_7 = \begin{bmatrix} -1 & 0 & 0 \\ 0 & -1 & 0 \\ 0 & 0 & 1 \end{bmatrix}$$ Interchange row 2 and row 3

$$A_8 = \begin{bmatrix} 1 & 0 & 0 \\ 0 & -1 & 0 \\ 0 & 0 & 1 \end{bmatrix}$$ Multiply row 1 by -1.

$$A_9 = \begin{bmatrix} 1 & 0 & 0 \\ 0 & 1 & 0 \\ 0 & 0 & 1 \end{bmatrix} = I_3$$ Multiply row 2 by -1.

Since $A_9 = I_3$ was obtained from A by a sequence of elementary row operations, A is row equivalent to I_3. Since A is row equivalent to I_3, we know A^{-1} exists. ■

The three elementary row operations can be combined to shorten the number of steps. Recall that the notation aRi + bRj means to replace the jth row with 'a' times row i plus 'b' times row j. When we write -2R2 + R1 we will replace row 1 by -2 times row 2 plus row 1. Remember the symbol '$\sim$' is row equivalence.

The sequence $A \sim A_1 \sim A_2 \sim A_3 \sim A_4 \sim A_5 \sim A_6 \sim A_7 \sim A_8 \sim A_9$ could be shortened to $A \sim A_2 \sim A_4 \sim A_6 \sim A_7 \sim A_9$ by combining some steps involving the same row. We would write

$$\begin{bmatrix} -1 & 1 & 1 \\ 1 & 4 & -5 \\ 1 & -2 & 0 \end{bmatrix} \begin{matrix} \\ R1 + R2 \\ R1 + R3 \end{matrix} \begin{bmatrix} -1 & 1 & 1 \\ 0 & 5 & -4 \\ 0 & -1 & 1 \end{bmatrix}$$

$$\begin{matrix} R3 + R1 \\ 5R3 + R2 \\ \\ \end{matrix} \begin{bmatrix} -1 & 0 & 2 \\ 0 & 0 & 1 \\ 0 & -1 & 1 \end{bmatrix} \begin{matrix} -2R2 + R1 \\ \\ -R2 + R3 \end{matrix} \begin{bmatrix} -1 & 0 & 0 \\ 0 & 0 & 1 \\ 0 & -1 & 0 \end{bmatrix}$$

$$R2 \leftrightarrow R3 \begin{bmatrix} -1 & 0 & 0 \\ 0 & -1 & 0 \\ 0 & 0 & 1 \end{bmatrix} \begin{matrix} -R1 \\ -R2 \\ \\ \end{matrix} \begin{bmatrix} 1 & 0 & 0 \\ 0 & 1 & 0 \\ 0 & 0 & 1 \end{bmatrix}$$

The row equivalence symbol '~' has been dropped and the matrices are separated by the row operation statements being used. We will adopt the procedure where the specific row operation used is written on the same line as the row being replaced.

Two new terms, **column clearing** and **pivot number,** will now be introduced. By clearing a column, we mean to use the elementary row operations in such a way as to make all of the entries in a specific column zero, except one. This nonzero entry is called a pivot number and is in the row being used to reduce the matrix.

In the above example the first series of elementary row operations cleared column one using -1 (in row 1) as the pivot number.

$$\begin{bmatrix} -1 & 1 & 1 \\ 1 & 4 & 5 \\ 1 & -2 & 0 \end{bmatrix} \begin{matrix} \\ R1 + R2 \\ R1 + R3 \end{matrix} \begin{bmatrix} -1 & 1 & 1 \\ 0 & 5 & -4 \\ 0 & -1 & 1 \end{bmatrix}.$$

The second series of row operations was used to clear column two using -1 (in row 3) as the pivot number.

$$\begin{bmatrix} -1 & 1 & 1 \\ 0 & 5 & -4 \\ 0 & -1 & 1 \end{bmatrix} \begin{matrix} R3 + R1 \\ 5R3 + R2 \\ \\ \end{matrix} \begin{bmatrix} -1 & 0 & 2 \\ 0 & 0 & 1 \\ 0 & -1 & 1 \end{bmatrix}.$$

This "column clearing" procedure is the major thrust of the elementary row operations and will be used repeatedly to solve many different kinds of problems. This procedure must be mastered.

The procedure we use here for finding the inverse of a matrix makes use of row reduction and row equivalent matrices. In any situation where row reduction is used (and we will use this technique again in other chapters), there is a great deal of arithmetic involved. In an attempt to reduce arithmetic errors, an arithmetic check column is introduced that allows us to uncover most of the arithmetic errors as we go. We return to the matrix of example 22 to illustrate the check column. Consider

$$\begin{bmatrix} -1 & 1 & 1 \\ 1 & 4 & -5 \\ 1 & -2 & 0 \end{bmatrix} \begin{matrix} 1 \\ 0 \\ -1 \end{matrix}$$

which is matrix A of example 22 with the arithmetic check column outside the matrix on the right.

The check column consists of the sums of the row entries for each row.

Row 1: $-1 + 1 + 1 = 1$.
Row 2: $1 + 4 - 5 = 0$.
Row 3: $1 - 2 + 0 = -1$.

As we use elementary row operations to reduce the matrix A we perform the same row operations on the check column. The row sums for the new matrix should equal the new values in the check column. The check column can be located anywhere, but we will write it outside the matrix to the right of the original matrix.

To illustrate the use of the check column, consider again the matrix of example 22.

$$\left[\begin{array}{rrr} -1 & 1 & 1 \\ 1 & 4 & -5 \\ 1 & -2 & 0 \end{array}\right] \begin{array}{r} 1 \\ 0 \\ -1 \end{array} \quad \begin{array}{r} \\ R1 + R2 \\ R1 + R3 \end{array} \left[\begin{array}{rrr} -1 & 1 & 1 \\ 0 & 5 & -4 \\ 0 & -1 & 1 \end{array}\right] \begin{array}{r} 1 \\ 1 \\ 0 \end{array}$$

$$\begin{array}{r} R3 + R1 \\ 5R3 + R2 \\ \\ \end{array} \left[\begin{array}{rrr} -1 & 0 & 2 \\ 0 & 0 & 1 \\ 0 & -1 & 1 \end{array}\right] \begin{array}{r} 1 \\ 1 \\ 0 \end{array} \quad \begin{array}{r} -2R2 + R1 \\ \\ -R2 + R3 \end{array} \left[\begin{array}{rrr} -1 & 0 & 0 \\ 0 & 0 & 1 \\ 0 & -1 & 0 \end{array}\right] \begin{array}{r} -1 \\ 1 \\ -1 \end{array}$$

And so on. We see at each step that the row sum equals the new row entry in the check column. The entries in the check column are obtained by performing the same row operations on this column that are used on the rest of the matrix.

If, in going from the second to the third matrix, we would have gotten

$$\left[\begin{array}{rrr} -1 & 0 & 2 \\ 0 & 0 & -1 \\ 0 & -1 & 1 \end{array}\right] \begin{array}{l} 1 \\ 1, \\ 0 \end{array}$$

we should see that the second row total does not equal the check column value, and hence there is an arithmetic error somewhere, telling us where to correct our error.

CAUTION: This check procedure is not perfect. If more than one arithmetic error is made, the errors may compensate, and the check will not catch it.

We now proceed with a method of computing the inverse of a matrix. Even though the procedure may seem artificial at this time, no attempt is made here to justify it mathematically. For now, it is important that you learn the procedure. A justification will be given in the next chapter. Consider the 2×2 matrices

$$A = \begin{bmatrix} 1 & 1 \\ 3 & 2 \end{bmatrix} \quad \text{and} \quad I_2 = \begin{bmatrix} 1 & 0 \\ 0 & 1 \end{bmatrix}.$$

First we **augment** A with I_2. The term augment means to combine into a single matrix. That is, we form the new matrix

$$[A|I_2] = \left[\begin{array}{cc|cc} 1 & 1 & 1 & 0 \\ 3 & 2 & 0 & 1 \end{array}\right]$$

by writing down A on the left and putting I_2 beside it. The vertical line between A and I_2 helps us keep the matrix A identified.

Use elementary row operations to reduce A to the identity I_2. That is, perform the row operations necessary to transform $[A|I_2]$ into the matrix $[I_2|B]$ if possible, where the row operations are done all the way across the augmented matrix. Once this is done, B can be shown to be the inverse of A. That is, A is invertible if and only if A is row equivalent to I_2.

■ **Example 23.** Find A^{-1} if $A = \begin{bmatrix} 1 & 1 \\ 3 & 2 \end{bmatrix}$.

Solution: Augmenting A with $I_2 = \begin{bmatrix} 1 & 0 \\ 0 & 1 \end{bmatrix}$ and adding the check column and using Gauss-Jordan reduction, we get

$$\left[\begin{array}{cc|cc} 1 & 1 & 1 & 0 \\ 3 & 2 & 0 & 1 \end{array}\right] \begin{array}{c} 3 \\ 6 \end{array} \; -3R1 + R2 \left[\begin{array}{cc|cc} 1 & 1 & 1 & 0 \\ 0 & -1 & -3 & 1 \end{array}\right] \begin{array}{c} 3 \\ -3 \end{array}$$

$$R2 + R1 \left[\begin{array}{cc|cc} 1 & 0 & -2 & 1 \\ 0 & -1 & -3 & 1 \end{array}\right] \begin{array}{c} 0 \\ -3 \end{array} \quad -R2 \left[\begin{array}{cc|cc} 1 & 0 & -2 & 1 \\ 0 & 1 & 3 & -1 \end{array}\right] \begin{array}{c} 0 \\ 3 \end{array}.$$

Then $B = \begin{bmatrix} -2 & 1 \\ 3 & -1 \end{bmatrix}$. To verify that $B = A^{-1}$, show that $AB = BA = I_2$.

$$AB = \begin{bmatrix} 1 & 1 \\ 3 & 2 \end{bmatrix} \begin{bmatrix} -2 & 1 \\ 3 & -1 \end{bmatrix}$$

$$= \begin{bmatrix} -2+3 & 1-1 \\ -6+6 & 3-2 \end{bmatrix} = \begin{bmatrix} 1 & 0 \\ 0 & 1 \end{bmatrix} \text{ and}$$

$$BA = \begin{bmatrix} -2 & 1 \\ 3 & -1 \end{bmatrix} \begin{bmatrix} 1 & 1 \\ 3 & 2 \end{bmatrix}$$

$$= \begin{bmatrix} -2+3 & -2+2 \\ 3-3 & 3-2 \end{bmatrix} = \begin{bmatrix} 1 & 0 \\ 0 & 1 \end{bmatrix}.$$

Hence $B = A^{-1}$. ■

■ **Example 24.** Find A^{-1} if $A = \begin{bmatrix} 2 & 3 \\ 1 & 4 \end{bmatrix}$.

Solution: Augmenting A with I_2 and the check column, we get

$$\left[\begin{array}{cc|cc} 2 & 3 & 1 & 0 \\ 1 & 4 & 0 & 1 \end{array}\right] \begin{array}{c} 6 \\ 6 \end{array}$$

$$\begin{array}{r} \\ -R1 + 2R2 \end{array} \left[\begin{array}{cc|cc} 2 & 3 & 1 & 0 \\ 0 & 5 & -1 & 2 \end{array}\right] \begin{array}{c} 6 \\ 6 \end{array}$$

$$\begin{array}{r} -3R2 + 5R1 \\ \\ \end{array} \left[\begin{array}{cc|cc} 10 & 0 & 8 & -6 \\ 0 & 5 & -1 & 2 \end{array}\right] \begin{array}{c} 12 \\ 6 \end{array}$$

$$\begin{array}{r} \frac{1}{10}R1 \\ \frac{1}{5}R2 \end{array} \left[\begin{array}{cc|cc} 1 & 0 & \frac{8}{10} & -\frac{6}{10} \\ 0 & 1 & -\frac{1}{5} & \frac{2}{5} \end{array}\right] \begin{array}{c} \frac{12}{10} \\ \frac{6}{5} \end{array}.$$

So $B = \begin{bmatrix} \frac{8}{10} & -\frac{6}{10} \\ -\frac{1}{5} & \frac{2}{5} \end{bmatrix} = \begin{bmatrix} \frac{4}{5} & -\frac{3}{5} \\ -\frac{1}{5} & \frac{2}{5} \end{bmatrix}.$

$$\begin{aligned}
AB &= \begin{bmatrix} 2 & 3 \\ 1 & 4 \end{bmatrix} \begin{bmatrix} \frac{4}{5} & -\frac{3}{5} \\ -\frac{1}{5} & \frac{2}{5} \end{bmatrix} \\
&= \begin{bmatrix} 2\cdot(\frac{4}{5}) + 3(-\frac{1}{5}) & 2(-\frac{3}{5}) + 3(\frac{2}{5}) \\ 1\cdot(\frac{4}{5}) + 4(-\frac{1}{5}) & 1(-\frac{3}{5}) + 4(\frac{2}{5}) \end{bmatrix} \\
&= \begin{bmatrix} \frac{8}{5} - \frac{3}{5} & -\frac{6}{5} + \frac{6}{5} \\ \frac{4}{5} - \frac{4}{5} & -\frac{3}{5} + \frac{8}{5} \end{bmatrix} \\
&= \begin{bmatrix} 1 & 0 \\ 0 & 1 \end{bmatrix} = I_2.
\end{aligned}$$

$$\begin{aligned}
BA &= \begin{bmatrix} \frac{4}{5} & -\frac{3}{5} \\ -\frac{1}{5} & \frac{2}{5} \end{bmatrix} \begin{bmatrix} 2 & 3 \\ 1 & 4 \end{bmatrix} \\
&= \begin{bmatrix} \frac{4}{5}(2) + (-\frac{3}{5})(1) & \frac{4}{5}(3) + (-\frac{3}{5})(4) \\ -\frac{1}{5}(2) + \frac{2}{5}(1) & -\frac{1}{5}(3) + \frac{2}{5}(4) \end{bmatrix} \\
&= \begin{bmatrix} \frac{8}{5} - \frac{3}{5} & \frac{12}{5} - \frac{12}{5} \\ -\frac{2}{5} + \frac{2}{5} & -\frac{3}{5} + \frac{8}{5} \end{bmatrix} \\
&= \begin{bmatrix} 1 & 0 \\ 0 & 1 \end{bmatrix} = I_2.
\end{aligned}$$

Then $A^{-1} = B = \begin{bmatrix} \frac{4}{5} & -\frac{3}{5} \\ -\frac{1}{5} & \frac{2}{5} \end{bmatrix}.$ ■

Earlier in this section it was pointed out that not all square matrices are invertible. The following example illustrates such a matrix and how our procedure tells us the matrix has no inverse.

■ **Example 25.** Given $A = \begin{bmatrix} 1 & 3 \\ 2 & 6 \end{bmatrix}$ find A^{-1} if it exists.

Solution:

$$\left[\begin{array}{cc|cc} 1 & 3 & 1 & 0 \\ 2 & 6 & 0 & 1 \end{array}\right] \begin{matrix} 5 \\ 9 \end{matrix} \; -2R1 + R2 \left[\begin{array}{cc|cc} 1 & 3 & 1 & 0 \\ 0 & 0 & -2 & 1 \end{array}\right] \begin{matrix} 5 \\ -1 \end{matrix}$$

In our procedure for inverting a matrix, the (2,2) entry in the final matrix is to be a 1. However, in this illustration, we cannot make this entry a 1 by using elementary row operations without affecting the 0 which we need in the (2,1) entry. Hence, we cannot reduce A to I_2 and therefore A is not invertible. ■

We restate, without proof, the following property.

> **Property 1.3** An $n \times n$ square matrix A is invertible if and only if A is row equivalent to I_n.

Before considering matrices larger than 2×2, a shortcut method for inverting 2×2 matrices can be considered. The following property provides this shortcut.

> **Property 1.4** If $A = \begin{bmatrix} a & b \\ c & d \end{bmatrix}$, then A is invertible if and only if $ad - bc \neq 0$. Furthermore, if $ad - bc \neq 0$, then
>
> $$A^{-1} = \frac{1}{ad - bc}\begin{bmatrix} d & -b \\ -c & a \end{bmatrix}.$$

Consider examples 23 and 24 of this section. Let $A = \begin{bmatrix} 1 & 1 \\ 3 & 2 \end{bmatrix}$. Then by property 1.4, since $ad - bc = 1 \cdot 2 - 1 \cdot 3 = -1 \neq 0$, we have $A^{-1} = \frac{1}{-1}\begin{bmatrix} 2 & -1 \\ -3 & 1 \end{bmatrix} = -1\begin{bmatrix} 2 & -1 \\ -3 & 1 \end{bmatrix} = \begin{bmatrix} -2 & 1 \\ 3 & -1 \end{bmatrix}$ which is what we found previously. Let $A = \begin{bmatrix} 2 & 3 \\ 1 & 4 \end{bmatrix}$. Then by the property 1.4, since $ad - bc = 2 \cdot 4 - 3 \cdot 1 = 5$, we have $A^{-1} = \frac{1}{5}\begin{bmatrix} 4 & -3 \\ -1 & 2 \end{bmatrix} = \begin{bmatrix} \frac{4}{5} & -\frac{3}{5} \\ -\frac{1}{5} & \frac{2}{5} \end{bmatrix}$ which is what we obtained by computation.

In example 25, where $A = \begin{bmatrix} 1 & 3 \\ 2 & 6 \end{bmatrix}$, we have $ab - bc = 1 \cdot 6 - 3 \cdot 2 = 0$, so by property 1.4, A is not invertible.

CAUTION: This shortcut procedure works only for 2×2 matrices.

Let us turn our attention now to the inverse of a 3 × 3 matrix.

■ Example 26. Let $A = \begin{bmatrix} -1 & 1 & 1 \\ 1 & 4 & -5 \\ 1 & -2 & 0 \end{bmatrix}$. Find A^{-1}.

Solution: Augment A with $I_3 = \begin{bmatrix} 1 & 0 & 0 \\ 0 & 1 & 0 \\ 0 & 0 & 1 \end{bmatrix}$. Use the check column and reduce A by Gauss-Jordan reduction.

$$\left[\begin{array}{rrr|rrr} -1 & 1 & 1 & 1 & 0 & 0 \\ 1 & 4 & -5 & 0 & 1 & 0 \\ 1 & -2 & 0 & 0 & 0 & 1 \end{array}\right] \begin{array}{r} 2 \\ 1 \\ 0 \end{array}$$

$$\begin{array}{r} \\ R1 + R2 \\ R1 + R3 \end{array} \left[\begin{array}{rrr|rrr} -1 & 1 & 1 & 1 & 0 & 0 \\ 0 & 5 & -4 & 1 & 1 & 0 \\ 0 & -1 & 1 & 1 & 0 & 1 \end{array}\right] \begin{array}{r} 2 \\ 3 \\ 2 \end{array}$$

$$\begin{array}{r} \\ R3 + R1 \\ 5R3 + R2 \end{array} \left[\begin{array}{rrr|rrr} -1 & 0 & 2 & 2 & 0 & 1 \\ 0 & 0 & 1 & 6 & 1 & 5 \\ 0 & -1 & 1 & 1 & 0 & 1 \end{array}\right] \begin{array}{r} 4 \\ 13 \\ 2 \end{array}$$

$$\begin{array}{r} -2R2 + R1 \\ \\ -R2 + R3 \end{array} \left[\begin{array}{rrr|rrr} -1 & 0 & 0 & -10 & -2 & -9 \\ 0 & 0 & 1 & 6 & 1 & 5 \\ 0 & -1 & 0 & -5 & -1 & -4 \end{array}\right] \begin{array}{r} -22 \\ 13 \\ -11 \end{array}$$

$$\begin{array}{r} -R1 \\ \\ -R3 \end{array} \left[\begin{array}{rrr|rrr} 1 & 0 & 0 & 10 & 2 & 9 \\ 0 & 0 & 1 & 6 & 1 & 5 \\ 0 & 1 & 0 & 5 & 1 & 4 \end{array}\right] \begin{array}{r} 22 \\ 13 \\ 11 \end{array}$$

$$\begin{array}{r} \\ \\ R2 \leftrightarrow R3 \end{array} \left[\begin{array}{rrr|rrr} 1 & 0 & 0 & 10 & 2 & 9 \\ 0 & 1 & 0 & 5 & 1 & 4 \\ 0 & 0 & 1 & 6 & 1 & 5 \end{array}\right] \begin{array}{r} 22 \\ 11 \\ 13 \end{array}$$

Hence $B = \begin{bmatrix} 10 & 2 & 9 \\ 5 & 1 & 4 \\ 6 & 1 & 5 \end{bmatrix}$.

The matrices A and B here are the same matrices that we used to begin this section, and we have already shown that $AB = BA = I_3$.

Therefore, if $A = \begin{bmatrix} -1 & 1 & 1 \\ 1 & 4 & -5 \\ 1 & -2 & 0 \end{bmatrix}$, $A^{-1} = \begin{bmatrix} 10 & 2 & 9 \\ 5 & 1 & 4 \\ 6 & 1 & 5 \end{bmatrix}$. ■

We now consider two additional examples.

■ **Example 27.** Let $A = \begin{bmatrix} 3 & 1 & 4 \\ 2 & 1 & 5 \\ 1 & 3 & -1 \end{bmatrix}$. Find A^{-1}.

Solution: The augmented matrix is

$$\left[\begin{array}{ccc|ccc} 3 & 1 & 4 & 1 & 0 & 0 \\ 2 & 1 & 5 & 0 & 1 & 0 \\ 1 & 3 & -1 & 0 & 0 & 1 \end{array}\right]\begin{array}{c} 9 \\ 9 \\ 4. \end{array}$$

In previous examples we have started our reduction using the (1,1) entry. Sometimes we may want to vary this procedure. In this example we have 1 as the entry in the (3,1), (1,2), and (2,2) positions. Let's take advantage of one of these positions to simplify some of the early arithmetic. We can fix it up later if necessary. We will begin with the (2,2) entry.

$$\left[\begin{array}{ccc|ccc} 3 & 1 & 4 & 1 & 0 & 0 \\ 2 & 1 & 5 & 0 & 1 & 0 \\ 1 & 3 & -1 & 0 & 0 & 1 \end{array}\right]\begin{array}{c} 9 \\ 9 \\ 4 \end{array}$$

$$\begin{array}{r} -R2 + R1 \\ \\ -3R2 + R3 \end{array}\left[\begin{array}{ccc|ccc} 1 & 0 & -1 & 1 & -1 & 0 \\ 2 & 1 & 5 & 0 & 1 & 0 \\ -5 & 0 & -16 & 0 & -3 & 1 \end{array}\right]\begin{array}{c} 0 \\ 9 \\ -23 \end{array}$$

$$\begin{array}{r} \\ -2R1 + R2 \\ 5R1 + R3 \end{array}\left[\begin{array}{ccc|ccc} 1 & 0 & -1 & 1 & -1 & 0 \\ 0 & 1 & 7 & -2 & 3 & 0 \\ 0 & 0 & -21 & 5 & -8 & 1 \end{array}\right]\begin{array}{c} 0 \\ 9 \\ -23 \end{array}$$

$$\begin{array}{r} -R3 + 21R1 \\ R3 + 3R2 \\ \\ \end{array}\left[\begin{array}{ccc|ccc} 21 & 0 & 0 & 16 & -13 & -1 \\ 0 & 3 & 0 & -1 & 1 & 1 \\ 0 & 0 & -21 & 5 & -8 & 1 \end{array}\right]\begin{array}{c} 23 \\ 4 \\ -23 \end{array}$$

$$\begin{array}{r} \frac{1}{21}R1 \\ \frac{1}{3}R2 \\ -\frac{1}{21}R3 \end{array}\left[\begin{array}{ccc|ccc} 1 & 0 & 0 & \frac{16}{21} & -\frac{13}{21} & -\frac{1}{21} \\ 0 & 1 & 0 & -\frac{1}{3} & \frac{1}{3} & \frac{1}{3} \\ 0 & 0 & 1 & -\frac{5}{21} & \frac{8}{21} & -\frac{1}{21} \end{array}\right]\begin{array}{c} \frac{23}{21} \\ \frac{4}{3} \\ \frac{23}{21} \end{array}$$

$$\text{So } B = \begin{bmatrix} \frac{16}{21} & -\frac{13}{21} & -\frac{1}{21} \\ -\frac{1}{3} & \frac{1}{3} & \frac{1}{3} \\ -\frac{5}{21} & \frac{8}{21} & -\frac{1}{21} \end{bmatrix} = \begin{bmatrix} \frac{16}{21} & -\frac{13}{21} & -\frac{1}{21} \\ -\frac{7}{21} & \frac{7}{21} & \frac{7}{21} \\ -\frac{5}{21} & \frac{8}{21} & -\frac{1}{21} \end{bmatrix}$$

$$= \left(\frac{1}{21}\right)\begin{bmatrix} 16 & -13 & -1 \\ -7 & 7 & 7 \\ -5 & 8 & -1 \end{bmatrix}.$$

We have done some additional work here to eliminate the fractions in the matrix itself. This is done to make computations easier when the matrix B is checked to see if $B = A^{-1}$ (i.e., show $AB = BA = I_3$). By factoring out the fraction it is possible to do the matrix multiplication using only integers and the fraction isn't involved until the final step. We will now show that $B = A^{-1}$.

$$BA = \left(\frac{1}{21}\right)\begin{bmatrix} 16 & -13 & -1 \\ -7 & 7 & 7 \\ -5 & 8 & -1 \end{bmatrix}\begin{bmatrix} 3 & 1 & 4 \\ 2 & 1 & 5 \\ 1 & 3 & -1 \end{bmatrix}$$

$$= \left(\frac{1}{21}\right)\begin{bmatrix} 21 & 0 & 0 \\ 0 & 21 & 0 \\ 0 & 0 & 21 \end{bmatrix} = \begin{bmatrix} 1 & 0 & 0 \\ 0 & 1 & 0 \\ 0 & 0 & 1 \end{bmatrix} = I_3.$$

$$\text{Also } AB = \begin{bmatrix} 3 & 1 & 4 \\ 2 & 1 & 5 \\ 1 & 3 & -1 \end{bmatrix}\left(\left(\frac{1}{21}\right)\begin{bmatrix} 16 & -13 & -1 \\ -7 & 7 & 7 \\ -5 & 8 & -1 \end{bmatrix}\right)$$

$$= \left(\frac{1}{21}\right)\left(\begin{bmatrix} 3 & 1 & 4 \\ 2 & 1 & 5 \\ 1 & 3 & -1 \end{bmatrix}\begin{bmatrix} 16 & -13 & -1 \\ -7 & 7 & 7 \\ -5 & 8 & -1 \end{bmatrix}\right) \quad \text{by property 1.2(d) of section 1.3}$$

$$= \left(\frac{1}{21}\right)\begin{bmatrix} 21 & 0 & 0 \\ 0 & 21 & 0 \\ 0 & 0 & 21 \end{bmatrix} = \begin{bmatrix} 1 & 0 & 0 \\ 0 & 1 & 0 \\ 0 & 0 & 1 \end{bmatrix} = I_3.$$

Therefore $B = A^{-1}$. By factoring out the fraction $\frac{1}{21}$, we didn't have to worry about multiplication and addition of fractions when we computed the matrix product. ■

■ Example 28. Let $A = \begin{bmatrix} 1 & 2 & 3 \\ 4 & 5 & 6 \\ 7 & 8 & 9 \end{bmatrix}$. Find A^{-1} if it exists.

Solution: The augmented matrix is

$$\left[\begin{array}{ccc|ccc} 1 & 2 & 3 & 1 & 0 & 0 \\ 4 & 5 & 6 & 0 & 1 & 0 \\ 7 & 8 & 9 & 0 & 0 & 1 \end{array}\right] \begin{array}{c} 7 \\ 16. \\ 25 \end{array}$$

Reducing, we obtain

$$\begin{array}{r} \\ -4R1 + R2 \\ -7R1 + R3 \end{array} \left[\begin{array}{ccc|ccc} 1 & 2 & 3 & 1 & 0 & 0 \\ 0 & -3 & -6 & -4 & 1 & 0 \\ 0 & -6 & -12 & -7 & 0 & 1 \end{array}\right] \begin{array}{c} 7 \\ -12 \\ -24 \end{array}$$

$$\begin{array}{r} 2R2 + 3R1 \\ \\ -2R2 + R3 \end{array} \left[\begin{array}{ccc|ccc} 3 & 0 & -3 & -5 & 2 & 0 \\ 0 & -3 & -6 & -4 & 1 & 0 \\ 0 & 0 & 0 & 1 & -2 & 1 \end{array}\right] \begin{array}{c} -3 \\ -12 \\ 0 \end{array}$$

Since the first three entries in row 3 are all 0, we cannot proceed further. We have already used row 1 and row 2 to clear the first two columns. Using either of these rows again will unclear a column we have already cleared. Hence we are not able to reduce the matrix A to the identity matrix I_3 and the inverse of A does not exist. In general, if a square matrix A contains a row or column of zeros, A is not row equivalent to I_n and is, therefore, not invertible. ■

There are several properties of the inverse of a matrix that are sometimes useful for computation purposes. Three of these properties are included here.

Property 1.5 Let A and B be invertible matrices of the same size. Then,

1. If k is a non-zero constant, (kA) is an invertible matrix and
$$(kA)^{-1} = k^{-1}A^{-1} = \frac{1}{k}A^{-1}.$$
2. $(A^{-1})^{-1} = A.$
3. (AB) is invertible and $(AB)^{-1} = B^{-1}A^{-1}.$

This last property says that the inverse of the product of two invertible matrices of the same size is the product of the inverses *in the opposite order*.

Recall here that matrix multiplication is not commutative, so in general $B^{-1}A^{-1} \neq A^{-1}B^{-1}$.

Consider now an example that uses property 1.5(1).

■ Example 29. Suppose $C = \begin{bmatrix} -\frac{1}{2} & \frac{1}{2} & \frac{1}{2} \\ \frac{1}{2} & 2 & -\frac{5}{2} \\ \frac{1}{2} & -1 & 0 \end{bmatrix}$ and we want to find C^{-1}.

Solution: Consider

$$C = \begin{bmatrix} -\frac{1}{2} & \frac{1}{2} & \frac{1}{2} \\ \frac{1}{2} & 2 & -\frac{5}{2} \\ \frac{1}{2} & -1 & 0 \end{bmatrix} = \begin{bmatrix} -\frac{1}{2} & \frac{1}{2} & \frac{1}{2} \\ \frac{1}{2} & \frac{4}{2} & -\frac{5}{2} \\ \frac{1}{2} & -\frac{2}{2} & 0 \end{bmatrix}$$

$$= \frac{1}{2}\begin{bmatrix} -1 & 1 & 1 \\ 1 & 4 & -5 \\ 1 & -2 & 0 \end{bmatrix}.$$

But $\begin{bmatrix} -1 & 1 & 1 \\ 1 & 4 & -5 \\ 1 & -2 & 0 \end{bmatrix}$ is the matrix A of example 26 of this section.

Then $C = \frac{1}{2}A$ where A is invertible, so $C^{-1} = \left(\frac{1}{2}A\right)^{-1} = \left(\frac{1}{2}\right)^{-1} A^{-1} = 2A^{-1}$.

And since, from example 26, $A^{-1} = \begin{bmatrix} 10 & 2 & 9 \\ 5 & 1 & 4 \\ 6 & 1 & 5 \end{bmatrix}$ we have

$$C^{-1} = 2\begin{bmatrix} 10 & 2 & 9 \\ 5 & 1 & 4 \\ 6 & 1 & 5 \end{bmatrix} = \begin{bmatrix} 20 & 4 & 18 \\ 10 & 2 & 8 \\ 12 & 2 & 10 \end{bmatrix}.$$

■

Perhaps the primary advantage of property 1.5(1) is that a matrix involving fractions can be rewritten as a constant times a matrix with integer entries. The computation can then be done with the matrix of integers. This procedure helps avoid messy fractions in the reduction process.

■ Example 30. Suppose $A = \begin{bmatrix} \frac{1}{2} & 0 & -1 \\ 2 & -1 & \frac{1}{2} \\ \frac{1}{2} & 1 & -5 \end{bmatrix} = \frac{1}{2}\begin{bmatrix} 1 & 0 & -2 \\ 4 & -2 & 1 \\ 1 & 2 & -10 \end{bmatrix}$. Find A^{-1}.

Solution: $A = \frac{1}{2}B$ where $B = \begin{bmatrix} 1 & 0 & -2 \\ 4 & -2 & 1 \\ 1 & 2 & -1 \end{bmatrix}$. Then if B is invertible, A is invertible and $A^{-1} = \left(\frac{1}{2}B\right)^{-1} = 2B^{-1}$.

Proceed by inverting B.

$$\left[\begin{array}{ccc|ccc} 1 & 0 & -2 & 1 & 0 & 0 \\ 4 & -2 & 1 & 0 & 1 & 0 \\ 1 & 2 & -10 & 0 & 0 & 1 \end{array}\right] \begin{array}{c} 0 \\ 4 \\ -6 \end{array}$$

$$\begin{array}{r} \\ -4R1 + R2 \\ -R1 + R3 \end{array} \left[\begin{array}{ccc|ccc} 1 & 0 & -2 & 1 & 0 & 0 \\ 0 & -2 & 9 & -4 & 1 & 0 \\ 0 & 2 & -8 & -1 & 0 & 1 \end{array}\right] \begin{array}{c} 0 \\ 4 \\ -6 \end{array}$$

$$\begin{array}{r} \\ \\ R2 + R3 \end{array} \left[\begin{array}{ccc|ccc} 1 & 0 & -2 & 1 & 0 & 0 \\ 0 & -2 & 9 & -4 & 1 & 0 \\ 0 & 0 & 1 & -5 & 1 & 1 \end{array}\right] \begin{array}{c} 0 \\ 4 \\ -2 \end{array}$$

$$\begin{array}{r} 2R3 + R1 \\ -9R3 + R2 \\ \\ \end{array} \left[\begin{array}{ccc|ccc} 1 & 0 & 0 & -9 & 2 & 2 \\ 0 & -2 & 0 & 41 & -8 & -9 \\ 0 & 0 & 1 & -5 & 1 & 1 \end{array}\right] \begin{array}{c} -4 \\ 22 \\ -2 \end{array}$$

$$\begin{array}{r} \\ -\frac{1}{2}R2 \\ \\ \end{array} \left[\begin{array}{ccc|ccc} 1 & 0 & 0 & -9 & 2 & 2 \\ 0 & 1 & 0 & -\frac{41}{2} & \frac{8}{2} & \frac{9}{2} \\ 0 & 0 & 1 & -5 & 1 & 1 \end{array}\right] \begin{array}{c} -4 \\ -11. \\ -2 \end{array}$$

Therefore,

$$B^{-1} = \begin{bmatrix} -9 & 2 & 2 \\ -\frac{41}{2} & \frac{8}{2} & \frac{9}{2} \\ -5 & 1 & 1 \end{bmatrix} \text{ and } A^{-1} = 2B^{-1} = \begin{bmatrix} -18 & 4 & 4 \\ -41 & 8 & 9 \\ -10 & 2 & 2 \end{bmatrix}.$$ ■

The next example illustrates the use of property 1.5(3).

■ **Example 31.** Let $A = \begin{bmatrix} 1 & 1 \\ 3 & 2 \end{bmatrix}$ and $B = \begin{bmatrix} 2 & 3 \\ 1 & 4 \end{bmatrix}$. Find $(AB)^{-1}$.

Solution: These are the matrices of examples 23 and 24 of this section. We found that

$$A^{-1} = \begin{bmatrix} -2 & 1 \\ 3 & -1 \end{bmatrix} \text{ and } B^{-1} = \begin{bmatrix} \frac{4}{5} & -\frac{3}{5} \\ -\frac{1}{5} & \frac{2}{5} \end{bmatrix} = \frac{1}{5}\begin{bmatrix} 4 & -3 \\ -1 & 2 \end{bmatrix}.$$

Hence $AB = \begin{bmatrix} 1 & 1 \\ 3 & 2 \end{bmatrix}\begin{bmatrix} 2 & 3 \\ 1 & 4 \end{bmatrix} = \begin{bmatrix} 3 & 7 \\ 8 & 17 \end{bmatrix}$ is invertible and $(AB)^{-1} = \begin{bmatrix} 3 & 7 \\ 8 & 17 \end{bmatrix}^{-1} = B^{-1}A^{-1} = \frac{1}{5}\begin{bmatrix} 4 & -3 \\ -1 & 2 \end{bmatrix}\begin{bmatrix} -2 & 1 \\ 3 & -1 \end{bmatrix} = \frac{1}{5}\begin{bmatrix} -17 & 7 \\ 8 & -3 \end{bmatrix}$. This can be verified by using the shortcut procedure for 2×2 matrices outlined earlier. That is

$$\begin{bmatrix} 3 & 7 \\ 8 & 17 \end{bmatrix}^{-1} = \frac{1}{(3 \cdot 17) - (7 \cdot 8)}\begin{bmatrix} 17 & -7 \\ -8 & 3 \end{bmatrix}$$

$$= -\frac{1}{5}\begin{bmatrix} 17 & -7 \\ -8 & 3 \end{bmatrix} = \frac{1}{5}\begin{bmatrix} -17 & 7 \\ 8 & -3 \end{bmatrix}.$$

■

This procedure is even more beneficial for matrices larger than 2×2 since no shortcut exists.

If A is a square matrix, then A^n is defined as $A \cdot A \cdot \ldots \cdot A$ (n factors). It can be shown that if A is invertible, then A^n is invertible and $(A^n)^{-1} = (A^{-1})^n$.

Exercise Set 1.4

1. Find the inverse of each matrix, if it exists.

a. $\begin{bmatrix} 2 & 3 \\ 1 & 2 \end{bmatrix}$ **b.** $\begin{bmatrix} 1 & 2 \\ -2 & 1 \end{bmatrix}$ **c.** $\begin{bmatrix} 3 & 1 \\ 1 & 1 \end{bmatrix}$

d. $\begin{bmatrix} 2 & -1 \\ -2 & 1 \end{bmatrix}$ **e.** $\begin{bmatrix} \frac{1}{2} & \frac{1}{4} \\ -\frac{1}{2} & \frac{1}{4} \end{bmatrix}$ **f.** $\begin{bmatrix} -2 & 2 \\ 3 & 1 \end{bmatrix}$

g. $\begin{bmatrix} 1 & -3 & 2 \\ 2 & 1 & 1 \\ 1 & 3 & 1 \end{bmatrix}$ **h.** $\begin{bmatrix} 9 & 8 & 7 \\ 6 & 5 & 4 \\ 3 & 2 & 1 \end{bmatrix}$

i. $\begin{bmatrix} 3 & -1 & 2 \\ -1 & 2 & 1 \\ 1 & -1 & 2 \end{bmatrix}$ **j.** $\begin{bmatrix} -4 & -1 & -1 \\ 5 & -2 & 3 \\ 2 & 1 & -1 \end{bmatrix}$

k. $\begin{bmatrix} 4 & 2 & -6 \\ 3 & -2 & 5 \\ -2 & -1 & 3 \end{bmatrix}$ **l.** $\begin{bmatrix} 3 & 4 & 2 \\ 2 & 1 & 2 \\ 1 & 1 & 2 \end{bmatrix}$

m. $\begin{bmatrix} 1 & -1 & 2 & 2 \\ 2 & 1 & 1 & 3 \\ -2 & 2 & -2 & -1 \\ 1 & 0 & 1 & 2 \end{bmatrix}$

2. Given $\begin{bmatrix} 1 & 2 & -1 \\ -1 & -1 & 2 \\ 1 & -1 & -3 \end{bmatrix}^{-1} = \begin{bmatrix} 5 & 7 & 3 \\ -1 & -2 & -1 \\ 2 & 3 & 1 \end{bmatrix}$ and

$\begin{bmatrix} 2 & 1 & 3 \\ 1 & 1 & 2 \\ 1 & 2 & 2 \end{bmatrix}^{-1} = \begin{bmatrix} 2 & -4 & 1 \\ 0 & -1 & 1 \\ -1 & 3 & -1 \end{bmatrix}$, find

a. $\begin{bmatrix} 5 & 7 & 3 \\ -1 & -2 & -1 \\ 2 & 3 & 1 \end{bmatrix}^{-1}$ **b.** $\begin{bmatrix} \frac{1}{2} & 1 & -\frac{1}{2} \\ -\frac{1}{2} & -\frac{1}{2} & 1 \\ \frac{1}{2} & -\frac{1}{2} & -\frac{3}{2} \end{bmatrix}^{-1}$

c. $\begin{bmatrix} 4 & 2 & 6 \\ 2 & 2 & 4 \\ 2 & 4 & 4 \end{bmatrix}^{-1}$ **d.** $\begin{bmatrix} \frac{5}{3} & \frac{7}{3} & 1 \\ -\frac{1}{3} & -\frac{2}{3} & -\frac{1}{3} \\ \frac{2}{3} & 1 & \frac{1}{3} \end{bmatrix}^{-1}$

e. $\left(\begin{bmatrix} 1 & 2 & -1 \\ -1 & -1 & 2 \\ 1 & -1 & -3 \end{bmatrix}\begin{bmatrix} 2 & 1 & 3 \\ 1 & 1 & 2 \\ 1 & 2 & 2 \end{bmatrix}\right)^{-1}$

f. $\left(\begin{bmatrix} 2 & 1 & 3 \\ 1 & 1 & 2 \\ 1 & 2 & 2 \end{bmatrix}\begin{bmatrix} 1 & 2 & -1 \\ -1 & -1 & 2 \\ 1 & -1 & -3 \end{bmatrix}\right)^{-1}$

g. $\left(\begin{bmatrix} 2 & -4 & 1 \\ 0 & -1 & 1 \\ -1 & 3 & -1 \end{bmatrix} \begin{bmatrix} 1 & 2 & -1 \\ -1 & -1 & 2 \\ 1 & -1 & -3 \end{bmatrix}\right)^{-1}$

3. Given $A = \begin{bmatrix} 3 & 4 \\ 5 & 7 \end{bmatrix}$ determine which of the following are true.

a. $(A^T)^{-1} = (A^{-1})^T$ **c.** $(A - I)^{-1} = A^{-1} - I$
b. $(A^2)^{-1} = (A^{-1})^2$ **d.** $(A^T - I)^{-1} = (A^{-1} - I)^T$

1.5 Shortcuts to Column Clearing (Optional)

The following shortcuts are combinations of the three known row reduction operations. Although the focus is on column clearing, it must be remembered that the operations are on rows. Once a method is selected to replace any one entry of a row, that method must be extended to all other entries of that row. In general, it is not necessary to use the same technique on the entire matrix. Different methods may be used on different rows. The choice of method depends on the entries and the preference of the solver.

We define the **product difference** of a 2×2 matrix as the product of the *pivot* number with the number diagonally opposite it minus the product of the two numbers on the other diagonal. The product difference depends on the four entries and the location of the pivot number. The following examples will illustrate. The pivot number has been circled for identification.

$$\begin{bmatrix} \textcircled{2} & 3 \\ 6 & 7 \end{bmatrix} = 14 - 18 = -4 \qquad \begin{bmatrix} 2 & \textcircled{3} \\ 6 & 7 \end{bmatrix} = 18 - 14 = +4$$

$$\begin{bmatrix} 2 & 3 \\ 6 & \textcircled{7} \end{bmatrix} = 14 - 18 = -4 \qquad \begin{bmatrix} 2 & 3 \\ \textcircled{6} & 7 \end{bmatrix} = 18 - 14 = +4$$

To find the 2×2 matrix whose product difference will produce the new entry, draw an imaginary arrow connecting the pivot number with the number to be replaced. The numbers on the corresponding opposite diagonal complete the matrix whose product difference is to be computed.

In actual practice, you will put your left index finger on the pivot number and your pencil on the number you are going to replace (or vice versa), make the mental multiplication, move your finger and pencil to the numbers on the opposite diagonal, compute that product, and subtract. In making this calculation, your finger and your pencil remain in their respective columns. The entries for an entire row are calculated in turn from left to right. Of course, the entry in the pivot column is zero and no calculation is necessary.

In the following example, product difference has been used on the second row to clear column two.

■ Example 32. Clear column 2 in the matrix $\begin{bmatrix} 2 & \textcircled{3} & 4 & 1 & 5 \\ 1 & 2 & 5 & 3 & 6 \end{bmatrix}$ to obtain $\begin{bmatrix} 2 & 3 & 4 & 1 & 5 \\ a & 0 & b & c & d \end{bmatrix}$.

Solution: Think $\begin{bmatrix} 2 & \textcircled{3} & 4 & 1 & 5 \\ 1 & 2 & 5 & 3 & 6 \end{bmatrix}$ and compute $a = \begin{bmatrix} 2 & \textcircled{3} \\ 1 & 2 \end{bmatrix} =$ $3 - 4 = -1$. Move to the next pair.

$\begin{bmatrix} 2 & \textcircled{3} & 4 & 1 & 5 \\ 1 & 2 & 5 & 3 & 6 \end{bmatrix}$ gives $b = \begin{bmatrix} \textcircled{3} & 4 \\ 2 & 5 \end{bmatrix} = 15 - 8 = 7$.

Next, $\begin{bmatrix} 2 & \textcircled{3} & 4 & 1 & 5 \\ 1 & 2 & 5 & 3 & 6 \end{bmatrix}$ and $c = \begin{bmatrix} \textcircled{3} & 1 \\ 2 & 3 \end{bmatrix} = 9 - 2 = 7$.

Likewise $\begin{bmatrix} 2 & \textcircled{3} & 4 & 1 & 5 \\ 1 & 2 & 5 & 3 & 6 \end{bmatrix}$ and $d = \begin{bmatrix} \textcircled{3} & 5 \\ 2 & 6 \end{bmatrix} =$ $18 - 10 = 8$.

The preceding description is for explanation only. The entire process would look like

$$\begin{bmatrix} 2 & \textcircled{3} & 4 & 1 & 5 \\ 1 & 2 & 5 & 3 & 6 \end{bmatrix} \to \begin{bmatrix} 2 & 3 & 4 & 1 & 5 \\ -1 & 0 & 7 & 7 & 8 \end{bmatrix}.$$ ■

■ **Example 33.** Reduce the matrix $\begin{bmatrix} 2 & -1 & 1 & 3 \\ 1 & 1 & 1 & 10 \\ 3 & -2 & 1 & 1 \end{bmatrix}$.

Solution:

$$\begin{bmatrix} \textcircled{2} & -1 & 1 & 3 \\ 1 & 1 & 1 & 10 \\ 3 & -2 & 1 & 1 \end{bmatrix} \overset{\textbf{a}}{\curvearrowright} \begin{bmatrix} 2 & -1 & 1 & 3 \\ 0 & \textcircled{3} & 1 & 17 \\ 0 & -1 & -1 & -7 \end{bmatrix}$$

$$\overset{\textbf{b}}{\curvearrowright} \begin{bmatrix} 6 & 0 & 4 & 26 \\ 0 & 3 & 1 & 17 \\ 0 & 0 & -2 & -4 \end{bmatrix} \overset{\textbf{c}}{\curvearrowright} \begin{bmatrix} 3 & 0 & 2 & 13 \\ 0 & 3 & 1 & 17 \\ 0 & 0 & \textcircled{1} & 2 \end{bmatrix}$$

$$\overset{\textbf{d}}{\curvearrowright} \begin{bmatrix} 3 & 0 & 0 & 9 \\ 0 & 3 & 0 & 15 \\ 0 & 0 & 1 & 2 \end{bmatrix} \overset{\textbf{e}}{\curvearrowright} \begin{bmatrix} 1 & 0 & 0 & 3 \\ 0 & 1 & 0 & 5 \\ 0 & 0 & 1 & 2 \end{bmatrix}$$

a. Select a pivot number and clear column 1.
b. Select a pivot number and clear column 2.
c. Reduce rows that have common factors ($\frac{1}{2}$R1 and $-\frac{1}{2}$R3)
d. Clear column 3.
e. Divide each row by its own pivot number ($\frac{1}{3}$R1 and $\frac{1}{3}$R2) and the matrix is in completely reduced form. ■

■ Example 34. Compute the inverse of the following matrix:

$$\begin{bmatrix} 2 & 3 & 4 \\ 1 & 3 & -1 \\ -2 & 0 & 4 \end{bmatrix}$$

Solution:

1. $$\left[\begin{array}{ccc|ccc} \textcircled{2} & 3 & 4 & 1 & 0 & 0 \\ 1 & 3 & -1 & 0 & 1 & 0 \\ -2 & 0 & 4 & 0 & 0 & 1 \end{array}\right] \begin{array}{c} 10 \\ 4 \\ 3 \end{array}$$

2. $$\left[\begin{array}{ccc|ccc} 2 & 3 & 4 & 1 & 0 & 0 \\ 0 & \textcircled{3} & -6 & -1 & 2 & 0 \\ 0 & 3 & 8 & 1 & 0 & 1 \end{array}\right] \begin{array}{c} 10 \\ -2 \\ 13 \end{array}$$

3. $$\left[\begin{array}{ccc|ccc} 2 & 0 & 10 & 2 & -2 & 0 \\ 0 & 3 & -6 & -1 & 2 & 0 \\ 0 & 0 & 14 & 2 & -2 & 1 \end{array}\right] \begin{array}{c} 12 \\ -2 \\ 15 \end{array}$$

4. $$\left[\begin{array}{ccc|ccc} 1 & 0 & 5 & 1 & -1 & 0 \\ 0 & 3 & -6 & -1 & 2 & 0 \\ 0 & 0 & \textcircled{14} & 2 & -2 & 1 \end{array}\right] \begin{array}{c} 6 \\ -2 \\ 15 \end{array}$$

5. $$\left[\begin{array}{ccc|ccc} 14 & 0 & 0 & 4 & -4 & -5 \\ 0 & 42 & 0 & -2 & 16 & 6 \\ 0 & 0 & 14 & 2 & -2 & 1 \end{array}\right] \begin{array}{c} 9 \\ 62 \\ 15 \end{array}$$

6. $$\left[\begin{array}{ccc|ccc} 1 & 0 & 0 & \frac{4}{14} & -\frac{4}{14} & -\frac{5}{14} \\ 0 & 1 & 0 & -\frac{2}{42} & \frac{16}{42} & \frac{6}{42} \\ 0 & 0 & 1 & \frac{2}{14} & -\frac{2}{14} & \frac{1}{14} \end{array}\right]$$

$$A^{-1} = \begin{bmatrix} \frac{4}{14} & -\frac{4}{14} & -\frac{5}{14} \\ -\frac{2}{42} & -\frac{16}{42} & \frac{6}{42} \\ \frac{2}{14} & -\frac{2}{14} & \frac{1}{14} \end{bmatrix}$$

$$A^{-1} = \left(\frac{1}{42}\right)\begin{bmatrix} 12 & -12 & -15 \\ -2 & 16 & 6 \\ 6 & -6 & 3 \end{bmatrix}$$

Comments:

1. Use the regular matrix setup, complete with arithmetic check column, and pick a pivot number to clear column 1.
2. Note: We have used product difference on row 2. However, row 3 was replaced simply by R1 + R3.
3. We circle a desired pivot number and clear column 2. Once again, simple addition, −R2 + R1 and −R2 + R3, is used. The idea is to use the simplest method.
4. Reduce row by common factor, $\frac{1}{2}$R1.
5. Circle the last pivot number and clear column 3.
6. Divide each row by its own pivot number $\frac{1}{14}$R1, $\frac{1}{42}$R2, & $\frac{1}{14}$R3 respectively. Most people omit this step and write A^{-1} directly from step 5. Factoring the common denominator usually results in the best form for computational purposes.

■

A second technique favored by some solvers is to write the pivot number and the number below the pivot number from the replacement row to the left of the matrix and use these to perform product differences directly from left to right as shown below.

$$\begin{array}{c} \textcircled{4} \\ 3 \end{array}\begin{bmatrix} 3 & 2 & \textcircled{4} & 5 & 7 \\ 2 & 1 & 3 & 2 & 0 \end{bmatrix} \rightarrow \begin{bmatrix} 3 & 2 & 4 & 5 & 7 \\ -1 & -2 & 0 & -7 & -21 \end{bmatrix}$$

This is exactly $-3R1 + 4R2$. Can you see that a further refinement is possible when the two numbers to the left have a common divisor?

If you wish to perform $-aR1 + bR2$ write

$$\begin{array}{c} \textcircled{b} \\ a \end{array}\begin{bmatrix} 3 & 2 & 1 & 4 \\ 0 & 1 & 2 & 3 \end{bmatrix}$$ and perform the product difference as before.

For example,

$$\begin{array}{c} 3 \\ 4 \end{array}\begin{bmatrix} 3 & 2 & \textcircled{12} & 2 & 4 \\ 5 & -7 & 16 & 3 & 1 \end{bmatrix} \rightarrow \begin{bmatrix} 3 & 2 & 12 & 2 & 4 \\ 3 & -29 & 0 & 1 & -13 \end{bmatrix}$$

NOTE: $12 \div 4 = 3$ and $16 \div 4 = 4$ and we have performed $-4R1 + 3R2$ which is an improvement over $-16R1 + 12R2$.

Exercise Set 1.5

1. Use product difference to clear column 1 in each of the following. Use the circled entry as the pivot number.

a. $\begin{bmatrix} \textcircled{3} & 2 & 1 & 4 \\ 2 & 3 & 4 & 1 \\ 0 & 2 & 1 & 3 \\ 1 & 1 & 3 & 2 \end{bmatrix}$ **b.** $\begin{bmatrix} 2 & -1 & 3 & 4 \\ 5 & 0 & 2 & 1 \\ \textcircled{3} & 2 & 1 & 4 \end{bmatrix}$

c. $\begin{bmatrix} \textcircled{4} & 3 & 4 & 1 & 2 & 3 \\ 6 & 1 & 2 & 3 & 4 & 7 \end{bmatrix}$

2. Compute the inverse of $\begin{bmatrix} 3 & -1 & 2 & 3 \\ 1 & 0 & 0 & 2 \\ 2 & 3 & 1 & 4 \\ -2 & 1 & 5 & 1 \end{bmatrix}$.

1.6 Chapter Review

Important Terms and Symbols in Chapter 1.

matrix (section 1.1)
([ ]) (section 1.1)
entries of a matrix (section 1.1)
size of a matrix (section 1.1)
square matrix (section 1.1)
equal matrices (section 1.1)
row matrix (section 1.1)
column matrix (section 1.1)
address of entry (section 1.1)
main diagonal (section 1.1)
identity matrices (section 1.1)
zero matrix (section 1.1)
transpose of a matrix (section 1.1)
matrix addition (section 1.2)
matrix subtraction (section 1.2)
scalar multiplication (section 1.2)
matrix multiplication (section 1.3)
inverse of a matrix (section 1.4)
invertible matrix (section 1.4)
elementary row operations (section 1.4)
equivalent matrices (section 1.4)
Gauss-Jordan reduction (section 1.4)
row equivalent (section 1.4)
completely reduced form (section 1.4)
column clearing (section 1.4)
pivot number (section 1.4)
augment a matrix (section 1.4)
product difference (section 1.5)

Review Exercises

1. Perform the matrix operations, if possible.

a. $\begin{bmatrix} 3 & 2 & -2 \\ 1 & 4 & 7 \end{bmatrix} + \begin{bmatrix} 2 & -2 & -1 \\ 3 & 1 & -2 \end{bmatrix}$

b. $\begin{bmatrix} 3 & -2 & 1 \\ 2 & 1 & 3 \\ 1 & 3 & -2 \end{bmatrix} - \begin{bmatrix} 3 & -2 & -1 \\ 4 & 2 & 1 \\ -1 & 3 & 2 \end{bmatrix}$

c. $\begin{bmatrix} 2 & -3 \\ 1 & 4 \\ -1 & 1 \end{bmatrix} \begin{bmatrix} 1 & 2 & -1 \\ 3 & -1 & 2 \end{bmatrix}$

d. $\begin{bmatrix} 3 & 2 & -2 \\ 1 & 4 & 1 \end{bmatrix} \begin{bmatrix} 1 & 2 \\ -2 & 1 \end{bmatrix}$

e. $\begin{bmatrix} 3 & -1 \\ 2 & 2 \end{bmatrix} \left(\begin{bmatrix} 2 & 1 & -1 \\ 3 & -1 & 2 \end{bmatrix} - 2 \begin{bmatrix} 1 & -1 & 1 \\ 2 & -2 & 3 \end{bmatrix} \right)$

f. $\begin{bmatrix} 3 & -1 & 2 \\ -2 & 1 & 3 \end{bmatrix}\left(\begin{bmatrix} 3 & -2 \\ 1 & 4 \\ -4 & 2 \end{bmatrix} + 3\begin{bmatrix} 1 & 3 \\ -2 & 2 \\ 3 & 1 \end{bmatrix}\right)$

2. Find the inverse of each, if it exists.

a. $\begin{bmatrix} 3 & -1 \\ 4 & 2 \end{bmatrix}$

b. $\begin{bmatrix} -1 & 4 \\ -2 & 6 \end{bmatrix}$

c. $\begin{bmatrix} 3 & -2 \\ -6 & 4 \end{bmatrix}$

d. $\begin{bmatrix} 6 & 5 \\ 1 & 2 \end{bmatrix}$

e. $\begin{bmatrix} 3 & -1 & 2 \\ 1 & -1 & 3 \\ 4 & 2 & 2 \end{bmatrix}$

f. $\begin{bmatrix} 4 & 1 & -3 \\ 1 & 2 & 1 \\ -2 & 3 & 1 \end{bmatrix}$

g. $\begin{bmatrix} 1 & -3 & -2 \\ 2 & 1 & -4 \\ -1 & -3 & 2 \end{bmatrix}$

h. $\begin{bmatrix} -2 & 1 & -3 \\ 1 & 2 & 1 \\ 4 & -2 & 6 \end{bmatrix}$

i. $\begin{bmatrix} 1 & -1 & 3 & 4 \\ 3 & 1 & 2 & 5 \\ -1 & 3 & 1 & 4 \\ 2 & 1 & 2 & 1 \end{bmatrix}$

j. $\begin{bmatrix} -2 & 1 & 3 & 1 \\ 1 & -3 & 2 & 4 \\ 0 & 1 & 3 & 2 \\ 2 & 1 & -1 & 1 \end{bmatrix}$

3. The new union contract calls for a 15% salary increase for all hourly employees. If the previous salary schedule was that given in the table below,
   a. What is the new salary schedule?
   b. Represent the actual increase by shift and job class as a matrix.

Job Class	*Shift* 1	2	3
I	8.40	8.60	9.00
II	8.60	8.80	9.40
III	8.80	9.20	9.80
IV	9.20	9.60	10.40

4. The company in Problem 3 has 1000 hourly employees on each shift divided into job classes in the same way—400 in class I, 300 in class II, 200 in class III, and 100 in class IV. Use matrices and matrix arithmetic to answer the following questions.
   a. What was the total hourly wage, by shift, prior to the increase?
   b. What is the total hourly wage after the increase?
   c. Assuming every hourly employee works a 40-hour week, what was the total weekly salary by shift prior to the increase?
   d. Same as **(c.)** but after the increase.

e. How much additional money per week is the new contract going to cost the company?

5. Overtime at the company in Problem 3 pays time and a half. Find the overtime salary schedule

a. before the increase.

b. after the increase.

6. Mr. Jones makes kitchen tables, chairs, and serving carts in his garage. The primary materials needed for construction are tubular chrome, formica, chrome trim, and vinyl. Each table requires 7 yards of tubular chrome, 2 square yards of formica, and 6 yards of chrome trim. A chair requires 2 yards of tubular chrome, 1 yard of trim, and 2 yards of vinyl, whereas a serving cart uses 3 yards of tubular chrome, 1 square yard of formica, and 2 yards of trim. He has orders for 6 tables, 30 chairs, and 4 serving carts.

a. Write the orders as a row matrix.

b. Write a matrix to represent the material needed for each item, where the rows represent the material used for a table, chair, and cart, respectively.

c. Use matrix multiplication to find out how much of each type of material Mr. Jones needs to fill his orders.

d. If you knew Mr. Jones's cost per unit of each kind of material, how could you use matrix arithmetic to find his total cost for materials?

7. Given

$$A = \begin{bmatrix} 2 & 1 & 3 \\ 1 & 1 & 1 \\ 3 & 6 & 1 \end{bmatrix}, \quad A^{-1} = \begin{bmatrix} -5 & 17 & -2 \\ 2 & -7 & 1 \\ 3 & -9 & 1 \end{bmatrix},$$

$$B = \begin{bmatrix} 5 & -6 & -13 \\ -4 & 5 & 11 \\ -2 & 2 & 5 \end{bmatrix}, \quad B^{-1} = \begin{bmatrix} 3 & 4 & -1 \\ -2 & -1 & -3 \\ 2 & 2 & 1 \end{bmatrix},$$

$$C = \begin{bmatrix} 2 & 1 & 0 \\ 5 & 2 & 3 \\ 7 & 3 & 2 \end{bmatrix}, \quad C^{-1} = \begin{bmatrix} -5 & -2 & 3 \\ 11 & 4 & -6 \\ 1 & 1 & -1 \end{bmatrix},$$

find

a. $(AB)^{-1}$. **b.** $(CA)^{-1}$. **c.** $(2B)^{-1}$. **d.** $\left(\frac{1}{3}C\right)^{-1}$.

e. $(AA^{-1})^{-1}$. **f.** $(AB^{-1})^{-1}$. **g.** $(A^{-1}B^{-1})^{-1}$. **h.** $(3A^{-1})^{-1}$.

i. $\left(\frac{1}{4}C^{-1}\right)^{-1}$. **j.** $3(A^{-1})^{-1}$.

2 Systems of Linear Equations

Introduction

The owner of a small candy store wants to prepare a special hard candy mix for Christmas sales. He has two types of candy available. One type is a regular hard candy mix that sells for \$.80 a pound. The other type is a ribbon candy that sells for \$1.20 a pound. The store owner would like to prepare 80 pounds of the mix and would like to sell the mix for \$.90 per pound. How much of each type of candy should be in the mix?

Suppose we let x be the number of pounds of regular mix and y the number of pounds of ribbon candy in the special mix. Then x and y must satisfy the equation $x + y = 80$. In addition, since the regular mix sells for \$.80 a pound and the ribbon candy sells for \$1.20 a pound, the revenue will be $.80x + 1.20y$. But the 80 pounds of mix is to sell for \$.90 a pound or a total of (\$.90) (80) = \$72 for 80 pounds. This gives a second equation, $.80x + 1.20y = 72$. We have, then, the two equations to solve.

$$\begin{cases} x + \quad y = 80 \\ .80x + 1.20y = 72 \end{cases}$$

The values we find for x and y must make both equations true.

This is an example of a **system of linear equations** in two variables (unknowns). A system of linear equations is a set of linear equations that are to be satisfied simultaneously. That is, values found as solutions must make every equation true. We will solve the candy store problem in section 2.1 when we begin to develop techniques used in solving systems of equations.

A **linear equation** is an equation of the form $a_1x_1 + a_2x_2 + \ldots + a_nx_n = b$ where $a_1, a_2, \ldots, a_n$, and b are constants (real numbers) and $x_1, x_2, \ldots, x_n$ are variables (unknowns). All variables have an exponent of one. In addition, mixed terms such as "x_1x_2" do not appear. Each term on the left is of the form a_ix_i where a_i is a constant. Although the constants can be any real numbers, in this book they will actually be rational numbers.

2.1 Systems of Linear Equations

In chapter 0 we discussed some properties of linear equations in two variables. In this section, a geometric interpretation of systems of linear equations in two variables will be considered. We also will consider, geometrically, equations in three variables with an initial approach to solving systems of equations. Geometric interpretations are not possible for linear equations with more than three variables. We will have to rely on algebraic techniques developed in the next section to solve systems with four or more variables.

Let us begin, then, by considering a system of two equations in two variables: $\begin{cases} a_1x + b_1y + c_1 = 0 \\ a_2x + b_2y + c_2 = 0 \end{cases}$. The graph of each of these equations is a line in the xy-plane. Figure 2.1 indicates the three ways that the two lines can be related geometrically.

In figure 2.1(a) the lines, called l_1 and l_2, are distinct lines that intersect in a unique point. The coordinates of this point give the solution to the system of equations.

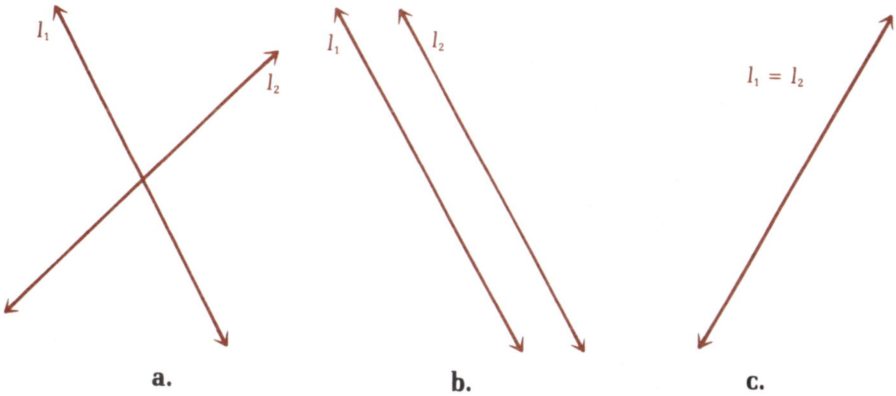

Figure 2.1

The lines in figure 2.1(b) are parallel, so there is no point that is on both lines. A system of equations of this type has no solution and is called an ***inconsistent system***.

If the system of equations has two (possibly) different forms for the same line, the graph would be like that in figure 2.1(c). In this case, every point on the line is a common point, and there are an infinite number of solutions. Such a system is called a ***dependent system***.

To solve a system of two linear equations in two variables, it is necessary to identify which of the three possibilities occurs. If the system has a unique solution, we want to know what that solution is.

The number of possibilities increases significantly when more than two equations are involved in the system. Consider, for example, three lines in a plane l_1, l_2, l_3. Assuming l_1 and l_2 intersect, as in figure 2.1(a), we have the possibilities indicated in figure 2.2.

We actually could indicate two more with l_3 parallel to l_2 (similar to figure 2.2(c)) or $l_3 = l_2$ (similar to figure 2.2(d)). Of the four (or six) pos-

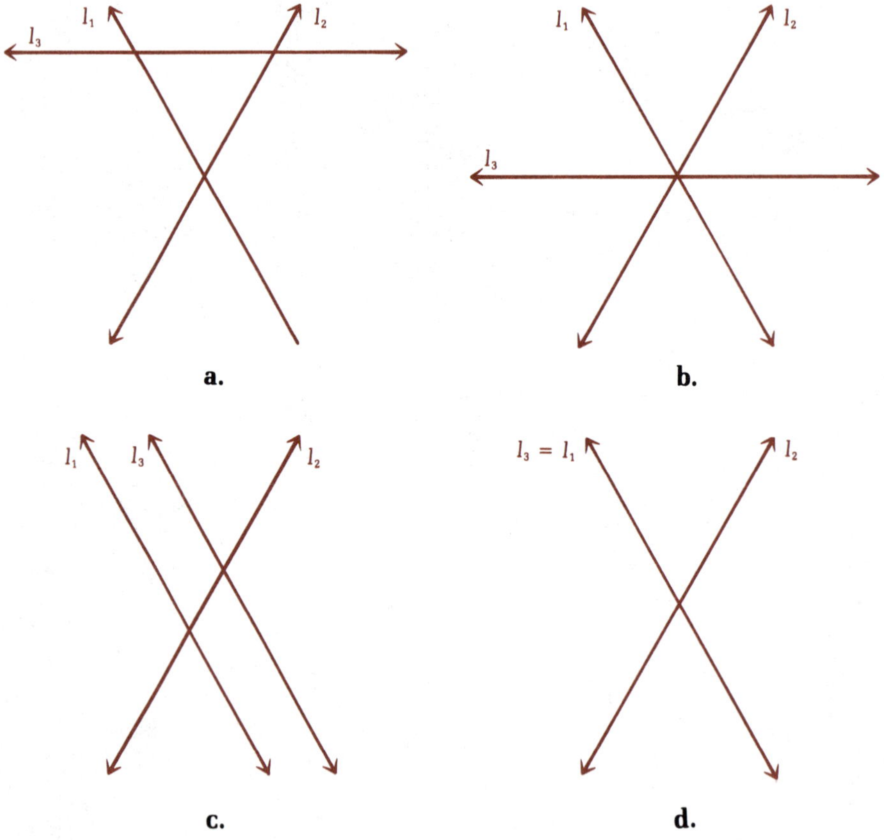

Figure 2.2

sibilities, only two (figures 2.2(b) and 2.2(d)) give solutions. In the remaining cases, there is no single point that is common to all three lines. If we include the possibilities generated by including a third line in the other two cases represented in figure 1, we only add one additional possible solution. If l_1, l_2, and l_3 are all the same line, we will have an infinite number of solutions.

Regardless of the number of equations in the system, there are only three possibilities for a solution. We always have no solution, exactly one solution, or an infinite number of solutions. In practical problems, however, restrictions sometimes limit us to only a few **acceptable solutions.** In many applications, negative values for the variables are not acceptable. Some problems also would not allow fractions in the solution.

Systems of equations in two (or more) variables arise in a variety of situations. A systematic procedure for solving problems involving systems of equations is desirable. A technique called **Gaussian elimination** will be developed here for solving systems of linear equations in two variables. Other techniques could be chosen. Gaussian elimination, however, has some advantages.

This technique is related to Gauss-Jordan reduction used in chapter 1 for inverting matrices. In fact, Gauss-Jordan reduction is an extension of Gaussian elimination or reduction. In addition, Gaussian elimination can be easily extended to systems with more than two variables. The extension requires only additional steps, not additional ideas.

Gaussian elimination is a procedure that converts a given system of equations into an **equivalent system** where at least some of the equations do not have all of the variables. The procedure is continued until at least one equation in the system has only one variable, if this is possible. We are then able to determine our solution.

Definition 2.1 Two systems of equations are equivalent if they have exactly the same solutions.

✓

NOTE: The equations in equivalent systems need not represent the same lines. In fact, normally they will not. The only requirement is that the lines represented by the systems must intersect in the same points.

■ Example 1. Consider the two systems of equations

$$\text{I}\begin{cases} x + y = 5 \\ x - y = 1 \end{cases} \quad \text{and} \quad \text{II}\begin{cases} x + y = 5 \\ x = 3 \end{cases}.$$

Solution:

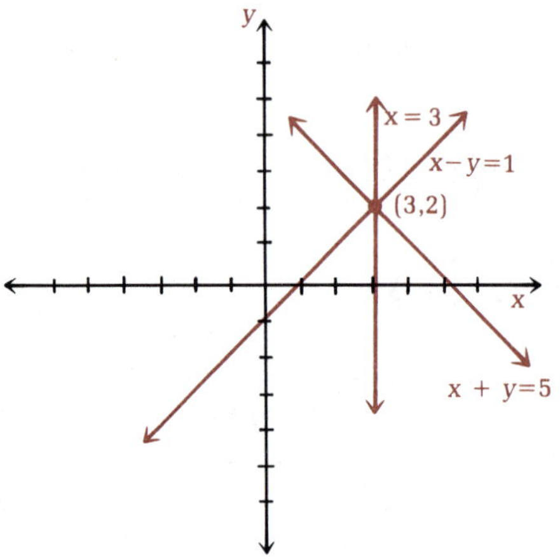

The graphs of the 3 lines ($x + y = 5$ is an equation in both systems) are placed on the same coordinate system to show that they all pass through the same point (3,2). ■

Using Gaussian elimination to find equivalent systems relies on three **elementary operations.** These are:

1. Multiply an equation by a nonzero constant.
2. Interchange two equations.
3. Add a nonzero multiple of one equation to a nonzero multiple of another equation.

✓ **NOTE:** In using operation 3, we replace one of the two equations by the combination of equations chosen.

The similarity between these elementary operations and the elementary row operations discussed in section 1.5 is not accidental. They are, in fact, the same operations performed on equations in a system instead of rows of a matrix.

■ Example 2. Use Gaussian elimination to solve the system $\begin{cases} x + y = 5 \\ 2x - 3y = 5 \end{cases}$.

Solution:

System	Graph	Elementary operations

$$\begin{cases} x + y = 5 \\ 2x - 3y = 5 \end{cases}$$

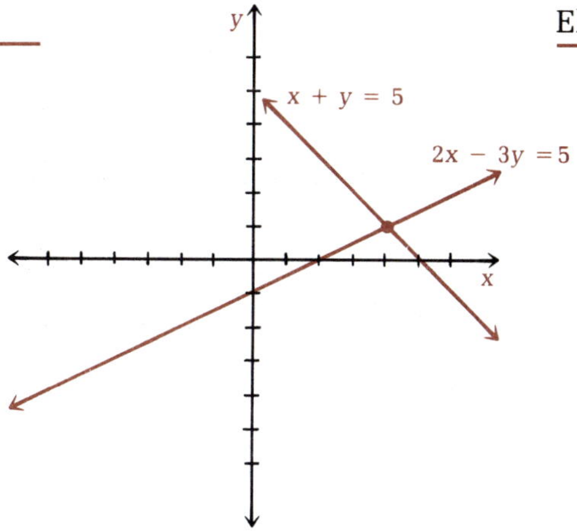

$$\begin{cases} x + y = 5 \\ 0 - 5y = -5 \end{cases}$$
or
$$\begin{cases} x + y = 5 \\ y = 1 \end{cases}$$

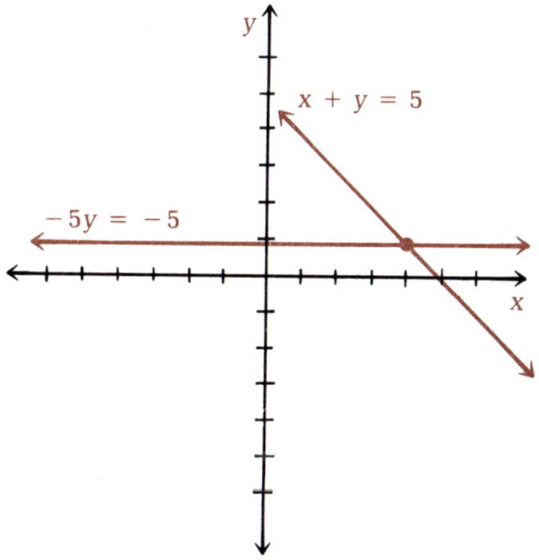

Replace equation 2 by $(-2 \times \text{equation } 1) + (\text{equation } 2)$.

$$\begin{cases} x = 4 \\ y = 1 \end{cases}$$

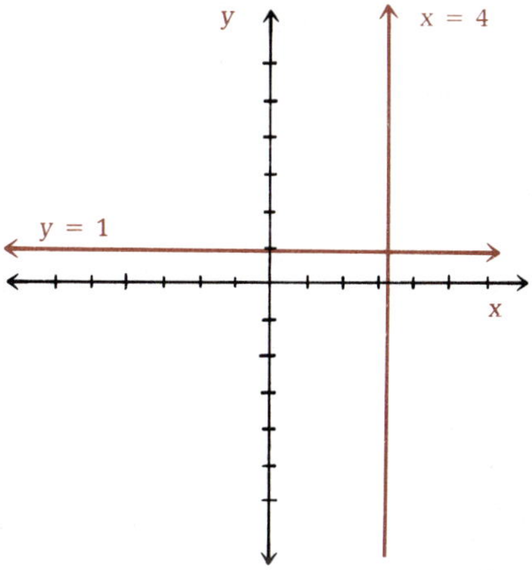

Multiply equation 2 by $-\frac{1}{5}$.

We now have an equation $(y = 1)$ that specifies a value for one of the variables. Using this value in the equation $x + y = 5$, we get $x + 1 = 5$ or $x = 4$. The solution is the point $(4,1)$. ■

■ **Example 3.** Consider now the candy store problem of the introduction. The system of equations to be solved is given by $\begin{cases} x + y = 80 \\ .80x + 1.20y = 72 \end{cases}$.

Solution:

System	Elementary Operations
$\begin{cases} x + y = 80 \\ .80x + 1.20y = 72 \end{cases}$	
$\begin{cases} x + y = 80 \\ 8x + 12y = 720 \end{cases}$	Multiply equation 2 by 10. (It is easier to work with integers than decimals.)
$\begin{cases} x + y = 80 \\ 4y = 80 \end{cases}$	Replace equation 2 by $(-8 \times \text{equation } 1) + (\text{equation } 2)$.
$\begin{cases} x + y = 80 \\ y = 20 \end{cases}$	Replace equation 2 by $\left(\frac{1}{4} \times \text{equation } 2\right)$.

Substituting $y = 20$ in the equation $x + y = 80$ gives $x + 20 = 80$, or $x = 60$. The candy store owner should mix 60 pounds of regular mix with 20 pounds of ribbon to get the 80 pounds of special mix. To check this answer, substitute $x = 60$ and $y = 20$ into the original equations.

$$\begin{cases} x + y = 80 \\ .80x + 1.20y = 72 \end{cases} \qquad \begin{cases} 60 + 20 = 80 \\ .80(60) + 1.20(20) = 48 + 24 = 72 \end{cases}$$

Since the values $x = 60$ and $y = 20$ satisfy the original equations, this solution is correct. ■

We now wish to consider an application of systems of linear equations in two variables. In this illustration, we consider a producer's willingness to supply a product and the consumer's demand for the product.

In a free market system there exists a strong interdependence between supply and demand. Graphs with a *p-axis* (price per unit) and a *q-axis* (quantity produced or consumed per week) are standard, but other units are sometimes used.

A supply-demand relationship can be looked at in two ways. From the point of view of the producer or supplier, we have a **supply curve** where the quantity of product supplied increases in direct relation with the price the consumer is willing to pay. From the point of view of the consumer, the opposite is true. As the price per unit decreases, the quantity the consumer is willing to buy will increase.

Geometrically, **supply curves** increase while **demand curves** decrease. Not all supply and demand curves are linear, but when they are, supply curves have positive slopes and demand curves have negative slopes.

The following figures show typical linear supply and demand curves. In keeping with current practice, the q-axis is always the horizontal axis.

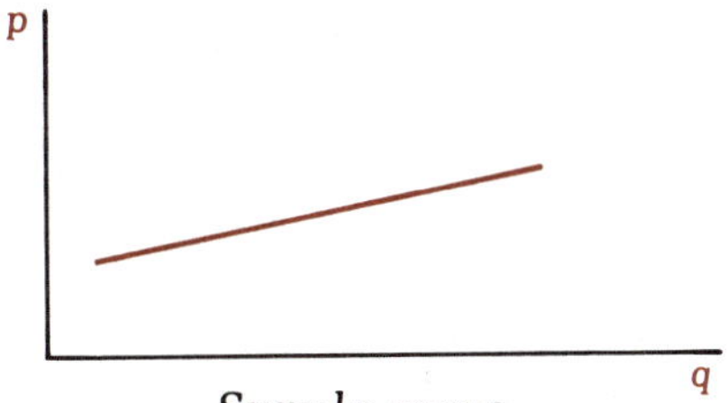

Supply curve

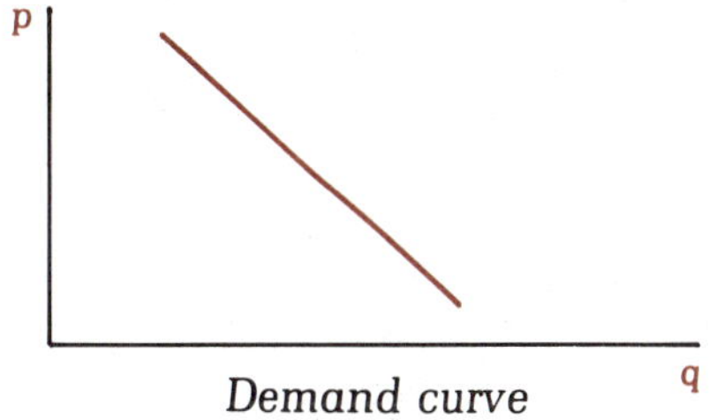

Demand curve

When both the supply and demand curves appear on the same graph, they will intersect at a point defining a specific price for, and quantity of, consumption. This point is referred to as the **point of equilibrium.**

The **equilibrium price** is, then, the price the consumer is willing to pay while using all the **equilibrium quantity.** This relationship between consumer and producer is an important one. To determine exactly the point of equilibrium, you need to solve the system of equations.

$$\begin{cases} \text{supply curve} \\ \text{demand curve} \end{cases}$$

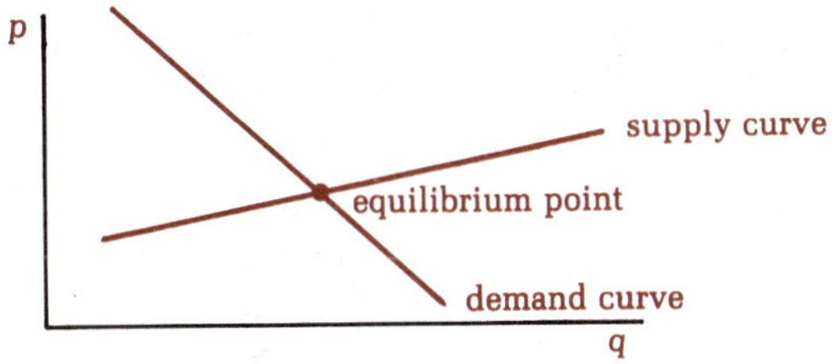

■ Example 4. For a certain product, the supply curve is given by $p = \dfrac{5}{1200}q + 5$ and the demand curve is given by $p = \dfrac{-7}{1200}q + 8$. Find the equilibrium price and quantity.

Solution: We are asked to find the point of equilibrium. We must solve the system of equations

$$\begin{cases} p = \dfrac{5}{1200}q + 5 \\ p = \dfrac{-7}{1200}q + 8, \end{cases}$$

or, equivalently

$$\begin{cases} \dfrac{-5}{1200}q + p = 5 \\ \dfrac{7}{1200}q + p = 8. \end{cases}$$

System	Elementary operations
$\begin{cases} \dfrac{-5}{1200}q + p = 5 \\ \dfrac{7}{1200}q + p = 8 \end{cases}$	
$\begin{cases} -5q + 1200p = 6000 \\ 7q + 1200p = 9600 \end{cases}$	Multiply each equation by 1200.
$\begin{cases} -5q + 1200p = 6000 \\ 12q = 3600 \end{cases}$	Add $(-1 \times$ equation 1) to (equation 2).
$\begin{cases} -5q + 1200p = 6000 \\ q = 300 \end{cases}$	Multiply equation 2 by $\dfrac{1}{12}$.

Hence the equilibrium quantity is 300. Substituting this value in equation 1 gives

$$\begin{aligned} -5(300) + 1200p &= 6000 \\ 1200p &= 7500 \\ p &= 6.25 \end{aligned}$$

The equilibrium price is \$6.25 and the point of equilibrium is (300, 6.25). ■

Another simple illustration of systems of equations involves the relationship between profit, revenue, and cost. Profit is defined as revenue minus cost. The point at which revenue equals cost is called the **break-even point.** It is determined by solving the system

$$\begin{cases} \text{revenue curve} \\ \text{cost curve} \end{cases}$$

for the value at which revenue equals cost.

■ Example 5. A farmer sells his chickens for \$2.00 each. His fixed cost is \$30 and it costs \$1.25 for each chicken to be fed until marketing. Find the revenue curve, the cost curve, and the break-even point.

Solution: Let x = the number of chickens. Then the revenue curve is $r = 2x$ (in dollars) and the cost curve is $c = 1.25x + 30$ (again in dollars). Solving

$$\begin{cases} r = 2x \\ c = 1.25x + 30 \end{cases}$$

when $r = c$ (revenue equals cost at the break-even point) we have

$$2x = 1.25x + 40$$
$$.75x = 30$$
$$x = \frac{30}{.75} = 40.$$

The farmer must sell 40 chickens just to break even. ■

Consider now the case of a system of linear equations in three variables. Geometrically, the graph of an equation of the form $ax + by + cz = d$ is a plane in three-space. If

$$\begin{cases} a_1x + b_1y + c_1z = d_1 \\ a_2x + b_2y + c_2z = d_2 \end{cases}$$

is a system of two equations in three-space, there are three possibilities for the geometric relationship of the two planes represented by the equations. The planes can be parallel with no common points, they can be identical and have all points in common, or they can intersect in a line. (See figure 2.3.)

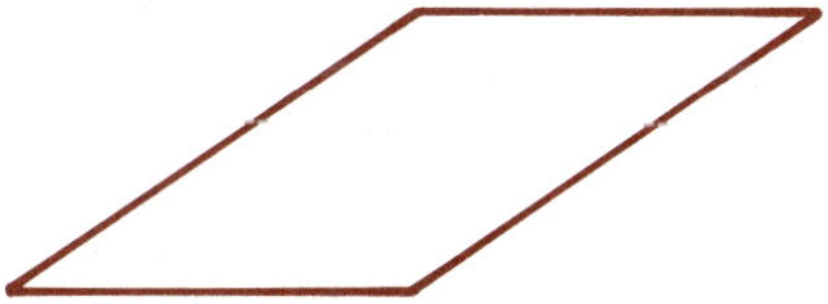

a. Planes are equal

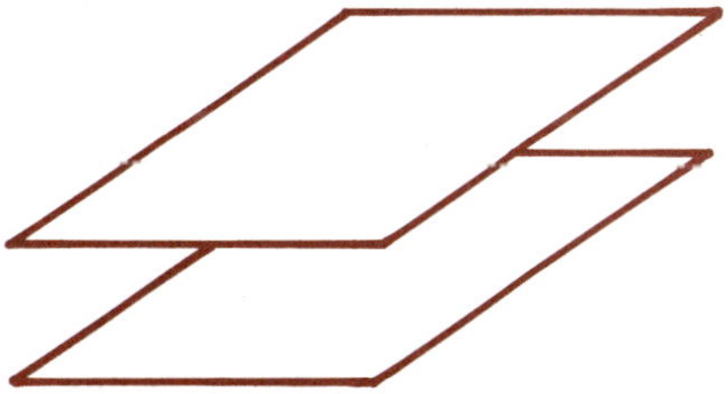

b. Planes are parallel

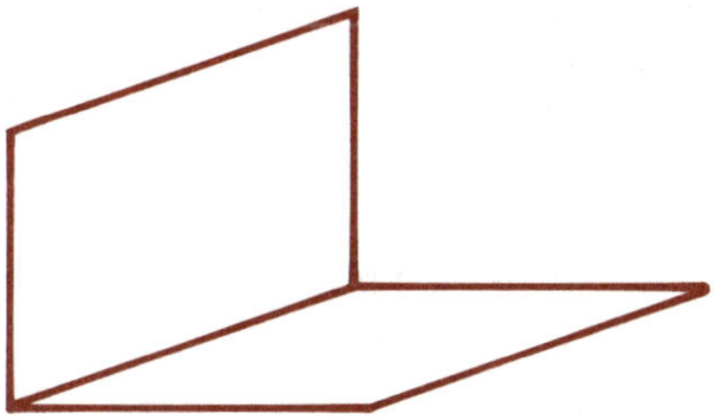

c. Planes intersect in a line

Figure 2.3

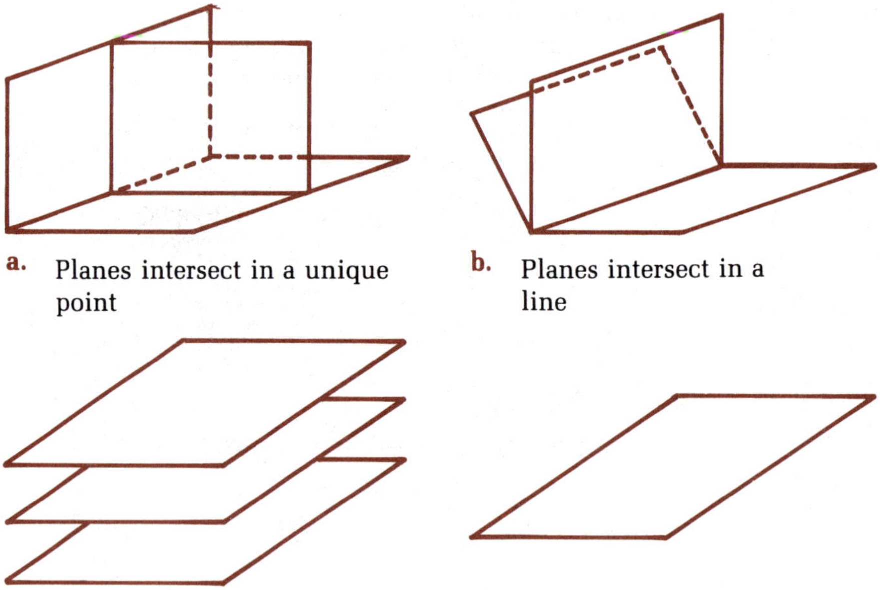

a. Planes intersect in a unique point

b. Planes intersect in a line

c. Planes parallel—no intersection

d. Same plane—all points in intersection

Figure 2.4

When a system of linear equations in three variables consists of three equations, the number of geometric relationships increases. However, only one new possibility for common points is introduced. This is the case where the three planes intersect in a unique point. (See figure 2.4.)

There are other possible geometric relationships that are not included in figure 2.3. Considering the other relationships will not alter the solution possibilities. A system of three linear equations in three variables will have no solutions, exactly one solution, a line of solutions, or a plane of solutions. The last two cases—a line or plane of solutions—both give an infinite number of points of intersection. In any system of linear equations, regardless of the number of equations or variables, there will always be no solution, exactly one solution, or an infinite number of solutions. However, in some application problems, restrictions occur that may limit the number of acceptable solutions.

Techniques used for solving systems of equations in three variables are the same as those used for systems of equations in two variables. The next three examples will illustrate the use of Gaussian elimination for systems of linear equations with three variables.

■ Example 6. Solve the system

$$\begin{cases} x + 2y + 3z = 3 \\ 2x + 5y + 3z = 2 \\ x \qquad\quad + 8z = 10 \end{cases}.$$

Solution:

System	Elementary operations
$\begin{cases} x + 2y + 3z = 3 \\ 2x + 5y + 3z = 2 \\ x \quad\quad + 8z = 10 \end{cases}$	Replace equation 2 by $(-2 \times$ equation 1$)$ + (equation 2), and replace equation 3 by (− equation 1) + (equation 3).
$\begin{cases} x + 2y + 3z = 3 \\ \quad\quad y - 3z = -4 \\ \quad -2y + 5z = 7 \end{cases}$	Replace equation 3 by $(2 \times$ equation 2$)$ + (equation 3).
$\begin{cases} x + 2y + 3z = 3 \\ \quad\quad y - 3z = -4 \\ \quad\quad\quad -z = -1 \end{cases}$	Since equation 3 contains only z, the value of z is now determined. $(z = 1)$

Substituting $z = 1$ into Equation 2 gives $y - 3(1) = -4$ or $y = -4 + 3 = -1$. Substituting $y = -1$ and $z = 1$ into equation 1, we get $x + 2(-1) + 3(1) = 3$ or $x - 2 + 3 = 3$. Then $x + 1 = 3$ or $x = 2$. The solution then is $x = 2$, $y = -1$, $z = 1$, or the point $(2, -1, 1)$.

To check this solution, these values are substituted for x, y, and z in the *original system* of equations:

$$\begin{cases} x + 2y + 3z = 3 \\ 2x + 5y + 3z = 2. \\ x \quad\quad + 8z = 10 \end{cases}$$

Letting $x = 2$, $y = -1$, and $z = 1$ gives

$$\begin{cases} 2 + 2(-1) + 3 \cdot 1 = 2 - 2 + 3 = 3 \\ 2 \cdot 2 + 5(-1) + 3 \cdot 1 = 4 - 5 + 3 = 2 \\ 2 \quad\quad + 8 \cdot 1 = 2 + 8 \quad = 10. \end{cases}$$

This solution works.

If no errors have been made in the reduction (elimination) steps, this is the only solution. ■

■ **Example 7.** Solve the system

$$\begin{cases} x + 6y + 4z = 3 \\ 2x + 4y - \;\; z = 5. \\ -x + 2y + 5z = 1 \end{cases}$$

Solution:

System	Elementary operations
$\begin{cases} x + 6y + 4z = 3 \\ 2x + 4y - \;\; z = 5 \\ -x + 2y + 5z = 1 \end{cases}$	Replace equation 2 by $(-2 \times$ equation 1$)$ + (equation 2), and replace equation 3 by (equation 1 + equation 3).

$$\begin{cases} x + 6y + 4z = 3 \\ \quad -8y - 9z = -1 \\ \quad\ \ 8y + 9z = 4 \end{cases}$$

Replace equation 3 by (equation 2 + equation 3).

$$\begin{cases} x + 6y + 4z = 3 \\ \quad -8y - 9z = -1 \\ \qquad\qquad\ 0 = 3 \end{cases}$$

The last equation (0 = 3) is not possible. Hence, there is no solution. ■

■ Example 8. Solve the system

$$\begin{cases} x + 6y + 3z = 3 \\ 2x + 4y - 2z = -2. \\ -x + 2y + 5z = 5 \end{cases}$$

Solution:

System	Elementary operations
$\begin{cases} x + 6y + 3z = 3 \\ 2x + 4y - 2z = -2 \\ -x + 2y + 5z = 5 \end{cases}$	Replace equation 2 by $(-2 \times$ equation 1) + (equation 2) and replace equation 3 by (equation 1 + equation 3).
$\begin{cases} x + 6y + 3z = 3 \\ \quad -8y - 8z = -8 \\ \quad\ \ 8y + 8z = 8 \end{cases}$	Multiply equation 2 by $-\frac{1}{8}$.
$\begin{cases} x + 6y + 3z = 3 \\ \qquad\ y + z = 1 \\ \quad\ \ 8y + 8z = 8 \end{cases}$	Replace equation 1 by $(-6 \times$ equation 2) + (equation 1) and equation 3 by $(-8 \times$ equation 2) + (equation 3).
$\begin{cases} x \quad - 3z = -3 \\ \ y + \ z = 1 \\ \ 0 + 0 \ = 0 \end{cases}$	Solve equation 1 for x and equation 2 for y.
$\begin{cases} x = -3 + 3z \\ y = 1 - z \\ 0 = 0 \end{cases}$	

The last equation, 0 = 0, is true for any values of x, y, and z. We obtain an infinite number of solutions where $x = -3 + 3z$, $y = 1 - z$, and z is any real number. Writing the solution as a point gives $(-3 + 3z, 1 - z, z)$, where z can be any real number. Unrestricted variables, like z in this example, are sometimes called **free** (or **independent**) **variables.** Since x and y both depend on the choice made for z, they are sometimes referred to as *dependent variables*. The choice of a free variable is not unique. We could have chosen x or y instead of z.

To get a **particular solution,** for example 8, a specific value is chosen for z and the corresponding values for x and y are determined. If we choose $z = 1$, then $x = -3 + 3 \cdot 1 = 0$ and $y = 1 - 1 = 0$. Then $x = 0$, $y = 0$, $z = 1$ is one solution for the system. Checking these values in the original system yields

$$\begin{aligned} 0 + 6 \cdot 0 + 3 \cdot 1 &= 3 \\ 2 \cdot 0 + 4 \cdot 0 - 2 \cdot 1 &= -2 \\ -0 + 2 \cdot 0 + 5 \cdot 1 &= 5 \end{aligned}$$

so this is one acceptable solution.

This is not, however, the only solution. If $z = 0$, then $x = -3 + 3 \cdot 0 = -3$ and $y = 1 - 0 = 1$. Checking these values in the system indicates (check it for yourself) that this is also an acceptable solution. ■

Exercise Set 2.1

For each of exercises 1–10, sketch a graph of the system and find the solution. In exercises 11–18 find the solution.

1. $\begin{cases} x + 2y = 5 \\ 2x - 3y = 3 \end{cases}$

2. $\begin{cases} 3x - 2y = 0 \\ 2x - 3y = -5 \end{cases}$

3. $\begin{cases} x - 2y = 5 \\ -2x + 4y = 3 \end{cases}$

4. $\begin{cases} -3x + 5y = 1 \\ 2x + 4y = 3 \end{cases}$

5. $\begin{cases} 3x - 2y = 1 \\ -6x + 4y = -2 \end{cases}$

6. $\begin{cases} x - 2y = -1 \\ 5x + 2y = 5 \end{cases}$

7. $\begin{cases} 3x - 2y = 1 \\ 5x + y = -7 \end{cases}$

8. $\begin{cases} 3x + 2y = -1 \\ 2x + y = -2 \end{cases}$

9. $\begin{cases} \frac{1}{2}x - \frac{1}{3}y = 1 \\ \frac{1}{4}x + \frac{2}{3}y = 3 \end{cases}$

10. $\begin{cases} 4x - 3y = 0 \\ 4x + 3y = 2 \end{cases}$

11. $\begin{cases} x + 2y - 3z = 1 \\ 2x + 2y - z = 5 \\ 2x + y + z = 6 \end{cases}$

12. $\begin{cases} x + 2y + z = 7 \\ 2x + 2y - z = -1 \\ x + 3y + z = -3 \end{cases}$

13. $\begin{cases} 2x + 3y - 2z = 3 \\ -x + y + z = 2 \\ 3x - 2y - 3z = -4 \end{cases}$

14. $\begin{cases} x + 2y + z = 2 \\ 2x + y - 4z = -2 \\ 3x + 3y + z = 3 \end{cases}$

15. $\begin{cases} 2x - 3y + 4z = 9 \\ 3x + 2y - z = -7 \\ x - 2y - 2z = -3 \end{cases}$

16. $\begin{cases} 4x + y + z = 2 \\ 4x + 11y - z = 4 \\ 8x + y + z = 3 \end{cases}$

17. $\begin{cases} 8x - 4y + 12z = 5 \\ x + 3y + z = 3 \\ 4x - 2y + 3z = 1 \end{cases}$

18. $\begin{cases} 2x + 3y - 2z = 1 \\ -x + y + z = 2 \\ 3x - 2y - 3z = -5 \end{cases}$

19. Suppose a manufacturer will supply $4p - 6$ units of a product at a set price p and consumers will buy $140 - 4p$ units at this price.

a. If the price were set at \$25 per unit would there be a shortage or a surplus?

b. For what value of p would the market be in equilibrium?

Given the supply-demand system of equations in exercises 20–22, identify which is the supply equation, which is the demand equation, and find the equilibrium point.

20. $\begin{cases} p = \dfrac{-q}{12} + 20 \\ p = \dfrac{q}{24} + 5 \end{cases}$ **21.** $\begin{cases} q - 20p = 80 \\ p + \quad q = 248 \end{cases}$ **22.** $\begin{cases} 6p + q = 51 \\ 40p - q = 18 \end{cases}$

In exercises 23 and 24, a system of revenue and cost curves is given. Find the break-even point, where q is the number of units involved.

23. $\begin{cases} r = 5q \\ c = 3q + 200 \end{cases}$ **24.** $\begin{cases} r = 30q + 120 \\ c = 20q + 200 \end{cases}$

25. A small business sells shoes at \$49.95 per pair and buys them for \$28 per pair. If the fixed cost assigned to this product is \$658.50 per week, how many pairs of shoes per week must be sold to just break even?

2.2 Matrices and Systems of Equations

In the preceding section, systems of linear equations involving two or three variables were considered with the corresponding geometric interpretation. When more than three variables are involved, it is no longer possible to represent the equations geometrically. This forces the use of different techniques for interpreting and solving systems of equations. The standard techniques make use of matrices and Gauss-Jordan reduction from chapter 1 or Gaussian reduction.

In Section 2.1 Gaussian elimination was used to solve some systems of equations. Gaussian reduction is a comparable technique used to reduce matrices.

Gauss-Jordan reduction was used in chapter 1 to reduce matrices to an identity matrix for the purpose of finding matrix inverses. Gaussian reduction is a modification of this technique but does not go as far in the reduction.

Gauss-Jordan reduction puts a matrix in completely reduced form (see section 1.4) so that it satisfies the following four conditions.

1. The first nonzero entry in each row is a 1.
2. The first nonzero entry in each row is to the right of the first nonzero entry in all preceding rows.
3. Each column containing the first nonzero entry of any row is cleared.
4. Any rows consisting entirely of zeros are placed at the bottom of the matrix.

Gaussian reduction places a matrix in reduced form where only conditions 1, 2, and 4 must be satisfied. That is, columns containing the first nonzero entry of a row do not have to be completely cleared. However, because of condition 2, any column containing a first nonzero row entry must have all zeros *below* this entry.

■ Example 9.

a. $$\begin{bmatrix} 1 & 1 & 1 & 1 & 2 \\ 0 & 1 & 0 & 1 & 3 \\ 0 & 0 & 1 & 1 & 4 \end{bmatrix}$$

This matrix is in reduced form, but not completely reduced form, since columns 2 and 3 are not cleared.

b. $$\begin{bmatrix} 1 & 1 & 1 & 1 & 3 \\ 0 & 1 & 2 & 2 & 1 \\ 0 & 1 & 1 & 1 & 2 \\ 0 & 0 & 1 & 1 & 2 \end{bmatrix}$$

This matrix is not in reduced form because row 3 violates condition 2. ■

When Gauss-Jordan reduction was used in chapter 1 for matrix inversion, it was necessary to reduce the matrix completely since it had to be reduced to I_n. When matrix reduction is used to solve systems of equations, complete reduction is not necessary. The matrix needs to be reduced only until it is possible to determine a solution. Hence, either Gaussian or Gauss-Jordan reduction can be used.

In this section, matrix reduction will be used to solve systems of equations. This technique will be illustrated in the following example where both methods, Gauss-Jordan and Gaussian reduction, are used. In subsequent examples, further modifications of these techniques will be illustrated.

To illustrate the technique used, consider example 4 of section 2.1.

■ Example 10. Solve the system

$$\begin{cases} x + 2y + 3z = 3 \\ 2x + 5y + 3z = 2 \\ x \qquad\quad + 8z = 10 \end{cases}.$$

When this system was solved in section 2.1, the elementary operations were performed to eliminate certain variables from some of the equations. The choice of elementary operations is dependent only on the respective coefficients. Changes will occur in the coefficients and in the constants on the right-hand side of the equations. Consider the coefficient matrix

$$A = \begin{bmatrix} 1 & 2 & 3 \\ 2 & 5 & 3 \\ 1 & 0 & 8 \end{bmatrix} \text{ and the column matrix of constants } B = \begin{bmatrix} 3 \\ 2 \\ 10 \end{bmatrix}.$$

Solution: Form a new matrix by augmenting A with the matrix B ($[A|B]$) (see chapter 1). Using Gauss-Jordan reduction, put the augmented matrix in completely reduced form.

$$[A|B] = \begin{array}{c} \\ \end{array}\left[\begin{array}{ccc|c} \overset{x}{1} & \overset{y}{2} & \overset{z}{3} & 3 \\ 2 & 5 & 3 & 2 \\ 1 & 0 & 8 & 10 \end{array}\right]$$

$$\begin{array}{r} \\ -2R + R2 \\ -R1 + R3 \end{array}\left[\begin{array}{ccc|c} \overset{x}{1} & \overset{y}{2} & \overset{z}{3} & 3 \\ 0 & 1 & -3 & -4 \\ 0 & -2 & 5 & 7 \end{array}\right]$$

$$\begin{array}{r} -2R2 + R1 \\ \\ 2R2 + R3 \end{array}\left[\begin{array}{ccc|c} \overset{x}{1} & \overset{y}{0} & \overset{z}{9} & 11 \\ 0 & 1 & -3 & -4 \\ 0 & 0 & -1 & -1 \end{array}\right]$$

$$\begin{array}{r} 9R3 + R1 \\ -3R3 + R2 \\ \\ \end{array}\left[\begin{array}{ccc|c} \overset{x}{1} & \overset{y}{0} & \overset{z}{0} & 2 \\ 0 & 1 & 0 & -1 \\ 0 & 0 & -1 & -1 \end{array}\right]$$

$$\begin{array}{r} \\ \\ -R3 \end{array}\left[\begin{array}{ccc|c} \overset{x}{1} & \overset{y}{0} & \overset{z}{0} & 2 \\ 0 & 1 & 0 & -1 \\ 0 & 0 & 1 & 1 \end{array}\right]$$

Each of the preceding matrices corresponds to a system of equations. The systems represented by these matrices are equivalent. The rows correspond to the equations in the system with the entries in columns 1, 2, and 3 as the coefficients of x, y, and z respectively. The entries in column 4 are the constants for the equations. For example, the third matrix

$$\left[\begin{array}{ccc|c} 1 & 0 & 9 & 11 \\ 0 & 1 & -3 & -4 \\ 0 & 0 & -1 & -1 \end{array}\right] \text{ corresponds to}$$

$$\begin{cases} 1\cdot x + 0\cdot y + 9\cdot z = 11 \\ 0\cdot x + 1\cdot y - 3\cdot z = -4 \\ 0\cdot x + 0\cdot y - 1\cdot z = -1 \end{cases} = \begin{cases} x + 9z = 11 \\ y - 3z = -4. \\ -z = -1 \end{cases}$$

The final matrix is in completely reduced form.
Row 1 corresponds to the equation $x + 0\cdot y + 0\cdot z = 2$ or $x = 2$.
Row 2 corresponds to the equation $0\cdot x + y + 0\cdot z = -1$ or $y = -1$.
Row 3 corresponds to the equation $0\cdot x + 0\cdot y + z = 1$ or $z = 1$.

The solution is, as before, $x = 2$, $y = -1$, $z = 1$ or the point $(2, -1, 1)$.

In matrix form we could write $\begin{bmatrix} x \\ y \\ z \end{bmatrix} = \begin{bmatrix} 2 \\ -1 \\ 1 \end{bmatrix}$. ■

When the final matrix is in completely reduced form, the solution can be read from this matrix. If Gaussian reduction had been chosen instead of Gauss-Jordan reduction, substitution could be used to complete the solution.

$$[A|B] = \begin{array}{c} \begin{array}{ccc} x & y & z \end{array} \\ \left[\begin{array}{ccc|c} 1 & 2 & 3 & 3 \\ 2 & 5 & 3 & 2 \\ 1 & 0 & 8 & 10 \end{array}\right] \end{array}$$

$$\begin{array}{r} \\ -2R1 + R2 \\ -R1 + R3 \end{array} \left[\begin{array}{ccc|c} 1 & 2 & 3 & 3 \\ 0 & 1 & -3 & -4 \\ 0 & -2 & 5 & 7 \end{array}\right]$$

$$\begin{array}{r} \\ \\ 2R2 + R3 \end{array} \left[\begin{array}{ccc|c} 1 & 2 & 3 & 3 \\ 0 & 1 & -3 & -4 \\ 0 & 0 & -1 & -1 \end{array}\right]$$

$$\begin{array}{r} \\ \\ -R3 \end{array} \left[\begin{array}{ccc|c} 1 & 2 & 3 & 3 \\ 0 & 1 & -3 & -4 \\ 0 & 0 & 1 & 1 \end{array}\right]$$

(Columns labelled x, y, z.)

Row 3 corresponds to the equation $z = 1$.

Row 2 corresponds to the equation $y - 3z = -4$ or $y = -4 + 3z = -4 + 3(1) = -4 + 3 = -1$.
Row 1 corresponds to the equation $x + 2y + 3z = 3$ or $x = 3 - 2y - 3z = 3 - 2(-1) - 3(1) = 3 + 2 - 3 = 2$. The solution is $x = 2$, $y = -1$, $z = 1$.

The choice of Gaussian or Gauss-Jordan reduction is not critical. The solution is easier to read if Gauss-Jordan reduction is used, but more reduction is needed. The shortcut procedure outlined in section 1.5 could also be used for the reduction. The best procedure for you is the one that makes you feel the most comfortable.

Since the technique used here is the same as that used for inverting matrices, any shortcut that worked before will still work. In fact, as we will see in the following paragraphs, there is actually more freedom when this technique is applied to systems of equations. In matrix inversion, row reduction is used to convert the augmented matrix $[A|I_n]$ into the matrix $[I_n|B]$. The final matrix *must* be in this form, so A **must** be reduced in I_n.

To solve a system of equations, the reduction only needs to be continued until a solution can be determined. Suppose A is the coefficient matrix of a system of n equations in n variables and B is the column matrix of constants. Then to solve the system, the augmented matrix $[A|B]$ is reduced using Gauss-Jordan (or Gaussian) reduction. If the final matrix is of the form $[I_n|C]$, the solution is given by the column matrix C. However, the reduction can be stopped at any step where the solution can be determined.

In particular, it is not necessary to have the first non zero entry of each row equal to 1. It also is not necessary to have the rows in the right order. Since the columns correspond to variables, solutions can be determined regardless of the row that non-zero entries are in.

We will consider several examples now to illustrate these ideas. The arithmetic check column used in chapter 1 can also be used here. We will use it in some of the examples.

■ **Example 11.** Solve the system $\begin{cases} x + y = 5 \\ 2x - 3y = 5 \end{cases}$.

This is the system that was solved in example 2 of section 2.1

Solution:

$$\left[\begin{array}{cc|c} 1 & 1 & 5 \\ 2 & -3 & 5 \end{array}\right] \begin{matrix} 7 \\ 4 \end{matrix}$$

$$\begin{matrix} \\ -2R1 + R2 \end{matrix} \left[\begin{array}{cc|c} 1 & 1 & 5 \\ 0 & -5 & -5 \end{array}\right] \begin{matrix} 7 \\ -10 \end{matrix}$$

$$\begin{array}{r} R2 + 5R1 \end{array} \left[\begin{array}{rr|r} 5 & 0 & 20 \\ 0 & -5 & -5 \end{array}\right] \begin{array}{r} 5 \\ -10 \end{array}$$

$$\begin{array}{r} \frac{1}{5}R1 \\ -\frac{1}{5}R2 \end{array} \left[\begin{array}{rr|r} 1 & 0 & 4 \\ 0 & 1 & 1 \end{array}\right] \begin{array}{r} 5 \\ 2 \end{array}$$

The solution, as before, is $x = 4$, $y = 1$ or the point (4,1). ■

■ Example 12. Use matrix techniques to solve the candy store problem of the introduction.

Solution: The system of equations is $\begin{cases} x + y = 80 \\ .80x + 1.20y = 72 \end{cases}$

$$\left[\begin{array}{rr|r} 1 & 1 & 80 \\ .80 & 1.20 & 72 \end{array}\right] \quad \begin{array}{r} \\ 5R2 \end{array} \left[\begin{array}{rr|r} 1 & 1 & 80 \\ 4 & 6 & 360 \end{array}\right]$$

$$\begin{array}{r} \\ -4R1 + R2 \end{array} \left[\begin{array}{rr|r} 1 & 1 & 80 \\ 0 & 2 & 40 \end{array}\right] \quad \begin{array}{r} -R2 + 2R1 \\ \\ \end{array} \left[\begin{array}{rr|r} 2 & 0 & 120 \\ 0 & 2 & 40 \end{array}\right]$$

Row 2 was multiplied by 5 in the first step to remove decimals and simplify computation. Then $x = \frac{120}{2} = 60$ and $y = \frac{40}{2} = 20$, so the owner should mix 60 pounds of regular hard candy with 20 pounds of ribbon candy. ■

■ Example 13. Solve the system

$$\begin{cases} 3x - 2y + z = 3 \\ 2x + 2y + z = -2. \\ x + 3y - z = 0 \end{cases}$$

Solution: Solution with check column:

$$\left[\begin{array}{rrr|r} 3 & -2 & 1 & 3 \\ 2 & 2 & 1 & -2 \\ 1 & 3 & -1 & 0 \end{array}\right] \begin{array}{r} 5 \\ 3 \\ 3 \end{array}$$

$$\begin{array}{r} \\ -R1 + R2 \\ R1 + R3 \end{array} \left[\begin{array}{rrr|r} 3 & -2 & 1 & 3 \\ -1 & 4 & 0 & -5 \\ 4 & 1 & 0 & 3 \end{array}\right] \begin{array}{r} 5 \\ -2 \\ 8 \end{array}$$

$$\begin{array}{r} 2R3 + R1 \\ -4R3 + R2 \\ \\ \end{array} \left[\begin{array}{rrr|r} 11 & 0 & 1 & 9 \\ -17 & 0 & 0 & -17 \\ 4 & 1 & 0 & 3 \end{array}\right] \begin{array}{r} 21 \\ -34 \\ 8 \end{array}$$

$$\begin{array}{r} \\ -\frac{1}{17}R2 \\ \\ \end{array}\left[\begin{array}{rrr|r} 11 & 0 & 1 & 9 \\ 1 & 0 & 0 & 1 \\ 4 & 1 & 0 & 3 \end{array}\right]\begin{array}{r} 21 \\ 2 \\ 8 \end{array}$$

$$\begin{array}{r} -11R2 + R1 \\ \\ -4R2 + R3 \end{array}\left[\begin{array}{rrr|r} 0 & 0 & 1 & -2 \\ 1 & 0 & 0 & 1 \\ 0 & 1 & 0 & -1 \end{array}\right]\begin{array}{r} -1 \\ 2 \\ 0 \end{array}$$

Since column 1 corresponds to x, column 2 to y, and column 3 to z, row 2 will determine the value of x, row 3 the value of y, and row 1 the value of z. The solution is given by $x = 1$, $y = -1$, $z = -2$. ■

■ **Example 14.** Solve the system

$$\begin{cases} x + y + z = 3 \\ 2x + y - 2z = -5. \\ -x + y - 3z = -1 \end{cases}$$

Solution: Solution with check column:

$$\left[\begin{array}{rrr|r} 1 & 1 & 1 & 3 \\ 2 & 1 & -2 & -5 \\ -1 & 1 & -3 & -1 \end{array}\right]\begin{array}{r} 6 \\ -4 \\ -4 \end{array}$$

$$\begin{array}{r} \\ -2R1 + R2 \\ R1 + R3 \end{array}\left[\begin{array}{rrr|r} 1 & 1 & 1 & 3 \\ 0 & -1 & -4 & -11 \\ 0 & 2 & -2 & 2 \end{array}\right]\begin{array}{r} 6 \\ -16 \\ 2 \end{array}$$

$$\begin{array}{r} R2 + R1 \\ \\ 2R2 + R3 \end{array}\left[\begin{array}{rrr|r} 1 & 0 & -3 & -8 \\ 0 & -1 & -4 & -11 \\ 0 & 0 & -10 & -20 \end{array}\right]\begin{array}{r} -10 \\ -16 \\ -30 \end{array}$$

$$\begin{array}{r} \\ -R2 \\ -\frac{1}{10}R3 \end{array}\left[\begin{array}{rrr|r} 1 & 0 & -3 & -8 \\ 0 & 1 & 4 & 11 \\ 0 & 0 & 1 & 2 \end{array}\right]\begin{array}{r} -10 \\ 16 \\ 3 \end{array}$$

$$\begin{array}{r} 3R3 + R1 \\ -4R3 + R2 \\ \\ \end{array}\left[\begin{array}{rrr|r} 1 & 0 & 0 & -2 \\ 0 & 1 & 0 & 3 \\ 0 & 0 & 1 & 2 \end{array}\right]\begin{array}{r} -1 \\ 4 \\ 3 \end{array}$$

The solution is $x = -2$, $y = 3$, $z = 2$ or the point $(-2,3,2)$. ■

■ Example 15. Solve $\begin{cases} 8x - 4y + 12z = 5 \\ x + 3y + z = 3. \\ 4x - 2y + 3z = 1 \end{cases}$

Solution:

$$\left[\begin{array}{ccc|c} 8 & -4 & 12 & 5 \\ 1 & 3 & 1 & 3 \\ 4 & -2 & 3 & 1 \end{array}\right]$$

$$\begin{array}{r} -8R2 + R1 \\ \\ -4R2 + R3 \end{array} \left[\begin{array}{ccc|c} 0 & -28 & 4 & -19 \\ 1 & 3 & 1 & 3 \\ 0 & -14 & -1 & -11 \end{array}\right]$$

$$\begin{array}{r} -2R3 + R1 \\ 3R3 + 14R2 \\ \\ \end{array} \left[\begin{array}{ccc|c} 0 & 0 & 6 & 3 \\ 14 & 0 & 11 & 9 \\ 0 & -14 & -1 & -11 \end{array}\right]$$

$$\begin{array}{r} \frac{1}{3}R1 \\ \\ \\ \end{array} \left[\begin{array}{ccc|c} 0 & 0 & 2 & 1 \\ 14 & 0 & 11 & 9 \\ 0 & -14 & -1 & -11 \end{array}\right]$$

$$\begin{array}{r} \\ -11R1 + 2R2 \\ R1 + 2R3 \end{array} \left[\begin{array}{ccc|c} 0 & 0 & 2 & 1 \\ 28 & 0 & 0 & 7 \\ 0 & -28 & 0 & -21 \end{array}\right]$$

NOTE: The check column is optional. We will not always use it. If you feel more comfortable using it, by all means do so.

The solution is $x = \dfrac{7}{28} = \dfrac{1}{4}$, $y = \dfrac{-21}{-28} = \dfrac{3}{4}$, $z = \dfrac{1}{2}$. ■

■ Example 16. Solve $\begin{cases} 2x - y + 2z = 4 \\ 3x + 6y - 3z = 5. \\ 3x + 3y + z = 7 \end{cases}$

Solution:

$$\left[\begin{array}{rrr|r} 2 & -1 & 2 & 4 \\ 3 & 6 & -3 & 5 \\ 3 & 3 & 1 & 7 \end{array}\right]$$

$$\begin{array}{r} \\ -3R1 + 2R2 \\ -3R1 + 2R3 \end{array} \left[\begin{array}{rrr|r} 2 & -1 & 2 & 4 \\ 0 & 15 & -12 & -2 \\ 0 & 9 & -4 & 2 \end{array}\right]$$

$$\begin{array}{r} R2 + 15R1 \\ \\ -3R2 + 5R3 \end{array} \left[\begin{array}{rrr|r} 30 & 0 & 18 & 58 \\ 0 & 15 & -12 & -2 \\ 0 & 0 & 16 & 16 \end{array}\right]$$

$$\begin{array}{r} \tfrac{1}{2}R1 \\ \\ \tfrac{1}{16}R3 \end{array} \left[\begin{array}{rrr|r} 15 & 0 & 9 & 29 \\ 0 & 15 & -12 & -2 \\ 0 & 0 & 1 & 1 \end{array}\right]$$

$$\begin{array}{r} -9R3 + R1 \\ 12R3 + R2 \\ \\ \end{array} \left[\begin{array}{rrr|r} 15 & 0 & 0 & 20 \\ 0 & 15 & 0 & 10 \\ 0 & 0 & 1 & 1 \end{array}\right]$$

So $x = \dfrac{20}{15} = \dfrac{4}{3},\ y = \dfrac{10}{15} = \dfrac{2}{3},\ z = 1$ ■

All of the previous examples have been similar in two respects. In each example, the number of equations has been the same as the number of variables. In addition, each system has had a unique solution. In the preceding section, systems with no solution or an infinite number of solutions were seen to exist. Systems with more equations than variables or more variables than equations also arise in practical problems.

Problems of these types can be handled using matrix reduction with no loss of efficiency. If a system has no solution or multiple solutions, the reduction will uncover this at some stage. Remember that the rows of the matrix at each stage of the reduction correspond to equations. Then if some row of one of the matrices has all entries equal to zero except the last one (the last entry in a row is in the column of constants), the corresponding equation will be of the form $0 \cdot x_1 + 0 \cdot x_2 + \ldots + 0 \cdot x_n = c$ where c is not zero. This simplifies to $0 = c$ which is a contradiction of $c \neq 0$. Hence, the system has no solution and is inconsistent.

The multiple solution situation is not quite as clear-cut. However, some general guidelines are possible. If the system contains fewer equations than variables, it is not possible to have a unique solution. Recall from section 2.1 that a linear equation in three variables represents a plane. Two planes do not intersect, intersect in a line (infinite number of solutions), or intersect in a plane (infinite number of solutions). The system will have no solution if, at some stage in the reduction, a row is obtained that corresponds to an equation of the form $0 = c$ where $c \neq 0$. That is, a row is obtained with zero as the coefficient of all variables but a constant value (last column) that is not zero. Otherwise, there will be multiple solutions.

If the number of equations is greater than or equal to the number of variables, any of the three solution possibilities—no solution, unique solution, multiple solutions—can occur. When multiple solutions occur in this situation, the final matrix in the reduction process will contain one or more rows in which every entry, including the last, is zero.

CAUTION: Be careful here. Having a row consisting entirely of zeros does not automatically imply multiple solutions.

The following examples will illustrate these various possibilities.

■ **Example 17.** Solve the system $\begin{cases} x - y = -5 \\ x + y = 1. \\ -2x + y = 7 \end{cases}$

Solution:

$$\left[\begin{array}{rr|r} 1 & -1 & -5 \\ 1 & 1 & 1 \\ -2 & 1 & 7 \end{array}\right] \begin{array}{r} -5 \\ 3 \\ 6 \end{array}$$

$$\begin{array}{r} \\ -R1 + R2 \\ 2R1 + R3 \end{array} \left[\begin{array}{rr|r} 1 & -1 & -5 \\ 0 & 2 & 6 \\ 0 & -1 & -3 \end{array}\right] \begin{array}{r} -5 \\ 8 \\ -4 \end{array}$$

$$\begin{array}{r} -R3 + R1 \\ 2R3 + R2 \\ \\ \end{array} \left[\begin{array}{rr|r} 1 & 0 & -2 \\ 0 & 0 & 0 \\ 0 & -1 & -3 \end{array}\right] \begin{array}{r} -1 \\ 0 \\ -4 \end{array}$$

Row 1 corresponds to the equation $x = -2$, and row 3 corresponds to $-y = -3$ or $y = 3$. Even though row 2 consists entirely of zeros, a unique solution $x = -2$, $y = 3$ is obtained. ■

■ **Example 18.** Solve the system $\begin{cases} x + y = 2 \\ -x + 2y = 1. \\ 2x - y = 4 \end{cases}$

Solution:

$$\left[\begin{array}{rr|r|r} 1 & 1 & 2 & 4 \\ -1 & 2 & 1 & 2 \\ 2 & -1 & 4 & 5 \end{array}\right]$$

$$\begin{array}{r} \\ R1 + R2 \\ -2R1 + R3 \end{array}\left[\begin{array}{rr|r|r} 1 & 1 & 2 & 4 \\ 0 & 3 & 3 & 6 \\ 0 & -3 & 0 & -3 \end{array}\right]$$

$$\begin{array}{r} -R2 + 3R1 \\ \\ R2 + R3 \end{array}\left[\begin{array}{rr|r|r} 3 & 0 & 3 & 6 \\ 0 & 3 & 3 & 3 \\ 0 & 0 & 3 & 3 \end{array}\right]$$

The last row corresponds to the equation $0 \cdot x + 0 \cdot y = 3$, or $0 = 3$. This is impossible. Hence, there is no solution. ■

■ **Example 19.** Solve the system $\begin{cases} 2x + 3y - 2z = 3 \\ -x + y + z = 2 \\ 3x - 2y - 3z = -4 \end{cases}$.

Solution:

$$\left[\begin{array}{rrr|r|r} 2 & 3 & -2 & 3 & 6 \\ -1 & 1 & 1 & 2 & 3 \\ 3 & -2 & -3 & -4 & -6 \end{array}\right]$$

$$\begin{array}{r} 2R2 + R1 \\ \\ 3R2 + R3 \end{array}\left[\begin{array}{rrr|r|r} 0 & 5 & 0 & 7 & 12 \\ -1 & 1 & 1 & 2 & 3 \\ 0 & 1 & 0 & 2 & 3 \end{array}\right]$$

$$\begin{array}{r} -5R3 + R1 \\ -R3 + R2 \\ \\ \end{array}\left[\begin{array}{rrr|r|r} 0 & 0 & 0 & -3 & -3 \\ -1 & 0 & 1 & 0 & 0 \\ 0 & 1 & 0 & 2 & 3 \end{array}\right]$$

Row 1 corresponds to $0 = -3$, which is impossible, so there is no solution. ■

■ **Example 20.** Solve the system $\begin{cases} 2x + 3y - 2z - 1 \\ -x + y + z = 2 \\ 3x - 2y - 3z = -5 \end{cases}$.

Solution:

$$\left[\begin{array}{ccc|c} 2 & 3 & -2 & 1 \\ -1 & 1 & 1 & 2 \\ 3 & -2 & -3 & -5 \end{array}\right]\begin{array}{c} 4 \\ 3 \\ -7 \end{array}$$

$$\begin{array}{r} 2R2 + R1 \\ \\ 3R2 + R3 \end{array}\left[\begin{array}{ccc|c} 0 & 5 & 0 & 5 \\ -1 & 1 & 1 & 2 \\ 0 & 1 & 0 & 1 \end{array}\right]\begin{array}{c} 10 \\ 3 \\ 2 \end{array}$$

$$\begin{array}{r} -5R3 + R1 \\ -R3 + R2 \\ \\ \end{array}\left[\begin{array}{ccc|c} 0 & 0 & 0 & 0 \\ -1 & 0 & 1 & 1 \\ 0 & 1 & 0 & 1 \end{array}\right]\begin{array}{c} 0 \\ 1 \\ 2 \end{array}$$

From row 2 we get $-x + z = 1$, or $-x = 1 - z$, or, finally, $x = -1 + z$. From row 3 we get $y = 1$. Then in the solution, $y = 1$ and $x = -1 + z$ where z is a free variable (z can be any real number). The solution is $x = -1 + z$, $y = 1$, $z = z$, or all points of the form $(-1 + z, 1, z)$ where z can be any real number. We could have chosen x as the free variable. Our solution would then have been $(x, 1, 1 + x)$. ■

■ **Example 21.** Solve $\begin{cases} x_1 + 3x_2 - 2x_3 + x_4 = 6 \\ 2x_1 + 5x_2 - 3x_3 + x_4 = 8. \\ x_1 + 3x_2 - x_3 + 3x_4 = 3 \end{cases}$

Solution:

$$\left[\begin{array}{cccc|c} 1 & 3 & -2 & 1 & 6 \\ 2 & 5 & -3 & 1 & 8 \\ 1 & 3 & -1 & 3 & 3 \end{array}\right]\begin{array}{c} 9 \\ 13 \\ 9 \end{array}$$

$$\begin{array}{r} \\ -2R1 + R2 \\ -R1 + R3 \end{array}\left[\begin{array}{cccc|c} 1 & 3 & -2 & 1 & 6 \\ 0 & -1 & 1 & -1 & -4 \\ 0 & 0 & 1 & 2 & -3 \end{array}\right]\begin{array}{c} 9 \\ -5 \\ 0 \end{array}$$

$$\begin{array}{r} 3R2 + R1 \\ \\ \\ \end{array}\left[\begin{array}{cccc|c} 1 & 0 & 1 & -2 & -6 \\ 0 & -1 & 1 & -1 & -4 \\ 0 & 0 & 1 & 2 & -3 \end{array}\right]\begin{array}{c} -6 \\ -5 \\ 0 \end{array}$$

$$\begin{array}{r} -R3 + R1 \\ -R3 + R2 \\ \\ \end{array}\left[\begin{array}{cccc|c} 1 & 0 & 0 & -4 & -3 \\ 0 & -1 & 0 & -3 & -1 \\ 0 & 0 & 1 & 2 & -3 \end{array}\right]\begin{array}{c} -6 \\ -5 \\ 0 \end{array}$$

No further reduction is possible since all three rows have been used to clear columns. Further reduction to clear column 4 will unclear one of the other columns. The three rows of this matrix correspond to the equations

$$\begin{cases} x_1 - 4x_4 = -3 \\ -x_2 - 3x_4 = -1 \\ x_3 + 2x_4 = -3 \end{cases} \quad \text{or} \quad \begin{cases} x_1 = -3 + 4x_4 \\ x_2 = 1 - 3x_4 \\ x_3 = -3 - 2x_4 \end{cases}$$

where x_4 is a free variable. The solution is $x_1 = -3 + 4x_4$, $x_2 = 1 - 3x_4$, $x_3 = -3 - 2x_4$, $x_4 = x_4$. Written as points, the solution is all points of the form $(-3 + 4x_4, 1 - 3x_4, -3 - 2x_4, x_4)$ where x_4 can be any real number. ■

To obtain the solution in a multiple solution problem, the general procedure is to solve the equations corresponding to the rows of the matrix for the *leading variable* (the first one appearing in the equation) in terms of the remaining variables and the constant. The variables corresponding to any uncleared columns will be *free variables*. Variables corresponding to cleared columns are dependent variables. They may be constants, or they may depend on the free variables. The following example will illustrate the process.

■ Example 22. Suppose each of the following matrices is the final matrix in the solution of a system of equations. Determine the solutions.

a. $\left[\begin{array}{cccc|c} 2 & -1 & 0 & 0 & 3 \\ 0 & 0 & 1 & 0 & 2 \\ 0 & 0 & 0 & 2 & 3 \end{array}\right]$ **b.** $\left[\begin{array}{cccc|c} 1 & 0 & 1 & 0 & 2 \\ 0 & 1 & 2 & 0 & 3 \\ 0 & 0 & 0 & 2 & 3 \\ 0 & 0 & 0 & 0 & 0 \end{array}\right]$

c. $\left[\begin{array}{ccccc|c} 3 & 0 & 1 & 0 & 1 & -1 \\ 0 & -2 & 2 & 0 & 1 & 3 \\ 0 & 0 & 0 & 2 & 0 & 3 \end{array}\right]$

Solutions:

a. $\left[\begin{array}{cccc|c} 2 & -1 & 0 & 0 & 3 \\ 0 & 0 & 1 & 0 & 2 \\ 0 & 0 & 0 & 2 & 3 \end{array}\right]$

There are 4 variables, say x_1, x_2, x_3, x_4, since there are five columns. The last column corresponds to the constants. We write the equations corresponding to the rows and solve for the leading variable.

$$\begin{cases} 2x_1 - x_2 = 3 \text{ or } 2x_1 = 3 + x_2 \text{ or } x_1 = \dfrac{3 + x_2}{2} \\ x_3 = 2 \\ 2x_4 = 3 \text{ or } x_4 = \dfrac{3}{2}. \end{cases}$$

The solution will be all points of the form $\left(\frac{3 + x_2}{2}, x_2, 2, \frac{3}{2}\right)$ where x_2 is a free variable and can be any real number.

b. $$\left[\begin{array}{cccc|c} 1 & 0 & 1 & 0 & 2 \\ 0 & 1 & 2 & 0 & 3 \\ 0 & 0 & 0 & 2 & 3 \\ 0 & 0 & 0 & 0 & 0 \end{array}\right]$$

There are 4 variables, say x_1, x_2, x_3, x_4. From the matrix we have

$$\begin{cases} x_1 + \quad x_3 = 2 \\ x_2 + 2x_3 = 3 \\ \qquad 2x_4 = 3 \\ \qquad\quad 0 = 0 \end{cases} \quad \text{or} \quad \begin{cases} x_1 = 2 - x_3 \\ x_2 = 3 - 2x_3 \\ x_4 = \frac{3}{2}. \end{cases}$$

In matrix form we have $\begin{bmatrix} x_1 \\ x_2 \\ x_3 \\ x_4 \end{bmatrix} = \begin{bmatrix} 2 - x_3 \\ 3 - 2x_3 \\ x_3 \\ \frac{3}{2} \end{bmatrix}$.

The solution consists of all points of the form $\left(2 - x_3, 3 - 2x_3, x_3, \frac{3}{2}\right)$ where x_3 is a free variable.

c. $$\left[\begin{array}{ccccc|c} 3 & 0 & 1 & 0 & 1 & -1 \\ 0 & -2 & 2 & 0 & 1 & 3 \\ 0 & 0 & 0 & 2 & 0 & 3 \end{array}\right]$$

There are five variables, say x_1, x_2, x_3, x_4, x_5. The rows of the matrix correspond to the equations

$$\begin{cases} 3x_1 + x_3 + x_5 = -1 \text{ or } 3x_1 = -1 - x_3 - x_5 \text{ or } x_1 = \dfrac{-1 - x_3 - x_5}{3} \\ -2x_2 + 2x_3 + x_5 = 3 \text{ or } -2x_2 = 3 - 2x_3 - x_5 \quad x_2 = \dfrac{-3 + 2x_3 + x_5}{2} \\ 2x_4 = 3 \text{ or } x_4 = \dfrac{3}{2} \end{cases}$$

The solution consists of all points of the form $\left(\frac{-1 - x_3 - x_5}{3}, \frac{-3 + 2x_3 + x_5}{2}, x_3, \frac{3}{2}, x_5\right)$ where both x_3 and x_5 are free variables. ■

Remember that to find the inverse of a square matrix A, it is necessary to reduce $[A|I_n]$ to the matrix $[I_n|B]$. That is, A must be reduced completely. To solve a system of equations having the matrix A as the coefficient matrix, it is not necessary to reduce A completely. The reduction only needs to be done until the solution can be obtained. However, when the solution is not unique, it is usually better to reduce completely.

Exercise Set 2.2

Find the solution for each system of equations, if it exists. If multiple solutions exist, give the general form.

1. $\begin{cases} 2x - 3y = 3 \\ x + 2y = 5 \end{cases}$

2. $\begin{cases} 2x - 3y = -5 \\ -x + 2y = 5 \end{cases}$

3. $\begin{cases} 2x - 4y = 3 \\ -x + 2y = 5 \end{cases}$

4. $\begin{cases} 2x + 4y = 3 \\ 3x - 5y = -1 \end{cases}$

5. $\begin{cases} x + 2y + z = 2 \\ -2x - y + 4z = 2 \\ 3x + 3y + z = 3 \end{cases}$

6. $\begin{cases} x_1 - 2x_2 + x_3 = 7 \\ 2x_1 + 2x_2 - x_3 = -1 \\ x_1 + 3x_2 + x_3 = -3 \end{cases}$

7. $\begin{cases} 2x_1 + 2x_2 - x_3 = 5 \\ x_1 + 2x_2 - 3x_3 = 1 \end{cases}$

8. $\begin{cases} 8x - 4y + 12z = 5 \\ x + 3y - 9z = 3 \\ 4x - 2y + 6z = 1 \end{cases}$

9. $\begin{cases} 4x + y - z = 2 \\ 4x + 11y - z = 4 \\ 8x + y + z = 3 \end{cases}$

10. $\begin{cases} 3x + 2y - 2z = -1 \\ 2x - 3y + 3z = -5 \\ 5x - 2y - z = -1 \end{cases}$

11. $\begin{cases} 6x + 2y - 2z = 4 \\ 4x - y + 3z = 9 \\ 3x + 2y + z = 7 \end{cases}$

12. $\begin{cases} 2x - 3y + 2z = 3 \\ x + y - z = 1 \\ x + 2y + 2z = -3 \end{cases}$

13. $\begin{cases} x - 2y + 3z = 2 \\ 2x + y - 2z = 1 \\ 3x - y + z = 3 \end{cases}$

14. $\begin{cases} 2x - 3y + 2z = 9 \\ 3x + 2z = 8 \\ 3y - 2z = -5 \end{cases}$

15. $\begin{cases} 2w + 3x - 2y + z = 8 \\ w - 2x + 2y + 2z = 2 \\ 3w - 2x + y + 2z = 6 \\ w + x + y + z = 4 \end{cases}$

16. $\begin{cases} 2w - 3x + y + 5z = 2 \\ 3w + 3x - 2y + z = 32 \\ 5w - x + 3y + 2z = 31 \\ 2w - 2x + 3y + 3z = 11 \end{cases}$

17. $\begin{cases} 3x + 2y + z = 15 \\ 2x - 3y + z = 1 \\ x + 2y + 2z = 6 \\ 2x - 2y - z = 5 \end{cases}$

18. $\begin{cases} 3x - y + 3z = 7 \\ 2x + 2y - 5z = 9 \\ 3x - 2y - z = 6 \\ 3x - 2y + 4z = 7 \end{cases}$

19. $\begin{cases} w + 2x - 5y + 4z = 2 \\ 3w + 3x - 7y + z = 3 \\ 2w + 3x + 3y + 2z = 5 \\ 3w + x + 2y - 2z = 3 \end{cases}$

20. $\begin{cases} 4w - 3x + y + 3z = 2 \\ 2w + x + y + z = 1 \\ 5w + 3x - y + z = 3 \\ w + 3x + y + 3z = 2 \end{cases}$

21. $\begin{cases} 3x + 2y - 2z = 10 \\ 2x - y + z = 2 \end{cases}$

22. $\begin{cases} 3x + 4y - 2z = 15 \\ 2x + 2y - 3z = 11 \end{cases}$

23. $\begin{cases} -2x + 4y + 5z = 4 \\ 3x - 6y - 7z = 3 \end{cases}$

24. $\begin{cases} 3x + 2y - 5z = 5 \\ x - y + 2z = 2 \\ 4x + y + z = 11 \\ -2x + 2y + z = 1 \end{cases}$

25. $\begin{cases} 3x + 2y + z = 10 \\ 2x + y + z = 7 \\ x + y + z = 5 \\ 2x + 2y - z = 4 \end{cases}$

26. $\begin{cases} 2x + y + z = 9 \\ x + 3y + z = 2 \\ 3x + 3y + z = 8 \\ 5x + 4y - z = 2 \end{cases}$

27. $\begin{cases} 3x + 2y + 3z = 12 \\ 2x + y + z = 7 \\ x + y + 2z = 5 \\ 2x + 3y + 7z = 13 \end{cases}$

28. $\begin{cases} x - 5y + 4z = 2 \\ 3x + 2y - 3z = 8 \\ 4x + y - 2z = 14 \\ -2x + 3y + 2z = 3 \end{cases}$

29. $\begin{cases} 2w + 3x - y + z = 10 \\ 5w - 2x + y - 2z = 9 \\ 3w + 2x + 2y + 3z = 24 \end{cases}$

30. $\begin{cases} 3w + 2x + 2y - 5z = -6 \\ w + x - 2y + z = -1 \\ 2w - 2x + 2y + z = 3 \end{cases}$

2.3 More on Systems of Equations

The preceding section involved the use of matrices and reduction techniques to solve systems of equations. This section will expand on some of those ideas and introduce some new ones. The focus will continue to be on systems of equations. The three main ideas of this section are:

A. **multiple systems** (more than one system with a common coefficient matrix),

B. **systems of equations** and the **inverse of a matrix**, and

C. **matrix equations to represent systems of equations.**

A. Multiple Systems of Equations

Consider the two systems of equations

$$\begin{cases} x - 3y + z = -4 \\ 2x + 2y - z = 10 \\ -x + y - z = -2 \end{cases} \quad \text{and} \quad \begin{cases} x - 3y + z = 4 \\ 2x + 2y - z = 7. \\ -x + y - z = -6 \end{cases}$$

Solving these systems using matrix methods would involve reducing the matrices

$$\left[\begin{array}{rrr|r} 1 & -3 & 1 & -4 \\ 2 & 2 & -1 & 10 \\ -1 & 1 & -1 & -2 \end{array}\right] \quad \text{and} \quad \left[\begin{array}{rrr|r} 1 & -3 & 1 & 4 \\ 2 & 2 & -1 & 7 \\ -1 & 1 & -1 & -6 \end{array}\right].$$

Examining these matrices closely reveals that the coefficient matrices are the same. The reduction steps involved in solving the two systems would probably be identical. Instead of solving the first system and then duplicating the steps to solve the second system, the systems can be solved simultaneously.

Since the coefficient matrices are the same, the augmented matrix $[A|B_1B_2]$ is formed where A is the common coefficient matrix, B_1 is the constant column of the first system, and B_2 is the constant column of the second system. The reduction is done and the solutions to the *first system* will correspond to the *first augmented column*. The solutions to the *second system* will correspond to the *second augmented column*. The following example will illustrate the method.

■ Example 23. Solve the systems of equations

$$\text{I} \begin{cases} x - 3y + z = -4 \\ 2x + 2y - z = 10 \\ -x + y - z = -2 \end{cases} \quad \text{and} \quad \text{II} \begin{cases} x - 3y + z = 4 \\ 2x + 2y - z = 7. \\ -x + y - z = -6 \end{cases}$$

Solution: Solution with check column:

The augmented matrix is

$$\left[\begin{array}{rrr|rr} 1 & -3 & 1 & -4 & 4 \\ 2 & 2 & -1 & 10 & 7 \\ -1 & 1 & -1 & -2 & -6 \end{array}\right] \begin{array}{r} -1 \\ 20. \\ -9 \end{array}$$

$$\begin{array}{r} \\ -2R1 + R2 \\ R1 + R3 \end{array} \left[\begin{array}{rrr|rr} 1 & -3 & 1 & -4 & 4 \\ 0 & 8 & -3 & 18 & -1 \\ 0 & -2 & 0 & -6 & -2 \end{array}\right] \begin{array}{r} -1 \\ 22 \\ -10 \end{array}$$

$$\begin{matrix} -3R3+2R1 \\ 4R3+R2 \\ \\ \end{matrix}\left[\begin{array}{ccc|cc|c} 2 & 0 & 2 & 10 & 14 & 28 \\ 0 & 0 & -3 & -6 & -9 & -18 \\ 0 & -2 & 0 & -6 & -2 & -10 \end{array}\right]$$

$$\begin{matrix} 2R2+3R1 \\ \\ \\ \end{matrix}\left[\begin{array}{ccc|cc|c} 6 & 0 & 0 & 18 & 24 & 48 \\ 0 & 0 & -3 & -6 & -9 & -18 \\ 0 & -2 & 0 & -6 & -2 & -10 \end{array}\right]$$

Further reduction could be done but is not necessary. The solution for system I (use the first augmented column) is given by $x = \frac{18}{6} = 3$, $y = \frac{-6}{-2} = 3$, and $z = \frac{-6}{-3} = 2$ or by the point (3,3,2). The solution for system II (use the second augmented column) is given by $x = \frac{24}{6} = 4$, $y = \frac{-2}{-2} = 1$, and $z = \frac{-9}{-3} = 3$ or by the point (4,1,3). ■

This technique can be used in any situation where several systems involving the same coefficient matrix are to be solved. However, if the number of systems is larger than the number of variables in the system, there is another method of solution that may be more economical. More will be said about this later.

■ Example 24. Solve the three systems of equations

$$\text{I}\begin{cases} 2x - y - 3z = 2 \\ -4x + 2y + 3z = -10 \\ 6x - 3y + z = 26 \end{cases} \quad \text{II}\begin{cases} 2x - y - 3z = 3 \\ -4x + 2y + 3z = -3 \\ 6x - 3y + z = -1 \end{cases} \quad \text{III}\begin{cases} 2x - y - 3z = 3 \\ -4x + 2y + 3z = -2. \\ 6x - 3y + z = 1 \end{cases}$$

Solution: The coefficient matrix is the same for each system, so we augment the matrix with the three columns and reduce.

$$\left[\begin{array}{ccc|cc|c} 2 & -1 & -3 & 2 & 3 & 3 & 6 \\ -4 & 2 & 3 & -10 & -3 & -2 & -14 \\ 6 & -3 & 1 & 26 & -1 & 1 & 30 \end{array}\right]$$

$$\begin{matrix} \\ 2R1+R2 \\ -3R1+R3 \end{matrix}\left[\begin{array}{ccc|ccc|c} 2 & -1 & -3 & 2 & 3 & 3 & 6 \\ 0 & 0 & -3 & -6 & 3 & 4 & -2 \\ 0 & 0 & 10 & 20 & -10 & -8 & 12 \end{array}\right]$$

$$\begin{matrix} -R2+R1 \\ \\ 10R2+3R3 \end{matrix}\left[\begin{array}{ccc|ccc|c} 2 & -1 & 0 & 8 & 0 & -1 & 8 \\ 0 & 0 & -3 & -6 & 3 & 4 & -2 \\ 0 & 0 & 0 & 0 & 0 & 16 & 16 \end{array}\right]$$

The augmented columns correspond to the three systems. From the final matrix, we see that systems I and II have multiple solutions (entire row of zeros) but system III has no solution (last row is all zeros except the last entry). Specifically, y is a free variable and the solution for system I is given by $2x - y = 8$, or $x = \dfrac{8 + y}{2}$, and $z = \dfrac{-6}{-3} = 2$, and y is arbitrary. The solution consists of all points of the form $\left(\dfrac{8 + y}{2}, y, 2\right)$.

System II has $2x - y = 0$, or $x = \dfrac{y}{2}$, and $z = \dfrac{3}{-3} = -1$, and y is a free variable. The solution consists of all points of the form $\left(\dfrac{y}{2}, y, -1\right)$.

For system III there is no solution since the last row corresponds to $0 = 16$, which isn't possible. ■

B. Systems of Equations and the Inverse of a Matrix

In chapter 1, Gauss-Jordan reduction was used to find the inverse of a square matrix. We are now in a position to consider why that technique works. Recall that the inverse of a square matrix A, if it exists, is a matrix B such that $AB = I_n$.

We will illustrate here with a 3×3 matrix A.

Let $A = \begin{bmatrix} a_{11} & a_{12} & a_{13} \\ a_{21} & a_{22} & a_{23} \\ a_{31} & a_{32} & a_{33} \end{bmatrix}$. We want to find a 3×3 matrix B so that

$AB = \begin{bmatrix} 1 & 0 & 0 \\ 0 & 1 & 0 \\ 0 & 0 & 1 \end{bmatrix}$. To simplify notation, let $B = \begin{bmatrix} x_1 & x_2 & x_3 \\ y_1 & y_2 & y_3 \\ z_1 & z_2 & z_3 \end{bmatrix}$.

$$\text{Then } AB = \begin{bmatrix} a_{11} & a_{12} & a_{13} \\ a_{21} & a_{22} & a_{23} \\ a_{31} & a_{32} & a_{33} \end{bmatrix} \begin{bmatrix} x_1 & x_2 & x_3 \\ y_1 & y_2 & y_3 \\ z_1 & z_2 & z_3 \end{bmatrix}$$

$$= \begin{bmatrix} a_{11}x_1 + a_{12}y_1 + a_{13}z_1 & a_{11}x_2 + a_{12}y_2 + a_{13}z_2 & a_{11}x_3 + a_{12}y_3 + a_{13}z_3 \\ a_{21}x_1 + a_{22}y_1 + a_{23}z_1 & a_{21}x_2 + a_{22}y_2 + a_{23}z_2 & a_{21}x_3 + a_{22}y_3 + a_{23}z_3 \\ a_{31}x_1 + a_{32}y_1 + a_{33}z_1 & a_{31}x_2 + a_{32}y_2 + a_{33}z_2 & a_{31}x_3 + a_{32}y_3 + a_{33}z_3 \end{bmatrix}$$

$$= \begin{bmatrix} 1 & 0 & 0 \\ 0 & 1 & 0 \\ 0 & 0 & 1 \end{bmatrix}.$$

Two matrices are equal when they are identical, so equating the first columns gives

$$\text{I}\quad \begin{cases} a_{11}x_1 + a_{12}y_1 + a_{13}z_1 = 1 \\ a_{21}x_1 + a_{22}y_1 + a_{23}z_1 = 0. \\ a_{31}x_1 + a_{32}y_1 + a_{33}z_1 = 0 \end{cases}$$

Equating second columns gives

$$\text{II}\quad \begin{cases} a_{11}x_2 + a_{12}y_2 + a_{13}z_2 = 0 \\ a_{21}x_2 + a_{22}y_2 + a_{23}z_2 = 1. \\ a_{31}x_2 + a_{32}y_2 + a_{33}z_2 = 0 \end{cases}$$

Equating third columns gives

$$\text{III}\quad \begin{cases} a_{11}x_3 + a_{12}y_3 + a_{13}z_3 = 0 \\ a_{21}x_3 + a_{22}y_3 + a_{23}z_3 = 0. \\ a_{31}x_3 + a_{32}y_3 + a_{33}z_3 = 1 \end{cases}$$

The three systems of equations have the same coefficient matrix, even though the variables are different. To solve the systems, reduce the augmented matrix of the system. This augmented matrix is

$$\left[\begin{array}{ccc|ccc} a_{11} & a_{12} & a_{13} & 1 & 0 & 0 \\ a_{21} & a_{22} & a_{23} & 0 & 1 & 0 \\ a_{31} & a_{32} & a_{33} & 0 & 0 & 1 \end{array}\right].$$

But this is just $[A|I_3]$, which is the matrix used in chapter 1 to find the inverse of A. In general, then, to invert the $n \times n$ matrix A, augment A with I_n and use Gauss-Jordan reduction. If A is not invertible, at least one of the above systems (I, II, or III) will not have a unique solution. This will show up in the reduction as it did in example 24 of this section.

C. Matrix Equations to Represent Systems of Equations

In the preceding section, systems of linear equations were represented by an augmented matrix, and then matrix reduction techniques were used to solve the system. Matrices can be used in another way to represent a system of equations. In fact, a matrix equation can be written to represent any given system of equations. As an illustration, consider the following system of three equations in four variables

$$\begin{cases} a_{11}x_1 + a_{12}x_2 + a_{13}x_3 + a_{14}x_4 = b_1 \\ a_{21}x_1 + a_{22}x_2 + a_{23}x_3 + a_{24}x_4 = b_2. \\ a_{31}x_1 + a_{32}x_2 + a_{33}x_3 + a_{34}x_4 = b_3 \end{cases}$$

Let $A = \begin{bmatrix} a_{11} & a_{12} & a_{13} & a_{14} \\ a_{21} & a_{22} & a_{23} & a_{24} \\ a_{31} & a_{32} & a_{33} & a_{34} \end{bmatrix}$ be the coefficient matrix for the system. Let $X = \begin{bmatrix} x_1 \\ x_2 \\ x_3 \\ x_4 \end{bmatrix}$ and $B = \begin{bmatrix} b_1 \\ b_2 \\ b_3 \end{bmatrix}$ be the column matrices of variables and constants, respectively. Consider the **matrix equation** $AX = B$.

$$AX = \begin{bmatrix} a_{11} & a_{12} & a_{13} & a_{14} \\ a_{21} & a_{22} & a_{23} & a_{24} \\ a_{31} & a_{32} & a_{33} & a_{34} \end{bmatrix} \begin{bmatrix} x_1 \\ x_2 \\ x_3 \\ x_4 \end{bmatrix}$$

$$= \begin{bmatrix} a_{11}x_1 + a_{12}x_2 + a_{13}x_3 + a_{14}x_4 \\ a_{21}x_1 + a_{22}x_2 + a_{23}x_3 + a_{24}x_4 \\ a_{31}x_1 + a_{32}x_2 + a_{33}x_3 + a_{34}x_4 \end{bmatrix} = \begin{bmatrix} b_1 \\ b_2 \\ b_3 \end{bmatrix} = B.$$

Equating corresponding entries of AX and B gives

$$\begin{cases} a_{11}x_1 + a_{12}x_2 + a_{13}x_3 + a_{14}x_4 = b_1 \\ a_{21}x_1 + a_{22}x_2 + a_{23}x_3 + a_{24}x_4 = b_2 \\ a_{31}x_1 + a_{32}x_2 + a_{33}x_3 + a_{34}x_4 = b_3 \end{cases}$$

which is the original system of equations.

Any system of equations can be written as a single matrix equation of the form $AX = B$, where A is the coefficient matrix of the system, X is the column matrix of variables, and B is the column matrix of constants. If the system of equations consists of m equations in n variables, the coefficient matrix A will be $m \times n$. The matrix X will be $n \times 1$, and the matrix B will be $m \times 1$. These dimension restrictions agree with those necessary for matrix products. That is, if A is $m \times n$ and X is $n \times 1$, then AX will be $m \times 1$, which is the dimension of B.

If A is a square matrix, a very useful result is sometimes available. Let $AX = B$ be the matrix form for a system of equations and suppose A is a square matrix. If A is $n \times n$, then X is $n \times 1$ and B is $n \times 1$. If A^{-1} exists, we have

$AX = B$	
$A^{-1}(AX) = A^{-1}B$	Multiply both sides on the *left* by A^{-1}.
$(A^{-1}A)X = A^{-1}B$	Property 1.2 (1), section 1.3
$I_nX = A^{-1}B$	Definition of A^{-1}
$X = A^{-1}B$	Definition of I_n

Since A is $n \times n$, A^{-1} is $n \times n$ and $A^{-1}B$ is $n \times 1$. Equating entries of X (the variables) and $A^{-1}B$ will give the solutions to the system.

CAUTION: Since, in general, $AB \neq BA$, the multiplication must be on the left.

■ Example 25. Solve the system of equations

$$\begin{cases} -x + y + z = 2 \\ x + 4y - 5z = 5 \\ x - 2y = -3 \end{cases}.$$

Solution: The coefficient matrix for the system is

$A = \begin{bmatrix} -1 & 1 & 1 \\ 1 & 4 & -5 \\ 1 & -2 & 0 \end{bmatrix}$. In example 26 in chapter 1 we found

$$A^{-1} = \begin{bmatrix} 10 & 2 & 9 \\ 5 & 1 & 4 \\ 6 & 1 & 5 \end{bmatrix}.$$

Then the solution to the system

$$AX = \begin{bmatrix} -1 & 1 & 1 \\ 1 & 4 & -5 \\ 1 & -2 & 0 \end{bmatrix} \begin{bmatrix} x \\ y \\ z \end{bmatrix} = \begin{bmatrix} 2 \\ 5 \\ -3 \end{bmatrix} = B \text{ is given by}$$

$$X = \begin{bmatrix} x \\ y \\ z \end{bmatrix} = A^{-1} \cdot \begin{bmatrix} 2 \\ 5 \\ -3 \end{bmatrix} = \begin{bmatrix} 10 & 2 & 9 \\ 5 & 1 & 4 \\ 6 & 1 & 5 \end{bmatrix} \begin{bmatrix} 2 \\ 5 \\ -3 \end{bmatrix} = \begin{bmatrix} 3 \\ 3 \\ 2 \end{bmatrix}.$$

Then $\begin{bmatrix} x \\ y \\ z \end{bmatrix} = \begin{bmatrix} 3 \\ 3 \\ 2 \end{bmatrix}$ or $x = 3$, $y = 3$, $z = 2$. ■

This technique can be useful in situations where there are multiple systems of equations with the same square coefficient matrix. If the coefficient matrix is A and the column matrices of constants are $B_1, B_2, B_3, \ldots, B_k$, there are k systems of equations ($AX = B_1$, $AX = B_2, \ldots, AX = B_k$) to solve. One procedure is to reduce the augmented matrix $[A|B_1B_2B_3 \ldots B_k]$ to find the solutions.

If, however, A^{-1} exists, the solutions will be determined by $A^{-1}B_1$, $A^{-1}B_2, \ldots, A^{-1}B_k$. If we let B be the matrix with $B_1, B_2, \ldots, B_k$ as its columns ($B = [B_1B_2 \ldots B_k]$), then $A^{-1}B$ will be a matrix whose k columns represent the solutions to the k systems of equations. If the number of systems (k) is larger than the number of rows in A, it may be quicker to use this technique. The following example will illustrate.

■ **Example 26.** Solve the six systems of equations $AX = B_i$ where

$$A = \begin{bmatrix} 1 & 1 & 1 \\ 2 & -1 & 2 \\ 2 & -1 & -1 \end{bmatrix},$$

$$B_1 = \begin{bmatrix} 5 \\ 1 \\ 7 \end{bmatrix}, B_2 = \begin{bmatrix} 3 \\ 3 \\ 0 \end{bmatrix}, B_3 = \begin{bmatrix} 5 \\ 7 \\ 1 \end{bmatrix}, B_4 = \begin{bmatrix} 4 \\ 8 \\ 5 \end{bmatrix}, B_5 = \begin{bmatrix} 1 \\ 8 \\ 5 \end{bmatrix}, B_6 = \begin{bmatrix} 2 \\ 1 \\ 7 \end{bmatrix}.$$

Solution: We will first compute A^{-1}, if it exists, and then use $A^{-1}B$ where $B = [B_1B_2B_3B_4B_5B_6]$.

$$\left[\begin{array}{ccc|ccc} 1 & 1 & 1 & 1 & 0 & 0 \\ 2 & -1 & 2 & 0 & 1 & 0 \\ 2 & -1 & -1 & 0 & 0 & 1 \end{array}\right] \begin{array}{c} 4 \\ 4 \\ 1 \end{array}$$

$$\begin{array}{r} \\ -2R1 + R2 \\ -2R1 + R3 \end{array} \left[\begin{array}{ccc|ccc} 1 & 1 & 1 & 1 & 0 & 0 \\ 0 & -3 & 0 & -2 & 1 & 0 \\ 0 & -3 & -3 & -2 & 0 & 1 \end{array}\right] \begin{array}{c} 4 \\ -4 \\ -7 \end{array}$$

$$\begin{array}{r} R2 + 3R1 \\ \\ -R2 + R3 \end{array} \left[\begin{array}{ccc|ccc} 3 & 0 & 3 & 1 & 1 & 0 \\ 0 & -3 & 0 & -2 & 1 & 0 \\ 0 & 0 & -3 & 0 & -1 & 1 \end{array}\right] \begin{array}{c} 8 \\ -4 \\ -3 \end{array}$$

$$\begin{array}{r} R3 + R1 \\ \\ \\ \end{array} \left[\begin{array}{ccc|ccc} 3 & 0 & 0 & 1 & 0 & 1 \\ 0 & -3 & 0 & -2 & 1 & 0 \\ 0 & 0 & -3 & 0 & -1 & 1 \end{array}\right] \begin{array}{c} 5 \\ -4 \\ -3 \end{array}$$

$$\begin{array}{r} \frac{1}{3}R1 \\ -\frac{1}{3}R2 \\ -\frac{1}{3}R3 \end{array} \left[\begin{array}{ccc|ccc} 1 & 0 & 0 & \frac{1}{3} & 0 & \frac{1}{3} \\ 0 & 1 & 0 & \frac{2}{3} & -\frac{1}{3} & 0 \\ 0 & 0 & 1 & 0 & \frac{1}{3} & -\frac{1}{3} \end{array}\right]$$

So $A^{-1} = \left(\dfrac{1}{3}\right)\begin{bmatrix} 1 & 0 & 1 \\ 2 & -1 & 0 \\ 0 & 1 & -1 \end{bmatrix}$ and $B = \begin{bmatrix} 5 & 3 & 5 & 4 & 1 & 2 \\ 1 & 3 & 7 & 8 & 8 & 1 \\ 7 & 0 & 1 & 5 & 5 & 7 \end{bmatrix}$.

$$A^{-1}B = \left(\frac{1}{3}\right)\begin{bmatrix} 1 & 0 & 1 \\ 2 & -1 & 0 \\ 0 & 1 & -1 \end{bmatrix}\begin{bmatrix} 5 & 3 & 5 & 4 & 1 & 2 \\ 1 & 3 & 7 & 8 & 8 & 1 \\ 7 & 0 & 1 & 5 & 5 & 7 \end{bmatrix}$$

$$= \left(\frac{1}{3}\right)\begin{bmatrix} 12 & 3 & 6 & 9 & 6 & 9 \\ 9 & 3 & 3 & 0 & -6 & 3 \\ -6 & 3 & 6 & 3 & 3 & -6 \end{bmatrix}$$

$$= \begin{bmatrix} 4 & 1 & 2 & 3 & 2 & 3 \\ 3 & 1 & 1 & 0 & -2 & 1 \\ -2 & 1 & 2 & 1 & 1 & -2 \end{bmatrix}.$$

Column 1 gives the solution $x = 4$, $y = 3$, $z = -2$ for $AX = B_1$. Column 3 gives the solution $x = 2$, $y = 1$, $z = 2$ for $AX = B_3$, and so on. That is, the ith column of $A^{-1}B$ gives the solution to the system $AX = B_i$.

An alternate procedure would have been to augment A with the six columns of constants for the six systems of equations and use row reduction on

$$\left[\begin{array}{ccc|cccccc} 1 & 1 & 1 & 5 & 3 & 5 & 4 & 1 & 2 \\ 2 & -1 & 2 & 1 & 3 & 7 & 8 & 8 & 1 \\ 2 & -1 & -1 & 7 & 0 & 1 & 5 & 5 & 7 \end{array}\right].$$

Either procedure is acceptable, although the use of A^{-1} requires row reduction across six columns whereas the other technique requires reduction across nine columns. ■

If the coefficient matrix for a system of equations is not invertible this technique will not work. Hence, whenever the coefficient matrix is not square or is square but not invertible, we are forced to augment with the constant columns and use standard row reduction.

In closing this section, we state more formally the result we have used in the previous two examples.

Property 2.1 If $AX = B$ is the matrix form of a system of n equations in n unknowns (A is square), then the system has a unique solution if and only if A is invertible. If A is invertible, the solution is given by $X = A^{-1}B$.

Property 2.1 is of no help if A is not invertible. It is necessary then to solve by other techniques to determine if there are multiple solutions or no solutions.

Exercise Set 2.3

1. For each of the following, write the matrix equation as a system of equations.

a. $$\begin{bmatrix} 3 & 1 & 4 \\ 1 & 2 & 3 \\ 2 & 1 & 4 \\ 1 & 3 & -1 \end{bmatrix}\begin{bmatrix} x \\ y \\ z \end{bmatrix} = \begin{bmatrix} 6 \\ -2 \\ 3 \\ 4 \end{bmatrix}$$

b. $$\begin{bmatrix} 3 & -1 & 2 \\ 4 & 1 & -2 \\ 1 & 4 & 5 \end{bmatrix}\begin{bmatrix} x \\ y \\ z \end{bmatrix} = \begin{bmatrix} 2 \\ 5 \\ 1 \end{bmatrix}$$

c. $$\begin{bmatrix} 2 & -1 & 3 & 4 \\ 1 & 1 & -1 & 3 \\ 4 & 2 & 1 & 1 \end{bmatrix}\begin{bmatrix} w \\ x \\ y \\ z \end{bmatrix} = \begin{bmatrix} 5 \\ 1 \\ 2 \end{bmatrix}$$

d. $$\begin{bmatrix} 2 & -1 & 4 & 3 & 1 \\ 3 & 1 & -1 & 2 & 1 \\ 3 & 4 & 1 & 1 & 2 \end{bmatrix}\begin{bmatrix} v \\ w \\ x \\ y \\ z \end{bmatrix} = \begin{bmatrix} -3 \\ 5 \\ 7 \\ 2 \end{bmatrix}$$

2. Solve the following three systems of equations.

$$\text{I}\quad \begin{cases} 6x + 2y - 2z = 4 \\ 4x - y + 3z = 11 \\ 3x + 2y + z = 10 \end{cases} \qquad \text{II}\quad \begin{cases} 6x + 2y - 2z = 20 \\ 4x - y + 3z = 13 \\ 3x + 2y + z = 14 \end{cases}$$

$$\text{III}\quad \begin{cases} 6x + 2y - 2z = 6 \\ 4x - y + 3z = 6 \\ 3x + 2y + z = 6 \end{cases}$$

3. Solve the matrix equation $AX = B$ for each of the following.

a. $$A = \begin{bmatrix} 3 & 2 & -2 \\ 2 & -3 & 3 \\ 5 & -2 & -1 \end{bmatrix},\ B = \begin{bmatrix} 5 \\ -1 \\ 0 \end{bmatrix}$$

b. $$A = \begin{bmatrix} 1 & -2 & 3 \\ 2 & 1 & -2 \\ 3 & -1 & 1 \end{bmatrix},\ B = \begin{bmatrix} 5 \\ 9 \\ 14 \end{bmatrix}$$

c. $A = \begin{bmatrix} 2 & 2 & -1 \\ 1 & 2 & -3 \\ 2 & 1 & 1 \end{bmatrix}$, $B = \begin{bmatrix} 7 \\ 4 \\ 6 \end{bmatrix}$

d. $A = \begin{bmatrix} 8 & -4 & 12 \\ 1 & 3 & -9 \\ 4 & -2 & 6 \end{bmatrix}$, $B = \begin{bmatrix} 12 \\ -2 \\ 6 \end{bmatrix}$

e. $A = \begin{bmatrix} 3 & 2 & 1 \\ 2 & 1 & 1 \\ 1 & 1 & 1 \\ 2 & 2 & -1 \end{bmatrix}$, $B = \begin{bmatrix} 10 \\ 7 \\ 5 \\ 4 \end{bmatrix}$

f. $A = \begin{bmatrix} 3 & -1 & 3 \\ 2 & 2 & -5 \\ 3 & -2 & -1 \\ 3 & -2 & 4 \end{bmatrix}$, $B = \begin{bmatrix} 7 \\ 9 \\ 6 \\ 7 \end{bmatrix}$

4. For each of the following, solve the matrix equation $AX = B$ for X. *Note*: The variable matrix X will be the same size as B.

a. $A = \begin{bmatrix} 2 & -3 \\ 1 & 2 \end{bmatrix}$, $B = \begin{bmatrix} 3 & 1 & -1 \\ 5 & 4 & 3 \end{bmatrix}$

b. $A = \begin{bmatrix} 2 & 4 \\ 3 & -5 \end{bmatrix}$, $B = \begin{bmatrix} 3 & 2 & -2 \\ -1 & -8 & 2 \end{bmatrix}$

c. $A = \begin{bmatrix} 1 & 2 & 1 \\ 3 & 0 & 2 \\ 3 & 3 & 1 \end{bmatrix}$, $B = \begin{bmatrix} 2 & 4 \\ 2 & 1 \\ 3 & 7 \end{bmatrix}$

d. $A = \begin{bmatrix} 2 & -3 & 2 \\ 3 & 0 & 2 \\ 0 & 3 & -2 \end{bmatrix}$, $B = \begin{bmatrix} 1 & 9 \\ 5 & 8 \\ 1 & -5 \end{bmatrix}$

5. Solve the matrix equation $AX = B$, where

$$A = \begin{bmatrix} 2 & 1 & 3 \\ 1 & 1 & 2 \\ 1 & 2 & 2 \end{bmatrix} \text{ and } B = \begin{bmatrix} 6 & 7 & 11 & 13 & 11 \\ 4 & 5 & 7 & 8 & 7 \\ 5 & 7 & 8 & 9 & 9 \end{bmatrix}.$$

6. Solve the matrix equation $AX = B$, where

$$A = \begin{bmatrix} 3 & 1 & 4 \\ 2 & 1 & 5 \\ 1 & 3 & -1 \end{bmatrix} \text{ and } B = \begin{bmatrix} 6 & 16 & 13 & 15 & 24 & 3 \\ 6 & 16 & 14 & 13 & 23 & 5 \\ -3 & 6 & 5 & 8 & 1 & 4 \end{bmatrix}.$$

7. Express each of the following as a single matrix equation.

a. $\begin{cases} x + 2y = 3 \\ 4x - 3y = 1 \end{cases}$ **b.** $\begin{cases} 3x - 4y + z = 8 \\ x + y - z = 3 \end{cases}$

c. $\begin{cases} 3x + 4y = 8 \\ x - y = 2 \\ 2x + 3y = 4 \end{cases}$

8. Express each of the problems in exercise set 2.2 as a single matrix equation.

2.4 Applications of Systems of Linear Equations

A major focus of this book, as indicated in the introduction to Chapter 1, is to consider how mathematical concepts or techniques can be used to solve applied problems. That is, this text is application-oriented.

This chapter began with a hypothetical problem that was solved in section 2.1 and again in section 2.2. This section will consider additional applied problems.

In general, when we speak of a "problem," we are not talking about a simple equation or other mathematical statement. A problem is stated (in words, a table, or other forms) in everyday language. Before the problem can be solved, it must be restated in mathematical language.

As we consider examples here, special attention will be given to the steps involved in "translating" everyday language into mathematical language. Many problem-solving difficulties occur because the problem is incorrectly translated into mathematics, not because of the solution procedure.

We hope that you are well-versed in the techniques used in the preceding sections of this chapter to solve systems of equations. If you are, you can concentrate your efforts here on "setting up the problem" (translating from everyday to mathematical language) to make use of these techniques and interpreting your answer in everyday language.

All of the problems in this section will involve solving systems of equations. The mathematical form of the problem, then, will be a system of equations. To write the system of equations will require identifying the variables (what x stands for, and so on), identifying the appropriate coefficients for the variables, and identifying the constants for the equations.

In some problems, a table can be constructed to represent the information. The table can be used to determine what the variables represent and how the equations are to be written. In the examples that follow, a table will be used whenever possible.

■ Example 27. A company has three factories (I, II, III), producing products A, B, and C. In one hour factory I produces 3 units of A, 1 unit of B, and 2 units of C. Factory II, in one hour, produces 2 units of A and 3 units of B. The hourly production for factory III is 1 unit of A, 1 unit of B, and 3 units of C. How many hours will each factory need to operate to fill an order for 70 units of A, 65 units of B, and 50 units of C?

Solution: The production table can be represented by

	Per Hour Production		
Factory	*A*	*B*	*C*
I	3	1	2
II	2	3	0
III	1	1	3
Requirements	70	65	50

Since the orders are for products A, B, and C, the requirements have been included as the last row in the table. We are asked to find the number of hours each factory should operate, so we let x be the number of hours factory I operates, y be the number of hours factory II operates, and z be the number of hours factory III operates.

To write the system of equations, we consider the following: Factory I produces 3 units of A per hour, so it produces $3x$ units of A in x hours; factory II produces 2 units of A per hour, so it produces $2y$ units of A in y hours; and factory III produces 1 unit of A per hour, so it produces $1 \cdot z = z$ units in z hours. We require 70 units of A, so $3x + 2y + z = 70$ is one of the equations in the system. Note that the coefficients 3, 2, and 1 and the constant 70 form the first column of the table.

By including the requirements in the table (in this case as a row), we get a direction for our equations. That is, if the requirements form a row in the table, the equations go down the columns. If, however, the requirements form a column, then the equations will go across the rows. In other words, if the requirements are correctly included in the table, the equations go toward the requirements.

Using an argument similar to the one in the preceding paragraph, the second column and third column give two additional equations. From column 2 we have $x + 3y + z = 65$, and from column 3 we get $2x + 3z = 50$.

The system of equations to be solved is

$$\begin{cases} 3x + 2y + z = 70 \\ x + 3y + z = 65 \\ 2x + 3z = 50 \end{cases}$$

The augmented matrix is

$$\left[\begin{array}{ccc|c} 3 & 2 & 1 & 70 \\ 1 & 3 & 1 & 65 \\ 2 & 0 & 3 & 50 \end{array}\right].$$

$$\begin{array}{r} \\ -R1 + R2 \\ -3R1 + R3 \end{array} \left[\begin{array}{ccc|c} 3 & 2 & 1 & 70 \\ -2 & 1 & 0 & -5 \\ -7 & -6 & 0 & -160 \end{array}\right]$$

$$\begin{array}{r} -2R2 + R1 \\ \\ 6R2 + R3 \end{array} \left[\begin{array}{ccc|c} 7 & 0 & 1 & 80 \\ -2 & 1 & 0 & -5 \\ -19 & 0 & 0 & -190 \end{array}\right]$$

$$\begin{array}{r} \\ \\ -\frac{1}{19}R3 \end{array} \left[\begin{array}{ccc|c} 7 & 0 & 1 & 80 \\ -2 & 1 & 0 & -5 \\ 1 & 0 & 0 & 10 \end{array}\right]$$

$$\begin{array}{r} -7R3 + R1 \\ 2R3 + R2 \\ \\ \end{array} \left[\begin{array}{ccc|c} 0 & 0 & 1 & 10 \\ 0 & 1 & 0 & 15 \\ 1 & 0 & 0 & 10 \end{array}\right]$$

The solution is $x = 10$, $y = 15$, $z = 10$. So factory I should operate for 10 hours, factory II for 15 hours, and factory III for 10 hours. ■

■ Example 28. The Cozee Glove Company manufactures three colors of men's leather driving gloves. For shipping purposes, the gloves are packed in cartons containing 12 dozen pairs (1 gross) each. The company has found that most of their distributors prefer mixed colors in a carton. They have determined that three types of cartons of mixed gloves work best. Type I cartons contain 3 dozen brown, 6 dozen black, and 3 dozen gray. Each type II carton contains 5 dozen brown, 4 dozen black, and 3 dozen gray. Each type III carton contains 4 dozen pairs of each color. How many cartons of each type should be sent to the Illinois distributor to fill an order for 200 dozen brown, 230 dozen black, and 170 dozen gray?

Solution: The given information, including the requirements, can be placed in a table.

	Carton Types			
Color	I	II	III	Requirements
Brown	3	5	4	200
Black	6	4	4	230
Gray	3	3	4	170

Since the requirements are given in terms of color, they go in this table as a column. The equations are determined by the rows of the table.

NOTE: We could have put the colors across the top and cartons down the left side. The equations would have been the same but would have come from the columns.

Letting x, y, z represent the number of cartons of types I, II, and III respectively, we obtain the system of equations $\begin{cases} 3x + 5y + 4z = 200 \\ 6x + 4y + 4z = 230. \\ 3x + 3y + 4z = 170 \end{cases}$

$$\left[\begin{array}{ccc|c} 3 & 5 & 4 & 200 \\ 6 & 4 & 4 & 230 \\ 3 & 3 & 4 & 170 \end{array}\right]$$

$$\begin{array}{r} \\ -2R1 + R2 \\ -R1 + R3 \end{array} \left[\begin{array}{ccc|c} 3 & 5 & 4 & 200 \\ 0 & -6 & -4 & -170 \\ 0 & -2 & 0 & -30 \end{array}\right]$$

$$\begin{array}{r} \\ -\frac{1}{2}R2 \\ -\frac{1}{2}R3 \end{array} \left[\begin{array}{ccc|c} 3 & 5 & 4 & 200 \\ 0 & 3 & 2 & 85 \\ 0 & 1 & 0 & 15 \end{array}\right]$$

$$\begin{array}{r} -5R3 + R1 \\ -3R3 + R2 \\ \\ \end{array} \left[\begin{array}{ccc|c} 3 & 0 & 4 & 125 \\ 0 & 0 & 2 & 40 \\ 0 & 1 & 0 & 15 \end{array}\right]$$

$$\begin{array}{r} -2R2 + R1 \\ \\ \\ \end{array} \left[\begin{array}{ccc|c} 3 & 0 & 0 & 45 \\ 0 & 0 & 2 & 40 \\ 0 & 1 & 0 & 15 \end{array}\right]$$

Then $x = \frac{45}{3} = 15$, $y = 15$, $z = \frac{40}{2} = 20$, so the company should ship 15 type I cartons, 15 type II cartons, and 20 type III cartons. ■

■ **Example 29.** The Cozee Glove Company (see example 28) receives orders from state distributors on a regular basis. During one month, they received orders from eight states as given in the table below.

	Orders							
Color	*IA*	*MO*	*WI*	*IN*	*MI*	*AR*	*TX*	*MN*
Brown	120	85	240	120	215	51	17	310
Black	140	90	280	120	260	56	18	380
Gray	100	65	200	120	185	41	13	270

How many cartons of each type should be sent to each state to fill these orders?

Solution: Management of Cozee Glove Company knows that they will be receiving orders on a regular basis. To facilitate filling the orders, they check the coefficient matrix to determine if it is invertible. If it is, they can keep the inverse on file, and as orders come in, they will multiply the inverse times the order to determine the appropriate number of cartons of each type to send. We will use that technique here. The coefficient matrix is

$$A = \begin{bmatrix} 3 & 5 & 4 \\ 6 & 4 & 4 \\ 3 & 3 & 4 \end{bmatrix}.$$

Augment A with I_3 and reduce.

$$\left[\begin{array}{ccc|ccc} 3 & 5 & 4 & 1 & 0 & 0 \\ 6 & 4 & 4 & 0 & 1 & 0 \\ 3 & 3 & 4 & 0 & 0 & 1 \end{array}\right]$$

$$\begin{array}{r} \\ -2R1 + R2 \\ -R1 + R3 \end{array} \left[\begin{array}{ccc|ccc} 3 & 5 & 4 & 1 & 0 & 0 \\ 0 & -6 & -4 & -2 & 1 & 0 \\ 0 & -2 & 0 & -1 & 0 & 1 \end{array}\right]$$

$$\begin{array}{r} 5R3 + 2R1 \\ -3R3 + R2 \\ \\ \end{array} \left[\begin{array}{ccc|ccc} 6 & 0 & 8 & -3 & 0 & 5 \\ 0 & 0 & -4 & 1 & 1 & -3 \\ 0 & -2 & 0 & -1 & 0 & 1 \end{array}\right]$$

$$\begin{matrix} 2R2 + R1 \\ \\ \\ \end{matrix}\left[\begin{array}{ccc|ccc} 6 & 0 & 0 & -1 & 2 & -1 \\ 0 & 0 & -4 & 1 & 1 & -3 \\ 0 & -2 & 0 & -1 & 0 & 1 \end{array}\right]$$

$$\begin{matrix} \\ \\ R2 \leftrightarrow R3 \end{matrix}\left[\begin{array}{ccc|ccc} 6 & 0 & 0 & -1 & 2 & -1 \\ 0 & -2 & 0 & -1 & 0 & 1 \\ 0 & 0 & -4 & 1 & 1 & -3 \end{array}\right]$$

$$\begin{matrix} \frac{1}{6}R1 \\ -\frac{1}{2}R2 \\ -\frac{1}{4}R3 \end{matrix}\left[\begin{array}{ccc|ccc} 1 & 0 & 0 & -\frac{1}{6} & \frac{2}{6} & -\frac{1}{6} \\ 0 & 1 & 0 & \frac{1}{2} & 0 & -\frac{1}{2} \\ 0 & 0 & 1 & -\frac{1}{4} & -\frac{1}{4} & \frac{3}{4} \end{array}\right]$$

$$\text{So } A^{-1} = \begin{bmatrix} -\frac{1}{6} & \frac{2}{6} & -\frac{1}{6} \\ \frac{1}{2} & 0 & -\frac{1}{2} \\ -\frac{1}{4} & -\frac{1}{4} & \frac{3}{4} \end{bmatrix} = \left(\frac{1}{12}\right)\begin{bmatrix} -2 & 4 & -2 \\ 6 & 0 & -6 \\ -3 & -3 & 9 \end{bmatrix}.$$

To fill the orders, $X = A^{-1}B$ where $X = \begin{bmatrix} x \\ y \\ z \end{bmatrix}$ and $B = [B_1 B_2 \ldots B_8]$ is the matrix of orders from the table.

$$\begin{bmatrix} x \\ y \\ z \end{bmatrix} = \frac{1}{12}\begin{bmatrix} -2 & 4 & -2 \\ 6 & 0 & -6 \\ -3 & -3 & 9 \end{bmatrix}\begin{bmatrix} 120 & 85 & 240 & 120 & 215 & 51 & 17 & 310 \\ 140 & 90 & 280 & 120 & 260 & 56 & 18 & 380 \\ 100 & 65 & 200 & 120 & 185 & 41 & 13 & 270 \end{bmatrix}$$

$$= \frac{1}{12}\begin{bmatrix} 120 & 60 & 240 & 0 & 240 & 40 & 12 & 360 \\ 120 & 120 & 240 & 0 & 180 & 60 & 24 & 240 \\ 120 & 60 & 240 & 360 & 240 & 48 & 12 & 360 \end{bmatrix}$$

$$= \begin{bmatrix} 10 & 5 & 20 & 0 & 20 & \frac{10}{3} & 1 & 30 \\ 10 & 10 & 20 & 0 & 15 & 5 & 2 & 20 \\ 10 & 5 & 20 & 30 & 20 & 4 & 1 & 30 \end{bmatrix}.$$

The orders are filled according to the following table.

	States							
Cartons	*IA*	*MO*	*WI*	*IN*	*MI*	*AR*	*TX*	*MN*
I	10	5	20	0	20	$\frac{10}{3}$	1	30
II	10	10	20	0	15	5	2	20
III	10	5	20	30	20	4	1	30

Two comments should be made here. For Indiana, the order was for the same number of pairs of each color. Since type III cartons are packed in this way, it is reasonable that this order, as well as any other of this kind, would be filled using only type III cartons.

The order for Arkansas includes $\frac{10}{3}$ or $3\frac{1}{3}$ cartons of type I. To fill the order exactly would require splitting a carton, which the company may not be willing to do. If they won't split a carton, there are two other options. They could send 4 type I cartons, thus sending more than the actual order, or they could send 3 type I cartons, thus not filling the order.

If 4 type I cartons are sent along with 5 type II and 4 type III, the actual number sent would be 53 dozen brown, 60 dozen black, and 43 dozen gray. This gives an excess of 2 dozen, 4 dozen, and 2 dozen, respectively.

Sending three type I cartons with 5 type II and 4 type III would provide 50 dozen brown, 54 dozen black, and 40 dozen gray. This would short Arkansas by 1 dozen, 2 dozen, and 1 dozen, respectively. ■

The preceding example is an illustration of **practical solutions** versus *general solutions*. In many applied situations, certain solutions are not feasible. For example, negative numbers frequently would not be reasonable in the solution of an applied problem. In some problems, fractions are not acceptable. In other problems fractions do not present a problem. The next four examples will help illustrate.

■ Example 30. The Dependo Battery Company manufactures batteries at three plants. The company has received an overseas order for 2700 size "C" batteries, 2700 size "D" batteries, and 210 6-volt batteries. How many hours of production will be required to fill the order if the per hour production is that given in the following table?

	Battery Type Per Hour		
Plant	C	D	6-Volt
X	300	300	20
Y	200	300	20
Z	300	100	20

Solution: The order requirement can be placed in the table as the bottom row.

	Battery Type		
Plant	C	D	6-Volt
X	300	300	20
Y	200	300	20
Z	300	100	20
Requirements	2700	2700	210

The requirements form a row so the columns form the equations in the system. Letting x, y, z be the time plants X, Y, Z operate, respectively, we obtain the system of equations

$$\begin{cases} 300x + 200y + 300z = 2700 \\ 300x + 300y + 100z = 2700. \\ 20x + 20y + 20z = 210 \end{cases}$$

Before writing the augmented matrix to solve the system, we divide equations 1 and 2 by 100 and equation 3 by 10 to reduce the size of the numbers for computation.

$$\left[\begin{array}{ccc|c} 3 & 2 & 3 & 27 \\ 3 & 3 & 1 & 27 \\ 2 & 2 & 2 & 21 \end{array}\right] \begin{array}{r} \\ -R1 + R2 \\ -2R1 + 3R3 \end{array} \left[\begin{array}{ccc|c} 3 & 2 & 3 & 27 \\ 0 & 1 & -2 & 0 \\ 0 & 2 & 0 & 9 \end{array}\right]$$

$$\begin{array}{r} -R3 + R1 \\ -R3 + 2R2 \\ \\ \end{array} \left[\begin{array}{ccc|c} 3 & 0 & 3 & 18 \\ 0 & 0 & -4 & -9 \\ 0 & 2 & 0 & 9 \end{array}\right]$$

$$\begin{array}{r} 3R2 + 4R1 \\ \\ \\ \end{array} \left[\begin{array}{ccc|c} 12 & 0 & 0 & 45 \\ 0 & 0 & -4 & -9 \\ 0 & 2 & 0 & 9 \end{array}\right]$$

Then $x = \frac{45}{12} = \frac{15}{4}$, $y = \frac{9}{2}$, $z = \frac{-9}{-4} = \frac{9}{4}$. Plant X operates $\frac{15}{4}$ hours, or 3 hours and 45 minutes. Plant Y operates $\frac{9}{2}$ hours, or 4 hours and 30 minutes. Plant Z operates $\frac{9}{4}$ hours, or 2 hours and 15 minutes.

The fractions are acceptable here since we can use fractional parts of an hour.

In addition, since plant X operates $\frac{15}{4}$ hours, it will produce $\frac{15}{4} \cdot 300 = 1125$ "C" batteries, $\frac{15}{4} \cdot 300 = 1125$ "D" batteries, and $\frac{15}{4} \cdot 20 = 75$ 6-volt batteries. Plant Y will produce $\frac{9}{2} \cdot 200 = 900$ "C" batteries, $\frac{9}{2} \cdot 300 = 1350$ "D" batteries, and $\frac{9}{2} \cdot 20 = 90$ 6-volt batteries. Production for plant Z will be $\frac{9}{4} \cdot 300 = 675$ "C" batteries, $\frac{9}{4} \cdot 100 = 225$ "D" batteries, and $\frac{9}{4} \cdot 20 = 45$ 6-volt batteries.

Then the actual production of the three plants is given in the following table.

	Production		
Plant	C	D	6-volt
X	1125	1125	75
Y	900	1350	90
Z	675	225	45
Total	2700	2700	210

■

■ Example 31. The Ace Electronics Company has three assembly lines for assembling three types of special radio equipment. The government has placed a special order for 27 units of type A, 27 units of type B, and 21 units of type C. The order needs to be filled immediately, so the president decides to have the three lines work overtime to fill the order. How long should each line operate to fill the order if the per hour production is that indicated by the following table?

	Radio Type		
Line	A	B	C
I	3	3	2
II	2	3	2
III	3	1	2

Solution: The orders are for types of radios, so we add these requirements to the table as a row.

	Radio Type		
Line	A	B	C
I	3	3	2
II	2	3	2
III	3	1	2
Requirements	27	27	21

Since the requirements form the last row of the table, the equations are formed by the columns. Letting x, y, and z represent the number of hours of operation for lines I, II, and III, respectively, we have

$$\begin{cases} 3x + 2y + 3z = 27 \\ 3x + 3y + z = 27 \\ 2x + 2y + 2z = 21 \end{cases}$$

Augmenting the coefficient matrix and solving gives

$$\left[\begin{array}{ccc|c} 3 & 2 & 3 & 27 \\ 3 & 3 & 1 & 27 \\ 2 & 2 & 2 & 21 \end{array}\right] \begin{array}{l} -3R3 + 2R1 \\ -3R3 + 2R2 \\ \end{array} \left[\begin{array}{ccc|c} 0 & -2 & 0 & -9 \\ 0 & 0 & -4 & -9 \\ 2 & 2 & 2 & 21 \end{array}\right]$$

$$\begin{array}{r} \\ \\ R1 + R3 \end{array} \left[\begin{array}{ccc|c} 0 & -2 & 0 & -9 \\ 0 & 0 & -4 & -9 \\ 2 & 0 & 2 & 12 \end{array}\right]$$

$$\begin{array}{r} \\ \\ R2 + 2R3 \end{array} \left[\begin{array}{ccc|c} 0 & -2 & 0 & -9 \\ 0 & 0 & -4 & -9 \\ 4 & 0 & 0 & 15 \end{array}\right].$$

The solution is $x = \frac{15}{4}$, $y = \frac{-9}{-2} = \frac{9}{2}$, and $z = \frac{-9}{-4} = \frac{9}{4}$. Line I should operate $\frac{15}{4}$ hours, or 3 hours and 45 minutes. Line II should operate $\frac{9}{2}$ hours, or 4 hours and 30 minutes. Line III should operate $\frac{9}{4}$ hours, or 2 hours and 15 minutes.

These solutions are the same as those for example 31, and it is still acceptable to split an hour using fractions. However, closer investigation reveals a problem. Since line I is to operate $\frac{15}{4}$ hours, this line will produce $\frac{15}{4} \cdot 3 = \frac{45}{4} = 11\frac{1}{4}$ type *A* radios. Line II will produce $\frac{9}{2} \cdot 2 = 9$ type *A* radios, and line III will produce $\frac{9}{4} \cdot 3 = \frac{27}{4} = 12\frac{3}{4}$ type *A* radios. Although it is true that $\frac{45}{4} + 9 + \frac{27}{4} = \frac{72}{4} + 9 = 18 + 9 = 27$, the fact that lines I and III produce partial radios presents a problem. In this situation, even though the fractions are acceptable in terms of hours, they lead to unrealistic solutions. ■

■ **Example 32.** The Acme Corporation needs to replace some of the machines used in one phase of their manufacturing process. The machines, which must be of a special design to handle three different products simultaneously, are available from three companies. Machines from Ace have the capability of handling 3 gizmos, 3 wizmos, and 2 wammos per hour. In one hour, machines from the Bango company can handle 2 gizmos, 3 wizmos, and 2 wammos. Machines from the Carmel company, while not as large, can still handle 3 gizmos, 1 wizmo, and 2 wammos per hour. In order to maintain a smooth, continuous production operation, Acme requires that

27 gizmos, 27 wizmos, and 21 wammos come through the machines each hour. How many machines should Acme buy from each company to fill their needs?

Solution: The following table will indicate the per hour capabilities together with company requirements.

	Product		
Machine	Gizmos	Wizmos	Wammos
Ace	3	3	2
Bango	2	3	2
Carmel	3	1	2
Requirements	27	27	21

Letting x be the number of Ace machines, y be the number of Bango machines, and z be the number of Carmel machines, we obtain the system of equations

$$\begin{cases} 3x + 2y + 3z = 27 \\ 3x + 3y + z = 27. \\ 2x + 2y + 2z = 21 \end{cases}$$

This is the same system of equations that was solved in example 32, above. The solution was given by $x = \frac{15}{4}$, $y = \frac{9}{2}$, and $z = \frac{9}{4}$. The company would need $3\frac{3}{4}$ Ace machines, $4\frac{1}{2}$ Bango machines, and $2\frac{1}{4}$ Carmel machines. This solution is not practical.

Some type of company decision needs to be made. Several possibilities exist, but they will have a definite effect on the rest of the manufacturing process. ■

■ **Example 33.** The Cozee Glove Company (see examples 28 and 29) has received an order from their Kansas distributor for 55 dozen pairs of brown gloves, 30 dozen pairs of black gloves, and 35 dozen pairs of gray gloves. How many cartons of each type should be sent to fill the order?

Solution: Using A^{-1} from example 27, the solution is

$$\begin{bmatrix} x \\ y \\ z \end{bmatrix} = \left(\frac{1}{12}\right)\begin{bmatrix} -2 & 4 & -2 \\ 6 & 0 & -6 \\ -3 & -3 & 9 \end{bmatrix}\begin{bmatrix} 55 \\ 30 \\ 35 \end{bmatrix}$$

$$= \left(\frac{1}{12}\right)\begin{bmatrix} -60 \\ 120 \\ 60 \end{bmatrix} = \begin{bmatrix} -5 \\ 10 \\ 5 \end{bmatrix}.$$

It is not possible for Cozee to send −5 cartons of type I. If the company sends 10 cartons of type II and 5 cartons of type III, Kansas will receive 70 dozen brown, 60 dozen black, and 50 dozen gray. This exceeds their order by a total of 60 dozen pairs of gloves. ■

The company cannot exactly fill the order without splitting cartons. In example 33, the Acme Company was not able to meet their requirements exactly. How are situations like this to be handled?

Several procedures could be used. One procedure that is sometimes used is to change all negative values to zero and round all positive fractions to the next higher whole number.

Using this procedure in example 33 would oversupply Kansas by 50 dozen pairs of gloves. There are other solutions (not obtained by the solution process) that would come closer to the actual order. For example, sending 11 cartons of type II would provide 55 dozen brown, 44 dozen black, and 33 dozen gray. This is 2 dozen gray pairs short but is only a total of 12 dozen pairs over the original order. Shipping 12 cartons of type II will meet all requirements with an excess of 5 dozen brown, 18 dozen black, and 1 dozen gray, for a total excess of 24 dozen pairs. You may be able to provide other solutions that are just as close, or closer, to the order than these.

Suppose in example 32 all of the fractions are rounded to the next highest whole number. The company would then purchase 4 machines from Ace, 5 machines from Bango, and 3 machines from Carmel. The per hour production would then be 31 gizmos, 30 wizmos, and 24 wammos. This would require that Acme increase production at both ends of the machine to handle the additional capability of the machines. The actual increase is 10 units. Can you find a whole number solution that meets the requirements but is closer to desired production than 10 units?

There are various ways to handle problems like those encountered in examples 32 and 33. Other information may be available to help find the "best" solution. In some problems, consideration of cost or other factors may be included.

In addition, some problems might not be stated in exact terms. That is, the problem might call for at least 25 or no more than 15. Some of these situations will be discussed in the next chapter.

In some applications, it is possible to obtain multiple solutions. It is still necessary to consider the practicality of possible solutions. Certain restrictions frequently limit the practical solutions to a finite number of possibilities.

■ Example 34. A college dean has been given permission to hire 10 full-time staff members for the new year, but he will be allowed to use only \$165,000 for yearly salaries. On the basis of the current job market, the dean knows he can hire beginning instructors for \$15,000 a year, experienced instructors for \$18,000 a year, and beginning assistant professors for \$21,000 a year.

How many people in each rank can the dean hire while using the full $165,000?

Solution: This example is similar to the one in the introduction to this chapter. One equation will be in terms of money and the other will not. A table is not appropriate in this problem.

Let x be the number of beginning instructors, y the number of experienced instructors, and z the number of beginning assistant professors. Then, since 10 staff members can be hired, one equation will be $x + y + z = 10$. The other equation, representing salaries, will be $15{,}000x + 18{,}000y + 21{,}000z = 165{,}000$. Before solving the system, we divide the second equation by 1000 to reduce the size of the numbers. The new system, then, is $\begin{cases} x + y + z = 10 \\ 15x + 12y + 21z = 165 \end{cases}$.

$$\left[\begin{array}{ccc|c} 1 & 1 & 1 & 10 \\ 15 & 18 & 21 & 165 \end{array}\right] \quad \begin{array}{r} \\ -15R1 + R2 \end{array} \left[\begin{array}{ccc|c} 1 & 1 & 1 & 10 \\ 0 & 3 & 6 & 15 \end{array}\right]$$

$$\begin{array}{r} -R2 + 3R1 \\ \\ \end{array} \left[\begin{array}{ccc|c} 3 & 0 & -3 & 15 \\ 0 & 3 & 6 & 15 \end{array}\right]$$

$$\begin{array}{r} \tfrac{1}{3}R1 \\ \tfrac{1}{3}R2 \end{array} \left[\begin{array}{ccc|c} 1 & 0 & -1 & 5 \\ 0 & 1 & 2 & 5 \end{array}\right]$$

The solution is $x = 5 + z$, $y = 5 - 2z$, $z = z$.

Since $x + y + z = 10$, z is restricted to values from 0 through 10. Other restrictions will be forced by the values for x and y. That is, x and y are also between 0 and 10. Then $x = 5 + z \le 10$, or $z < 5$, and $y = 5 - 2z \ge 0$, so $5 \ge 2z$, or $z \le \frac{5}{2}$. Hence z is restricted to the values 0, 1, or 2. (Remember, the dean has to hire full-time staff.)

There are, then, three acceptable solutions:

$z = 0$ implies $x = 5$, $y = 5$
$z = 1$ implies $x = 6$, $y = 3$
$z = 2$ implies $x = 7$, $y = 1$.

The solutions, then, are given by

1. Five beginning instructors and 5 experienced instructors,
2. Six beginning instructors, 3 experienced instructors, and 1 beginning assistant professor, or
3. Seven beginning instructors, 1 experienced instructor, and 2 beginning assistant professors. ■

■ **Example 35.** The management of Sav a Bundle Supermarkets, Inc. has decided to prepare Easter baskets consisting of jelly beans, small chocolate eggs, chocolate rabbits, and creme-filled chocolate eggs. The baskets are to be sold in stores in the Southwest, Midwest, and West. Because of regional price variations, the price of a basket will vary from region to region, but management wants the baskets to contain the same items. The products to be included are all individual items except the jelly beans, which come in 6-ounce packages. Management wants the desired selling price to be $9.60 in the Southwest, $8.80 in the Midwest, and $11.80 in the West. How many units of each item should be included if the individual item prices are those given in the table?

	Item			
Region	*Jelly beans*	*Chocolate eggs*	*Rabbits*	*Creme eggs*
Southwest	.40	.60	1.40	.30
Midwest	.40	.50	1.30	.30
West	.50	.70	1.70	.40

Solution: Let w be the number of packages of jelly beans, x the number of chocolate eggs, y the number of chocolate rabbits, and z the number of creme eggs.

The system of equations, determined by the costs for the three regions,

$$\text{is} \begin{cases} .40w + .60x + 1.40y + .30z = 9.60 \\ .40w + .50x + 1.30y + .30z = 8.80. \\ .50w + .70x + 1.70y + .40z = 11.80 \end{cases}$$

To remove decimals and simplify computation, we will multiply each equation by 10.

$$\left[\begin{array}{cccc|c} 4 & 6 & 14 & 3 & 96 \\ 4 & 5 & 13 & 3 & 88 \\ 5 & 7 & 17 & 4 & 118 \end{array}\right]$$

$$\begin{array}{r} \\ -R1 + R2 \\ -5R1 + 4R3 \end{array} \left[\begin{array}{cccc|c} 4 & 6 & 14 & 3 & 96 \\ 0 & -1 & -1 & 0 & -8 \\ 0 & -2 & -2 & 1 & -8 \end{array}\right]$$

$$\begin{array}{r} 6R2 + R1 \\ \\ -2R2 + R3 \end{array} \left[\begin{array}{cccc|c} 4 & 0 & 8 & 3 & 48 \\ 0 & -1 & -1 & 0 & -8 \\ 0 & 0 & 0 & 1 & 8 \end{array}\right]$$

$$\begin{array}{r} -3R3 + R1 \\ -R2 \\ \\ \end{array} \left[\begin{array}{cccc|c} 4 & 0 & 8 & 0 & 24 \\ 0 & 1 & 1 & 0 & 8 \\ 0 & 0 & 0 & 1 & 8 \end{array}\right]$$

$$\begin{array}{r} \frac{1}{4}R1 \\ \\ \\ \end{array} \left[\begin{array}{cccc|c} 1 & 0 & 2 & 0 & 6 \\ 0 & 1 & 1 & 0 & 8 \\ 0 & 0 & 0 & 1 & 8 \end{array}\right]$$

The solution is given by $w = 6 - 2y$, $x = 8 - y$, $y = y$, $z = 8$. Each basket will have 8 creme-filled eggs. The variable y can be any whole value subject to $w \geq 0$ and $x \geq 0$. Then $w = 6 - 2y \geq 0$ gives $6 \geq 2y$, or $y \leq 3$, and $x = 8 - y \geq 0$ gives $8 \geq y$. Since $y \leq 3$ implies $y \leq 8$, the acceptable values for y are 0, 1, 2, and 3. The actual possibilities for Easter baskets with the desired selling prices are given in the following table:

	Basket			
Item	*1*	*2*	*3*	*4*
Jelly beans	6	4	2	0
Chocolate eggs	8	7	6	5
Rabbits	0	1	2	3
Creme eggs	8	8	8	8

■

Exercise Set 2.4

1. How many gallons of $\frac{1}{2}$% butterfat milk and 4% butterfat whole milk should be mixed to get 50 gallons of 2% butterfat milk?
2. A man has $30,000 to invest, part at 6% in a tax-free municipal bond and part at 14%, of which $\frac{1}{7}$ must be paid to the government in taxes. If he wishes the two investments to yield exactly the same net returns, how should he invest?
3. A rancher needs 24 seed bulls and has $50,000 to spend on this project. He can buy type *A* bulls at $3500, type *B* bulls at $1500, and type *C* bulls at $7500. What are his options?
4. A chemist needs 40 milliliters of a 25% acid solution. In the lab she has a 20% acid solution and a 40% acid solution. How much of each type should she use to get the desired solution?
5. A company has factories I, II, and III producing products *A*, *B*, and *C*. In one hour factory I produces 8 units of *A*, 7 units of *B*, and 3 units of *C*. Factory II produces 4 units of *A*, 5 units of *B*, and 8 units of *C*. Factory III produces 7 units of *A*, 3 units of *B*, and 2 units of *C*. The company is required to produce 1350 units of *A*, 1100 units of *B*, and 800 units of *C*.
   a. Construct a data table for this problem.
   b. Let x = number of hours of production for factory I
   y = number of hours of production for factory II
   z = number of hours of production for factory III
   and set up a system of equations.

c. Solve the system of equations and interpret, in words, the solution.

6. A total of 12,000 tickets were sold for a concert at a college. The tickets cost \$5 for local college students, \$6 for other students, and \$8 for the general public. If the revenue was \$70,000 and the number of tickets sold to local college students was three times the number sold to the general public, how many people of each category bought tickets?

7. John has been instructed by his doctor to take 10 units of vitamin A, 9 units of vitamin E, and 19 units of vitamin C per day. John has three brands of vitamins at home. Brand X has 2 units of A, 3 units of E, and 5 units of C per tablet. Each brand Y tablet has 1 unit of A, 3 units of E, and 4 units of C, whereas each brand Z tablet has 1 unit of A and 1 unit of C. How many tablets of each type should John take to fulfill his requirement?

8. The personnel manager is requested to hire 22 new employees for office work. He can hire a file clerk for \$800 a month, a secretary for \$1000 a month, and a stenographer for \$1200 a month. How can he spend the \$20,000 allotted?

9. The company purchasing agent is required to buy 30 new automobiles. She decides on three models: model A, which costs \$6000 per unit; model B at \$8000 per unit; and model C at \$10,000 per unit. Discuss how the agent can spend \$210,000 for the 30 autos.

10. A child needs to buy 32 favors as door prizes for his party guests. He has only \$8 to spend, and he can buy suckers at 10¢ each, gum at 20¢ a package, peanuts at 30¢ a bag, and candy bars for 40¢ each. Discuss the possibilities.

11. An oil company has four refineries that produce four specialized products. Refinery I produces 3 units of oil, 1 unit of lubricant, and 5 units of transmission fluid per hour. Refinery II produces 4 units of oil, 5 units of lubricant, and 5 units of transmission fluid per hour. In one hour, refinery III produces 5 units of oil, 6 units of lubricant, and 6 units of transmission fluid while refinery IV produces 2 units of oil and 5 units of transmission fluid. Each of the 4 refineries produces 1 unit of solvent per hour. How many hours should each refinery operate to fill an order for 69 units of oil, 49 units of lubricant, 113 units of transmission fluid, and 22 units of solvent?

12. How many hours should each refinery in exercise 11 operate to fill the following orders?

	Orders							
Product	*1*	*2*	*3*	*4*	*5*	*6*	*7*	*8*
Oil	67	42	345	335	210	140	220	400
Lubricant	55	36	245	275	180	80	160	400
Trans. fluid	109	63	565	545	315	225	320	550
Solvent	21	12	110	105	60	45	60	100

13. A company packages four products (A, B, C, D) in four types of cartons (I, II, III, IV) in the following combinations:

Carton Type	Number per Carton *A*	*B*	*C*	*D*
I	3	3	3	3
II	4	4	4	0
III	12	0	0	0
IV	0	12	0	0

a. Write and solve a system of equations to fill an order for a units of A, b units of B, c units of C, and d units of D. (Your solution will be in terms of a, b, c, d.)

b. Use the result of part **(a)** to fill the following orders.

Product	Order Number 1	2	3	4	5	6	7	8	9	10
A	12	36	41	64	42	20	780	240	100	0
B	12	48	53	124	44	30	300	360	100	120
C	12	0	29	64	37	50	180	120	20	120
D	12	0	9	24	20	10	60	48	10	72

14. You are given the following instructions when asked to buy 100 sandwiches for the family picnic. Buy the same number of ham as ham and cheese sandwiches, and buy equal numbers of ham salad and peanut butter and jelly sandwiches. Ham and cheese sandwiches cost 90¢ each, ham cost 80¢ each, ham salad cost 70¢ each, and peanut butter and jelly cost 60¢ each. You are given \$80 to buy sandwiches with. Can you exactly meet all requirements?

15. A company markets prepackaged plaster mix in 10 pound boxes. The boxes are of type A, B, C, D, and E. Each contains sand (s), cement (c), and a texture compound (t), according to the table below.

Contents in Pounds	Box Type A	B	C	D	E
s	2	3	4	1	7
c	3	3	3	3	3
t	5	4	3	6	0

The company has found the 10 pound boxes very marketable. In order to give the customer a wider choice of finishes, new finishes can be made up by mixing various combinations and more important to the customers, if the supplier is temporarily out of one type of box, that type may be duplicated with certain other combinations.

a. Suppose you need ten boxes (100 pounds) of plaster mix type *A* but the store is out. What other combinations can you use?

b. You need 20 boxes of type *C*. What other combinations could you use?

c. Can box types *A*, *B*, and *C* be made up from boxes of types *D* and *E* alone? Explain.

d. Can boxes of type *A* and *C* only be used to replace *B*?

16. A factory has three assembly lines for making four types of machines. How long should each line operate to fill an order for 40 type *A*, 30 type *B*, 24 type *C*, and 34 type *D* machines if the per hour production is that indicated in the table?

	Machine Type			
Line	*A*	*B*	*C*	*D*
I	2	3	2	1
II	2	2	0	0
III	1	1	4	4

17. A feed dealer has three types of liquid food supplements that he mixes to meet farmers' requirements. Each supplement contains minerals, vitamins, and protein, plus filler to make 12-ounce bottles. The composition of the critical ingredients is given in the following table, in ounces per bottle.

Type	*Minerals*	*Vitamins*	*Protein*
I	4	2	6
II	2	2	4
III	2	4	4

How many bottles of each type should be mixed to fill the following requirements? (The total amount of supplement is not critical.)

	Farmer								
Ingredient	1	2	3	4	5	6	7	8	9
Minerals	80	84	80	86	90	76	77	79	60
Vitamins	84	80	80	72	74	80	81	78	120
Protein	140	144	140	146	150	136	137	139	120

18. A novelty shop has four popular inexpensive party favors, *A*, *B*, *C*, and *D*. The owner wants to prepare boxes of 40 party favors using a combination of the four. Since favors B and D are frequently used together, she wants to use the same number of each of these. The prices of A, B, C, and D are 10¢, 20¢, 40¢, and 10¢, respectively, and she wants to sell the boxes of 40 for \$6.50. Find all of the possible combinations the owner can prepare.

2.5 Input-Output Analysis

In 1973 Wassily W. Leontief won a Nobel prize for his work in applied economics. Leontief's methods were designed to solve problems of increasing and decreasing supply and demand on a system where a strong interdependence exists between the various sectors of the system. This means that part of the production (*output*) is consumed in the production process (*input*), hence the term **input-output analysis.**

Consider a bad cook who burns 10% of the cookies he bakes. Then if we wish to establish a relationship between the production and the final demand, we would need to determine the relation between x, the number of cookies baked, and y, the number of good cookies needed. That is

$$\begin{bmatrix}\text{Total value}\\ \text{of output}\end{bmatrix} = \begin{bmatrix}\text{Value consumed in}\\ \text{production (input)}\end{bmatrix} + \begin{bmatrix}\text{Value consumed by}\\ \text{final demand}\end{bmatrix}$$

or simply

$$\begin{aligned} x &= .1x + y \\ 1x - .1x &= y \\ (1 - .1)x &= y \\ x &= (1 - .1)^{-1}y \\ x &= (.9)^{-1}y \\ x &= \frac{10y}{9} \end{aligned}$$

Although this simple algebra may seem tedious and formal, we are doing the algebra in this way to establish a pattern. If we make the basic assumption that the input percentages do not change (in this case 10%), then the final equation $x = \frac{10y}{9}$ will always result in knowing where to set the production to meet the final demand. For example, if we require 90 good cookies, we let $y = 90$ and x then equals 100. If we need 130 good cookies, we would set the production level at $\frac{10(130)}{9}$, or 145, since parts of cookies don't count.

We will now consider a simple two-sector problem. A farmer raises corn and cattle where

1. 10% of the dollar value of the cattle produced in a year is budgeted to pay expenses incurred in the cattle production.
2. 50% of the dollar value of the cattle is used to offset the cost of producing corn.
3. 20% of the dollar value of the corn is used to feed the cattle.
4. 10% of the dollar value of the corn is used to buy seed, fertilizer, and so on, i.e., input into the grain production process.

The following table brings all of these factors together.

	Input	
Production	*Cattle*	*Corn*
Cattle	.1	.2
Corn	.5	.1

The matrix $L = \begin{bmatrix} .1 & .2 \\ .5 & .1 \end{bmatrix}$ is called the input-output matrix. If it is required to have a gross profit (final demand) of \$4000 from cattle and \$6000 from corn, what must the production level be? We have two relations to consider:

1. $\begin{bmatrix} \text{Total value} \\ \text{of cattle} \\ \text{production} \end{bmatrix} = \begin{bmatrix} \text{Input into} \\ \text{cattle} \\ \text{production} \end{bmatrix} + \begin{bmatrix} \text{Input into} \\ \text{corn} \\ \text{production} \end{bmatrix} + \begin{bmatrix} \text{Final} \\ \text{Demand} \\ \end{bmatrix}$

2. $\begin{bmatrix} \text{Total value} \\ \text{of corn} \\ \text{produced} \end{bmatrix} = \begin{bmatrix} \text{Input into} \\ \text{cattle} \\ \text{production} \end{bmatrix} + \begin{bmatrix} \text{Input into} \\ \text{corn} \\ \text{production} \end{bmatrix} + \begin{bmatrix} \text{Final} \\ \text{Demand} \\ \end{bmatrix}$

Hence if we let

$$x_1 = \text{value of cattle}$$

$$x_2 = \text{value of corn}$$

then we have the system

$$x_1 = .1x_1 + .2x_2 + 4{,}000$$

$$x_2 = .5x_1 + .1x_2 + 6{,}000$$

If $X = \begin{bmatrix} x_1 \\ x_2 \end{bmatrix}$, called the production matrix,

$L = \begin{bmatrix} .1 & .2 \\ .5 & .1 \end{bmatrix}$, called the input-output matrix or sometimes called a consumption matrix, and

$D = \begin{bmatrix} 4{,}000 \\ 6{,}000 \end{bmatrix}$, called the final demand matrix, then

$$X = LX + D$$
$$X - LX = D$$
$$IX - LX = D \qquad \text{(In this case } I \text{ is } I_2.\text{)}$$
$$(I - L)X = D$$
$$X = (I - L)^{-1}D$$

The matrix $(I - L)^{-1}$ is referred to as the Leontief matrix and remains unchanged whereas the final demand D may vary. The central idea here is that once the internal consumption percentages are known, they tend to remain fixed, and to compute production levels necessary to meet new final demands it is only necessary to compute the matrix $(I - L)^{-1} \cdot D$.

To solve our problem

$$I - L = \begin{bmatrix} 1 & 0 \\ 0 & 1 \end{bmatrix} - \begin{bmatrix} .1 & .2 \\ .5 & .1 \end{bmatrix} = \begin{bmatrix} .9 & -.2 \\ -.5 & .9 \end{bmatrix}$$

$$(I - L)^{-1} = \left((.1)\begin{bmatrix} 9 & -2 \\ -5 & 9 \end{bmatrix}\right)^{-1} = 10\begin{bmatrix} 9 & -2 \\ -5 & 9 \end{bmatrix}^{-1} = \frac{10}{71}\begin{bmatrix} 9 & 2 \\ 5 & 9 \end{bmatrix}$$

$$\text{Hence, } X = (I - L)^{-1} \cdot D = \frac{10}{71}\begin{bmatrix} 9 & 2 \\ 5 & 9 \end{bmatrix}\begin{bmatrix} 4{,}000 \\ 6{,}000 \end{bmatrix} = \begin{bmatrix} 6{,}760.56 \\ 10{,}422.54 \end{bmatrix}$$

The production level is then

$$x_1 = \$6{,}760.56 = \text{total value of cattle}$$
$$x_2 = \$10{,}422.54 = \text{total value of corn.}$$

Checking we have,

Income from cattle = \$6,760.56, but this production level had expenses of 10% of \$6,760.56 and 20% of \$10,422.54.

$$(6{,}760.56) - (.1)(6{,}760.56) - (.2)(10{,}422.54) = \$4{,}000.$$

And similarly,

$$(10{,}422.54) - (.5)(6{,}760.56) - (.1)(10{,}422.54) = \$6{,}000.$$

■ Example 36. If a final demand $D = \begin{bmatrix} 14{,}200 \\ 7{,}100 \end{bmatrix}$ is used, then the production

$$X = (I - L)^{-1}D \text{ is } \frac{10}{71}\begin{bmatrix} 9 & 2 \\ 5 & 9 \end{bmatrix}\begin{bmatrix} 14{,}200 \\ 7{,}100 \end{bmatrix}$$

$$= 1000\begin{bmatrix} 9 & 2 \\ 5 & 9 \end{bmatrix}\begin{bmatrix} 2 \\ 1 \end{bmatrix} = \begin{bmatrix} 20{,}000 \\ 19{,}000 \end{bmatrix}.$$

Whether or not it is possible for this farm to produce \$20,000 worth of cattle and \$19,000 worth of corn is not the question that is answered here. What is answered is that if these internal consumption percentages remain unchanged, then these values—\$20,000 and \$19,000—are the respective gross productions necessary to meet this final demand. ■

■ Example 37. Given the following data, compute the input-output matrix L.

	Input				
Output	*A*	*B*	*C*	*Final Demand*	*Production Level*
A	100	80	240	80	500
B	50	40	60	250	400
C	0	120	180	300	600

Solution: Since each entry in column 1 is the production of A consumed by the sectors A, B, and C, respectively, we divide each of these entries by the production level of sector $A = 500$. Similarly, the entries in column B are divided by 400 and those in column C by 600.

$$L = \begin{bmatrix} \frac{100}{500} & \frac{80}{400} & \frac{240}{600} \\ \frac{50}{500} & \frac{40}{400} & \frac{60}{600} \\ \frac{0}{500} & \frac{120}{400} & \frac{180}{600} \end{bmatrix} = \begin{bmatrix} .2 & .2 & .4 \\ .1 & .1 & .1 \\ 0 & .3 & .3 \end{bmatrix}$$ ■

■ Example 38. Find the production level necessary to meet the final demand of $D = \begin{bmatrix} 100 \\ 200 \\ 300 \end{bmatrix}$, using the matrix L from the last example. Check the results.

Solution:

$$(I - L)^{-1} \cdot D = \begin{bmatrix} .8 & -.2 & -.4 \\ -.1 & .9 & -.1 \\ 0 & -.3 & .7 \end{bmatrix}^{-1}\begin{bmatrix} 100 \\ 200 \\ 300 \end{bmatrix}$$

$$= \begin{bmatrix} 1.3216 & .5727 & .8370 \\ .1542 & 1.2335 & .2643 \\ .0661 & .5286 & 1.5419 \end{bmatrix}\begin{bmatrix} 100 \\ 200 \\ 300 \end{bmatrix} = \begin{bmatrix} 497.8 \\ 341.4 \\ 574.9 \end{bmatrix}$$

A computer program, INVMAT, was used to compute $(I - L)^{-1}$.

$$497.8 - (.2)(497.8) - (.2)(341.4) - (.4)(574.9) = 100$$
$$341.4 - (.1)(497.8) - (.1)(341.4) - (.1)(574.9) = 200$$
$$574.9 - 0 - (.3)(341.4) - (.3)(574.9) = 300$$

Exercise Set 2.5

1. Compute $I - L$ when L equals

a. $\begin{bmatrix} \frac{1}{2} & \frac{1}{3} \\ \frac{1}{4} & \frac{1}{5} \end{bmatrix}$ **b.** $\begin{bmatrix} .1 & .2 & .3 \\ .4 & .3 & .2 \\ .2 & .2 & 0 \end{bmatrix}$ **c.** $\begin{bmatrix} .1 & .2 & .1 & .2 \\ .1 & .3 & 0 & .3 \\ .2 & .1 & .2 & .1 \\ .1 & .2 & .3 & .2 \end{bmatrix}$

2. Given $I - L = \begin{bmatrix} .9 & -.1 & -.2 & -.3 \\ -.4 & .95 & -.1 & 0 \\ -.2 & -.3 & .9 & 0 \\ 0 & -.4 & -.05 & .95 \end{bmatrix}$ find L.

3. Given that the Leontief matrix $(I - L)^{-1}$ in problem 2 is

$$\begin{bmatrix} 1.3992 & .4553 & .3861 & .4419 \\ .6445 & 1.3006 & .2990 & .2035 \\ .5258 & .5347 & 1.2966 & .1660 \\ .2990 & .5758 & .1942 & 1.1471 \end{bmatrix}$$

find the level of production necessary to meet a final demand of

a. $\begin{bmatrix} 1{,}000 \\ 1{,}000 \\ 1{,}000 \\ 2{,}000 \end{bmatrix}$ **b.** $\begin{bmatrix} 30{,}000 \\ 20{,}000 \\ 10{,}000 \\ 20{,}000 \end{bmatrix}$ **c.** $\begin{bmatrix} 5 \\ 7 \\ 3 \\ 4 \end{bmatrix}$

4. Given the following input-output matrix

	Input			
Output	A	B	Final Demand	Production Level
A	80	240	180	500
B	240	180	80	500

find the production level necessary to change the final demand for A from 180 to 200 and for B from 80 to 100.

5. Given the following input-output matrix, find the production level necessary to change the final demand to $\begin{bmatrix} 450 \\ 220 \end{bmatrix}$.

	A	B	Final Demand
A	300	200	500
B	100	200	200

6. Given the input-output matrix below, find the production level necessary to change the final demand to

a. $\begin{bmatrix} 600 \\ 400 \\ 300 \end{bmatrix}$ **b.** $\begin{bmatrix} 750 \\ 500 \\ 400 \end{bmatrix}$ **c.** $\begin{bmatrix} 800 \\ 600 \\ 300 \end{bmatrix}$

	x	y	z	Final Demand
x	200	100	100	600
y	100	50	50	300
z	100	100	100	200

2.6 Chapter Review

Important Terms and Symbols in Chapter 2.

system of linear equations (introduction)
linear equation (introduction)
inconsistent system (section 2.1)
dependent system (section 2.1)
acceptable solution (section 2.1)
Gaussian elimination (section 2.1)
equivalent systems (section 2.1)
elementary operations (section 2.1)
supply curve (section 2.1)
demand curve (section 2.1)
point of equilibrium (section 2.1)
equilibrium price (section 2.1)
equilibrium quantity (section 2.1)
break-even point (section 2.1)
free variable, independent variable (section 2.1)
particular solution (section 2.1)
leading variables (section 2.2)
multiple systems of equations (section 2.3)
systems of equations and the inverse of a matrix (section 2.3)
matrix equations to represent systems of equations (section 2.3)
matrix equation (section 2.3)
practical solutions (section 2.4)
input-output analysis (section 2.5)

Review Exercises

In exercises 1–20 solve the system of equations.

1. $\begin{cases} 3x - y = 1 \\ x + y = 7 \end{cases}$

2. $\begin{cases} \frac{1}{3}x + \frac{1}{4}y = 8 \\ \frac{1}{5}x + \frac{1}{6}y = 5 \end{cases}$

3. $\begin{cases} 2x - y + z = 2 \\ x - y + z = 0 \\ 3x - 2y + 2z = 2 \end{cases}$

4. $\begin{cases} 2x + 3y + z = -5 \\ y + 2z = 3 \\ 4x - y - 4z = 1 \end{cases}$

5. $\begin{cases} 2x + y - z + w = 4 \\ -x - y + z + w = 4 \\ 3x - y + z - 3w = 0 \\ y - z + w = 0 \end{cases}$

6. $\begin{cases} x + y + z + w = 26 \\ x - y + z - w = 2 \\ x + y - z + w = 14 \\ 2x + y - 2z - w = 6 \end{cases}$

7. $\begin{cases} 2x - 3y = 5 \\ x + 2y = 6 \\ 2x + y = 9 \end{cases}$

8. $\begin{cases} x + 3y = 4 \\ x - y = -1 \\ 3x + y = 2 \end{cases}$

9. $\begin{cases} 3x - 2y = 5 \\ 2x + 3y = 12 \\ x + y = 2 \end{cases}$

10. $\begin{cases} 3x + 2y + z = 3 \\ 2x - y + 3z = 2 \\ 2x + y + z = 2 \end{cases}$

11. $\begin{cases} 4x + 2y + z = 3 \\ 3x + 3y = 3 \\ x + y - 2z = 2 \end{cases}$

12. $\begin{cases} 2x + 3y - 4z = 0 \\ -x - y + 2z = -1 \\ 3x + y - 6z = 7 \end{cases}$

13. $\begin{cases} 3x - 6y + 9z = 17 \\ 2x + y + z = 9 \\ -x + 2y - 3z = -7 \end{cases}$

14. $\begin{cases} x - 4y + 2z = 8 \\ 2x + 2y - z = -9 \\ -x + 2y - z = -3 \end{cases}$

15. $\begin{cases} x + y + z = 9 \\ y + z + w = 10 \\ x + z + w = 12 \\ x + y + w = 11 \end{cases}$

16. $\begin{cases} 3x + 2y - z + 3w = 8 \\ 2x + 6y - 3z + w = 2 \\ x - 4y + 2z - w = 6 \\ x - 2y + z - 2w = 3 \end{cases}$

17. $\begin{cases} 2x + 3y - 2z = 1 \\ x + y + z = 3 \end{cases}$

18. $\begin{cases} 2x - 3y + 4z = -9 \\ x - y + 2z = -4 \end{cases}$

19. $\begin{cases} 3x + 2y + 4z - w = 1 \\ 2x + y + z + w = 2 \\ x - 2y + 2z + w = -5 \end{cases}$

20. $\begin{cases} 2v + 3w - 2x + y + z = 5 \\ v + 2w + x - y - z = 2 \\ v + w - x + y + z = 3 \\ 3v - 2w + x + y - z = 5 \end{cases}$

In exercises 21–26, a system of supply and demand equations is given. For each exercise, tell which is the supply equation, which is the demand equation, and find the point of equilibrium.

21. $\begin{cases} p = -\dfrac{1}{20}q + 100 \\ p = \dfrac{1}{200}q + 1 \end{cases}$

22. $\begin{cases} p = -\dfrac{1}{25}q + 14 \\ p = \dfrac{1}{25}q - 10 \end{cases}$

23. $\begin{cases} p = \dfrac{1}{24}q - 600 \\ p = -\dfrac{1}{400}q + 62.5 \end{cases}$

24. $\begin{cases} p = \dfrac{1}{500}q + 0.1 \\ p = -\dfrac{1}{25}q + 208 \end{cases}$

25. $\begin{cases} p = \dfrac{1}{200}q - \dfrac{1}{40} \\ p = -\dfrac{1}{20}q + \dfrac{39}{2} \end{cases}$

26. $\begin{cases} p = \dfrac{1}{40}q - 150 \\ p = -\dfrac{1}{60}q + \dfrac{1265}{12} \end{cases}$

In exercises 27–32, find the break-even point for the system of revenue and cost equations.

27. $\begin{cases} r = 25\,q \\ c = 15\,q + 200 \end{cases}$

28. $\begin{cases} r = 650\,q \\ c = 450\,q + 2000 \end{cases}$

29. $\begin{cases} r = 18\,q \\ c = 7\,q + 2200 \end{cases}$

30. $\begin{cases} r = 15\,q \\ c = 9\,q + 180 \end{cases}$

31. $\begin{cases} r = 9.75\,q \\ c = 5.25\,q + 4500 \end{cases}$

32. $\begin{cases} r = 12.45\,q \\ c = 6.95\,q + 110 \end{cases}$

In exercises 33–36, solve the matrix equation $AX = B$ for X. *Note*: The matrix X will have the same dimension as B.

33. $A = \begin{bmatrix} 2 & -3 \\ 2 & -4 \end{bmatrix}, \quad B = \begin{bmatrix} 1 & 2 & -1 \\ 2 & 4 & 2 \end{bmatrix}$

34. $A = \begin{bmatrix} 1 & 2 \\ 2 & 5 \end{bmatrix}, \quad B = \begin{bmatrix} 3 & 1 & 3 & 2 \\ 1 & 4 & 2 & 1 \end{bmatrix}$

35. $A = \begin{bmatrix} 1 & 1 & 1 \\ 2 & -1 & 2 \\ 2 & -1 & -1 \end{bmatrix}, \quad B = \begin{bmatrix} 5 & 3 & 2 & 1 & 4 \\ 1 & -1 & 1 & 1 & 4 \\ 7 & 4 & 5 & 3 & 2 \end{bmatrix}$

36. $A = \begin{bmatrix} 10 & 2 & 9 \\ 5 & 1 & 4 \\ 6 & 1 & 5 \end{bmatrix}, \quad B = \begin{bmatrix} 50 & 25 & 42 & 21 & 41 & 40 \\ 24 & 12 & 20 & 10 & 19 & 19 \\ 29 & 14 & 24 & 12 & 23 & 23 \end{bmatrix}$

37. In order to remain active after retiring, three former carpenters set up a shop in Mr. Smith's garage to build single-unit bird houses, multiple-unit bird houses, and bird feeders. After developing their plans and building a few of each item, they discover that Mr. Smith can build 2 single-unit houses, 1 multiple-unit house, and 3 feeders per day. Mr. Brown can build 1 single-unit house, 1 multiple-unit house, and 6 feeders per day. Mr. Allen can build 3 single-unit houses, 1 multiple-unit house, and 1 feeder per day. How many days should each man work to fill a weekly order for 18 single-unit houses, 9 multiple-unit houses, and 30 feeders?

38. After Mr. Smith, Mr. Brown, and Mr. Allen get their business established (see problem 37), they begin to take weekly orders. How many days should each man work each week to fill the following orders?

	Week					
Product	1	2	3	4	5	6
Single-unit houses	11	10	13	13	17	23
Multiple-unit houses	5	5	6	7	8	11
Feeders	14	16	18	26	24	34

39. A veterinarian has told a man that his dog needs to take food supplements to regain strength after an accident. The dog needs 20 units of protein, 16 units of iron, and 10 units of calcium each day. The man has three mineral supplements, in tablet form, to use. Supplement *A* has 2 units of protein, 1 unit of iron, and 1 unit of calcium per tablet; *B* has 4 units of protein, 2 units of iron, and 2 units of calcium per tablet; and *C* has 2 units of protein, 3 units of iron, and 1 unit of calcium per tablet. How many tablets of each supplement should the man give his dog each day to meet requirements exactly?

40. A greeting card company has made special arrangements with a service organization to sell last year's leftover Christmas cards. The company packs them in cartons, each containing 12 boxes of cards. Three types of cartons are available, each containing a mix of religious, humorous, and standard cards. Each type *A* carton contains 3 boxes of religious, 2 boxes of humorous, and 7 boxes of standard cards. Each type *B* carton has 1 box of religious, 1 box of humorous, and 10 boxes of standard cards. Type *C* cartons contain 3 boxes of religious, 1 box of humorous, and 8 boxes of standard cards. Orders are received from each region of the country in terms of the number of boxes of each type desired. How many cartons of each type should the company send to fill the following regional orders?

Card Type	New England	Mid-Atlantic	Southeast	North Central	Midwest	Rocky Mountains	Southwest	West
Religious	1600	1400	1600	1400	2000	1400	1100	1900
Humorous	900	800	800	900	900	700	700	1200
Standard	4700	5000	4800	4900	4900	5100	4200	7700

41. A small nation has a simple two-sector economy consisting only of business and industry. If the input-output matrix is given by

$$L = \begin{array}{c} \\ B \\ I \end{array} \begin{array}{c} \begin{array}{cc} B & I \end{array} \\ \begin{bmatrix} .2 & .4 \\ .3 & .2 \end{bmatrix} \end{array}$$

find the production levels necessary to meet a final demand of \$2,600,000 for business and \$3,900,000 for industry.

42. The following table gives the final demand for business and industry for the small country of problem 41 for a six-year period. Find production levels necessary to meet each year's demand. The demands are in millions of dollars, and L is the same as in problem 41.

	Year					
Sector	*1*	*2*	*3*	*4*	*5*	*6*
Business	5.2	6.5	7.8	10.4	14.3	16.9
Industry	6.5	6.5	8.45	11.7	16.9	22.1

43. An emerging nation has divided its economy into three sectors—agriculture, industry, and business. The input-output matrix for the economy is given by

$$L = \begin{array}{c} \\ A \\ I \\ B \end{array} \begin{array}{c} \begin{array}{ccc} A & I & B \end{array} \\ \begin{bmatrix} 0.5 & 0.3 & 0.1 \\ 0.1 & 0.4 & 0.1 \\ 0.2 & 0.2 & 0.5 \end{bmatrix} \end{array}.$$

Find production levels necessary to meet a final demand of \$105 million for agriculture, \$126 million for industry, and \$84 million for business for this year.

44. Assuming that the input-output matrix L for the nation in problem 43 remains the same, find production levels necessary to meet the yearly final demands given in the table, where demands are in billions of dollars.

	Year					
Sector	*1*	*2*	*3*	*4*	*5*	*6*
Agriculture	5.25	7.35	9.45	9.45	11.55	15.75
Industry	6.30	8.40	10.50	11.55	13.65	17.85
Business	4.20	7.35	8.40	9.45	13.65	16.80

3 Linear Programming

Introduction

Many problems in quantitative methods involve specific restrictions. Some of the more common restrictions are time, number of employees, machine time, and availability of material. Our object is to maximize or minimize some quantity—frequently profit, revenue, or cost—while production levels stay within the bounds forced by the restrictions. Many of these problems can be placed in the general category called **linear programming** problems. As a simple illustration, consider the following.

For their Junior Achievement project, Sue and Bill decide to make and sell kites. They have chosen kite building as a project because Sue's father has donated 600 feet of bamboo stored in his garage and Bill has obtained 60 yards of faded silk from his Aunt May.

Together, Bill and Sue design a box kite that requires 24 feet of bamboo and 2 yards of silk to construct and a standard kite that requires 8 feet of bamboo and 1 yard of silk to construct. They expect to sell all the kites they can build at a profit of \$2 for each box kite and \$1.50 for each standard kite. How many of each type of kite should they build?

The problem facing Bill and Sue is a typical linear programming problem. How can they best utilize their resources to make the most money?

The restrictions on manufacture forced by the limited quantities of bamboo and silk available are called *constraints*. The profit, to be maximized, is expressed by the linear equation $p = 1.5x + 2y$, where x = number of standard kites built and sold and y = number of box kites built and sold. This equation is referred to as the *objective equation* or statement.

After thinking about a production schedule for a long time, Bill and Sue list some of their possibilities and examine the expected profit. Their list (incomplete) is as follows:

	x	y	$x + 2y$	$8x + 24y$			$1.5x + 2y$
	Number of Standard Kites	*Number of Box Kites*	*Amount of Silk Used*	*Amount of Bamboo Used*	*Amount of Silk Left*	*Amount of Bamboo Left*	*Profit*
Plan 1	10	20	50	560	10	40	\$55
Plan 2	10	21	52	584	6	16	\$57
Plan 3	24	17	58	600	2	0	\$70

Bill and Sue decide to use Plan 3 because almost all of the raw material will be used. Their profit is \$70, which may or may not be their maximum possible profit. Any unused amount of silk or bamboo will more formally be referred to as *slack* or *surplus.*

It will be useful to restate some of the more common terms used in this chapter. The restrictions that force limits on the production process are called **constraints.** The set of all values (points) defined by the constraints (points that satisfy all of the constraints) is called the **feasible region.**

This feasible region is apart from, and unchanged by, any objective statement that may be made. In fact, it is quite common, once a feasible region is known, to ask several questions about it. These questions are referred to as **objective statements.** Typical objective statements are the following: "How do we generate more revenue?"; "What is our maximum profit?"; "How can we reduce cost?"; and "What is the lowest cost?". In general, how do we husband our resources for optimal results?

Although the variety of problems is unlimited, the largest single group involves linear equations and inequalities. Linear programming is the general category for these problem-solving techniques.

3.1 Linear Inequalities

Consider the simple case of a single variable $x = 5$. This value divides the number line into three distinct sets:

1. The half-line $\{x|x < 5\}$.
2. $\{5\}$.
3. The half-line $\{x|x > 5\}$.

On the number line, the point $x = 5$ becomes the boundary between all numbers less than 5 and all numbers greater than 5.

In a similar way, a linear equation in two variables divides the plane into three sets. For example, the straight line $2x + y = 8$ is the boundary between the **half-planes** $2x + y > 8$ and $2x + y < 8$.

In figure 1 the graph of $2x + y = 8$ is the boundary between $2x + y > 8$, the shaded half-plane, and $2x + y < 8$, the non-shaded half-plane. To sketch or graph the inequality $2x + y < 8$, we first graph the boundary line $2x + y = 8$, choose a point not on the boundary, and evaluate $2x + y$ at this point. We call this our test point.

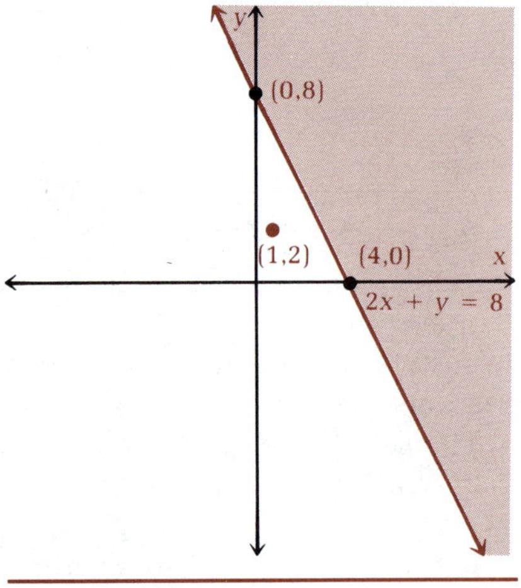

Figure 3.1

Any point not on the boundary may be selected as a test point. In figure 3.1 we arbitrarily pick (1,2). The variable $2x + y$ evaluated at this point becomes $2(1) + 2 = 4$, which indicates that the point (1,2) belongs to the half-plane $2x + y < 8$. The points in the shaded half-plane have x and y values for which $2x + y > 8$.

Conventionally, we shade the half-plane that is being discarded. The unshaded half-plane in figure 3.1 is the graph of $2x + y \leq 8$. Although some problems do exist with strict inequalities, all problems in this text will include the boundary.

The **graph of a system of inequalities** is the region of the plane common to all of the inequalities. In sketching a system of inequalities, we shade the *unused half-planes*, always discarding these shaded regions. The resulting feasible region will be the unshaded portion of the plane.

Consider figure 3.2. The feasible region is the polygonal region OACB defined by the following three restrictions:

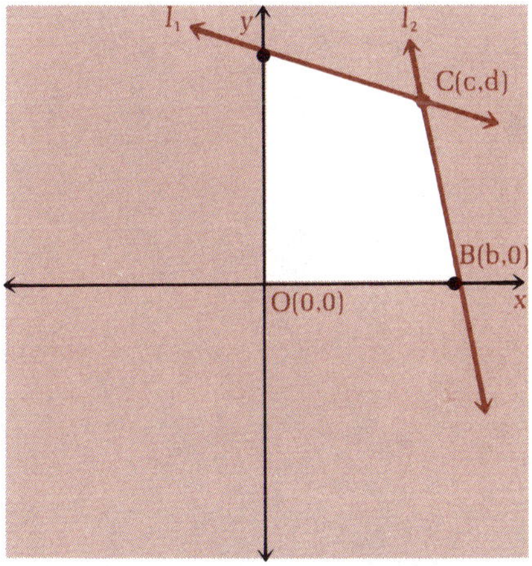

Figure 3.2

1. $x \geq 0$ and $y \geq 0$.
2. The half-plane below line l_1.
3. The half-plane below line l_2.

The **corner points** are O, A, C, and B and are the points on the boundary that are determined by the intersections of the boundary lines.

■ Example 1. Sketch the region defined by the following system and identify the corner points.

Solution:

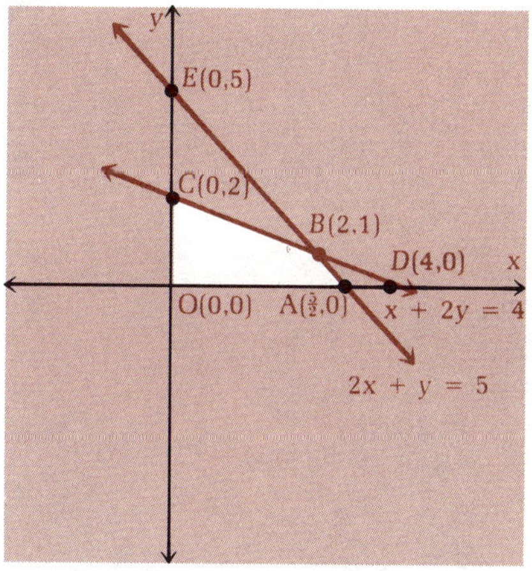

Figure 3.3

$$\begin{cases} x \geq 0,\ y \geq 0 \\ 2x + y \leq 5 \\ x + 2y \leq 4 \end{cases}$$

The restrictions $x \geq 0$ and $y \geq 0$ limit us to the first quadrant. Begin by sketching the boundary lines $2x + y = 5$ and $x + 2y = 4$. Use (0,0) to test $2x + y \leq 5$ and $x + 2y \leq 4$ to determine which half-plane to shade and discard. Find the corner points by solving the appropriate system of equations, i.e.

$$\begin{cases} 2x + y = 5 \\ x + 2y = 4 \end{cases} \quad \begin{cases} x = 0 \\ y = 0 \end{cases} \quad \begin{cases} 2x + y = 5 \\ y = 0 \end{cases} \quad \begin{cases} x + 2y = 4 \\ x = 0 \end{cases}$$

for the respective corner points $B(2,1)$, $O(0,0)$, $A\left(\frac{5}{2},0\right)$ and $C(0,2)$. ■

■ Example 2. Sketch the region bounded by the system

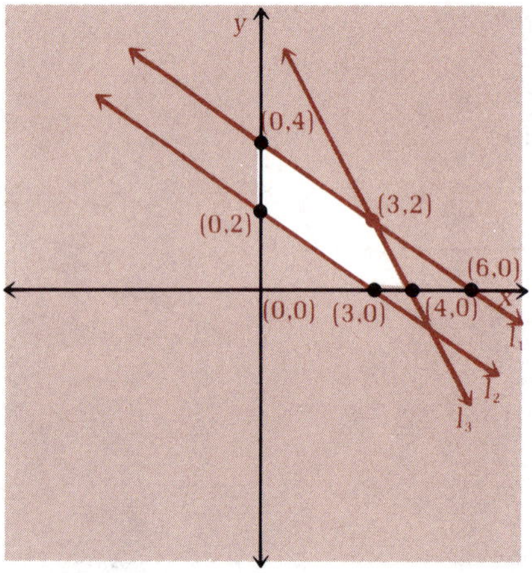

Figure 3.4

$$\begin{cases} x \geq 0,\ y \geq 0 \\ 2x + 3y \leq 12,\ l_1 \\ 2x + 3y \geq 6,\ l_2 \\ 2x + y \leq 8,\ l_3 \end{cases}$$

Solution: Start by shading out all but the first quadrant. Draw lines l_1, l_2, and l_3 using their respective x-axis and y-axis intercepts.

	Boundary Line	Axis Intercepts
Line 1	$2x + 3y = 12$	(6,0), (0,4)
Line 2	$2x + 3y = 6$	(3,0), (0,2)
Line 3	$2x + 8y = 8$	(4,0), (0,8)

Use (0,0) to test $2x + 3y \leq 12$. Since $2(0) + 3(0) \leq 12$, we shade and discard the half-plane above l_1. Using (0,0) to test $2x + 3y \geq 6$, we discard the half-plane below l_2 since $2(0) + 3(0) \ngeq 6$. Using (0,0) to test $2x + y \leq 8$ we find $2(0) + 0 \leq 8$ and again we discard the half-plane that does not contain the test point. Solving for the intersection of l_1 and l_3 $\begin{cases} 2x + 3y = 12 \\ 2x + y = 8, \end{cases}$ we locate the point (3,2) which completes the identification of the corner points.

Although the lines l_2 and l_3 are boundary lines, their intersection $\left(\frac{9}{2}, -1\right)$ is not a corner point. All corner points must be on the boundary of the feasible region. ■

We observe that as more constraints are sketched, the feasible region becomes smaller. In a practical sense we need to shade only that portion of the half-plane that has never before been shaded. Once a portion of the plane is shaded, it is excluded forever from the feasible region. Hence it is not necessary to shade a region more than once. Our focus is always on the unshaded portion of the plane.

■ Example 3. Identify the system of constraints defined by the non-shaded region in figure 3.5.

Solution:

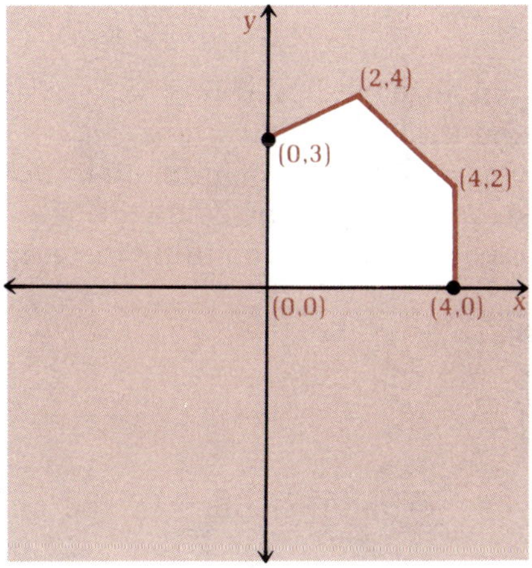

Figure 3.5

The unshaded region lies in the first quandrant, so we start with $x \geq 0$, $y \geq 0$. The other three boundary lines are easily expressed as equations using the formula $\dfrac{y - y_1}{x - x_1} = \dfrac{y_2 - y_1}{x_2 - x_1}$.

The line determined by (0,3) and (2,4) is $x - 2y = -6$. The line determined by (2,4) and (4,2) is $x + y = 6$. The line determined by (4,2) and (4,0) is $x = 4$.

Use (0,0), always a good test point, to determine which half-plane to define. Since our test point is in the feasible region we select the half-plane that contains it. For example, $x - 2y = -6$ defines two bounded half-planes, $x - 2y \leq -6$ and $x - 2y \geq -6$. We substitute (0,0) and discard the half-plane it doesn't satisfy. After testing each boundary line in turn, our system of constraints is

$$\begin{cases} x \geq 0,\ y \geq 0 \\ x - 2y \geq -6 \\ x + y \leq 6 \\ x \leq 4 \end{cases}$$

■

■ Example 4. Find the system of constraints defined by the graph in figure 3.6.

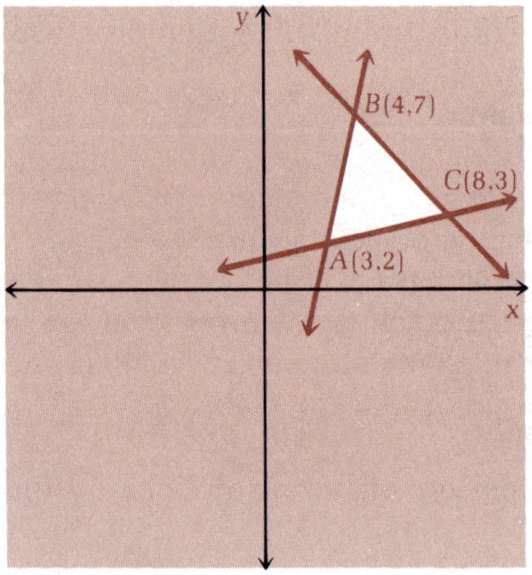

Figure 3.6

Solution: The equation of the line through A and B is $5x - y = 13$, through A and C is $-x + 5y = 7$, and through B and C is $x + y = 11$. Using (0,0) as a test point we wish to discard the half-plane containing (0,0) for the boundary AB and AC but keep the half-plane below BC which does contain the test point. The system of constraints defining the graph in figure 3.6 is

$$\begin{cases} 5x - y \geq 13 \\ -x + 5y \geq 7 \\ x + y \leq 11 \end{cases}$$

■

■ Example 5. Identify the system of constraints defined by the unbounded region in figure 3.7.

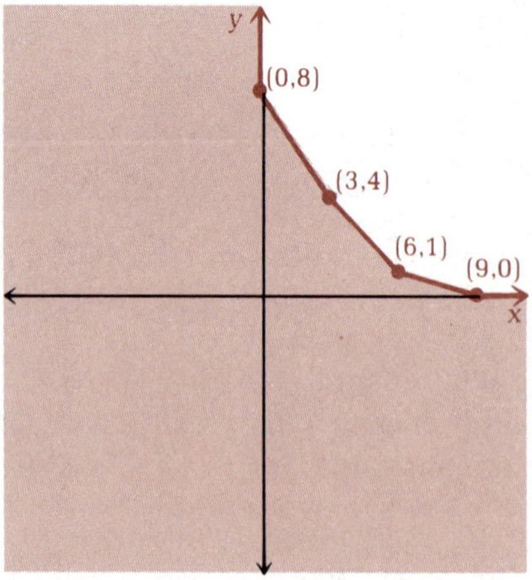

Figure 3.7

Solution: $x \geq 0$, $y \geq 0$ defines the first quadrant. The line $4x + 3y = 24$ is the boundary defined by the two points (0,8) and (3,4). $x + y = 7$ is the boundary line defined by (3,4) and (6,1) while $x + 3y = 9$ is the boundary line defined by (6,1) and (9,0). Using (0,0) as a test point, we have the system

$$\begin{cases} x \geq 0,\ y \geq 0 \\ 4x + 3y \geq 24 \\ x + y \geq 7 \\ x + 3y \geq 9 \end{cases}$$

Exercise Set 3.1

In problems 1–5, graph the region defined by the system of constraints and identify the corner points.

1. $\begin{cases} x \geq 0,\ y \geq 0 \\ x + y \leq 8 \end{cases}$

2. $\begin{cases} x \geq 0,\ y \geq 0 \\ 2x + y \leq 13 \\ x + y \leq 8 \end{cases}$

3. $\begin{cases} x + y \geq 12 \\ 3x - y \geq 16 \\ x \geq 0,\ y \geq 0 \end{cases}$

4. $\begin{cases} x \geq 0,\ y \geq 0 \\ x + 2y \geq 4 \\ 2x - 5y \geq -15 \\ x + y \leq 10 \\ x - y \leq 6 \end{cases}$

5. $\begin{cases} x \geq 3,\ y \geq 0 \\ x + y \geq 11 \\ 2x + y \geq 16 \\ x - 2y \geq 4 \end{cases}$

In problems 6–10, identify the system of constraints defined by the non-shaded regions.

6.

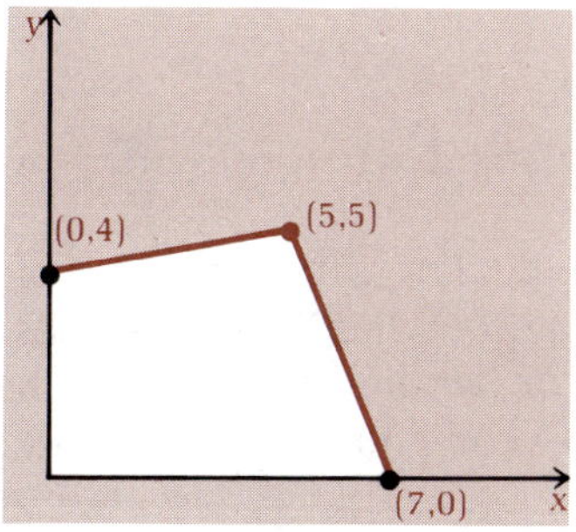

7.

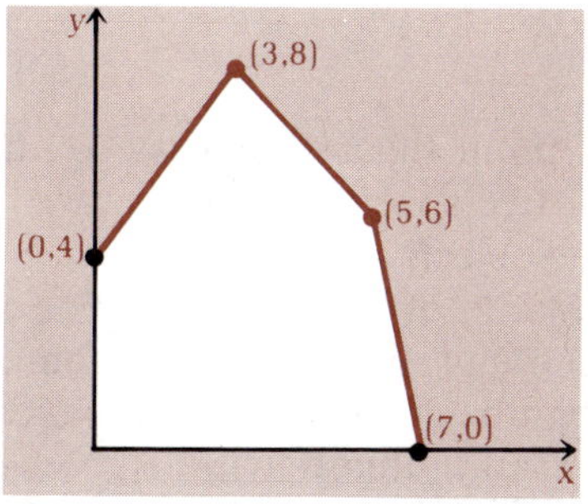

8.

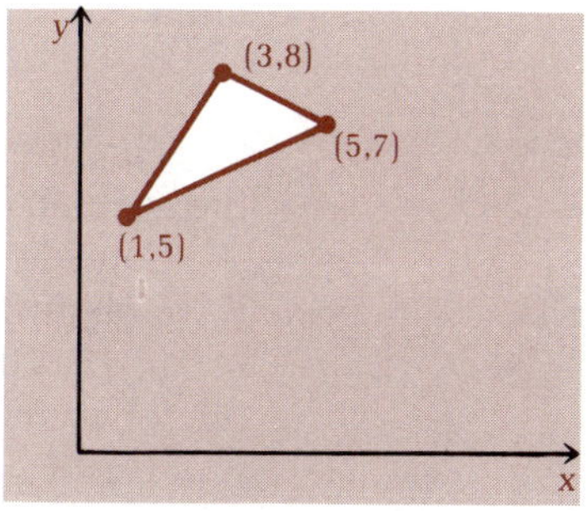

9.

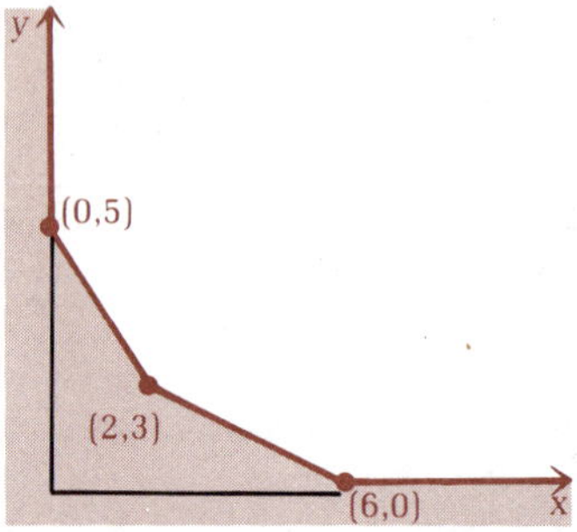

10.

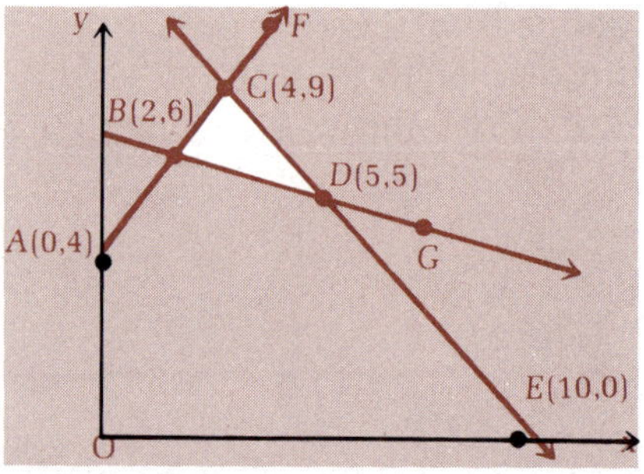

a. the region *OACE*

b. the region *OABDE*

c. the region *BCD*

d. the unbounded region *FCDG*

3.2 A Geometric Approach to Optimization

In linear programming problems, the goal is to optimize an objective statement subject to whatever restrictions exist. This will almost always take the form of **maximizing** or **minimizing** a linear **objective equation** subject to values found in the feasible region. For this reason, the following theorem is the basis of all techniques designed to solve linear programming problems.

The Corner Point Theorem The linear objective statement will have its optimal values at the corner points of the feasible region whenever the feasible region is bounded. If a feasible region is unbounded, the optimal values may not exist, but if they do, they exist at the corner points of the feasible region.

NOTE: The optimal, or extreme, value will be the maximum or minimum.

The corner point theorem is easy to understand when presented in the geometry of two-space.

Consider the objective statement $P = 2x + 5y$. For different values of P we obtain a family of parallel lines. For example when P takes values of 10, 15, 20, 30, and 40 we have the family of parallel lines shown in the following figure. Of course these are only five members of an infinite number of lines, all mutually parallel.

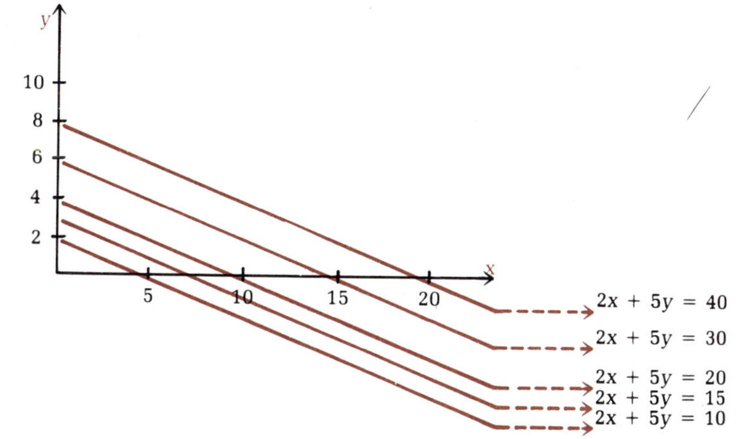

If a feasible region, say

$$\begin{cases} x \geq 0,\ y \geq 0 \\ x - 5y \geq -15 \\ 8x + 5y \leq 60 \end{cases},$$

is superimposed on this family of lines (see figure below)

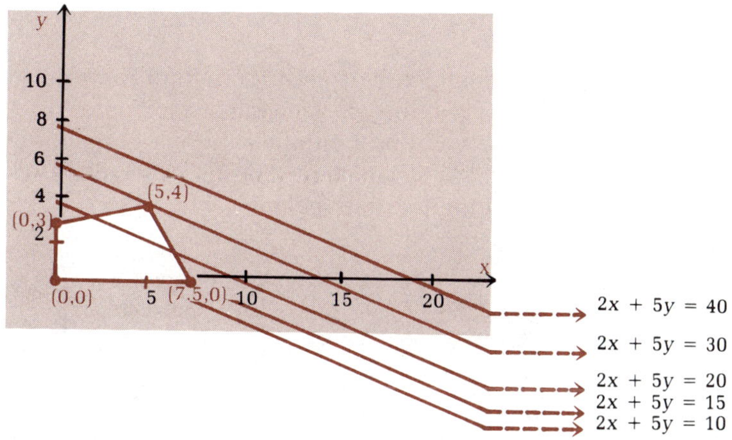

we observe that the value $P = 30$ is the maximum possible. If P is larger than 30 the straight line defined for that value of P would fall completely outside the feasible region, as does $2x + 5y = 40$.

At least in two space the following statements should be intuitively clear.

1. For a bounded feasible region, any linear objective statement will have both a minimum and a maximum value on that region.

2. An unbounded region may or may not have optimal values for a specific linear objective statement. An example (involving an unbounded feasible region) is shown below where each and every possible member of the family of lines defined by the objective statement would intersect the feasible region.

Maximize $P = x - y$ subject to $\begin{cases} x \geq 0, y \geq 0 \\ x + y \geq 5 \\ 3x - 4y \geq 18 \end{cases}$.

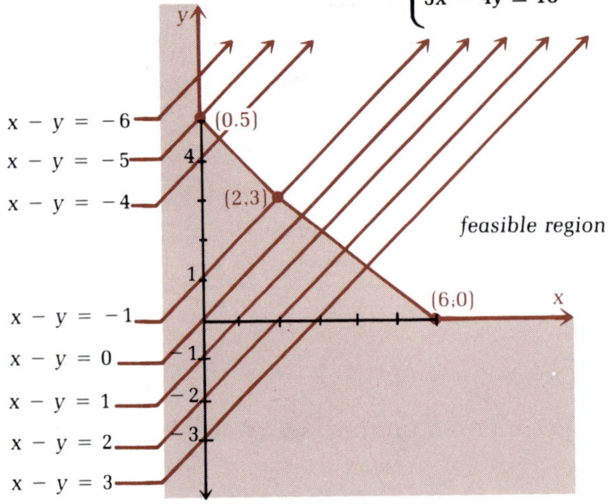

3. While some corner point (vertex) of the feasible region is always involved in finding extrema (maximum or minimum values), that corner point may not be a unique solution. Consider the figure below where the objective statement has the same slope as one of the boundary lines of the feasible region.

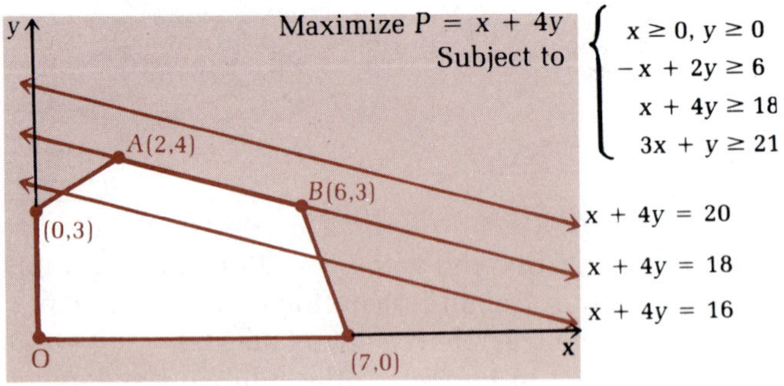

In this linear programming problem any point on the line segment $\overline{AB}$ would produce a maximum. ■

■ Example 6. Maximize $P = x + y$ and minimize $C = 40 + 2x - 5y$ subject to the restrictions

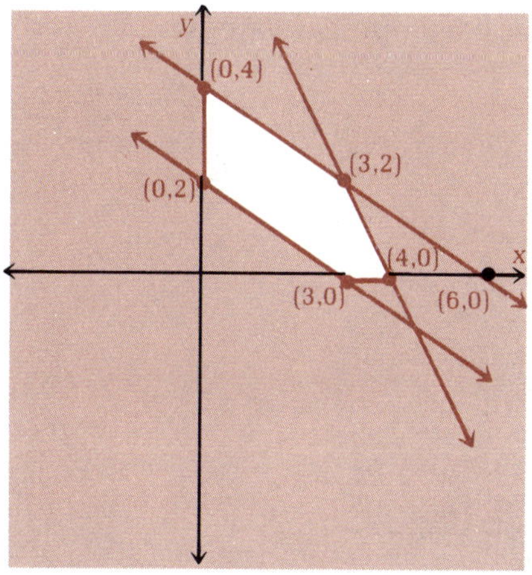

$$\begin{cases} x \geq 0,\ y \geq 0 \\ 2x + 3y \leq 12 \\ 2x + 3y \geq 6 \\ 2x + y \leq 8 \end{cases}.$$

Figure 3.8

Solution: This is the same system found in example 2 of the last section. The feasible region and corner points are shown in figure 3.8. To find the values of x and y in the feasible region that will make $P = x + y$ a maximum, we need only evaluate this *objective equation* at the corner points and identify the largest value. Similarly, evaluate $C = 40 + 2x - 5y$ at these corner points and locate the minimum.

Corner Points	Value of $P = x + y$	Value of $C = 40 + 2x - 5y$
(0,2)	2	30
(0,4)	4	20**
(3,2)	5*	36
(4,0)	4	48
(3,0)	3	46

*We see that P has a maximum of 5 when $x = 3$ and $y = 2$.
**C has a minimum of 20 when $x = 0$ and $y = 4$.

■

■ Example 7. Subject to the following constraints

$$\begin{cases} x \geq 0,\ y \geq 0 \\ 4x + y \geq 8 \\ 3x + 2y \geq 11 \\ x + 5y \geq 8 \end{cases}$$

a. minimize $A = 2x + 3y$.
b. minimize $B = 8x - y$.
c. maximize $B = 8x - y$.

Solution: The graph of the feasible region with the corner points is shown below in figure 3.9. The objective equations A and B have been evaluated at the corner points and these values are shown in the accompanying table. Since we are dealing with an unbounded feasible region, some of our objective equations may not have optimal values.

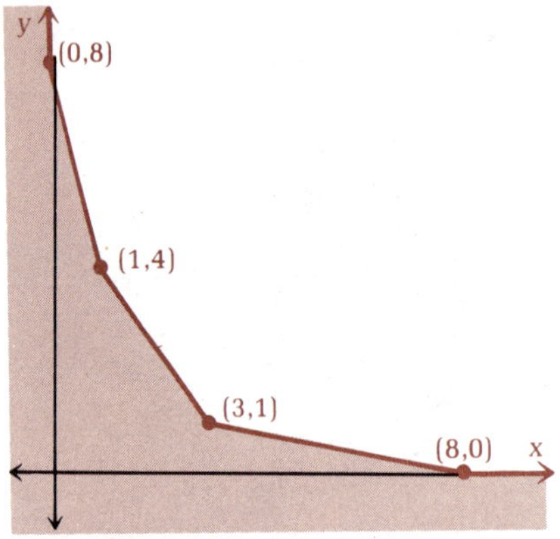

Figure 3.9

Corner Point	A	B
(0,8)	24	−8
(1,4)	14	4
(3,1)	9	23
(8,0)	16	64

a. The minimum value of A is 9 at (3,1).
b. B does not have a minimum value, since there are many points, (1,10) for one, that will make $B = -9$ a value lower than any corner point value.
c. B does not have a maximum value. We test (10,2), a point in the feasible region, and $B = 78$, again a value greater than any found at the corner points. ■

When the feasible region is unbounded, it is usually advantageous to evaluate a few points in the feasible region other than the corner points. If any of these test points gives an objective equation value greater than the largest or smaller than the least corner point value, then the respective extrema do not exist. Great care should be taken with unbounded feasible regions.

The following is a geometric test to determine if an optimal value exists for an unbounded region.

> If the feasible region is unbounded, and if there exists a line parallel to the objective equation set equal to zero that lies entirely outside the feasible region, then an optimal value will exist. This optimal value may be a maximum or a minimum but not both. Conversely, if all lines parallel to the objective equation intersect the feasible region, then optimal values do not exist.

In example 7, any line with a positive slope must cut the feasible region. Then objective equation $B = 8x - y$ which has slope of $+8$ implies B has no optimal values. $A = 2x + 3y$ when set equal to zero gives the line $2x + 3y = 0$ and this line falls entirely outside the feasible region. Hence, equation A does have an optimal value. In this case A has a minimum value. Therefore, it cannot have a maximum value.

Exercise Set 3.2

1. Maximize $P = 4x + 2y$ subject to $\begin{cases} x \geq 0,\ y \geq 0 \\ 5x + 2y \leq 30. \\ 3x + 4y \leq 32 \end{cases}$

2. Minimize $C = 40 - 2x + y$ subject to $\begin{cases} x + y \leq 10 \\ 0 \leq x \leq 6\ . \\ 0 \leq y \leq 4 \end{cases}$

3. Maximize $Z = 4x + 3y$ and minimize $T = 8 + 3x - 4y$ subject to
$\begin{cases} x - \quad y \leq 2 \\ x + \quad y \geq 10. \\ x + 2y \leq 16 \end{cases}$

4. Maximize $A = 2x + y$ subject to $\begin{cases} x \geq 0,\ y \geq 0 \\ x - y \leq 4. \\ x + y \geq 6 \end{cases}$

5. Minimize $B = 2x - y$ subject to the constraints in problem 4.
6. Minimize $C = 3x + 2y$ subject to the constraints in problem 4.
7. Maximize $D = 2x - y$ subject to the constraints in problem 4.
8. Maximize and minimize $A = 40 + 2x - y$ subject to

$$\begin{cases} 2x - 3y \leq 16 \\ 2x - 3y \geq 6 \\ 3x - \quad y \leq 14^{\cdot} \\ x - 2y \leq 10 \end{cases}$$

9. Find the maximum and minimum values of $R = 120 - .01x + 3y$

$$\text{subject to} \begin{cases} x \geq 0,\ y \geq 0 \\ x + y \leq 100 \\ 2x + y \leq 194 \\ x - 5y \geq 20 \end{cases}.$$

3.3 Linear Programming Applications

In the last section we were very careful to consider only problems in mathematical form. Problems from the real world are generally stated in a less stylized manner. The purpose of this section is to aid the student in converting word problems into mathematical language. We will state a linear programming problem, solve the problem, and generalize our technique into comments that should aid the reader in setting up future problems.

Linear programming problems become complex when a large number of unknowns and constraints are involved, and new techniques are needed to solve them. Some of these new tools will be developed later in this text, whereas others are beyond the scope of this book; however, the one common link in all word problems is data analysis. That is, we must convert from everyday language to mathematical language.

■ Example 8. Ten women find that there exists a ready market for quilts. They decide to form a company and produce two types of quilts: model B, a small baby quilt that requires 1 hour of sewing machine time, 2 hours of handstitching time, and 1 hour of embroidery work; and a regular model R quilt that requires 4 hours of sewing machine time and 1 hour of handstitching time but uses no embroidery work.

The women set up their factory in the basement of Mrs. Smythe's house and decide to work 8-hour days. After a work pattern has been established, it becomes apparent that 40 hours per day of sewing machine time is available (they own only five machines). In addition, 24 hours of handstitching time and 10 hours of embroidery work are the maximum for these operations. The profit for baby quilts is $24 each, whereas the regular type quilt produces a profit of only $18. What should be the women's production schedule to maximize their profit? What is that profit?

Solution *With comments:*

1. If possible, organize data into a table.

1.

	Model B	Model R	Limits
Machine time	1 hr	4 hr	40 hr
Handstitching time	2 hr	1 hr	24 hr
Embroidery time	1 hr	–	10 hr
Profits per item	\$24	\$18	

2. Identify the unknown values you are trying to find as literal numbers.

2. Let x = number of model B quilts, y = number of model R quilts, and profit, $P = 24x + 18y$.

3. Find the known bounds or limits of the given restrictions. Write a system of constraints. These constraints are identified from the table as the row or column that contains exactly one bound. Each bound defines a constraint.

3. The sewing machine can be used at most 40 hours. Hence 40 is a bound, and row one identifies the constraint $x + 4y \leq 40$. Similarly, $2x + y \leq 24$ and $x \leq 10$. Our feasible region is defined as

$$\begin{cases} x \geq 0, y \geq 0 \\ x + 4y \leq 40 \\ 2x + y \leq 24 \\ x \leq 10. \end{cases}$$

4. Graph the feasible region and identify the corner points.

4.

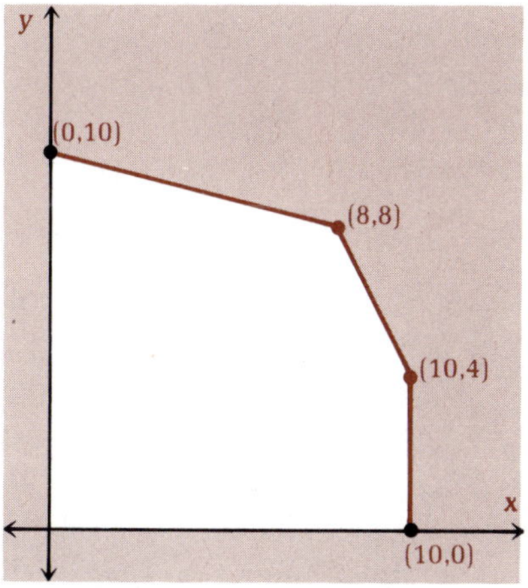

5. Evaluate the objective equation at the corner points and select the optimal value if it exists.

5.

Corner Point	$P = 24x + 18y$
(0,0)	0
(0,10)	180
(8,8)	336
(10,4)	312
(10,0)	240

The optimal value is 336. The women, therefore, schedule their work to produce 8 type *B* quilts and 8 type *R* quilts per day for a total profit of $336.

Putting $(x,y) = (8,8)$ into the constraints and checking,

$$x + 4y \leq 40 \text{ becomes } 8 + 4(8) = 40 \leq 40,$$
$$2x + y \leq 24 \text{ becomes } 2(8) + 8 = 24 \leq 24,$$
and
$$x \leq 10 \text{ becomes } 8 = 8 \leq 10.$$

The optimal solution uses all sewing machine time and handstitching time available. However, not all embroidery time is used. Using the full ten hours of embroidery time will decrease profit. ■

■ **Example 9.** A farm of 360 acres is to be planted. By hedging his crops, the farmer can assure a profit of $120 per acre of beans and $60 an acre for corn. These profits include government parity, which carries with it two restrictions. At least one third of the acreage must be planted, and the number of acres of beans cannot exceed one third the sum of the number of acres of corn plus 600. Once in the program, the farmer can "land-bank" (let land lie fallow and not plant) as many acres as he wishes at a profit of $70 per acre. How should the farmer plant to assure a maximum profit? What will be his profit?

Solution *Using the comments:*

1. Some problems do not lend themselves to tables. This is one.
2. Let x = the number of acres of corn, y = the number of acres of beans, and $360 - [x + y]$ = the number of acres not to be planted.

$$P = 60x + 120y + 70(360 - [x + y])$$
$$P = 50y - 10x + 25{,}200$$

3. Constraint 1: size of farm $\quad x + y \leq 360$

   Constraint 2: number of acres that must be planted $\quad x + y \geq 120$

   Constraint 3: formula for number of acres of beans $\quad y \leq \dfrac{600 + x}{3}$

4.

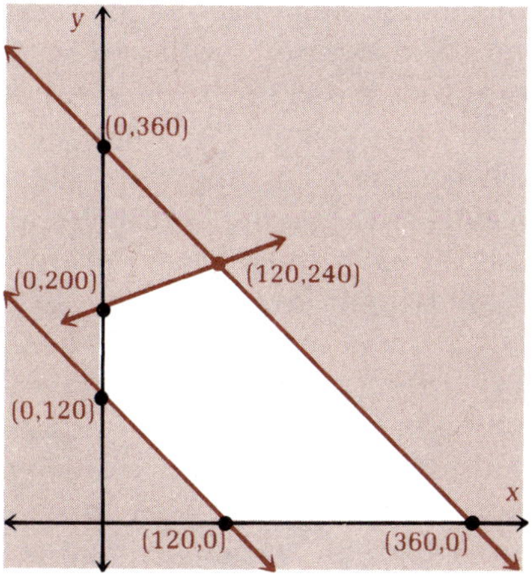

Corner Points	Profit
(0,200)	35,200
(0,120)	31,200
(120,240)	36,000
(360,0)	21,600
(120,0)	24,000

Although the farmer can maximize his profit to the optimal value of \$36,000 by planting 120 acres of corn and 240 acres of beans, he may decide to put 160 acres in the land bank and plant only 200 acres of beans. If the farmer decided not to plant these 160 acres, his profit is reduced by only \$800. ■

Exercise Set 3.3

1. A small boat builder makes two types of crafts. These boats pass through three stages in their construction. Type A, the basic model, requires 2 hours of basic construction, 3 hours of assembly, and 1 hour for trim and finishing. Type B, a deluxe model, needs 3 hours for basic construction, 4 hours of assembly, and 1 hour in trim and finishing. The profit per boat is \$120 for each type A and \$150 for each type B. The factory, limited by skilled labor and space, has available 36 hours per day for basic construction, 48 hours for assembly, and 14 hours for trim and finishing.

   a. Complete the table below.

	Total Time		
	Available	*Model A*	*Model B*
Basic construction			
Assembly			
Trim and finishing			
Profit per item			

   b. Let x = number of model A boats and y = number of model B boats. Write a system of constraints and an objective equation for profit.

c. Graph the feasible region and identify the corner points.
d. Find the maximum profit and set the production schedule.

2. If the profit were \$120 per model *A* boat and \$110 for each model *B* boat, would this change the production schedule? How?

3. The boat company in problem 1 has a large operational loan due shortly and must maximize revenue to meet the cash shortfall. The boats sell for \$500 and \$800, respectively.
a. Find maximum revenue.
b. What is the new production schedule?
c. What is the amount of the extra revenue over a 28-day period?

4. A cargo plane has a maximum cargo space of 3040 cubic feet. The load limit for cargo is listed at 10,200 pounds. Available for transport are two kinds of freight: type *A* boxes, which consume 32 cubic feet of space and weigh 30 pounds; and boxes of type *B*, which take up only 8 cubic feet of space but weigh 90 pounds each. The revenue is \$4 per box of either type. In what way does the dispatcher have the plane loaded to maximize revenue?

5. In marketing a new soft drink product, a company needs to add as much flavor as possible. Flavoring agent RH5 gives 30 flavor units per loaf, whereas SS3 gives 40 flavor units per loaf. These products also contain artificial preservatives and sweeteners, which company policy limits. Subject to the following restrictions, how is flavor maximized?

	RH5 Per Loaf	*SS3 Per Loaf*	*Company Limits*
Benzoic acid	.1 oz.	.3 oz.	4.2 oz.
Methylparaben	.1 oz.	.1 oz.	1.8 oz.
Sodium saccharin	2 oz.	1 oz.	26 oz.
Red dye 30	.001 mg.	.002 mg.	.016 mg.
Flavor units per loaf	30 f.u.	40 f.u.	

6. A truck can carry three kinds of boxes limited only by size and weight. Set up a profit objective statement and a feasible region from the data given in the following table. Do not solve.

Types of Boxes	x	y	z	*Maximum Amount*
Weight	1	2	3	5000
Size	4	4	2	4000
Profit per box (¢)	8	6	8	

7. Two kinds of 10-ounce food packages can be purchased for WA KEN BOW summer camp. The two types of food packs contain the following nutritional composition in ounces.

	Carbohydrate	*Fat*	*Protein*	*Filler*
Type I	2	3	4	1
Type II	3	1	3	3

The cost of type I is 80¢ per package. The cost of type II is 65¢ per package. Find the minimum cost if the daily requirement is 500 ounces of carbohydrates, 200 ounces of fat, and 900 ounces of protein.

8. If in problem 7 a third type of package were available that contained 4 ounces of carbohydrates, 2 ounces of fat, 3 ounces of protein, and 1 ounce of filler and sold for 70¢ per package:
   - **a.** Write the objective statement. (Hint: It contains three variables.)
   - **b.** Find the feasible region expressed as a system of inequalities.

3.4 The Simplex Method

Consider the following linear programming problem. Maximize $P = x + y + z$ subject to the constraints

$$\begin{cases} x \geq 0,\ y \geq 0,\ z \geq 0 \\ 6x + y + 4z \leq 48 \\ -2x + 2y + z \leq 12. \\ 12x + 16y + z \leq 96 \\ 3x + 4y + 2z \leq 31 \end{cases}$$

Each of these constraints is a plane in three-space and the feasible region is the interior of an irregular polyhedron (see figure 3.10).

Using the corner point theorem we can maximize our objective equation, $P = x + y + z$, in the usual way, picking the corner point that gives the largest value.

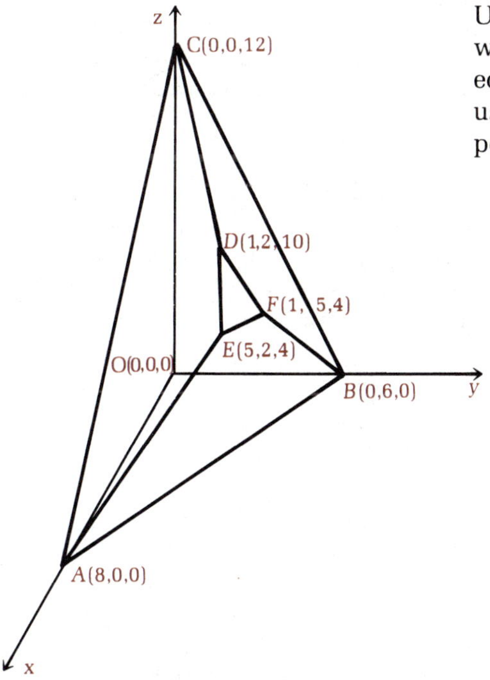

Corner Point (x,y,z)	Objective Value $P = x + y + z$
O(0,0,0)	0
A(8,0,0)	8
B(0,6,0)	6
C(0,0,12)	12
D(1,2,10)	13* (max.)
E(5,2,4)	11
F(1,5,4)	10

Figure 3.10

Although this solution seems straightforward enough, it should be noted that the feasible region is bounded by seven planes. Each three-plane combination determines a point, and most of these points are not corner points of the feasible region. In fact, there are 35 possible points in this problem, but only 7 are corner points for the feasible region.

If problems become more complex, and they do, the geometric approach becomes unusable. It's just as well, since there exists a relatively simple algorithm involving nothing more complicated than elementary row reduction for matrices that will find the best corner and identify the maximum value. The algorithm is known as the **simplex method.**

To begin our study, we must first identify the restrictions on the type of problem this algorithm will solve. As with many mathematical problems, we will later find ways around some of these restrictions. But for now, all problems to be solved by the simplex method must satisfy the following **initial conditions.**

1. All variables must be nonnegative, i.e., $x \geq 0$, $y \geq 0$, $z \geq 0$, and so on.
2. All constraints involving linear combinations of the variables must be less than or equal to a positive bound. For example, $2x + 4y - 3z \leq 15$ is an acceptable constraint, but $2x - y + z \geq 4$ is excluded because of the "greater than" symbol, as is $-2x + y - z \leq -4$ for the negative bound.
3. The objective equation is to be maximized.

In exercise set 3.2, problem 1 does satisfy these conditions, whereas the remaining problems do not. The adjustments necessary to solve problems with mixed constraints will be considered in the next section. The minor adjustments necessary to minimize the objective equation will be discussed in section 3.6.

There are three important steps involved in the use of the simplex method. These steps will be discussed separately in the beginning, and then the actual procedure will be summarized in a flow chart. The steps are as follows:

1. Setting up the simplex tableau (matrix).
2. "Reading" the tableau. (At each stage of the simplex algorithm, the tableau can be "read" to provide information about the values of the variables and the objective equation.)
3. Changing the tableau. (Several changes may be necessary to obtain an optimal solution.)

1. Setting Up the Tableau

Suppose you were going to make kites out of silk and bamboo and you have on hand 60 yards of silk (one yard wide) and 600 running feet of bamboo. If you make a standard kite, you will use 8 feet of bamboo and 1 yard of silk. If you make a box kite, you will use 24 feet of bamboo and 2 yards of silk. On the sale of the kites, you realize a profit of $1.50 per standard kite and $2.00 per box kite.

	Silk	*Bamboo*	*Profit Per Kite*
Standard kite	1 yd.	8 ft.	$1.50
Box kite	2 yds.	24 ft.	$2.00
On hand	60 yds.	600 ft.	

Let x be the number of standard kites and y be the number of box kites to be made. Then $x + 2y \le 60$ is a constraint on kite making due to the limited amount of silk. If you decide to make 10 standard kites and 20 box kites you will use only 50 yards of silk and have 10 yards of surplus. Let s be the number of yards of surplus silk. Then $x + 2y + s = 60$. The number of yards of silk used to make standard kites plus the number of yards of silk used to make box kites plus the number of yards of silk left over account for a total of 60 yards. Similarly $8x + 24y + t = 600$ is a statement about bamboo where t is the surplus bamboo. These variables s and t are called *slack variables* and they represent the "what's left" in most problems. The profit equation is given by $P = 1.5x + 2y$.

We wish, then, to maximize $P = 1.5x + 2y$ subject to

$$\begin{cases} x \ge 0,\ y \ge 0 \\ x + 2y \le 60 \\ 8x + 24y \le 600 \end{cases}$$

. The corresponding system of equations,

$$\begin{cases} x + 2y + s = 60 \\ 8x + 24y + t = 600 \\ 1.5x + 2y - P = 0 \end{cases}$$

leads to the matrix

$$\begin{array}{cccccc} x & y & s & t & P & \\ \end{array}$$
$$\begin{bmatrix} 1 & 2 & 1 & 0 & 0 & 60 \\ 8 & 24 & 0 & 1 & 0 & 600 \\ 1.5 & 2 & 0 & 0 & -1 & 0 \end{bmatrix},$$

where rows one, two, and three are as follows:

row one is 1 2 1 0 0 60 from $x + 2y + s + 0 + 0 = 60$,

row two is 8 24 0 1 0 600 from $8x + 24y + 0 + t + 0 = 600$,

row three is 1.5 2 0 0 −1 0 from $1.5x + 2y + 0 + 0 - P = 0$.

The simplex tableau is set up in matrix form as follows.

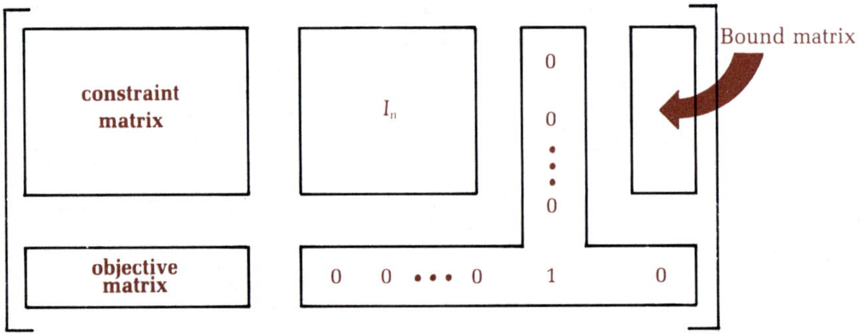

Figure 3.11

The identity matrix, I_n, is chosen to correspond to the number of constraints. That is, n is the number of rows in the **constraint matrix.**

■ Example 10. In the following linear programming problems, identify

1. The constraint matrix,
2. I_n (n refers to the number of bounded constraints),
3. The bound matrix,
4. The objective matrix,

and set up the simplex tableau according to Figure 3.11.

Solution:

Problem	*Parts*	*Simplex Tableau*
a. Maximize $P = 4x + 2y$ subject to $\begin{cases} x \geq 0,\ y \geq 0 \\ 5x + 2y \leq 30 \\ 3x + 4y \leq 32 \end{cases}$	Constraint matrix $\begin{bmatrix} 5 & 2 \\ 3 & 4 \end{bmatrix}$ $I_2 = \begin{bmatrix} 1 & 0 \\ 0 & 1 \end{bmatrix}$ Bound matrix $\begin{bmatrix} 30 \\ 32 \end{bmatrix}$ Objective matrix $[4 \quad 2]$	$\begin{bmatrix} 5 & 2 & 1 & 0 & 0 & 30 \\ 3 & 4 & 0 & 1 & 0 & 32 \\ 4 & 2 & 0 & 0 & -1 & 0 \end{bmatrix}$

b. Maximize $P = 4x - y$ subject to

$$\begin{cases} x \geq 0,\ y \geq 0 \\ x + 2y \leq 6 \\ 2x + y \leq 6 \\ 3x + 2y \leq 8 \end{cases}$$

Constraint matrix $\begin{bmatrix} 1 & 2 \\ 2 & 1 \\ 3 & 2 \end{bmatrix}$

$I_3 = \begin{bmatrix} 1 & 0 & 0 \\ 0 & 1 & 0 \\ 0 & 0 & 1 \end{bmatrix}$

Bound matrix $\begin{bmatrix} 6 \\ 6 \\ 8 \end{bmatrix}$

Objective matrix $[4 \quad -1]$

$$\begin{bmatrix} 1 & 2 & 1 & 0 & 0 & 0 & 6 \\ 2 & 1 & 0 & 1 & 0 & 0 & 6 \\ 3 & 2 & 0 & 0 & 1 & 0 & 8 \\ 4 & -1 & 0 & 0 & 0 & -1 & 0 \end{bmatrix}$$

c. Maximize $P = x + y + z$ subject to

$$\begin{cases} x \geq 0,\ y \geq 0,\ z \geq 0 \\ 6x + y + 4z \leq 48 \\ -2x + 2y + z \leq 12 \\ 12x + 16y + z \leq 96 \\ 3x + 4y + 2z \leq 31 \end{cases}$$

Constraint matrix $\begin{bmatrix} 6 & 1 & 4 \\ -2 & 2 & 1 \\ 12 & 16 & 1 \\ 3 & 4 & 2 \end{bmatrix}$

$I_4 = \begin{bmatrix} 1 & 0 & 0 & 0 \\ 0 & 1 & 0 & 0 \\ 0 & 0 & 1 & 0 \\ 0 & 0 & 0 & 1 \end{bmatrix}$

Bound matrix $\begin{bmatrix} 48 \\ 12 \\ 96 \\ 31 \end{bmatrix}$

Objective matrix $[1 \quad 1 \quad 1]$

$$\begin{bmatrix} 6 & 1 & 4 & 1 & 0 & 0 & 0 & 0 & 48 \\ -2 & 2 & 1 & 0 & 1 & 0 & 0 & 0 & 12 \\ 12 & 16 & 1 & 0 & 0 & 1 & 0 & 0 & 96 \\ 3 & 4 & 2 & 0 & 0 & 0 & 1 & 0 & 31 \\ 1 & 1 & 1 & 0 & 0 & 0 & 0 & -1 & 0 \end{bmatrix}$$

d. Maximize $p = x_1 + 2x_2 + x_3 + 3x_4$ subject to

$$\begin{cases} x_1, x_2, x_3, x_4 \geq 0 \\ 2x_1 + x_2 + x_3 - x_4 \leq 6 \\ x_1 - x_2 + 2x_3 + x_4 \leq 4 \end{cases}$$

Constraint matrix $\begin{bmatrix} 2 & 1 & 1 & -1 \\ 1 & -1 & 2 & 1 \end{bmatrix}$

$I_2 = \begin{bmatrix} 1 & 0 \\ 0 & 1 \end{bmatrix}$

Bound matrix $\begin{bmatrix} 6 \\ 4 \end{bmatrix}$

Objective matrix
$[1 \quad 2 \quad 1 \quad 3]$

$$\begin{bmatrix} 2 & 1 & 1 & -1 & 1 & 0 & 0 & 6 \\ 1 & -1 & 2 & 1 & 0 & 1 & 0 & 4 \\ 1 & 2 & 1 & 3 & 0 & 0 & -1 & 0 \end{bmatrix}$$

■

Notice that the simplex tableau is a matrix of linear equations. The identity matrix, I_n, represents the coefficients of the slack variables. It is not the intent of this book to explain exactly the inner workings of the simplex method. We will concern ourselves with using the algorithm to solve problems.

2. "Reading" the Simplex Tableau (Matrix)

The solving of problems using the simplex method involves changing the form of the simplex tableau according to an exact procedure. Before we discuss the procedure for changing the simplex matrix it is important that you can *"read" the simplex matrix*. Each variable in the problem is represented by a column. In the preceding kite problem, column 1 identifies the x coefficients, column 2 the y coefficients, column 3 the coefficients of the slack variable for the first constraint, column 4 the coefficients of the slack variable for the second constraint, and column 5 the coefficients of the profit variable P, our objective value.

In any simplex matrix, we can read the values of the variables by

a. assigning zero to the variable identified by any column that contains two or more nonzero entries, i.e., assign zero to any uncleared column or

b. in any cleared column (a column with exactly one nonzero entry), the value of the variable identified by that column is the ratio of the last row entry divided by the nonzero column entry.

NOTE: We will know we have obtained an optimal solution when the bottom row contains no positive numbers.

The following examples should help to explain. We will discuss where these matrices came from later.

■ Example 11. Read the following simplex matrix.

$$\begin{array}{ccccccccc} x & y & z & s_1 & s_2 & s_3 & s_4 & P & \\ \end{array}$$
$$\begin{bmatrix} 14 & 0 & 7 & 2 & -1 & 0 & 0 & 0 & 84 \\ -2 & 2 & 1 & 0 & 1 & 0 & 0 & 0 & 12 \\ 28 & 0 & -7 & 0 & -8 & 1 & 0 & 0 & 0 \\ 7 & 0 & 0 & 0 & -2 & 0 & 1 & 0 & 7 \\ 4 & 0 & 1 & 0 & -1 & 0 & 0 & -2 & -12 \end{bmatrix}$$

column 1 uncleared: $x = 0$

column 2 cleared: $y = \dfrac{12}{2} = 6$

column 3 uncleared: $z = 0$

column 4 cleared: $s_1 = \dfrac{84}{2} = 42$

column 5 uncleared: $s_2 = 0$

column 6 cleared: $s_3 = \dfrac{0}{1} = 0$

column 7 cleared: $s_4 = \dfrac{7}{1} = 7$

column 8 cleared: $P = \dfrac{-12}{-2} = 6$

This means that, at the point (0,6,0), the value of P is 6. However, since the bottom row contains some positive numbers, this is not the maximum value for P. ■

■ Example 12. Read the following simplex matrix.

$$\begin{array}{ccccccccc} x & y & z & s_1 & s_2 & s_3 & s_4 & P & \\ \end{array}$$
$$\begin{bmatrix} 0 & 0 & 7 & 2 & 3 & 0 & -2 & 0 & 70 \\ 0 & 7 & 0 & -1 & 0 & 0 & 2 & 0 & 14 \\ 0 & 0 & 0 & 2 & 3 & 1 & -6 & 0 & 42 \\ 7 & 0 & 0 & 0 & 2 & 0 & 1 & 0 & 7 \\ 0 & 0 & 0 & -2 & -2 & 0 & -2 & -14 & -182 \end{bmatrix}$$

$x = \dfrac{7}{7} = 1$

$y = \dfrac{14}{7} = 2$

$z = \dfrac{70}{7} = 10$

$s_1 = s_2 = s_4 = 0$

$s_3 = \dfrac{42}{1} = 42$

and $P = \dfrac{-182}{-14} = 13$

Since the bottom row contains no positive entries, this value for P is a maximum. This is the last matrix in the simplex solution to the problem solved geometrically at the beginning of this section. We have the additional information here that $s_1 = s_2 = s_4 = 0$ and $s_3 = 42$, i.e., constraints 1, 2, and 4 have no slack, but constraint 3 ($12x + 16y + z \leq 96$) has 42 surplus units. Checking (1,2,10) we see $12(1) + 16(2) + 10 = 12 + 32 + 10 = 54$, which is exactly 42 short of 96. ■

3. Changing the Simplex Tableau

The simplex tableau is a matrix corresponding to a system of linear equations. The procedure used for **changing the tableau** is the elementary row reduction technique used in chapter 2 to solve systems of linear equations. However, there are three notable differences from standard row reduction.

1. The last row of the simplex tableau corresponds to the objective equation and is *never* used for clearing columns.
2. The choice of a pivot number to be used for clearing a column is not arbitrary.
3. When a linear combination of two rows is used, the row being replaced is never multiplied by a negative number. That is, if row j is to be replaced by $aRi + bRj$, b must be a positive number.

Three steps are involved in changing the simplex matrix. These are the following:

1. Select a column with a positive last entry. This choice is free, but after working several problems, you will become aware that some choices are better than others. It is generally best to choose the column with the largest last entry.
2. Once a column is chosen, determine the pivot number. Divide each entry, except the one in the bottom row, into the bound entry in the last column. The **pivot number** is a positive entry that gives the smallest nonnegative quotient.
3. Use the **pivot row**, the row containing the pivot number, to clear the column using elementary row operations, with one exception. Never multiply any row except the pivot row by a negative number.

This process is repeated as long as positive entries remain in the last row. When the bottom row no longer contains positive entries, the final, optimal solution is obtained.

To see why the pivot number is chosen as a positive entry giving the smallest non-negative quotient, consider the following problem.

Maximize $P = x + y + z$ subject to

$$\begin{cases} x \geq 0,\ y \geq 0,\ z \geq 0 \\ 6x + y + 4z \leq 48 \\ 2x + 2y + z \leq 12 \\ 12x + 16y + z \leq 96 \\ 3x + 4y + 2z \leq 31. \end{cases}$$

The simplex matrix is given by (see example 10c)

$$\begin{bmatrix} 6 & 1 & 4 & 1 & 0 & 0 & 0 & 0 & 48 \\ -2 & 2 & 1 & 0 & 1 & 0 & 0 & 0 & 12 \\ 12 & 16 & 1 & 0 & 0 & 1 & 0 & 0 & 96 \\ 3 & 4 & 2 & 0 & 0 & 0 & 1 & 0 & 31 \\ 1 & 1 & 1 & 0 & 0 & 0 & 0 & -1 & 0 \end{bmatrix}$$

When we read this simplex matrix, we obtain $x = y = z = 0$ and $P = 0$. The slack variables have values $s_1 = 48$, $s_2 = 12$, $s_3 = 96$, and $s_4 = 31$. However, P does not involve the slack variables. The value of P can be increased by increasing the value of any one of the initial variables, x, y, or z. An increase in one of the initial variables will cause a decrease in at least one of the slack variables.

We need to determine how much we can increase an initial variable and still satisfy the constraints. The first four rows of the matrix are the constraints

$$\begin{aligned} 6x + y + 4z + s_1 &= 48 \\ -2x + 2y + z + s_2 &= 12 \\ 12x + 16y + z + s_3 &= 96 \\ 3x + 4y + 2z + s_4 &= 31. \end{aligned}$$

The slack variables have been included since they are affected by a change in one of the initial variables.

We choose one of the initial variables we wish to increase, keeping the remaining variables at zero. Suppose we choose z to increase, leaving $x = y = 0$. From the first constraint, we see that increasing z by 12 units (from 0 to 12) and making $s_1 = 0$ will give

$$6 \cdot 0 + 0 + 4 \cdot 12 + 0 = 48.$$

Hence an increase of 12 units in z satisfies constraint one. The second constraint will also be satisfied by a 12 unit increase in z, making $s_2 = 0$.

For the third constraint, we can increase z by 96 units while for constraint four we can increase z by $15\frac{1}{2}$ units. To determine the amount by which z can be increased, we let all other variables have value 0 and solve for z. That is,

$$\begin{aligned} 0 + 0 + 4z + 0 &= 48 \text{ or } z = \frac{48}{4} = 12, \\ 0 + 0 + z + 0 &= 12 \text{ or } z = 12, \\ 0 + 0 + z + 0 &= 96 \text{ or } z = 96, \\ 0 + 0 + 2z + 0 &= 31 \text{ or } z = \frac{31}{2} = 15\frac{1}{2}. \end{aligned}$$

The amount by which z can be increased for each constraint is the ratio of the number in the bound column (right column) to the coefficient of z (entry in the z column).

Since the new value of z (amount by which z is increased) must satisfy all of the constraints, we must choose the smallest increase, which is the smallest ratio. For this problem, an increase in z of more than 12 units (96 or $15\frac{1}{2}$) will not satisfy the first two constraints. Hence we would choose a 12 unit increase. That makes row 1 or row 2 the pivot row with the corresponding z column entry (4 or 1) as the pivot number for clearing the column.

We use 0 for the value of the slack so that z is increased by the largest possible amount. Had we chosen $s_1 = 12$ instead of $s_1 = 0$, we would have had

$$\begin{aligned} 0 + 0 + 4z + 12 &= 48, \text{ or} \\ 4z &= 36, \text{ or} \\ z &= 9. \end{aligned}$$

This is not the largest amount by which z could be increased. When determining the amount by which a variable can be increased, we always use 0 for the slack variable values.

Suppose we had chosen to increase x, keeping $y = z = 0$. We then look at the equations

$$\begin{aligned} 6x + 0 + 0 + 0 &= 48, \\ -2x + 0 + 0 + 0 &= 12, \\ 12x + 0 + 0 + 0 &= 96, \\ 3x + 0 + 0 + 0 &= 31. \end{aligned}$$

Solving these equations for x, we would obtain $x = 8$, $x = -6$, $x = 8$, and $x = 10\frac{1}{3}$ respectively. The smallest of these values is -6. However, this is a decrease in x, not an increase, and would result in a decrease in the value of P from $P = 0$ to $P = -6$. The value $x = -6$ also violates the initial constraint $x \geq 0$. Hence, we would choose the increase of 8 units using row 1 or row 3 as the pivot row. The pivot number would be either the 6 in row 1 or the 12 in row 3. It doesn't matter which of the two we use when the ratios are the same. Choosing the smaller of the two might simplify the arithmetic but might also increase the number of steps.

Choosing a pivot number for continuing the reduction is based on this same procedure. We look at the ratio of the bound number to the column entry in the column we are clearing. The pivot number is a number producing the smallest non-negative ratio.

As the reduction is continued and additional initial variables are increased, the values of variables that have already been increased may well decrease. However, the value of P will either stay the same or increase. Since we are seeking the maximum value for P, we want to continue to change values of the initial variables until we locate a corner of the feasible region where P is a maximum. It doesn't matter if some of the values begin to decrease somewhat.

All of the preceding information is summarized in the following flow chart.

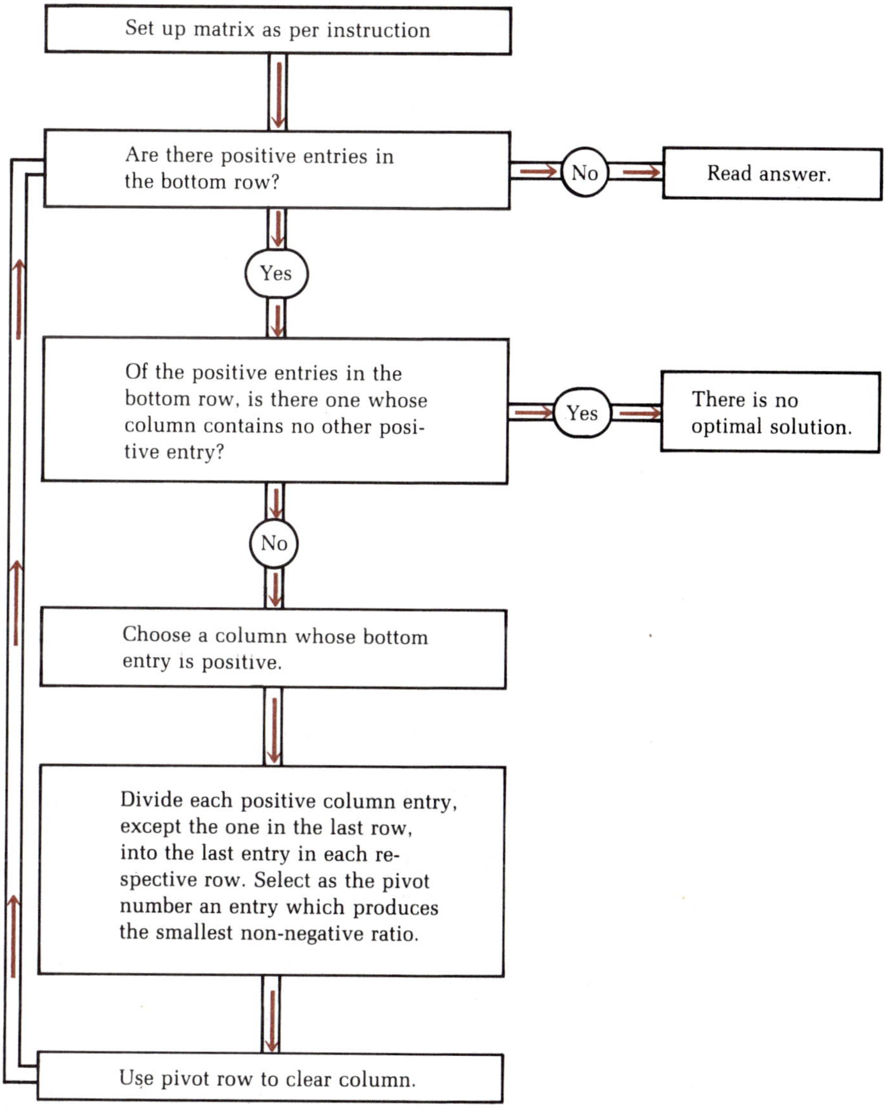

The following examples will illustrate the use of the **flow chart.**

■ **Example 13.** Maximize $p = 2x + 3y$ subject to $\begin{cases} x \geq 0,\ y \geq 0 \\ 2x + \ \ y \leq 5. \\ x + 2y \leq 4 \end{cases}$

Solution:

Tableau 1
(the set-up matrix)

$$\begin{bmatrix} 2 & 1 & 1 & 0 & 0 & 5 \\ 1 & \textcircled{2} & 0 & 1 & 0 & 4 \\ 2 & 3 & 0 & 0 & -1 & 0 \end{bmatrix}$$

Step 1. Set the problem up in simplex form. There are two positive entries in the last row and both have positive entries above them. Choose a column with a positive last entry (column 2 was picked).

Step 2. Check the ratios $5 \div 1 = 5$ and $4 \div 2 = 2^*$.
*Determine the proper pivot number which is 2.

Step 3. Clear column 2. Since there still exists a positive number in the last row, we repeat the process.

In column 1 of tableau 2, $6 \div 3 = 2^*$ and $4 \div 1 = 4$
*Select 3 as the pivot number and clear column 1 to obtain tableau 3.

Tableau 2

$$\begin{bmatrix} \textcircled{3} & 0 & 2 & -1 & 0 & 6 \\ 1 & 2 & 0 & 1 & 0 & 4 \\ 1 & 0 & 0 & -3 & -2 & -12 \end{bmatrix}$$

Tableau 3

$$\begin{bmatrix} 3 & 0 & 2 & -1 & 0 & 6 \\ 0 & 6 & -2 & 4 & 0 & 6 \\ 0 & 0 & -2 & -8 & -6 & -42 \end{bmatrix}$$

Step 4. Since the bottom row of tableau 3 contains no positive entries, the solution is complete. We read the matrix to obtain $x = 2$, $y = 1$, $p = 7$ and slack variables $s_1 = s_2 = 0$.

Geometrically (see Figure 3.12), we were at the corner (0,0) in the set-up matrix. The first change (read tableau 2) moved us along the *y*-axis (column 2) to the corner *A*(0,2). The last change (read tableau 3) moved us along the edge of the line segment $\overline{AB}$ to the corner point *B*(2,1) and identified this as the optimal solution. If we had chosen to clear column 1 first, we would have first moved to point *C*, and then in the second change to point *B* along the line segment $\overline{BC}$.

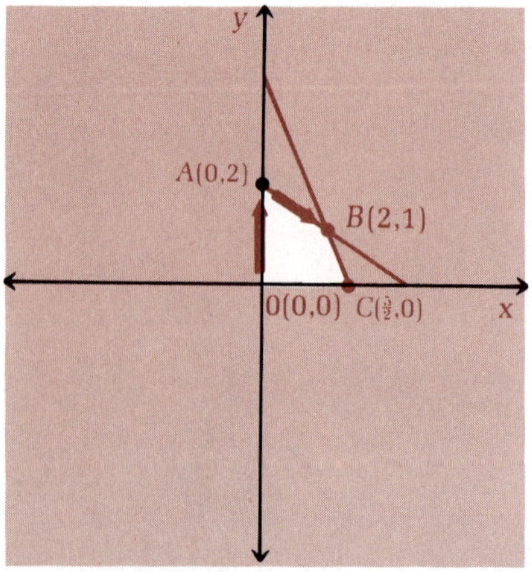

Figure 3.12 ■

■ Example 14. Maximize $P = x + y + z$ subject to

$$\begin{cases} x \geq 0,\ y \geq 0,\ z \geq 0 \\ 6x + y + 4z \leq 48 \\ -2x + 2y + z \leq 12. \\ 12x + 16y + z \leq 96 \\ 3x + 4y + 2z \leq 31 \end{cases}$$

Solution: This problem was solved earlier using the corner point theorem. Simplex tableau 1 is set up below. Column 2 is randomly selected, and the pivot number is circled.

Tableau 1

$$\begin{bmatrix} 6 & 1 & 4 & 1 & 0 & 0 & 0 & 0 & 48 \\ -2 & \textcircled{2} & 1 & 0 & 1 & 0 & 0 & 0 & 12 \\ 12 & 16 & 1 & 0 & 0 & 1 & 0 & 0 & 96 \\ 3 & 4 & 2 & 0 & 0 & 0 & 1 & 0 & 31 \\ 1 & 1 & 1 & 0 & 0 & 0 & 0 & -1 & 0 \end{bmatrix}$$

In simplex tableau 2 below, column 2 has been cleared. We read the matrix and find we are at (0,6,0) where $P = 6$. The last row has some positive entries, so we must continue the process.

Tableau 2

$$\begin{bmatrix} 14 & 0 & \textcircled{7} & 2 & -1 & 0 & 0 & 0 & 84 \\ -2 & 2 & 1 & 0 & 1 & 0 & 0 & 0 & 12 \\ 28 & 0 & -7 & 0 & -8 & 1 & 0 & 0 & 0 \\ 7 & 0 & 0 & 0 & -2 & 0 & 1 & 0 & 7 \\ 4 & 0 & 1 & 0 & -1 & 0 & 0 & -2 & -12 \end{bmatrix}$$

We chose to clear column 3 next, even though column 1 has the largest last entry. We avoid, whenever possible, a pivot number with a zero bound. Choosing a pivot number with a zero bound may be necessary sometimes. However, since the zero bound will not improve the value of the objective equation, we will use it only if we have to. Remember, pivot numbers are positive. Hence -7 is not a candidate, even though it produces the smallest ratio. The ratios for the two candidates are equal. Either choice will unclear a column that has already been cleared. We chose 7 so we don't unclear column 2. We are sometimes forced to unclear columns because we must pick a pivot number with the smallest ratio.

Tableau 3

$$\begin{bmatrix} 14 & 0 & 7 & 2 & -1 & 0 & 0 & 0 & 84 \\ -28 & 14 & 0 & -2 & 8 & 0 & 0 & 0 & 0 \\ 42 & 0 & 0 & 2 & -9 & 1 & 0 & 0 & 84 \\ \textcircled{7} & 0 & 0 & 0 & -2 & 0 & 1 & 0 & 7 \\ 14 & 0 & 0 & -2 & -6 & 0 & 0 & -14 & -168 \end{bmatrix}$$

We have moved to the corner (0,0,12) where $p = \dfrac{-168}{-14} = 12$, but we need to continue the simplex method by clearing column 1 since it still has a positive entry in the last row.

Tableau 4

$$\begin{bmatrix} 0 & 0 & 7 & 2 & 3 & 0 & 2 & 0 & 70 \\ 0 & 14 & 0 & -2 & 0 & 0 & 4 & 0 & 28 \\ 0 & 0 & 0 & 2 & 3 & 1 & -6 & 0 & 42 \\ 7 & 0 & 0 & 0 & -2 & 0 & 1 & 0 & 7 \\ 0 & 0 & 0 & -2 & -2 & 0 & -2 & -14 & -182 \end{bmatrix}$$

Look at that beautiful last row! We are finished. Read

$$x = \frac{7}{7} = 1, \qquad y = \frac{28}{14} = 2, \qquad z = \frac{70}{7} = 10,$$

$$\text{and } P = \frac{-182}{-14} = 13.$$

With reference to figure 3.10 in section 3.4, we moved along the edges of the polyhedron from $O(0,0,0)$ to $B(0,6,0)$ to $C(0,0,12)$ to $D(1,2,10)$ where the objective equation was maximum. One last word of advice—use check columns if you are unsure of your arithmetic (see section 1.4).

When the objective equation contains a constant term—for example, if $P = 10 + 4x + 6y$ is the objective equation to maximize—we could enter -10 as the very last (southeast corner) entry of our set-up matrix and proceed as usual. Alternately, we could maximize $2x + 3y$ and compute $P = 10 + 2(2x + 3y)$ as a final step. ■

■ Example 15. Maximize $P = 10 + 4x + 6y$ subject to

$$\begin{cases} x \geq 0,\ y \geq 0 \\ 2x + y \leq 18 \\ x + 2y \leq 16 \\ x + y \leq 10 \end{cases}.$$

Solution:

$$\begin{bmatrix} 2 & 1 & 1 & 0 & 0 & 0 & 18 \\ 1 & 2 & 0 & 1 & 0 & 0 & 16 \\ 1 & 1 & 0 & 0 & 1 & 0 & 10 \\ 4 & 6 & 0 & 0 & 0 & -1 & -10 \end{bmatrix} \quad 4x + 6y - P = -10$$

$$\begin{matrix} -R2 + 2R1 \\ \\ -R2 + 2R3 \\ -3R2 + R4 \end{matrix} \begin{bmatrix} 3 & 0 & 2 & -1 & 0 & 0 & 20 \\ 1 & 2 & 0 & 1 & 0 & 0 & 16 \\ 1 & 0 & 0 & -1 & 2 & 0 & 4 \\ 1 & 0 & 0 & -3 & 0 & -1 & -58 \end{bmatrix}$$

$$\begin{matrix} -3R3 + R1 \\ -R3 + R2 \\ \\ -R3 + R4 \end{matrix} \begin{bmatrix} 0 & 0 & 2 & 2 & -6 & 0 & 8 \\ 0 & 2 & 0 & 2 & -2 & 0 & 12 \\ 1 & 0 & 0 & -1 & 2 & 0 & 4 \\ 0 & 0 & 0 & -2 & -2 & -1 & -62 \end{bmatrix}$$

Read the matrix to obtain

$$x = 4,\ y = \frac{12}{2} = 6,\ s_1 = \frac{8}{2} = 4,\ s_2 = s_3 = 0, \text{ and } P = 62.$$

Checking

$2(4) + 6 = 14 \leq 18 \qquad (s_1 = 18 - 14 = 4)$

$4 + 2(6) = 12 \leq 12 \qquad (s_2 = 12 - 12 = 0)$

$4 + 6 = 10 \leq 10 \qquad (s_3 = 10 - 10 = 0)$

$P = 10 + 4(4) + 6(6) = 10 + 16 + 36 = 62$ ■

Exercise Set 3.4

1. Set up the simplex tableau for the following:

   a. Maximize $P = 2x + y$ subject to

   $$\begin{cases} x \geq 0,\ y \geq 0 \\ 3x - 2y \leq 14 \\ x + y \leq 8 \end{cases}$$

   b. Maximize $R = 2x - y + z$ subject to

   $$\begin{cases} x \geq 0,\ y \geq 0,\ z \geq 0 \\ x + y + z \leq 8 \\ 2x - y - z \leq 10 \\ x + y \leq 5 \end{cases}$$

c. Maximize $P = 2x_1 + x_2$
subject to $x_1, x_2 \geq 0$ and

$$\begin{cases} x_1 + x_2 \leq 4 \\ 2x_1 - x_2 \leq 8 \\ 3x_1 + 4x_2 \leq 15 \end{cases}$$

d. Maximize $Z = x_1 + x_2 + x_3$
subject to $x_1, x_2, x_3 \geq 0$ and

$$\begin{cases} 2x_1 - x_2 + x_3 \leq 8 \\ x_1 + x_2 - x_3 \leq 5 \end{cases}$$

2. Why do each of the following fail to satisfy the initial conditions for the simplex method?

a. Maximize $P = x + y$
subject to

$$\begin{cases} 2x + y \leq 5 \\ x - y \geq 6 \end{cases}$$

b. Maximize $P = 2x - 3y$
subject to

$$\begin{cases} x \geq 0,\ y \geq 0 \\ 3x + y \leq 10 \\ -x + y \leq -6 \end{cases}$$

3. Identify the correct pivot number in

a. column 1 b. column 2 c. column 3

$$\begin{bmatrix} 3 & 1 & 2 & 1 & 0 & 0 & 5 \\ 1 & 2 & 3 & 0 & 1 & 0 & 2 \\ 3 & 4 & 0 & 0 & 0 & -1 & 0 \end{bmatrix}$$

4. Clear column 1 in the matrix in problem 3.
5. Use the simplex method to solve the following problems.

a. Maximize $P = 2x + 3y$
subject to $x \geq 0,\ y \geq 0$
and

$$\begin{cases} x + y \leq 6 \\ -x + 2y \leq 6 \end{cases}$$

b. Maximize $R = 9 + 5x + 7y$
subject to $x \geq 0,\ y \geq 0$ and

$$\begin{cases} 2x + y \leq 13 \\ x + y \leq 8 \end{cases}$$

c. Maximize $P = x + y + z$
subject to $x, y, z, \geq 0$ and

$$\begin{cases} 2x + y - z \leq 8 \\ x - y + 3z \leq 10 \end{cases}$$

d. Maximize $P = 3x - y + 19$
subject to $x \geq 0,\ y \geq 0$ and

$$\begin{cases} 3x + y \leq 12 \\ 2x - y \leq 8 \\ x + 3y \leq 15 \end{cases}$$

6. Solve the kite problem given at the beginning of this section.
7. Solve problem 6 of exercise set 3.3.
8. Solve problem 8 of exercise set 3.3.

9. Maximize $P = 8x + 5y + 3z$ subject to $x,y,z \geq 0$ and

$$\begin{cases} 6x + y + 4z \leq 48 \\ -2x + 2y + z \leq 12 \\ 12x + 16y + z \leq 96 \\ 3x + 4y + 2z \leq 31 \end{cases}$$

10. Maximize $P = 50 + x + 2y$ subject to $x \geq 0$, $y \geq 0$ and $3x + 2y \leq 14$

11. Maximize $P = x - y - z$ subject to the constraints

$$\begin{cases} x,y,z \geq 0 \\ x + 10y + z \leq 100 \\ 2x - 100y + 3z \leq 150 \end{cases}$$

12. Maximize $R = 3x + y + 3z$ subject to $x,y,z \geq 0$ and

$$\begin{cases} x + 10y + z \leq 100 \\ 2x - 100y + 3z \leq 150 \end{cases}$$

3.5 More on the Simplex Method

When linear programming problems do not satisfy the initial requirements for using the simplex method, it *may* still be possible to find an *acceptable initial matrix* where the simplex method can be carried out. This adjustment will be referred to as *Phase I*, and it needs to be considered when the following conditions exist:

1. The problem contains $\geq$ constraints.
2. The problem contains negative bounds. For example, $-2x + 4y \leq -8$.
3. One or more of the constraints is an equation.

The flow chart which follows will transform a matrix of constraints into an acceptable initial matrix where the simplex method can be used. It should be noted here that a simple adjustment will also allow for *minimizing the objective statement*. Section 3.6 addresses itself to that adjustment.

In most computer systems, "software packages" are available for solving simplex problems that include **minimization** problems and problems with **modified**, or **mixed, constraints,** referred to here as Phase I.

NOTE: Following the **flow chart for Phase I,** some "points to remember" have been stated. It is extremely important to observe the restrictions on clearing and multiplying by -1.

Phase I

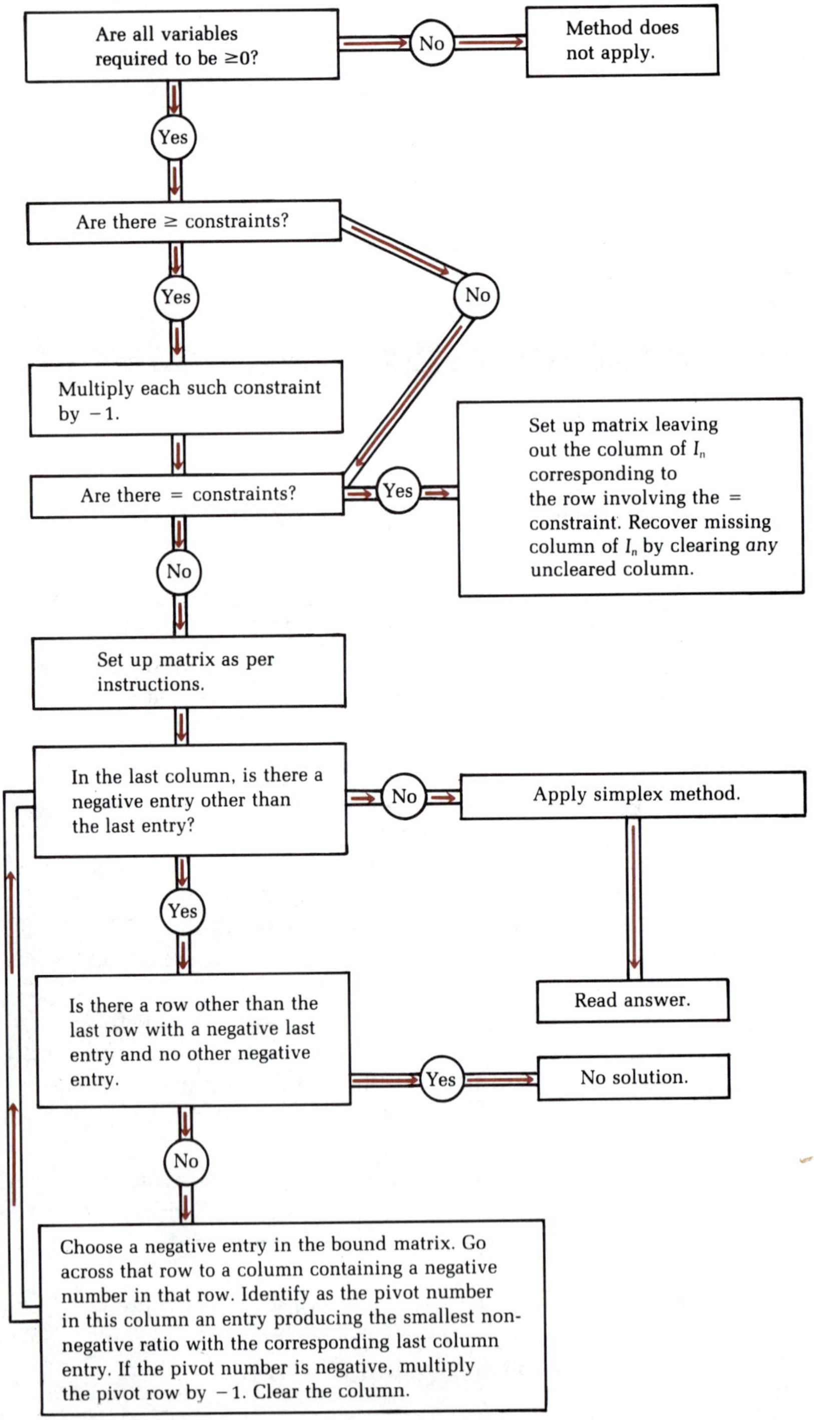
Are all variables required to be ≥0?
No
Method does not apply.
Yes
Are there ≥ constraints?
No
Yes
Multiply each such constraint by −1.
Are there = constraints?
Yes
Set up matrix leaving out the column of I_n corresponding to the row involving the = constraint. Recover missing column of I_n by clearing *any* uncleared column.
No
Set up matrix as per instructions.
In the last column, is there a negative entry other than the last entry?
No
Apply simplex method.
Read answer.
Yes
Is there a row other than the last row with a negative last entry and no other negative entry.
Yes
No solution.
No
Choose a negative entry in the bound matrix. Go across that row to a column containing a negative number in that row. Identify as the pivot number in this column an entry producing the smallest non-negative ratio with the corresponding last column entry. If the pivot number is negative, multiply the pivot row by −1. Clear the column.

Points to Remember

1. If the pivot number is negative, change the sign of each entry in the pivot row.
2. You cannot change the signs in any row other than the pivot row. Since the purpose of this section is to eliminate negative entries from the last column, it is tempting to multiply rows by -1 at random. *This will not work! Don't do it!* The fact that you may sometimes get a correct answer only compounds this type of error.
3. The pivot number is always a column entry that gives the least non-negative ratio in the column being cleared. If an entry produces a zero ratio, this is always the least non-negative ratio. However, we would prefer to avoid pivot numbers with zero ratios since they may not improve the value of the objective variable. If possible, another column should be selected to clear. If this is not possible, the entry producing the zero ratio must be used.
4. When choices exist in choosing which column to clear, look ahead for possible difficulties. You may be able to save a step. If no choice is clear-cut, select the column with the largest last entry to clear first.

■ Example 16. Maximize $A = 2x + y$ subject to

$$\begin{cases} x \geq 0,\ y \geq 0 \\ x - y \leq 4. \\ x + y \geq 6 \end{cases}$$

NOTE: This is problem 4 in exercise set 3.2.

Solution:

$$\begin{bmatrix} 1 & -1 & 1 & 0 & 0 & 4 \\ -1 & \enclose{circle}{-1} & 0 & 1 & 0 & -6 \\ 2 & 1 & 0 & 0 & -1 & 0 \end{bmatrix}$$

Comments

Change $x + y \geq 6$ to $-x - y \leq -6$ and set up the matrix as per instructions. Next, go down the last column to (-6) and across to (-1) in column 2, and check ratios. There is only one candidate for a positive ratio. Circle the pivot number.

$$\begin{bmatrix} 1 & -1 & 1 & 0 & 0 & 4 \\ 1 & \enclose{circle}{1} & 0 & -1 & 0 & 6 \\ 2 & 1 & 0 & 0 & -1 & 0 \end{bmatrix}$$

Change signs in pivot row because pivot number was negative. Clear column 2.

$$\begin{bmatrix} 2 & 0 & 1 & -1 & 0 & 10 \\ 1 & 1 & 0 & -1 & 0 & 6 \\ 1 & 0 & 0 & 1 & -1 & -6 \end{bmatrix}$$

There is no negative entry in the last column, so we can apply the regular simplex method. Column 4 assures us that no solution exists, since there is no possible pivot number in column 4. ■

■ **Example 17.** Maximize $P = 3x + 5y$ subject to

$$\begin{cases} x \geq 0,\ y \geq 0 \\ -x + 3y \leq 16 \\ x + 4y \geq 26 \\ 2x + y \leq 24 \end{cases}.$$

Solution:

$$\begin{array}{r} R3 + 2R1 \\ R3 + 2R2 \\ \\ -3R3 + 2R4 \end{array} \begin{bmatrix} -1 & 3 & 1 & 0 & 0 & 0 & 16 \\ -1 & -4 & 0 & 1 & 0 & 0 & -26 \\ \textcircled{2} & 1 & 0 & 0 & 1 & 0 & 24 \\ 3 & 5 & 0 & 0 & 0 & -1 & 0 \end{bmatrix}$$

Comments

The (-26) in the last column indicates we are not "on the feasible region." Using Phase I, we are directed to go down the last column to -26 and across the second row to -1 (we could also have stopped in column 2); identify and circle the proper pivot number in column 1 to use for clearing column 1.

$$\begin{array}{r} R2 + R1 \\ \\ R2 + 7R3 \\ R2 + R4 \end{array} \begin{bmatrix} 0 & 7 & 2 & 0 & 1 & 0 & 56 \\ 0 & \textcircled{-7} & 0 & 2 & 1 & 0 & -28 \\ 2 & 1 & 0 & 0 & 1 & 0 & 24 \\ 0 & 7 & 0 & 0 & -3 & -2 & -72 \end{bmatrix}$$

Once again, go down the last column to -28 and across row 2 to -7. Identify the pivot number in column 2 and clear. Remember to change all signs in the pivot row, since the pivot number is negative.

$$\begin{bmatrix} 0 & 0 & 2 & \textcircled{2} & 2 & 0 & 28 \\ 0 & 7 & 0 & -2 & -1 & 0 & 28 \\ 14 & 0 & 0 & 2 & 8 & 0 & 140 \\ 0 & 0 & 0 & 2 & -2 & -2 & -100 \end{bmatrix}$$

There is no negative number in the last column. We are ready for regular simplex methods. Since there is a positive number in the last row, column 4 must be cleared.

$$\begin{array}{r} \\ R1 + R2 \\ -R1 + R3 \\ -R1 + R4 \end{array} \begin{bmatrix} 0 & 0 & 2 & 2 & 2 & 0 & 28 \\ 0 & 7 & 2 & 0 & 1 & 0 & 56 \\ 14 & 0 & -2 & 0 & 6 & 0 & 112 \\ 0 & 0 & -2 & 0 & -4 & -2 & -128 \end{bmatrix}$$

Read the answer.

$x = \dfrac{112}{14} = 8,$

$y = \dfrac{56}{7} = 8,$

$s_1 = s_3 = 0,\ s_2 = 14$, and

$P = \dfrac{-128}{-2} = 64.$ ■

If the four matrices in example 17 are read in the regular way, following along with figure 3.13 below, our first matrix put us at the point (0,0). This point is off the feasible region, as implied by a negative bound.

Our first move put us at the point (12,0), again off the feasible region, as indicated by a negative bound.

Our second move took us to the point (10,4), which is a corner of the feasible region. We can now employ the regular simplex method, since all bounds are positive.

Our third move took us to the best corner point (8,8), where we were able to read the objective equation's maximum value, as well as slack variable values.

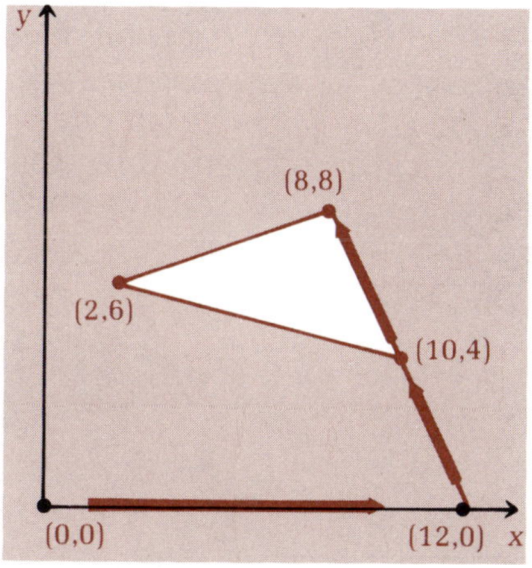

Figure 3.13

In a linear programming problem, if the feasible region is defined by a system of constraints containing one or more equations, the feasible region will remain a collection of points, usually infinite. The corner point theorem and the simplex method may still be used to evaluate the objective equation for maximum values.

In the next example we will state and solve a simple linear programming problem in two-space, first using geometry, then algebra, and finally using our more powerful simplex matrix tool.

■ Example 18. Maximize $P = 5x + 3y$ subject to $\begin{cases} x \geq 0,\ y \geq 0 \\ x + \ \ y \leq 8 \\ 2x + \ \ y \leq 12 \\ 4x + 3y = 26 \end{cases}$.

Solution:

1. Geometrically our feasible region defined by the inequalities

$$\begin{cases} x \geq 0,\ y \geq 0 \\ x + y \leq 8 \\ 2x + y \leq 12 \end{cases}$$ is the polygon *OADH*.

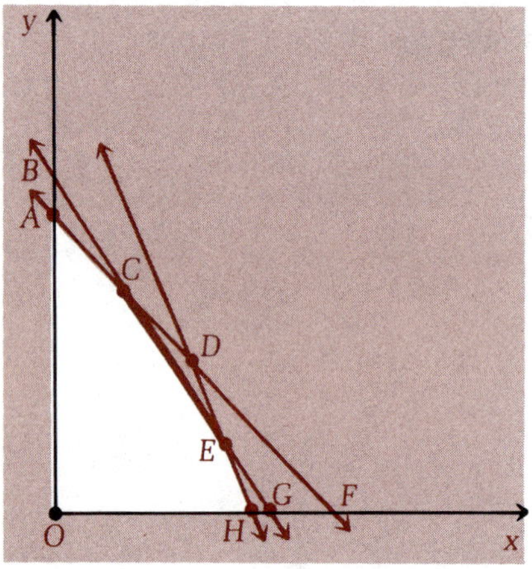

O(0,0) E(5,2)
A(0,8) F(8,0)
$B(0,\frac{26}{3})$ $G(\frac{13}{2},0)$
C(2,6) H(6,0)
D(4,4)

Figure 3.14

The additional constraint is the line $4x + 3y = 26$ which intersects this polygon in the line segment $\overline{CE}$. Hence our feasible region is the collection of points on $\overline{CE}$ and the end points, *C* and *E*, are the corner points of the region. At the corner points *C*(2,6) and *E*(5,2), *P* takes on values of 28 and 31 respectively and by the corner point theorem $P = 31$ is maximum.

2. Algebraically, solving the equation $4x + 3y = 26$ for *y*, we get $y = \dfrac{26 - 4x}{3}$. Then $x + y \leq 8$ becomes $x + \dfrac{26 - 4x}{3} \leq 8$ or $x \geq 2$. The constraint $2x + y \leq 12$ becomes $2x + \dfrac{26 - 4x}{3} \leq 12$ or $x \leq 5$. As a result, we have a set of x values where $2 \leq x \leq 5$. The end points are, of course, (2,6) and (5,2), and again by the corner point theorem we get a maximum of $P = 31$.

3. Now we present the matrix form with comments. Since there is an equality constraint, an adjustment in the set-up procedure is necessary, as indicated in the Phase I flow chart. There is no slack variable for the third constraint since it is already an equality. However, with three constraints, we should work with the identity matrix I_3. To resolve the situation, we leave out the third column (the equality constraint is the third constraint) of I_3 in the simplex matrix.

$$\begin{bmatrix} 1 & 1 & 1 & 0 & 0 & 8 \\ 2 & 1 & 0 & 1 & 0 & 12 \\ 4 & \textcircled{3} & 0 & 0 & 0 & 26 \\ 5 & 3 & 0 & 0 & -1 & 0 \end{bmatrix}$$

Comments
This corresponds to $x = 0$, $y = 0$, $s_1 = 8$, $s_2 = 12$, and $P = 0$.
The point (0,0) is not on the feasible region (see figure 3.14).

In order for the simplex method to work, we need the identity matrix I_n where n is the number of constraints. Our problem has three constraints, so we need I_3. The first step in solving the problem is to clear a column, other than the last column, to **recover the missing column** of I_3. There are only two columns (1 and 2) that are candidates in this example. We have chosen to use column 2, using the pivot number 3. Since we need the column $\begin{bmatrix} 0 \\ 0 \\ 1 \\ 0 \end{bmatrix}$, we must use row 3. Ratios are not considered when selecting the pivot number to recover a missing column of I_n.

$$\begin{matrix} -R3 + 3R1 \\ -R3 + 3R2 \\ \\ -R3 + R4 \end{matrix} \begin{bmatrix} \textcircled{-1} & 0 & 3 & 0 & 0 & -2 \\ 2 & 0 & 0 & 3 & 0 & 10 \\ 4 & 3 & 0 & 0 & 0 & 26 \\ 1 & 0 & 0 & 0 & -1 & -26 \end{bmatrix}$$

We read $x = 0$, $y = \frac{26}{3}$, $s_1 = -\frac{2}{3}$, $s_2 = \frac{10}{3}$, and $P = 26$.

The point $\left(0, \frac{26}{3}\right)$ (point B in figure 3.14) is not on the feasible region as is indicated by the -2 in the last column.

NOTE: Column 2 is not the third column of I_3. Multiplying row 3 by $\frac{1}{3}$ would correct this but would introduce unnecessary fractions. We are satisfied to have the non-zero entry in the right row (row 3), regardless of what the number is.

With a negative entry in the last column, we use Phase I and continue. The only negative entry in row 1 (the row with the negative entry in the last column) is in column 1, so this column needs to be cleared. The minimum positive ratio is obtained using the -1 in row 1, so -1 is the pivot number. Since the pivot number is negative, all signs in the pivot row are changed.

$$\begin{array}{r} -R1 \\ 2R1+R2 \\ 4R1+R3 \\ R1+R4 \end{array}\begin{bmatrix} 1 & 0 & -3 & 0 & 0 & 2 \\ 0 & 0 & \textcircled{6} & 3 & 0 & 6 \\ 0 & 3 & 12 & 0 & 0 & 18 \\ 0 & 0 & 3 & 0 & -1 & -28 \end{bmatrix}$$

This corresponds to $x = 2$, $y = 6$, $s_1 = 0$, $s_2 = 2$ and $P = 28$. The point (2,6) is on the feasible region (point C in figure 3.14) as indicated by all positive bounds in the last column.

All bounds are now positive, so the regular simplex method can be applied. Column 3 must be cleared since it has a positive entry in the last row. The smallest ratio indicates the new pivot number should be the 6 in row 2.

$$\begin{array}{r} R2+2R1 \\ \\ -2R2+R3 \\ -R2+2R4 \end{array}\begin{bmatrix} 2 & 0 & 0 & 3 & 0 & 10 \\ 0 & 0 & 6 & 3 & 0 & 6 \\ 0 & 3 & 0 & -6 & 0 & 6 \\ 0 & 0 & 0 & -3 & -2 & -62 \end{bmatrix}$$

Reading this matrix gives $x = 5$, $y = 2$, $s_1 = 1$, $s_2 = 0$ and $P = 31$.

This is the final matrix and gives the maximum value $P = 31$ when $x = 5$ and $y = 2$. This solution agrees with those obtained previously by geometric and algebraic techniques. ■

In the preceding example, the first two methods may seem simpler. However, as the size of our problems increase (more variables and/or more constraints) the first two methods become nightmares whereas the simplex method increases only moderately. In fact, the simplex method does not become more difficult, only longer. Two examples follow.

■ **Example 19.** Maximize $P = 7x - 2y + 3z + 4w$ subject to $x,y,z,w \geq 0$ and the following three constraints:

$$\begin{cases} 2x + y + z + w \leq 18 & (1) \\ x + y + 2z - w \geq 10. & (2) \\ x + y + z + w = 8 & (3) \end{cases}$$

Solutions:

$$\begin{bmatrix} 2 & 1 & 1 & 1 & 1 & 0 & 0 & 18 \\ -1 & -1 & -2 & 1 & 0 & 1 & 0 & -10 \\ 1 & 1 & 1 & \textcircled{1} & 0 & 0 & 0 & 8 \\ 7 & -2 & 3 & 4 & 0 & 0 & -1 & 0 \end{bmatrix}$$

Comments
The initial matrix is set up after multiplying constraint (2) by -1. Select any column, say column 4, circle the pivot number, and recover the missing column of I_3.

$$\begin{array}{r} -R3+R1 \\ -R3+R2 \\ \\ -4R3+R4 \end{array}\left[\begin{array}{cccccccc} 1 & 0 & 0 & 0 & 1 & 0 & 0 & 10 \\ -2 & -2 & -3 & 0 & 0 & 1 & 0 & -18 \\ \textcircled{1} & 1 & 1 & 1 & 0 & 0 & 0 & 8 \\ 3 & -6 & -1 & 0 & 0 & 0 & -1 & -32 \end{array}\right]$$

The −18 in the last column implies that an adjustment is necessary. Apply Phase I. Each of the first three columns has a negative entry in row 2. Column 1 is chosen here in order to get rid of the plus 3 in the last row. This may save a step.

$$\begin{array}{r} -R3+R1 \\ 2R3+R2 \\ \\ -3R3+R4 \end{array}\left[\begin{array}{cccccccc} 0 & -1 & -1 & -1 & 1 & 0 & 0 & 2 \\ 0 & 0 & \textcircled{-1} & 2 & 0 & 1 & 0 & -2 \\ 1 & 1 & 1 & 1 & 0 & 0 & 0 & 8 \\ 0 & -9 & -4 & -3 & 0 & 0 & -1 & -56 \end{array}\right]$$

With a −2 in the last column, Phase I is applied again. There is no choice for a pivot number since the only negative entry in row 2 is in column 3, and this entry gives the smallest positive ratio. Change the signs in the pivot row and clear column 3.

$$\begin{array}{r} -R2+R1 \\ -R2 \\ R2+R3 \\ -4R2+R4 \end{array}\left[\begin{array}{cccccccc} 0 & -1 & 0 & -3 & 1 & -1 & 0 & 4 \\ 0 & 0 & 1 & -2 & 0 & -1 & 0 & 2 \\ 1 & 1 & 0 & 3 & 0 & 1 & 0 & 6 \\ 0 & -9 & 0 & -11 & 0 & -4 & -1 & -48 \end{array}\right]$$

We are finished. Read the simplex matrix to obtain $x = 6, y = 0, z = 2, w = 0, s_1 = 4, s_2 = 0$, and $P = 48$.

Arithmetic check of original problem

$P = 7x - 2y + 3z + 4w$ — $7(6) - 2(0) + 3(2) + 4(0) = 48$

1. $2x + y + z + w \leq 18$ — $2(6) + 0 + (2) + 0 = 14 \leq 18$; $s_1 = 4$

2. $x + y + 2z - w \geq 10$ — $(6) + 0 + 2(2) - 0 = 10 \geq 10$; $s_2 = 0$

3. $x + y + z + w = 8$ — $6 + 0 + 2 + 0 = 8$. ■

■ Example 20. Maximize $P = 4x_1 + 3x_2 + x_3 + x_4 + x_5$ subject to $x_1, x_2, x_3, x_4, x_5 \geq 0$

$$\text{and}\begin{cases} 3x_1 - 2x_2 + x_3 + x_4 - 5x_5 \leq 12 \\ x_2 + 3x_3 + x_4 \geq 16 \\ x_1 + x_2 + x_3 = 8 \\ x_3 + x_4 + x_5 = 10 \end{cases}$$

Solution: The arithmetic check column has been included here to illustrate that it also works for simplex problems.

$$\left[\begin{array}{ccccccccc} 3 & -2 & 1 & 1 & -5 & 1 & 0 & 0 & 12 \\ 0 & -1 & -3 & -1 & 0 & 0 & 1 & 0 & -16 \\ 1 & 1 & 1 & 0 & 0 & 0 & 0 & 0 & 8 \\ 0 & 0 & 1 & 1 & \textcircled{1} & 0 & 0 & 0 & 10 \\ 4 & 3 & 1 & 1 & 1 & 0 & 0 & -1 & 0 \end{array}\right] \begin{array}{c} 11 \\ -20 \\ 11 \\ 13 \\ 9 \end{array} \quad \text{recover} \quad \left[\begin{array}{c} 0 \\ 0 \\ 0 \\ 1 \\ 0 \end{array}\right]$$

$$\begin{array}{r} 5R4 + R1 \\ \\ \\ \\ -R4 + R5 \end{array} \left[\begin{array}{ccccccccc} 3 & -2 & 6 & 6 & 0 & 1 & 0 & 0 & 62 \\ 0 & -1 & -3 & -1 & 0 & 0 & 1 & 0 & -16 \\ 1 & \textcircled{1} & 1 & 0 & 0 & 0 & 0 & 0 & 8 \\ 0 & 0 & 1 & 1 & 1 & 0 & 0 & 0 & 10 \\ 4 & 3 & 0 & 0 & 0 & 0 & 0 & -1 & -10 \end{array}\right] \begin{array}{c} 76 \\ -20 \\ 11 \\ 13 \\ -4 \end{array} \quad \text{recover} \quad \left[\begin{array}{c} 0 \\ 0 \\ 1 \\ 0 \\ 0 \end{array}\right]$$

$$\begin{array}{r} 2R3 + R1 \\ R3 + R2 \\ \\ \\ -3R3 + R5 \end{array} \left[\begin{array}{ccccccccc} 5 & 0 & 8 & 6 & 0 & 1 & 0 & 0 & 78 \\ 1 & 0 & -2 & \textcircled{-1} & 0 & 0 & 1 & 0 & -8 \\ 1 & 1 & 1 & 0 & 0 & 0 & 0 & 0 & 8 \\ 0 & 0 & 1 & 1 & 1 & 0 & 0 & 0 & 10 \\ 1 & 0 & -3 & 0 & 0 & 0 & 0 & -1 & -34 \end{array}\right] \begin{array}{c} 98 \\ -9 \\ 11 \\ 13 \\ -37 \end{array} \quad \text{apply Phase I}$$

$$\begin{array}{r} 6R2 + R1 \\ -R2 \\ \\ R2 + R4 \\ \\ \end{array} \left[\begin{array}{ccccccccc} 11 & 0 & -4 & 0 & 0 & 1 & 6 & 0 & 30 \\ -1 & 0 & 2 & 1 & 0 & 0 & -1 & 0 & 8 \\ 1 & 1 & 1 & 0 & 0 & 0 & 0 & 0 & 8 \\ \textcircled{1} & 0 & -1 & 0 & 1 & 0 & 1 & 0 & 2 \\ 1 & 0 & -3 & 0 & 0 & 0 & 0 & -1 & -34 \end{array}\right] \begin{array}{c} 44 \\ 9 \\ 11 \\ 4 \\ -37 \end{array} \quad \text{apply the simplex method.}$$

$$\begin{array}{r} -11R4 + R1 \\ R4 + R2 \\ -R4 + R3 \\ \\ -R4 + R5 \end{array} \left[\begin{array}{ccccccccc} 0 & 0 & 7 & 0 & -11 & 1 & -5 & 0 & 8 \\ 0 & 0 & 1 & 1 & 1 & 0 & 0 & 0 & 10 \\ 0 & 1 & 2 & 0 & -1 & 0 & -1 & 0 & 6 \\ 1 & 0 & -1 & 0 & 1 & 0 & 1 & 0 & 2 \\ 0 & 0 & -2 & 0 & -1 & 0 & -1 & -1 & -36 \end{array}\right] \begin{array}{c} 0 \\ 13 \\ 7 \\ 4 \\ -41 \end{array}$$

Read the answer. $x_1 = 2$, $x_2 = 6$, $x_3 = 0$, $x_4 = 10$, $x_5 = 0$, $s_1 = 8$, $s_2 = 0$, and $P = 36$. ■

Exercise Set 3.5

1. Maximize $R = 4x + 5y$

$$\text{subject to} \begin{cases} x \geq 0,\ y \geq 0 \\ x + 2y \leq 8\,. \\ 2x - y \geq 6 \end{cases}$$

2. Maximize $P = 3x - 2y$

$$\text{subject to} \begin{cases} x \geq 0,\ y \geq 0 \\ 2x + y \geq 8 \\ x \leq 10 \\ y - x \leq 4 \end{cases}$$

3. Maximize $P = 4x + 3y$

$$\text{subject to} \begin{cases} x \geq 0,\ y \geq 0 \\ x + y \geq 3 \\ 2x + y \geq 4 \\ 2x + 5y \leq 10 \end{cases}.$$

4. Maximize $C = 2x + 3y$

$$\text{subject to} \begin{cases} x \geq 0,\ y \geq 0 \\ 9x + 5y \geq 45 \\ 2x + 6y \geq 12 \\ 2x - y \leq 4 \end{cases}.$$

5. Maximize $P = 4x + 5y + 12$

$$\text{subject to} \begin{cases} x \geq 0,\ y \geq 0 \\ -x + y \leq 2 \\ 3x + 2y \leq 14 \\ x + y \geq 8 \end{cases}.$$

6. Maximize $P = 4x + 5y + 12$

$$\text{subject to} \begin{cases} x \geq 0,\ y \geq 0 \\ -x + y \leq 2 \\ 3x + 2y \leq 24 \\ x + y \geq 6 \end{cases}.$$

7. A major food processing company has been contracted by the government to prepare food packets for space flights. Each packet must weigh exactly 16 ounces, must contain at least 12 units of iron and 32 units of protein, and must not contain more than 42 units of fat. The company has two food types available to mix for the packets. Type *A* contains 3 units of iron, 1 unit of protein, and 2 units of fat per ounce. Type *B* contains 1 unit of iron, 5 units of protein, and 3 units of fat per ounce. The company's profit will be 3¢ per ounce of type *A* and 5¢ per ounce of type *B* in the packets.

a. How many ounces of each should be in the packets to maximize profit?

b. What is this profit per packet?

8. The government has decided to reduce the allowable fat content for the food packets (see problem 7) to no more than 38 units. In addition, because of a shortage of some foods, the company profit for type *B* food has decreased to 2¢ per ounce. They cannot increase prices because of contract specifications. How should the packets be made to maximize profit, and what will the profit per packet be?

9. Recover the missing column or columns of I_n in each of the following simplex matrices.

$$\begin{bmatrix} 2 & 3 & 4 & 1 & 0 & 0 & 5 \\ 1 & -2 & 3 & 0 & 1 & 0 & 4 \\ 3 & 3 & 2 & 0 & 0 & 0 & 6 \\ 2 & 3 & 4 & 0 & 0 & -1 & 0 \end{bmatrix} \quad \begin{bmatrix} 2 & 3 & 1 & 0 & 0 & 5 \\ 4 & 2 & 0 & 1 & 0 & 8 \\ 3 & -1 & 0 & 0 & 0 & 9 \\ 1 & 1 & 0 & 0 & -1 & 0 \end{bmatrix}$$

10. Maximize $P = 3x + 2y$

$$\text{subject to} \begin{cases} x \geq 0,\ y \geq 0 \\ x + 4y \geq 8 \\ x + y = 5 \\ 2x + y \leq 7 \end{cases}$$

11. Maximize $R = 4x + 3y$

$$\text{subject to} \begin{cases} x \geq 0,\ y \geq 0 \\ x + y = 3 \\ 2x + 6y \leq 6 \\ 3x - y \leq 14 \end{cases}$$

12. Maximize $P = x + 8y + 5z$ subject to

$$\begin{cases} x \geq 0,\ y \geq 0,\ z \geq 0 \\ x + y + z \leq 8 \\ -x + 2y + z \leq 2 \\ 3x - 3y = 4 \end{cases}.$$

13. Maximize $P = 5x + 10y$ subject to

$$\begin{cases} x \geq 0,\ y \geq 0 \\ x + y \leq 20 \\ 2x - y \geq 10 \\ 3x - 2y = 10 \\ -x + 2y = 10 \end{cases}.$$

3.6 Minimization

In any finite collection of numbers, it is possible to find the minimum value by changing the signs of all the numbers and selecting the maximum. For example, if we wish to find the smallest number in the set $A = \{-2,-1,5,8,12\}$ we could select the largest entry of the set $B = \{2,1,-5,-8,-12\}$. Since 2 is the largest entry of B, -2 is the smallest entry of A.

This principle and the corner point theorem can be used with the simplex method to minimize objective equations. If P is the objective equation and $A,B,C,\ldots,N$ are the corner points of the feasible region, a set of numbers $X = \{P_A,P_B,P_C,\ldots,P_N\}$ can be generated where $P_A,P_B,P_C,\ldots,P_N$ are the values of P at the corner points $A,B,C,\ldots,N$ respectively. The minimum value in X will be the negative of the maximum value in $Y = \{-P_A,-P_B,-P_C,\ldots,-P_N\}$. That is, if $-P_K$ is the maximum value in the set Y, $-(-P_K) = P_K$ is the minimum value in X.

In order to minimize P, we need only maximize $-P$. If $P = a_1x_1 + a_2x_2 + \ldots + a_nx_n$ is the objective equation, multiply both sides by (-1) to get $-P = -a_1x_1 - a_2x_2 - \ldots - a_nx_n$. For a maximization problem, the objective equation $P = a_1x_1 + a_2x_2 + \ldots + a_nx_n$ is rewritten $a_1x_1 + a_2x_2 + \ldots + a_nx_n - P = 0$. This equation becomes the last row of the simplex matrix.

For a minimization problem, $P = a_1x_1 + a_2x_2 + \ldots + a_nx_n$ is first rewritten as $-P = -a_1x_1 - a_2x_2 - \ldots - a_nx_n$ and then as $-a_1x_1 - a_2x_2 - \ldots - a_nx_n + P = 0$. This final equation is the last row in the simplex matrix for minimization. Note here that the entry in the last row corresponding to P is $+1$ instead of -1. In the final matrix, the last entry in the last column will also be positive, so the value of P will still correspond to a positive number.

In summary if we wish to minimize an objective equation, the following step is added at the beginning of the flow chart for Phase I.

Multiply the objective equation by -1 and proceed as usual.

■ Example 21. Minimize and maximize $C = 3x + 2y$ subject to

$$\begin{cases} x \geq 0, y \geq 0 \\ x + y \leq 1 \\ 5y - x \geq 7 \\ 5x - y \geq 13 \end{cases}.$$

(SET-UP)

Minimize

$$\begin{bmatrix} 1 & 1 & 1 & 0 & 0 & 0 & 11 \\ 1 & -5 & 0 & 1 & 0 & 0 & -7 \\ -5 & 1 & 0 & 0 & 1 & 0 & -13 \\ -3 & -2 & 0 & 0 & 0 & 1 & 0 \end{bmatrix}$$

Maximize

$$\begin{bmatrix} 1 & 1 & 1 & 0 & 0 & 0 & 11 \\ 1 & -5 & 0 & 1 & 0 & 0 & -7 \\ -5 & 1 & 0 & 0 & 1 & 0 & -13 \\ 3 & 2 & 0 & 0 & 0 & -1 & 0 \end{bmatrix}$$

note difference

PHASE I

Minimize:

$$\begin{bmatrix} 6 & 0 & 5 & 1 & 0 & 0 & 48 \\ -1 & 5 & 0 & -1 & 0 & 0 & 7 \\ -24 & 0 & 0 & 1 & 5 & 0 & -72 \\ -17 & 0 & 0 & -2 & 0 & 5 & 14 \end{bmatrix}$$

Maximize:

$$\begin{bmatrix} 6 & 0 & 5 & 1 & 0 & 0 & 48 \\ -1 & 5 & 0 & -1 & 0 & 0 & 7 \\ -24 & 0 & 0 & 1 & 5 & 0 & -72 \\ 17 & 0 & 0 & 2 & 0 & -5 & -14 \end{bmatrix}$$

PHASE I

Minimize:

$$\begin{bmatrix} 0 & 0 & 20 & 5 & 5 & 0 & 120 \\ 0 & 120 & 0 & -25 & -5 & 0 & 240 \\ 24 & 0 & 0 & -1 & -5 & 0 & 72 \\ 0 & 0 & 0 & -65 & -85 & 120 & 1560 \end{bmatrix}$$

Maximize:

$$\begin{bmatrix} 0 & 0 & 20 & 5 & 5 & 0 & 120 \\ 0 & 120 & 0 & -25 & -5 & 0 & 240 \\ 24 & 0 & 0 & -1 & -5 & 0 & 72 \\ 0 & 0 & 0 & 65 & 85 & 120 & 1560 \end{bmatrix}$$

Answer (Minimize)

$x = 3$ $\quad s_1 = 6$

$y = 13$ $\quad s_2 = s_3 = 0$ $\quad C = 13$

SIMPLEX

$$\begin{bmatrix} 0 & 0 & 20 & 5 & 5 & 0 & 120 \\ 0 & 120 & 20 & -20 & 0 & 0 & 360 \\ 24 & 0 & 20 & 4 & 0 & 0 & 192 \\ 0 & 0 & -340 & -20 & 0 & -120 & -3600 \end{bmatrix}$$

Answer (Maximize)

$x = 8$ $\quad s_1 = s_2 = 0$

$y = 3$ $\quad s_3 = 24$ $\quad C = 30$ ■

■ **Example 22.** Minimize $P = 8 + 2x - y$ subject to

$$\begin{cases} x \geq 0,\ y \geq 0 \\ x + y \leq 10 \\ x \leq 6 \\ y \leq 4 \end{cases}.$$

Solution:

Comment

1. $\begin{bmatrix} 1 & 1 & 1 & 0 & 0 & 0 & 10 \\ 1 & 0 & 0 & 1 & 0 & 0 & 6 \\ 0 & \textcircled{1} & 0 & 0 & 1 & 0 & 4 \\ -2 & 1 & 0 & 0 & 0 & 1 & +8 \end{bmatrix}$

1. The last row has been replaced by $-P$, i.e., for maximum we would have used $2x - y - P = -8$. We now use $-2x + y + P = +8$.

2. $\begin{bmatrix} 1 & 0 & 1 & 0 & -1 & 0 & 6 \\ 1 & 0 & 0 & 1 & 0 & 0 & 6 \\ 0 & 1 & 0 & 0 & 1 & 0 & 4 \\ -2 & 0 & 0 & 0 & -1 & 1 & 4 \end{bmatrix}$

2. Apply the simplex method and read the answer:

$x = 0$, $y = 4$, $s_1 = 6$, $s_2 = 6$, $s_3 = 0$, and $P = 4$. ■

NOTE: Notice the positive entries in the last two columns of the last row. In minimization, these *need not be negative* to assure us that we have reached an optimal solution.

■ **Example 23.** Minimize $P = 4x_1 + 3x_2 + x_3 + x_4 + x_5$ subject to each $x_i \geq 0$ and

$$\begin{cases} 3x_1 - 2x_2 + x_3 + x_4 - 5x_5 \leq 12 \\ x_2 + 3x_3 + x_4 \geq 16 \\ x_1 + x_2 + x_3 = 8 \\ x_3 + x_4 + x_5 = 10 \end{cases}.$$

Solution:

$$\left[\begin{array}{cccccccc|c} 3 & -2 & 1 & 1 & -5 & 1 & 0 & 0 & 12 \\ 0 & -1 & -3 & -1 & 0 & 0 & 1 & 0 & -16 \\ 1 & 1 & 1 & 0 & 0 & 0 & 0 & 0 & 8 \\ 0 & 0 & 1 & 1 & \textcircled{1} & 0 & 0 & 0 & 10 \\ -4 & -3 & -1 & -1 & -1 & 0 & 0 & 1 & 0 \end{array}\right] \begin{array}{c} 11 \\ -20 \\ 11 \\ 13 \\ -9 \end{array} \quad \text{Recover} \begin{bmatrix} 0 \\ 0 \\ 0 \\ 1 \\ 0 \end{bmatrix}$$

Column 5 contains the most zeros.

$$\begin{array}{r} 5R4 + R1 \\ \\ \\ \\ R4 + R5 \end{array} \left[\begin{array}{cccccccc|c} 3 & -2 & 6 & 6 & 0 & 1 & 0 & 0 & 62 \\ 0 & -1 & -3 & -1 & 0 & 0 & 1 & 0 & -16 \\ 1 & 1 & \textcircled{1} & 0 & 0 & 0 & 0 & 0 & 8 \\ 0 & 0 & 1 & 1 & 1 & 0 & 0 & 0 & 10 \\ -4 & -3 & 0 & 0 & 0 & 0 & 0 & 1 & 10 \end{array}\right] \begin{array}{c} 76 \\ -20 \\ 11 \\ 13 \\ 4 \end{array} \quad \text{Recover} \begin{bmatrix} 0 \\ 0 \\ 1 \\ 0 \\ 0 \end{bmatrix}$$

Choose column 3 to remove negative number in last column.

$$\begin{array}{r} -6R3 + R1 \\ 3R3 + R2 \\ \\ -R3 + R4 \\ \\ \end{array} \begin{bmatrix} -3 & -8 & 0 & 6 & 0 & 1 & 0 & 0 & 14 \\ 3 & 2 & 0 & -1 & 0 & 0 & 1 & 0 & 8 \\ 1 & 1 & 1 & 0 & 0 & 0 & 0 & 0 & 8 \\ -1 & -1 & 0 & 1 & 1 & 0 & 0 & 0 & 2 \\ -4 & -3 & 0 & 0 & 0 & 0 & 0 & 1 & 10 \end{bmatrix} \begin{array}{l} 10 \\ 13 \\ 11 \\ 2 \\ 4 \end{array}$$

Since the last column contains no negatives and the last row contains no positives except the last two, we are finished.

Reading the answer from this matrix gives

$x_1 = 0, x_2 = 0, x_3 = 8, x_4 = 0, x_5 = 2,$

$s_1 = 14, s_2 = 8,$ and $P = 10.$

Checking this answer gives
$P = 4(0) + 3(0) + 8 + 0 + 2 = 10$

1. $0 - 0 + 8 + 0 - 10 = -2 \leq 12$ and $12 - (-2) = 14 = s_1$
2. $0 + 24 + 0 = 24 \geq 16$ and $24 - 16 = 8 = s_1$
3. $0 + 0 + 8 = 8$
4. $8 + 0 + 2 = 10.$ ■

Exercise Set 3.6

1. Maximize $P = 2x + y$ and minimize $C = 40 - x - 2y$

$$\text{subject to} \begin{cases} x \geq 0, y \geq 0 \\ -2x + 3y \geq 1 \\ y \leq 5x - 4 \\ 3x + 2y \leq 18 \end{cases}$$

2. Maximize $Z = 3x + 7y$

$$\text{subject to} \begin{cases} 2x + 3y \leq 400 \\ x + y = 150 \\ x \geq 0, y \geq 0 \end{cases}$$

3. Minimize $R = 2x + 3y$

$$\text{subject to} \begin{cases} x + 4y \geq 8 \\ x + y = 5 \\ 2x + y \geq 7 \end{cases}$$

4. Maximize $M = 4x + 3y$

$$\text{subject to} \begin{cases} x + y \leq 5 \\ 2x - 3y = 12 \\ x - y \geq 1 \\ x \geq 0, y \geq 0 \end{cases}$$

5. Maximize $M = 4x + 3y$

$$\text{subject to} \begin{cases} x + y \le 5 \\ 2x - 3y = 6 \\ x - y \ge 1 \\ x \ge 0, y \ge 0 \end{cases}$$

6. Find the minimum cost to produce a daily meal for the dogs at the city pound if the state requires a minimum of 50 lbs. of carbohydrates, a minimum of 90 lbs. of meat protein but allows only a maximum of 40 lbs. of soybean meal protein for the number of dogs in residence. The dog catcher can select from three types of state approved dog food which have the following nutritional composition.

	Carbohydrates	*Meat*	*Soybean*	*Filler*
Dog food type I	20%	30%	40%	10%
Dog food type II	30%	10%	30%	30%
Dog food type III	40%	20%	30%	10%

Each dog food type is packed in 10 lb. bags for easy handling and cost 80¢, 68¢, and 72¢ per package respectively.

7. Minimize $T = 8x + 12y$

$$\text{subject to} \begin{cases} x \ge 0, y \ge 0 \\ x + y \ge 3 \\ 2x + 6y \ge 8 \\ 3x + 3y \ge 8 \end{cases}$$

8. Make up a word problem to fit the numbers in problem 7.
9. A chemical company needs to package two caustic compounds, 3RG2 and 6BR7. One pound of 3RG2 contains 1 thorm of BY5, 2 glubs of Nh3R, 3 ounces tryzine and 1 klute of N4/3G whereas one pound of 6BR7 contains 1 thorm BY5, 6 glubs of Nh3R, 3 ounces tryzine and 5 klutes of N4/3G.

   The compounds must be packaged separately and mixed at the job site in liquid form. For safe handling, each pound of 3RG2 must be mixed with 8 quarts of water, and each pound of 6BR7, being stronger, requires 12 quarts of water for dilution. How should the company package these two compounds to minimize the amount of water to be used if the final mix must equal or exceed 3 thorms of BY5, 8 glubs of Nh3R, and 8 ounces of tryzine and must contain exactly 5 klutes of N4/3G.
10. A company has two warehouses, *A* and *B*. Warehouse *A* has 25 golf carts, and *B* has 35 golf carts. Southern Country Club orders 20 carts, and Bogey Hills orders 30. The shipping cost from *A* to Southern Country Club is \$3 per cart and to Bogey Hills is \$5 per cart. It costs

$5 per cart to ship from *B* to Southern and $6 per cart from *B* to Bogey Hills. How many carts should be shipped from each warehouse to each golf course so that the total shipping cost is a minimum?

11. Three vice-presidents are able to complete certain jobs at different rates according to the following table.

	Answer Letters	*Make Cost Estimate*	*Make Telephone Calls*	*Overtime Cost per Hour*
Vice-president I	2	3	4	$30
Vice-president II	3	1	3	$40
Vice-president III	4	0	4	$40

That is, vice-president I can, in one hour, answer 2 letters, make 3 cost estimates, and complete 4 telephone calls, and so on. Work is piling up, and at least 12 letters, 8 cost estimates, and 24 phone calls must be made in overtime. Who should be given the overtime, how much should they be given, and what will be the total cost to the company?

12. Work problem 11 if the overtime requirement were changed to read "work is piling up, and at least 20 letters, 6 cost estimates, and 28 phone calls need to be made."

3.7 Minimization and the Dual

In the preceding section we looked at minimization in terms of maximizing the negative of the objective function. There is an alternate procedure that can be used for some minimization problems. This technique is particularly useful for simplex solutions of game theory problems discussed in chapter 9 (section 9.3).

The technique involves the use of the **dual** for minimization problems in *standard form*. Recall that a maximization problem in standard form is one in which all variables must be ≥ 0, all constraints must involve $\leq$, and all bounds are ≥ 0. A minimization problem is in standard form if all constraints involve $\geq$.

Consider a maximization problem in standard form. We can use a technique similar to that used in section 2.4 for systems of equations to write the problem in matrix form. To illustrate, consider the following example.

■ Example 24 Write the following linear programming problems in matrix form.

a. Maximize $P = 4x + 2y$ subject to

$$\begin{cases} x \geq 0,\ y \geq 0 \\ 5x + 2y \leq 30 \\ 3x + 4y \leq 32 \end{cases}.$$

b. Maximize $P = x + y + z$ subject to

$$\begin{cases} x \geq 0,\ y \geq 0,\ z \geq 0 \\ 6x + y + 4z \leq 48 \\ -2x + 2y + z \leq 12 \\ 12x + 16y + z \leq 96 \\ 3x + 4y + 2z \leq 31 \end{cases}.$$

Solutions: **a.** Let $C = \begin{bmatrix} 5 & 2 \\ 3 & 4 \end{bmatrix}$ be the constraint matrix,

let $X = \begin{bmatrix} x \\ y \end{bmatrix}$ be the variable matrix,

let $B = \begin{bmatrix} 30 \\ 32 \end{bmatrix}$ be the bound matrix,

let $A = [4 \quad 2]$ be the objective matrix, and

let $O = \begin{bmatrix} 0 \\ 0 \end{bmatrix}$ be the appropriate zero matrix. The problem can then be written as follows:

$$\text{Maximize } P = AX \text{ subject to } \begin{cases} X \geq O \\ CX \leq B \end{cases}.$$

b. Let $C = \begin{bmatrix} 6 & 1 & 4 \\ -2 & 2 & 1 \\ 12 & 16 & 1 \\ 3 & 4 & 2 \end{bmatrix}$, $X = \begin{bmatrix} x \\ y \\ z \end{bmatrix}$, $B = \begin{bmatrix} 48 \\ 12 \\ 96 \\ 31 \end{bmatrix}$,

$A = [1 \quad 1 \quad 1]$, and $O = \begin{bmatrix} 0 \\ 0 \\ 0 \end{bmatrix}$. The problem can then be written as follows:

$$\text{Maximize } P = AX \text{ subject to } \begin{cases} X \geq O \\ CX \leq B \end{cases}.$$

■

Minimization problems in standard form can also be written in matrix form. Consider the following illustration.

$$\text{Minimize } P = 30x + 32y \text{ subject to } \begin{cases} x \geq 0,\ y \geq 0 \\ 5x + 3y \geq 4 \\ 2x + 4y \geq 2 \end{cases}.$$

Letting $C = \begin{bmatrix} 5 & 3 \\ 2 & 4 \end{bmatrix}$, $X = \begin{bmatrix} x \\ y \end{bmatrix}$, $B = \begin{bmatrix} 4 \\ 2 \end{bmatrix}$, $A = [30 \quad 32]$, and $O = \begin{bmatrix} 0 \\ 0 \end{bmatrix}$, we have the following:

$$\text{Minimize } P = AX \text{ subject to } \begin{cases} X \geq O \\ CX \geq B \end{cases}.$$

CAUTION: Since the matrix form of the constraints involves only one symbol ($\leq$ or $\geq$), we cannot have mixed constraints. All constraints must be of the same form. This may require that some constraints be multiplied by -1 to reverse the inequality. We will not consider any equality constraints.

We are now ready to consider the **dual.** Corresponding to any linear programming problem in standard form is a linear programming problem called the dual. The dual of a maximization problem is a minimization problem, and the dual of a minimization problem is a maximization problem.

It is this latter situation that is of interest to us here. That is, we will solve a minimization problem by maximizing the dual. The techniques involved in the simplex method will still be used with two exceptions. The use of the dual will change the initial set-up of the simplex tableau and the way in which the solution is read.

We first consider what the dual of a minimization problem is and how the tableau is set up. Consider the following minimization problem in matrix form.

$$\text{Minimize } P = AX \text{ subject to } \begin{cases} X \geq O \\ CX \geq B \end{cases}$$

where C is $m \times n$, X is $n \times 1$, B is $m \times 1$, A is $1 \times n$, and O is $n \times 1$. The dual maximization problem is as follows:

$$\text{Maximize } R = B^TY \text{ subject to } \begin{cases} Y \geq O \\ C^TY \leq A^T \end{cases}$$

where C^T, B^T, and A^T are the transposes of the original matrices C, B, and A, respectively; Y is a variable matrix of size $n \times 1$; and O is the $m \times 1$ zero matrix.

The simplex tableau is set up according to the maximization problem, using the transpose of the original constraint matrix as the new constraint matrix, the transpose of the original bound matrix as the new objective matrix, and the transpose of the original objective matrix as the new bound matrix. Note here that the matrix Y is of size $m \times 1$, and the original variable matrix X is of size $n \times 1$. That is, X and Y may not have the same number of variables. This will not be a problem, since the solutions are read in a different way.

The solutions will appear at the bottom of the columns where the identity matrix was originally. If C is $m \times n$, then C^T is $n \times m$, and we will use the $n \times n$ identity matrix in the initial tableau. There are, then, n columns corresponding to the n original variables in X. We will now consider several examples to illustrate setting up the tableau and reading the solutions.

The tableau will be set up according to the following: Let C be the $m \times n$ constraint matrix, X be the $n \times 1$ variable matrix, B be the $m \times 1$ bound matrix, and A be the $1 \times n$ objective matrix. The initial tableau for the dual will be

$$\left[\begin{array}{c|c|c|c} C^T & I_n & \begin{matrix}0\\0\\\vdots\\0\end{matrix} & A^T \\ \hline B^T & 0\ \ 0 \ldots 0 & -1 & 0\end{array}\right]$$

We now consider how to read the solutions to the original problem when using the dual. The solutions for standard simplex problems were determined using the last column of the final matrix. That is, the optimal values for the initial variables, the slack or surplus variables, and the objective variable were quotients of last column values with corresponding cleared column entries.

When the dual is used, the solutions are determined using the entries in the last row of the final matrix. The values of the surplus variables will be determined from the last row entries below the constraint matrix columns, the values of the initial variables are determined using the last row entries under the identity matrix, and the value of the objective variable is determined using the last row entry in the last column. The actual values of the variables are found by dividing the last row by the entry in the last row that is in the next-to-last column.

Suppose the last row of a minimization problem with three initial variables and three initial constraints is

$$[-8 \quad 0 \quad 0 \quad -16 \quad 0 \quad -8 \quad -4 \quad -24].$$

Since there were three initial variables and three initial constraints, the constraint matrix and its transpose are both 3×3. Hence the first three entries correspond to surplus variables and the next three to the initial variables. We divide by the next-to-last entry in this row to obtain the values of the variables

$$s_1 = \frac{-8}{-4} = 2, \quad s_2 = \frac{0}{-4} = 0, \quad s_3 = \frac{0}{-4} = 0,$$

$$x = \frac{-16}{-4} = 4, \quad y = \frac{0}{-4} = 0, \quad z = \frac{-8}{-4} = 2,$$

and $C = \dfrac{-24}{-4} = 6.$

The following example will provide additional illustrations of reading the answer from the final (dual) simplex matrix.

■ Example 25. Each of the following matrices is the final matrix in a simplex minimization problem using the dual. Read the solutions.

a.
$$\begin{bmatrix} 0 & 1 & 2 & 1 & -1 & 0 & 0 & 1 \\ 1 & 0 & 0 & -1 & 2 & 0 & 0 & 1 \\ 0 & 0 & 2 & -1 & -1 & 1 & 0 & 2 \\ 0 & 0 & -1 & -1 & -2 & 0 & -1 & -7 \end{bmatrix}$$

b.
$$\begin{bmatrix} 0 & 2 & 1 & -2 & -1 & 0 & 3 \\ 3 & -1 & 0 & 1 & 2 & 0 & 5 \\ 0 & -2 & 0 & -4 & -6 & -2 & -20 \end{bmatrix}$$

c.
$$\begin{bmatrix} 0 & 1 & 2 & 1 & -1 & 0 & 0 & 1 \\ 1 & 0 & 0 & -1 & 2 & 0 & 0 & 1 \\ 0 & 0 & 2 & -1 & -1 & 1 & 0 & 2 \\ 0 & 0 & -1 & -1 & -2 & 0 & -2 & -7 \end{bmatrix}$$

d.
$$\begin{bmatrix} 8 & 7 & 14 & 0 & 5 & 3 & 0 & 0 & 13 \\ 3 & 0 & 7 & 7 & 1 & 2 & 0 & 0 & 4 \\ -6 & 0 & -11 & 0 & -3 & -2 & 1 & 0 & 2 \\ -15 & 0 & -28 & 0 & -26 & -10 & 0 & -7 & -62 \end{bmatrix}$$

Solutions: Before determining solutions, we must determine how many initial variables and how many slack variables we have. This is done by determining the number of rows in the matrix. If the matrix has m rows, then the identity matrix used is $(m - 1) \times (m - 1)$. Then there are $m - 1$ initial variables. For this example, a, c, and d have three initial variables and b has two. We now look at the bottom row and identify the entries that will be used to determine the solutions.

a. Three initial variables:

$$\begin{bmatrix} 0 & 0 & -1 & -1 & -2 & 0 & \textcircled{-1} & -7 \end{bmatrix}$$
$$\quad s_1 \quad s_2 \quad s_3 \quad x \quad y \quad z \qquad\quad C$$

To find the values of the variables, divide each entry by -1, the next-to-last entry in the last row.

$$s_1 = s_2 = \frac{0}{-1} = 0, \qquad s_3 = \frac{-1}{-1} = 1,$$
$$x = \frac{-1}{-1} = 1, \qquad y = \frac{-2}{-1} = 2,$$
$$z = \frac{0}{-1} = 0, \qquad \text{and } C = \frac{-7}{-1} = 7.$$

Since the surplus variables are not our major interest here, we have as a solution $C = 7$ when $x = 1$, $y = 2$, and $z = 0$.

b. Two initial variables:

$$\begin{bmatrix} 0 & -2 & 0 & -4 & -6 & \textcircled{-2} & -20 \end{bmatrix}$$
$$\quad s_1 \quad s_2 \quad s_3 \quad x \quad y \qquad\quad C$$

The solution is

$$C = \frac{-20}{-2} = 10 \text{ when } x = \frac{-4}{-2} = 2, \qquad y = \frac{-6}{-2} = 3,$$
$$s_1 = s_3 = 0, \qquad \text{and } s_2 = \frac{-2}{-2} = 1.$$

c. Three initial variables:

$$\begin{bmatrix} 0 & 0 & -1 & -1 & -2 & 0 & \textcircled{-2} \end{bmatrix}$$
$$\quad s_1 \quad s_2 \quad s_3 \quad x \quad y \quad z$$

The solution is

$$C = \frac{-7}{-2} = \frac{7}{2} \text{ when } x = \frac{-1}{-2} = \frac{1}{2}, \quad y = \frac{-2}{-2} = 1, \text{ and } z = 0.$$

Note that $s_1 = s_2 = 0$ and $s_3 = \frac{1}{2}$.

d. Three initial variables:

$$\begin{bmatrix} -15 & 0 & -28 & 0 & -26 & -10 & 0 & \textcircled{-7} & -62 \end{bmatrix}$$
$$\quad s_1 \quad s_2 \quad s_3 \quad s_4 \quad x \quad y \quad z \qquad\quad C$$

The solution is

$$C = \frac{62}{7} \text{ when } x = \frac{26}{7}, \qquad y = \frac{10}{7}, \text{ and } z = 0.$$

Note that $s_1 = \frac{15}{7}$, $s_3 = \frac{28}{7} = 4$, and $s_2 = s_4 = 0$. ■

■ Example 26. Minimize $P = 6x + 8y$ subject to

$$\begin{cases} x, y \geq 0 \\ 3x + 2y \geq 12 \\ x + 2y \geq 4 \end{cases}.$$

Solution: $C = \begin{bmatrix} 3 & 2 \\ 1 & 2 \end{bmatrix}$, $X = \begin{bmatrix} x \\ y \end{bmatrix}$, $B = \begin{bmatrix} 12 \\ 4 \end{bmatrix}$, $A = [6 \quad 8]$, $C^T = \begin{bmatrix} 3 & 1 \\ 2 & 2 \end{bmatrix}$.

Initial Tableau	*Comments*
$C^T \quad B^T \quad I_2 \quad A^T$ $\begin{bmatrix} \textcircled{3} & 1 & 1 & 0 & 0 & 6 \\ 2 & 2 & 0 & 1 & 0 & 8 \\ 12 & 4 & 0 & 0 & -1 & 0 \end{bmatrix}$	Pick pivot number and clear column 1.
$\begin{bmatrix} 3 & 1 & 1 & 0 & 0 & 6 \\ 0 & 4 & -2 & 3 & 0 & 12 \\ 0 & 0 & -4 & 0 & -1 & -24 \end{bmatrix}$ $s_1 \quad s_2 \quad x \quad y \quad\quad P$	Since there are no positives in the last row, we are ready to read the answer.

To obtain x, y, and P, divide the entries in the last row by -1 (the entry in the next-to-last column) to get

$$x = \frac{-4}{-1} = 4, \; y = \frac{0}{-1} = 0, \; P = \frac{-24}{-1} = 24.$$ ■

NOTE: The matrix reduction is the same as for the simplex, using regular simplex or Phase I. The differences are in the set-up and reading of the solution.

It certainly is not necessary to write the constraint matrix and its transpose and the other matrices before setting up the simplex tableau. We can actually begin by writing the constraints as columns with the bounds in the last row. That is, the first constraint becomes the first column, the second constraint is the second column, and so on. We then fill in the identity matrix, the last two columns (the last column with the objective matrix), and the bottom row.

■ Example 27. Set up the simplex tableau to minimize $C = 9x + 10y$ subject to

$$\begin{cases} x \geq 0, y \geq 0 \\ 2x + 5y \geq 24 \\ 3x + 4y \geq 28 \end{cases}.$$

Solution: The first constraint becomes the first column, and the second constraint is the second column.

$$\begin{array}{c} \\ \\ \end{array}\qquad \overset{I_2}{} \qquad\qquad \text{Objective matrix} \downarrow$$

$$\left[\begin{array}{cc:cc:cc} 2 & 3 & 1 & 0 & 0 & 9 \\ 5 & 4 & 0 & 1 & 0 & 10 \\ 24 & 28 & 0 & 0 & -1 & 0 \end{array}\right]$$

↑ First constraint (first column); ↑ Second constraint (second column)

■

■ **Example 28.** Minimize $C = 9x + 10y$ subject to

$$\begin{cases} x \geq 0, y \geq 0 \\ 2x + 5y \geq 24 \\ 3x + 4y \geq 28 \end{cases}.$$

Solution:

$$\begin{bmatrix} 2 & 3 & 1 & 0 & 0 & 9 \\ 5 & \textcircled{4} & 0 & 1 & 0 & 10 \\ 24 & 28 & 0 & 0 & -1 & 0 \end{bmatrix}$$

$$\begin{bmatrix} -7 & 0 & 4 & -3 & 0 & 6 \\ 5 & 4 & 0 & 1 & 0 & 10 \\ -11 & 0 & 0 & -7 & -1 & -70 \end{bmatrix}$$
$$\quad s_1 \qquad s_2 \qquad x \qquad y \qquad\qquad\quad C$$

Hence $x = \dfrac{0}{-1} = 0$, $y = \dfrac{-7}{-1} = 7$, and $C = \dfrac{-70}{-1} = 70$. ■

■ **Example 29.** Minimize $C = 4x + 2y + 3z$ subject to

$$\begin{cases} x, y, z \geq 0 \\ 5x + y + 2z \geq 6 \\ 2x + y + 4z \geq 2 \\ 3x + 2y + z \geq 5 \end{cases}.$$

Solution:

$$\begin{bmatrix} \textcircled{5} & 2 & 3 & 1 & 0 & 0 & 0 & 4 \\ 1 & 1 & 2 & 0 & 1 & 0 & 0 & 2 \\ 2 & 4 & 1 & 0 & 0 & 1 & 0 & 3 \\ 6 & 2 & 5 & 0 & 0 & 0 & -1 & 0 \end{bmatrix}$$

$$\begin{bmatrix} 5 & 2 & 3 & 1 & 0 & 0 & 0 & 4 \\ 0 & 3 & 7 & -1 & 5 & 0 & 0 & 6 \\ 0 & 16 & -1 & -2 & 0 & 5 & 0 & 7 \\ 0 & -2 & 7 & -6 & 0 & 0 & -5 & -24 \end{bmatrix}$$

$$\begin{bmatrix} 35 & 5 & 0 & 10 & -15 & 0 & 0 & 10 \\ 0 & 3 & 7 & -1 & 5 & 0 & 0 & 6 \\ 0 & 115 & 0 & -15 & 5 & 35 & 0 & 55 \\ 0 & -5 & 0 & -5 & -5 & 0 & -5 & -30 \end{bmatrix}$$

$$\begin{matrix} s_1 & s_2 & s_3 & x & y & z & & C \end{matrix}$$

Then

$$x = \frac{-5}{-5} = 1,\ y = \frac{-5}{-5} = 1,\ z = \frac{0}{-5} = 0,\ \text{and } C = \frac{-30}{-5} = 6.$$

■

■ **Example 30.** Minimize $C = 3x + 2y$ subject to

$$\begin{cases} x \geq 0,\ y \geq 0 \\ x + y \leq 11 \\ -x + 5y \geq 7 \\ 5x - y \geq 13 \end{cases}.$$

Solution: To use the dual, we must have all constraints as $\geq$, so we first multiply constraint 1 $(x + y \leq 11)$ by -1 to obtain $-x - y \geq -11$ and proceed.

$$\begin{bmatrix} -1 & -1 & 5 & 1 & 0 & 0 & 3 \\ -1 & 5 & -1 & 0 & 1 & 0 & 2 \\ -11 & 7 & 13 & 0 & 0 & -1 & 0 \end{bmatrix}$$

$$\begin{bmatrix} -1 & -1 & 5 & 1 & 0 & 0 & 3 \\ -6 & 24 & 0 & 1 & 5 & 0 & 13 \\ -42 & 48 & 0 & -13 & 0 & -5 & -39 \end{bmatrix}$$

$$\begin{bmatrix} -30 & 0 & 120 & 25 & 5 & 0 & 85 \\ -6 & 24 & 0 & 1 & 5 & 0 & 13 \\ -30 & 0 & 0 & -15 & -10 & -5 & -65 \end{bmatrix}$$

$$\begin{matrix} s_1 & s_2 & s_3 & x & y & & C \end{matrix}$$

Then $x = \dfrac{-15}{-5} = 3,\ y = \dfrac{-10}{-5} = 2,$ and $C = \dfrac{-65}{-5} = 13.$

■

■ **Example 31.** Minimize $C = 6x - y$ subject to

$$\begin{cases} x \geq 0, y \geq 0 \\ x - y \leq 4 \\ x + y \geq 6 \\ x + 3y \leq 16 \end{cases}.$$

Solution: Since constraints 1 and 3 involve $\leq$, we first multiply both of them by -1 to get the new system of constraints

$$\begin{cases} -x + y \geq -4 \\ x + y \geq 6 \\ -x - 3y \geq -16 \end{cases}.$$

The initial tableau is

$$\begin{bmatrix} -1 & 1 & -1 & 1 & 0 & 0 & 6 \\ 1 & 1 & \textcircled{-3} & 0 & 1 & 0 & -1 \\ -4 & 6 & -16 & 0 & 0 & -1 & 0 \end{bmatrix}.$$

Since the last column contains a negative entry, we begin with Phase 1 and proceed. The appropriate pivot number has been circled.

$$\begin{bmatrix} -4 & \textcircled{2} & 0 & 3 & -1 & 0 & 19 \\ -1 & -1 & 3 & 0 & -1 & 0 & 1 \\ -28 & 2 & 0 & 0 & -16 & -3 & 16 \end{bmatrix}$$

$$\begin{bmatrix} -4 & 2 & 0 & 3 & -1 & 0 & 19 \\ -6 & 0 & 6 & 3 & -3 & 0 & 21 \\ -24 & 0 & 0 & -3 & -15 & -3 & -3 \end{bmatrix}$$

$$\quad s_1 \quad s_2 \quad s_3 \quad x \quad y \quad \quad C$$

Then $x = \dfrac{-3}{-3} = 1$, $y = \dfrac{-15}{-3} = 5$, and $C = \dfrac{-3}{-3} = 1$. ■

Exercise Set 3.7

In exercises 1–8, set up the initial simplex tableau of the dual.

1. Minimize $c = x + 2y$ subject to

$$\begin{cases} x \geq 0, y \geq 0 \\ 2x + 3y \geq 18 \\ 3x + 2y \geq 12 \end{cases}.$$

2. Minimize $c = 3x + 4y$ subject to

$$\begin{cases} x \geq 0, y \geq 0 \\ 3x + y \geq 9 \\ x + 2y \geq 8 \\ 3x + 3y \geq 13 \end{cases}.$$

3. Minimize $c = 3x + 2y + z$ subject to

$$\begin{cases} x \geq 0, y \geq 0, z \geq 0 \\ 3x + 2y - z \geq 11 \\ 2x + 3y + z \geq 15 \\ 3x + 2y + 2z \geq 18 \end{cases}.$$

4. Minimize $c = x + y + 2z$ subject to

$$\begin{cases} x \geq 0, y \geq 0, z \geq 0 \\ 3x + y + z \geq 10 \\ 2x + 2y - z \geq 10 \\ 3x + y + 2z \geq 15 \end{cases}.$$

5. Minimize $c = 2x + 3y$ subject to

$$\begin{cases} x \geq 0, y \geq 0 \\ 2x + 3y \geq 15 \\ 3x + 2y \geq 11 \\ x + y \leq 12 \end{cases}.$$

6. Minimize $c = x + 4y$ subject to

$$\begin{cases} x \geq 0, y \geq 0 \\ 2x + y \leq 18 \\ 3x + 3y \geq 10 \\ 4x + y \geq 9 \end{cases}.$$

7. Minimize $c = 3x + 2y - z$ subject to

$$\begin{cases} x \geq 0, y \geq 0, z \geq 0 \\ 2x + 3y + 3z \leq 30 \\ 3x + y + z \leq 28 \\ x + 2y + 4z \geq 12 \end{cases}.$$

8. Minimize $c = 2x - y + 3z$ subject to

$$\begin{cases} x \geq 0, y \geq 0, z \geq 0 \\ x + y + z \leq 25 \\ 3x + 2y + 2z \geq 60 \\ 2x + y - z \leq 27 \end{cases}.$$

In exercises 9–16, the simplex tableau is the final tableau in a simplex minimization problem using the dual. Determine the solution.

9. $$\begin{bmatrix} 0 & 1 & -1 & 1 & 0 & 3 \\ 2 & 0 & -2 & -1 & 0 & 8 \\ 0 & 0 & -1 & -1 & -1 & -3 \end{bmatrix}$$

10. $$\begin{bmatrix} 4 & 0 & 1 & 3 & 0 & 12 \\ 0 & 3 & -2 & -1 & 0 & 12 \\ 0 & 0 & -2 & -3 & -1 & -13 \end{bmatrix}$$

11. $$\begin{bmatrix} 1 & 0 & 3 & -1 & 2 & 0 & 11 \\ 2 & 4 & 0 & 2 & 2 & 0 & 15 \\ -3 & 0 & 0 & -8 & -12 & -4 & -52 \end{bmatrix}$$

12. $$\begin{bmatrix} 0 & -3 & 2 & 1 & -2 & 0 & 5 \\ 1 & 2 & 0 & -1 & -1 & 0 & 15 \\ 0 & -3 & 0 & -9 & -21 & -3 & -30 \end{bmatrix}$$

13.
$$\begin{bmatrix} 0 & 3 & 2 & 0 & -1 & 0 & 5 \\ 0 & 0 & -1 & 3 & 4 & 0 & 4 \\ 2 & 0 & 1 & 0 & 2 & 0 & 15 \\ 0 & 0 & -10 & 0 & -25 & -5 & -60 \end{bmatrix}$$

14.
$$\begin{bmatrix} 0 & 0 & 4 & -2 & 3 & 0 & 6 \\ 2 & 0 & 0 & 1 & 2 & 0 & 8 \\ 0 & 3 & 0 & -3 & -3 & 0 & 12 \\ 0 & 0 & 0 & -9 & -11 & -4 & -62 \end{bmatrix}$$

15.
$$\begin{bmatrix} 3 & 0 & 0 & 0 & -1 & 1 & -3 & -1 & 0 & 2 \\ 0 & 0 & 0 & 2 & 2 & 1 & 2 & -1 & 0 & 3 \\ 0 & 2 & 0 & 0 & 1 & -2 & 3 & 3 & 0 & 4 \\ 0 & 0 & 3 & 0 & -1 & -3 & -2 & -2 & 0 & 1 \\ 0 & 0 & 0 & 0 & -15 & -10 & -25 & -5 & -5 & -55 \end{bmatrix}$$

16.
$$\begin{bmatrix} 3 & 0 & 0 & 2 & -3 & 0 & 3 & -1 & 0 & 4 \\ 2 & 0 & 3 & 0 & 3 & 0 & -3 & -3 & 0 & 5 \\ -2 & 0 & 0 & 0 & 2 & 2 & 2 & 2 & 0 & 8 \\ 3 & 3 & 0 & 0 & -3 & 0 & -3 & -1 & 0 & 12 \\ -7 & 0 & 0 & 0 & -9 & 0 & -15 & -12 & -3 & -57 \end{bmatrix}$$

In exercises 17–28, solve using the dual.

17. Minimize $C = 6x + 3y$
subject to

$$\begin{cases} x \geq 0, y \geq 0 \\ x + y \geq 2 \\ 4x + y \geq 4 \end{cases}.$$

18. Minimize $C = 5x + 4y$
subject to

$$\begin{cases} x \geq 0, y \geq 0 \\ 3x + y \geq 11 \\ 2x + 3y \geq 12 \end{cases}.$$

19. Minimize $C = 18x + 2y$
subject to

$$\begin{cases} x \geq 0, y \geq 0 \\ x + y \geq 4 \\ 3x - y \geq 4 \end{cases}.$$

20. Minimize $C = x + 4y$
subject to

$$\begin{cases} x \geq 0, y \geq 0 \\ 2x + y \geq 4 \\ -x + y \geq 1 \end{cases}.$$

21. Minimize $C = 6x + 3y$ subject to

$$\begin{cases} x \geq 0, y \geq 0 \\ -x + y \geq 1 \\ x + y \leq 2 \\ 16x - y \geq 4 \end{cases}.$$

22. Minimize $C = 10x + 17y$ subject to

$$\begin{cases} x \geq 0, y \geq 0 \\ 2x + y \geq 16 \\ x + 2y \geq 11 \\ x + y \leq 9 \end{cases}.$$

23. Minimize $C = 16x + 8y + 18z$ subject to

$$\begin{cases} x \geq 0, y \geq 0, z \geq 0 \\ 3x - 2y + 2z \geq 4 \\ -2x + y + 2z \geq 5 \\ 2x + 4y + 3z \geq 5 \end{cases}.$$

24. Minimize $C = 9x + 24y + 15z$ subject to

$$\begin{cases} x \geq 0, y \geq 0, z \geq 0 \\ x + y + z \geq 7 \\ x + 2y + z \geq 5 \\ -x + y \geq 6 \end{cases}.$$

25. Minimize $C = 8x - y$ subject to

$$\begin{cases} x \geq 0, y \geq 0 \\ 3x + y \geq 16 \\ x + 2y \leq 22 \\ 2x - y \leq 4 \end{cases}.$$

26. Minimize $C = -x + 4y$ subject to

$$\begin{cases} x \geq 0, y \geq 0 \\ 2x - y \geq 1 \\ -x + 2y \geq 7 \\ x + y \leq 14 \end{cases}.$$

27. Minimize $C = 2x - y + 3z$ subject to

$$\begin{cases} x \geq 0, y \geq 0, z \geq 0 \\ x + y + z \leq 15 \\ 2x + 3y - z \geq 20 \\ 3x - y + z \geq 10 \end{cases}.$$

28. Minimize $C = x + 2y - z$ subject to

$$\begin{cases} x \geq 0, y \geq 0, z \geq 0 \\ 3x + 2y + z \leq 30 \\ 2x + 3y + z \geq 40 \\ x + 3y + z \geq 45 \end{cases}.$$

3.8 Chapter Review

Important Terms and Symbols in Chapter 3.

linear programming (introduction)
constraints (introduction)
feasible region (introduction)
objective statements (introduction)
half-plane (section 3.1)
test point (section 3.1)
graphs of inequalities (section 3.1)
corner points (section 3.1)
corner point theorem (section 3.2)
maximizing and minimizing (section 3.2)
objective equation (section 3.2)

simplex (simplex method) (section 3.4)
- initial conditions (section 3.4)
- simplex tableau (matrix) (section 3.4)
- objective matrix (section 3.4)
- bound matrix (section 3.4)
- constraint matrix (section 3.4)
- "reading" the tableau (section 3.4)
- changing the tableau (section 3.4)

pivot number, pivot row (section 3.4)
flow chart for simplex (section 3.4)
flow chart for Phase I (section 3.5)
- minimization (section 3.5)
- mixed constraints (section 3.5)
- recover missing column (section 3.5)

dual (section 3.7)

Review Exercises

1. Graph the feasible region defined by the system of constraints

$$\begin{cases} x \geq 0, \; y \geq 0 \\ 2x - y \leq 8 \\ 4y - 3x \leq 8 \end{cases}$$

and label the corner points.

2. Write the system of constraints that determines the following feasible region.

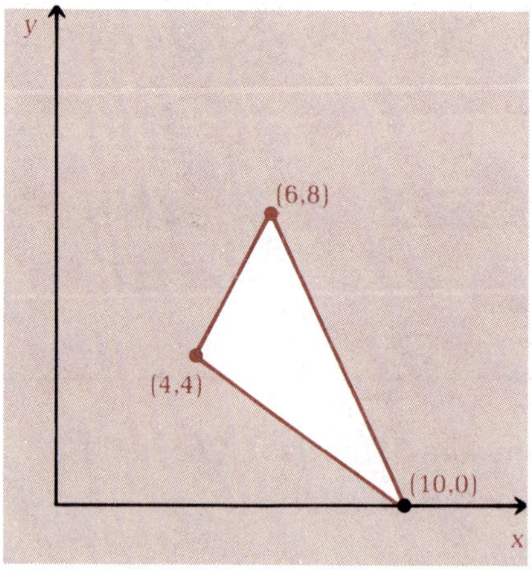

3. Graph the unbounded feasible region.

$$x \geq 0, \; y \geq 0 \qquad x + 2y \geq 12$$
$$x + y \geq 10 \qquad 2x + y \geq 13$$

4. Which of the following points are in the half-plane $2x + 5y \leq 38$?
a. (0,0) **b.** (8,2) **c.** (10,5)
d. (4,6) **e.** (5,5) **f.** (19,2)

5. Maximize $P = 3x - 2y$ subject to

$$\begin{cases} x, y \geq 0 \\ x + y \leq 8 \\ 2x - 3y \leq 1 \end{cases}$$

by graphing the feasible regions to determine the corner points.

6. Minimize $C = 80 + 3x - 2y$ subject to

$$\begin{cases} x, y \geq 0 \\ x + y \leq 8\,. \\ 2x - 3y \leq 1 \end{cases}$$

7. Use simplex to solve problem 5 and determine the values of the slack variables.

8. Use simplex to solve problem 6.

9. Use simplex to maximize $A = 3x - 2y$ subject to

$$\begin{cases} x \geq 0,\ y \geq 0 \\ x + y \leq 8 \\ 2x - 3y \leq 1 \\ 6x + y \geq 13 \end{cases}.$$

10. Use simplex to minimize $A = 3x \quad 2y$ subjcct to

$$\begin{cases} x \geq 0,\ y \geq 0 \\ x + y \leq 8 \\ 2x - 3y \leq 1 \\ 6x + y \geq 13 \end{cases}.$$

11. Set up the initial simplex tableau necessary to maximize $R = 2x - 3y + 4z$ subject to

$$\begin{cases} x \geq 0,\ y \geq 0,\ z \geq 0 \\ x + y + 2z \geq 6 \\ x - y + 2z \geq 4\,. \\ x + 2y + 3z \leq 9 \\ x + y + z = 5 \end{cases}$$

12. Set up the initial simplex tableau necessary to minimize $C = 3x - 2y - z$ subject to the constraints in problem 11.

13. Maximize $P = x - y + 3z$ subject to

$$\begin{cases} x, y, z \geq 0 \\ x + y \leq 12 \\ y - z \leq 4 \\ x - y + z \leq 1 \end{cases}.$$

14. Maximize $B = 2x + y + z$ subject to the constraints in problem 13.

15. Minimize $C = 2x - y + 3z$ subject to

$$\begin{cases} x, y, z \geq 0 \\ x + y \geq 3 \\ y + z \geq 5 \\ x + z \geq 7 \end{cases}.$$

16. Minimize $C = 2x + y + 3z$ subject to

$$\begin{cases} x, y \geq 0 \\ x + y \geq 3 \\ y + z \geq 5 \\ x + z \geq 7 \end{cases}.$$

17. A company has three factories (I, II, III) producing products A, B, and C. In one hour factory I produces 3 units of A, 1 unit of B, and 2 units of C. Factory II, in one hour, produces 2 units of A and 3 units of B. The hourly production for factory III is 1 unit of A, 1 unit of B, and 3 units of C. The company needs to fill an order of at least 70 units of A, at least 65 units of B, and at least 50 units of C. Provided the cost per hour for factory I is \$210, for factory II is \$210, and for factory III is \$280, how do we minimize our production cost? What is our production cost? Have we overproduced, and if so by how much?

18. A chemical plant has two processes it can employ to refine raw ore. Process I will produce, on a daily basis, 400 pounds of copper, 30 tons of slag, and 30 ounces of gold, whereas process II will recover, daily, 100 pounds of copper, 40 tons of slag, and 100 ounces of gold. There was a time when the process selected was determined only by the market value of the recovered metals. But now the EPA (Environmental Protection Agency) has stiff fines for the discharge of polluted water into a nearby river. Unfortunately, process I discharges 400 gallons of polluted water per day into the river, and process II discharges 300 gallons per day.

The company decides it needs to produce at least 1400 pounds of copper, 300 tons of slag, and 480 ounces of gold.

a. How should it operate its refinery to minimize the run-off?
b. Are there any surpluses?
c. What amount of polluted water was discharged?

4 Transportation and Assignment Problems

Introduction

An important application of linear programming is the **transportation problem.** This type of problem was designed in the early 1940s. In its simplest form it is designed to minimize shipping cost from points of manufacture to points of sale. The transportation problem gives rise to a model that will fit many problems, and any of these problems will have its solution in the same algorithm that solves the transportation problem.

Since the problem and its solution are fairly new to the business world, the types of problems that fit this model are still growing and will continue to grow as more and more middle management people become aware of the possibilities. In a few years maybe you will find a problem unique to your career field that can be solved by this technique.

Under certain conditions a computer may be needed to solve these problems completely, but in many cases they can be handled by hand in a reasonable time. In any case, the ability to recognize that a particular type of problem has an accessible solution may be the most important aspect of this chapter.

4.1 Shipping Schedules

Since transportation problems are linear programming problems, they can be solved using the simplex method developed in chapter 3. However, certain difficulties arise.

In this chapter, we will develop a new model for solving transportation problems. Before developing the model, we will state and solve a transportation problem using the simplex method of chapter 3. As you study this illustration, you should begin to see where some of the difficulties arise. These will be specifically indicated at the end of the problem.

Suppose a company has 30 snowblowers in their St. Louis warehouse and 20 in their Tulsa warehouse. They receive an order for 15 snowblowers from a store in Kansas City and 25 snowblowers from a store in Wichita. Shipping costs are \$10 per unit from St. Louis to Kansas City, \$15 per unit from St. Louis to Wichita, \$12 per unit from Tulsa to Kansas City, and \$12 per unit from Tulsa to Wichita. How should they ship the snowblowers to minimize their cost?

In order to solve this problem using the simplex method, we need to write constraints and an objective equation. A diagram will be helpful.

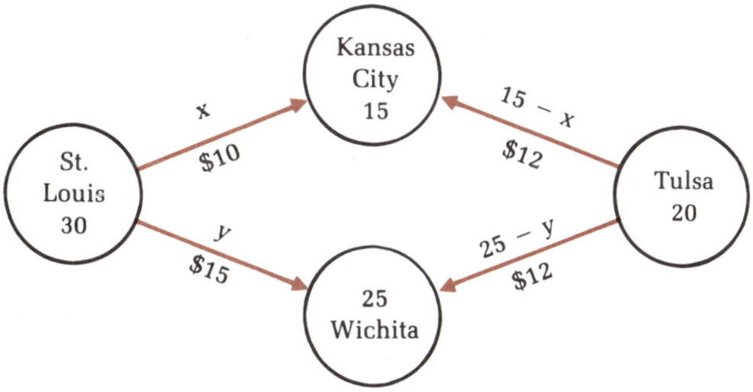

We label the diagram to include the supply in St. Louis and Tulsa and the demand from Kansas City and Wichita, together with the shipping costs. Let x be the number of units to be shipped from St. Louis to Kansas City and y the number from St. Louis to Wichita. Then $15 - x$ and $25 - y$ will be the number of units shipped from Tulsa to Kansas City and Wichita, respectively.

We now begin to write constraints. Since St. Louis has 30 units available, $x \leq 30$ and $y \leq 30$. However, because of the orders from Kansas City and Wichita, we can say $x \leq 15$ and $y \leq 25$, so we don't need the $x \leq 30$ and $y \leq 30$. For shipments from Tulsa, we have both $15 - x \leq 15$ and

$15 - x \le 20$ and $25 - y \le 25$ and $25 - y \le 20$. Simplifying these four inequalities gives

$$\begin{aligned} 15 - x \le 15 &\to \quad 0 \le x \text{ or } x \ge 0, \\ 15 - x \le 20 &\to -5 \le x \text{ or } x \ge -5, \\ 25 - y \le 25 &\to \quad 0 \le y \text{ or } y \ge 0, \text{ and} \\ 25 - y \le 20 &\to \quad 5 \le y \text{ or } y \ge 5. \end{aligned}$$

The only one of these that needs to be used in the simplex matrix is the last one, $y \ge 5$. St. Louis has only 30 units, so $x + y \le 30$. Since the orders total 40 and Tulsa has only 20 units, $x + y \ge 20$. Using the same reasoning,

$$\begin{aligned} (15 - x) + (25 - y) &\le 20 \text{ and} \\ (15 - x) + (25 - y) &\ge 10. \end{aligned}$$

Simplifying gives

$$\begin{aligned} -x - y &\le -20 \text{ or } x + y \ge 20 \text{ and} \\ -x - y &\ge -30 \text{ or } x + y \le 30. \end{aligned}$$

Again, no new constraint is obtained. We have, then, the following constraints.

1. $x \le 15$
2. $y \le 25$
3. $y \ge 5$
4. $x + y \le 30$
5. $x + y \ge 20$

However, constraints 1 and 5 together will force constraint 3 to be satisfied. The system of constraints is, then,

$$\begin{cases} x \ge 0, y \ge 0 \\ x \le 15 \\ y \le 25 \\ x + y \le 30 \\ x + y \ge 20 \end{cases}$$

The cost equation will be

$$\begin{aligned} C &= 10x + 15y + 12(15 - x) + 12(25 - y) \\ &= 10x + 15y + 180 - 12x + 300 - 12y \\ &= 480 - 2x + 3y \end{aligned}$$

The simplex tableau and solution follow.

$$\begin{bmatrix} 1 & 0 & 1 & 0 & 0 & 0 & 0 & 15 \\ 0 & 1 & 0 & 1 & 0 & 0 & 0 & 25 \\ 1 & 1 & 0 & 0 & 1 & 0 & 0 & 30 \\ -1 & \textcircled{-1} & 0 & 0 & 0 & 1 & 0 & -20 \\ 2 & -3 & 0 & 0 & 0 & 0 & 1 & 480 \end{bmatrix}$$

$$\begin{bmatrix} \textcircled{1} & 0 & 1 & 0 & 0 & 0 & 0 & 15 \\ -1 & 0 & 0 & 1 & 0 & 1 & 0 & 5 \\ 0 & 0 & 0 & 0 & 1 & 1 & 0 & 10 \\ 1 & 1 & 0 & 0 & 0 & -1 & 0 & 20 \\ 5 & 0 & 0 & 0 & 0 & -3 & 1 & 540 \end{bmatrix}$$

$$\begin{bmatrix} 1 & 0 & 1 & 0 & 0 & 0 & 0 & 15 \\ 0 & 0 & 1 & 1 & 0 & 1 & 0 & 20 \\ 0 & 0 & 0 & 0 & 1 & 1 & 0 & 10 \\ 0 & 1 & -1 & 0 & 0 & -1 & 0 & 5 \\ 0 & 0 & -5 & 0 & 0 & -3 & 1 & 465 \end{bmatrix}$$

Reading the answer gives $x = 15$ and $y = 5$, with $C = \$465$. Then $15 - x = 0$ and $25 - y = 20$. Hence, St. Louis ships 15 units to Kansas City and 5 to Wichita. Tulsa ships the other 20 units to Wichita. We can then write a **shipping schedule** in table form as

From	**To:** *Kansas City*	*Wichita*
St. Louis	15	5
Tulsa	0	20

The objection to this technique is that as the number of sources of supply or demand or both increases, it becomes much more difficult to write the constraints. If we are to be sure that we have obtained the optimum value, we must have all constraints. Extra constraints increase the computation in reducing the simplex matrix and should be avoided.

The model to be developed to handle transportation problems in this chapter does not require that we write the constraints. The new model is much easier to use than the simplex method and can be used for other types of problems as well.

■ Example 1. A shipping schedule must be produced that will move 230 items from three warehouse locations to five stores according to the following tables. In this example, note that supply and demand are equal.

Warehouse	*Supply*
1	40
2	60
3	130

Store	*Orders (Demand)*
A	30
B	60
C	80
D	20
E	40

Solution: The following **matrix chart** brings both of these tables together where the rows represent supply and the columns represent demand.

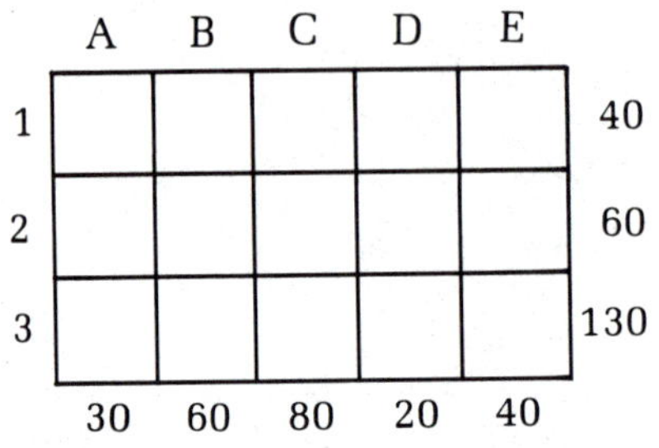

The row and column totals are included as a part of the chart. In order to fill in the chart and arrive at an acceptable shipping schedule, we use a technique called the **northwest corner method.** Starting in the northwest (upper-left) corner, fill in the smaller of the row or column total. In this example, this is the 30 in the first column.

The first column is now completed, and we move to column 2(B), still in row 1. The column total is 60, but we have only 10 left in row 1. We enter the 10, completing row 1. Move down to row 2 and continue. Each time we fill a square, we complete the requirements of either a row or a column or both. The last entry, in the southeast (lower-right) corner, completes the last row and last column of the chart and produces our initial shipping schedule, as given in the following chart and table.

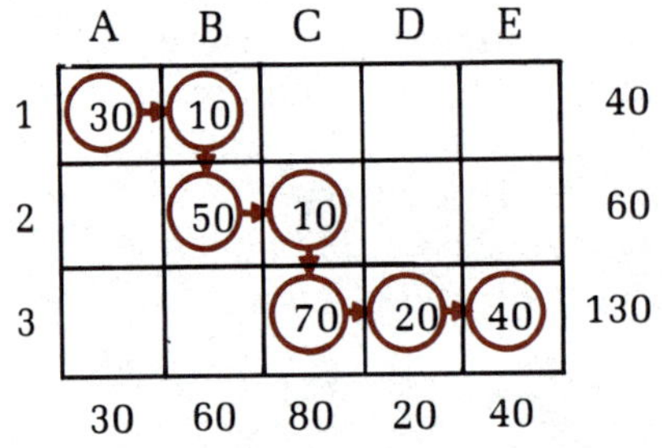

Shipping Schedule

From	**To:** A	B	C	D	E
1	30	10	—	—	—
2	—	50	10	—	—
3	—	—	70	20	40

The entries in the chart are circled to distinguish them from other numbers that will be inserted later. ■

■ Example 2. Complete the following chart using the northwest corner method.

	A	B	C	D	
1					30
2					30
3					20
4					10
5					60
	20	40	70	20	

	A	B	C	D	
1	20	10			30
2		30			30
3			20		20
4			10		10
5			40	20	60
	20	40	70	20	

When 30 was entered in square (2,2), the requirements for both row 2 and column B are satisfied. Square (3,3) becomes the northwest corner of the uncompleted rows and columns. Continue the process as before by entering the maximum possible amount, 20, in square (3,3). The final entry in the southeast corner completes the chart. It should be noted that every entry completes the requirement for some row or some column. ■

■ Example 3. The United States Army operates several training centers. Five of these are located at Fort Knox, KY; Fort Jackson, SC; Fort Sill, OK; Fort Benning, GA; and Fort Dix, NJ. After completing the training, the Army plans to reassign the trainees to seven line divisions, as indicated in the following tables. Use the northwest corner method to construct an initial shipping schedule.

Centers	*Number of Trainees*
Fort Knox	431
Fort Jackson	285
Fort Sill	643
Fort Benning	127
Fort Dix	114
Total	1600

Units	*Allotted Number of Trainees*
82nd Airborne, Fort Benning, GA	300
4th Division, Fort Carson, CO	200
1st Cavalry, Fort Hood, TX	200
7th Infantry, Fort Ord, CA	100
5th Infantry, Fort Polk, LA	400
9th Infantry, Fort Lewis, WA	200
2nd Division, Korea	200
Total	1600

Solution: We make the chart and use the northwest corner method.

	82nd	4th	1st	7th	5th	9th	2nd	
Fort Knox	(300)	(131)						431
Fort Jackson		(69)	(200)	(16)				285
Fort Sill				(84)	(400)	(159)		643
Fort Benning						(41)	(86)	127
Fort Dix							(114)	114
	300	200	200	100	400	200	200	

In table form, the shipping schedule would look like the following.

From	To:	82nd GA	4th CO	1st TX	7th CA	5th LA	9th WA	2nd Korea
Fort Knox, KY		300	131	—	—	—	—	—
Fort Jackson, SC		—	69	200	16	—	—	—
Fort Sill, OK		—	—	—	84	400	159	—
Fort Benning, GA		—	—	—	—	—	41	86
Fort Dix, NJ		—	—	—	—	—	—	114 ■

Although the shipping schedule is a correct solution to the problem as presented, a closer look would reveal that some of the personnel are being moved unnecessarily long distances. Better shipping schedules that are more cost efficient can be found, and we will address this problem in the next sections.

Exercise Set 4.1

1. Use the northwest corner method to find a shipping schedule for the following problems.

	1	2	3	4	
A					100
B					100
C					200
D					300
	150	250	100	200	

	1	2	3	4	5	
A						100
B						100
C						100
D						100
	100	50	50	50	150	

	1	2	3	
A				100
B				50
C				50
D				40
E				30
F				30
	100	100	100	

	1	2	3	4	5	6	
A							120
B							130
C							150
D							200
E							300
F							100
	120	180	200	190	160	150	

2. The following tables contain the necessary information to construct a shipping schedule to supply the orders for stores *A, B, C, D* & *E* from factories I, II, and III. Construct a shipping schedule.

Factory	*Production*
I	200
II	300
III	200

Stores	*Orders*
A	112
B	179
C	131
D	108
E	170

4.2 Loops

Closed paths making only 90° turns at exactly two squares in every row and in every column entered are the only paths under consideration. Paths may be as simple as *abhg* in figure 4.1 or more complicated, like *cdxyolqr* in figure 4.2, which seems to cross itself before closing.

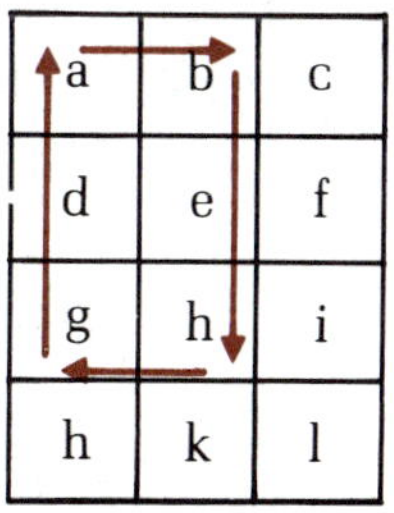

Figure 4.1

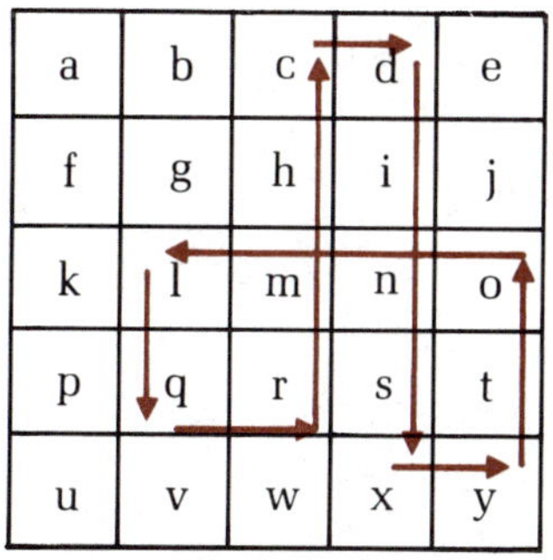

Figure 4.2

A **loop** will be defined here as a closed path where exactly one of the corner squares involved is blank.

A loop has exactly two corner squares in each row and exactly two corner squares in each column it enters. A loop may pass through several squares while making its path, but only the squares at the corners of the loop are of importance to us. We alternately assign add and subtract labels to these corner squares. If we choose an amount less than or equal to the smallest number in the subtract squares, we can increase the add squares and decrease the subtract squares by this amount while maintaining the exact row and column requirements unaltered. The rows and columns will have been simultaneously increased and decreased by the same amount.

a (20) S	b (10) A	c	d	30
e	f (20) S	g (10) A	h	30
i A	j	k (40) S	l	40
m	n	o	p (50)	50
20	30	50	50	

Chart 4.1

a (5) S	b (25) A	c	d	30
e	f (5) S	g (25) A	h	30
i (15) A	j	k (25) S	l	40
m	n	o	p (50)	50
20	30	50	50	

Chart 4.2

In chart 4.1, we wish to change the configuration by adding 15 units to square *i*. The loop—and there is only one—is the path *ikgfba*. The blank square, in this case *i*, is always labeled an add square. We enter *A*'s and *S*'s alternately around the loop.

Chart 4.2 is the result of adding 15 to the *A* squares and taking 15 away from the *S* squares.

NOTE: The sum of the entries in each row and in each column remains unchanged.

In practical application, if it were desirable to move 15 units around this loop, it would be even more cost effective to move the maximum number of 20 units around this loop. The result would give chart 4.3.

a	b (30) S	c	d A	30
e	f	g (30)	h	30
i (20)	j	k (20)	l	40
m	n A	o	p (50) S	50
20	30	50	50	

Chart 4.3

	b		d (30)	30
		(30)		30
(20)		(20)		40
	n (30)		p (20)	50
20	30	50	50	

Chart 4.4

Assigning *A* and *S* labels to the path *nbdp* in chart 4.3 and moving the maximum possible 30 units produces the configuration in chart 4.4.

It becomes apparent that a large number of "solutions" are possible using this technique. When additional constraints, usually shipping cost, are imposed on our model, some of the shipping schedules become more cost effective than others. These solutions are called optimal solutions and we seek them lovingly.

It was determined in 1959 by Simonnard and Hadley that the number of entries needed to determine the final best solution was exactly $m + n - 1$ where m is the number of rows and n is the number of columns.

If more than $m + n - 1$ entries are used there will exist at least one path with no blank squares. A new configuration could be made reducing the amount of the entry in the least desirable square until one square in that path becomes blank. If fewer than $m + n - 1$ entries are used, it is generally impossible to proceed to the most desirable solution because some possibilities are eliminated.

> **Rule 1:** The number of entries at every stage of the solution must equal the number of rows plus the number of columns minus 1: $(m + n - 1)$.

In order to satisfy rule 1, it may be necessary to insert zeros in some of the blanks.

> **Case 1.** When making the initial chart, if an entry completes the requirements for both a row and a column simultaneously, enter a zero in an adjacent blank.
>
> **Case 2.** When pushing numbers through loops, if more than one square becomes empty, enter a zero in all but one.

■ Example 4. Chart 4.4 should have a zero in either square *o* or square *l*. (This zero will also appear in chart 4.3.) Chart 4.3 would also require a zero in either square *a* or square *f*. Whichever is selected, the other would be left blank.

When a **proper initial assignment,** having $m + n - 1$ entries has been made, each blank square will identify a unique loop, and filling that blank is called **making a pattern change.** ■

■ Example 5. Use the northwest corner method to make a proper initial solution to the problem below.

	1	2	3	4	5	
A						50
B						40
C						60
D						60
	30	20	100	50	10	

→

	1	2	3	4	5	
A	30	20				50
B		0	40			40
C			60	0		60
D				50	10	60
	30	20	100	50	10	

■

■ **Example 6.** Identify all the loops in the following tableau.

	1	2	3	4	5	
A	(50) a	(60) b	c	d	e	110
B	f	(30) g	h	i	j	30
C	k	(10) l	(50) m	(40) n	o	100
D	p	q	r	(30) s	(30) t	60
	50	100	50	70	30	

Solution: Each blank square defines exactly one loop. The first letter identifies that loop.

1. *cblm*
2. *dbln*
3. *eblnst*
4. *fgba*
5. *hmlg*
6. *inlg*
7. *jglnst*
8. *klba*
9. *onst*
10. *psnlba*
11. *qsnl*
12. *rsnm* ■

■ **Example 7.** Identify the loop for square *w* in chart 4.5 and perform a pattern change.

(30) a	b	c	(20) d	e	50
f	(30) g	(10) h	i	j	40
k	(60) l	m	n	(0) o	60
p	q	(10) r	(40) s	t	50
u	v	w	x	(50) y	50
30	90	20	60	50	

(30) a	b	c	(20) d	e	50
f	(40) g	h	i	j	40
k	(50) l	m	n	(10) o	60
p	q	(10) r	(40) s	t	50
u	v	(10) w	x	(40) y	50
30	90	20	60	50	

Chart 4.5

Solution: The loop *whgloy* is the one in which *w*, g, and *o* are the add squares, and *h*, *l*, and *y* are the subtract squares. Since the smallest value in all the subtract squares is 10, the pattern change reflected in chart 4.5 involves adding 10 units to squares *w*, g, and *o* while subtracting 10 units from squares *h*, *l*, and *y*. Square *h* is now blank, and the number of entries remains the same in both charts. ($m + n - 1 = 5 + 5 - 1 = 9$). ■

NOTE: While loops are unique, the direction around any loop may be either clockwise or counterclockwise. It makes no difference. For example, in chart 4.5, the loop *whgloy* could have been *wyolgh*.

Exercise Set 4.2

1. In a 7 by 5 tableau how many entries are required for a proper initial assignment?
2. In the following chart all supply and demands are satisfied. Insert circled zeros necessary to have a proper initial assignment.

(50)					50
	(30)	(20)	(10)		60
				(40)	40
				(50)	50
50	30	20	10	90	

3. In the chart below identify the loops for blanks (1), (15), (6), (23), (36) and (11).

1	(40) 2	3	4	5	6	40
7	8	(60) 9	10	11	12	60
13	14	15	16	(60) 17	(0) 18	60
(10) 19	20	21	22	23	(70) 24	80
25	(20) 26	27	28	(30) 29	30	50
31	32	(30) 33	(10) 34	35	36	40
(50) 37	38	39	(40) 40	41	42	90
60	60	90	50	90	70	

4. Make a pattern change in the chart in problem 3 for blanks
   a. 22 **b.** 31 **c.** 25 **d.** 11 **e.** 7 **f.** 12

4.3 Transportation Problems

The classic problem consists of three kinds of information: **1.** supply or production, **2.** demand or orders, and **3.** shipping costs. We then find a shipping schedule that exhausts supply, meets demand, and succeeds at a minimum cost. The following example will illustrate how the problem should be approached. The three types of information necessary are listed in the following tables.

■ Example 8.

Supply

Factory	Production
A	100
B	200
C	50

Demand

Store	Orders
1	70
2	80
3	100
4	100

Shipping Cost (in Dollars)

From	To: 1	2	3	4
A	3	2	2	3
B	4	3	5	2
C	3	2	1	2

Solution: All the information is brought together in the following chart. It should be noted that the last column is the supply, the last row is the demand, and the upper right-hand entries are the respective shipping costs. The circled numbers provide an initial solution using the northwest corner method.

	1	2	3	4	
A	[3] (70)	[2] (30)	[2]	[3]	100
B	[4]	[3] (50)	[5] (100)	[2] (50)	200
C	[3]	[2]	[1]	[2] (50)	50
	70	80	100	100	

Chart 4.6

The shipping cost for this schedule is 70(3) + 30(2) + 50(3) + 100(5) + 50(2) + 50(2) = $1,120. ■

To find the cheapest total shipping cost, we should perform selected pattern changes that will produce new shipping schedules, each reducing the total shipping cost.

Each of the unfilled squares in our chart will have a **cost factor.** The cost factor is the amount of change per unit in the total cost if that blank were filled by performing a pattern change on the unique loop determined by that blank. The cost factor is computed by alternately adding and subtracting the shipping cost around the loop. The cost factors have been added to our chart, and the table explains how each was computed.

	1	2	3	4	
A	[3] (70) *a*	[2] (30) *b*	[2] −2 *c*	[3] +2 *d*	100
B	[4] 0 *e*	[3] (50) *f*	[5] (100) *g*	[2] (50) *h*	200
C	[3] −1 *i*	[2] −1 *j*	[1] −4 *k*	[2] (50) *l*	50
	70	80	100	100	

Chart 4.7

Blank	*Loop Used*	*Arithmetic Sum Equals Cost Factor*
c	*cbfg*	2 − 2 + 3 − 5 = −2
d	*dbfh*	3 − 2 + 3 − 2 = +2
e	*efba*	4 − 3 + 2 − 3 = 0
i	*ilhfba*	3 − 2 + 2 − 3 + 2 − 3 = −1
j	*jlhf*	2 − 2 + 2 − 3 = −1
k	*klhg*	1 − 2 + 2 − 5 = −4

	1	2	3	4	
A	[3] (70) *a*	[2] (30) *b*	[2] −2 *c*	[3] +2 *d*	100
B	[4] 0 *e*	[3] (50) *f*	[5] (50) *g*	[2] (100) *h*	200
C	[3] +3 *i*	[2] +3 *j*	[1] (50) *k*	[2] +4 *l*	50
	70	80	100	100	

Chart 4.8

Blank *k* has a (−4) cost factor, and if we perform a pattern change on loop *klhg*, we can push a maximum of 50 units through that loop for a gain of (+4)(50) or $200. The total shipping cost is now 3(70) + 2(30) + 3(50) + 5(50) + 2(100) + 1(50) = $920, which was exactly what we expected.

Cost factors are now recomputed for the blanks in chart 4.8 as follows:

Blank	*Loop Used*	*Algebraic Sum Equals Cost Factor*
c	*cbfg*	2 − 2 + 3 − 5 = −2
d	*dbfh*	3 − 2 + 3 − 2 = +2
e	*efba*	4 − 3 + 2 − 3 = 0
i	*ikgfba*	3 − 1 + 5 − 3 + 2 − 3 = +3
j	*jkgf*	2 − 1 + 5 − 3 = +3
l	*lhgk*	2 − 2 + 5 − 1 = +4

	1	2	3	4	
A	[3] (70) *a*	[2] +2 *b*	[2] (30) *c*	[3] +4 *d*	100
B	[4] −2 *e*	[3] (80) *f*	[5] (20) *g*	[2] (100) *h*	200
C	[3] +1 *i*	[2] +3 *j*	[1] (50) *k*	[2] +4 *l*	50
	70	80	100	100	

Chart 4.9

The only gain now is to fill *c* and since the largest amount that can be pushed through loop *cbfg* is 30 units, our total shipping cost will be reduced by $60. Chart 4.9 reflects this pattern change.

a [3] (50)	*b* [2] 0	*c* [2] (50)	*d* [3] +2	100
e [4] (20)	*f* [3] (80)	*g* [5] +2	*h* [2] (100)	200
i [3] +1	*j* [2] +1	*k* [1] (50)	*l* [2] +2	50
70	80	100	100	

Chart 4.10

Total shipping cost now is 3(70) + 2(30) + 3(80) + 5(20) + 2(100) + 1(50) = 860 or $860. Recomputing the cost factors finds *e* with a −2 and pushing the maximum 20 units through loop *egca*, we further reduce our shipping cost by $40. Chart 4.10 is the result of this pattern change.

The total shipping cost now equals 3(50) + 2(50) + 4(20) + 3(80) + 2(100) + 1(50) = $820. Computing the cost factors in chart 4.10 reveals no negative numbers, and since no negative cost factors remain, we have reached the lowest possible total shipping cost of $820. This is an optimal solution.

The zero cost factor in square *b* would allow us to perform a pattern change on loop *baef*, resulting in the same total shipping cost of $820, producing an alternate shipping schedule that may be important to management. Hence it is a part of the answer to our problem.

Shipping Schedule I

	1	2	3	4
A	50	0	50	0
B	20	80	0	100
C	0	0	50	0

Shipping Schedule II

	1	2	3	4
A	0	50	50	0
B	70	30	0	100
C	0	0	50	0

■

Points to Remember

1. Only blank squares are assigned cost factors.
2. The blank in a loop is always the first letter in the path pattern, e.g., if *rstuvw* is a path, only the *r* square is empty.
3. Each loop contains an even number of squares: two in each row and each column the path enters.
4. After a pattern change has been made, at least some of the cost factors will have changed, and all of them must therefore be reexamined.
5. It is not required to compute all cost factors before making a pattern change. Any pattern change using a loop with a negative cost factor will result in lowering the total shipping cost. See section 4.5 for more detail.

■ Example 9. The Ace Car Rental Company has an excess of 150 cars in four of its locations, and these cars are needed elsewhere. The following tables give the complete information. The shipping cost is the amount necessary to move a single car from where it stands to a new location and is expressed in dollars and rounded to the nearest $10.

Location	*Surplus Number of Cars*
Houston	30
Dallas	30
New Orleans	40
Kansas City	50

Location	*Allotted Number of Cars*
Oklahoma City	20
Baton Rouge	30
Denver	50
Tulsa	50

Transport Cost in Dollars

From	**To:** *Okla. City*	*Baton Rouge*	*Denver*	*Tulsa*
Houston	40	20	60	50
Dallas	30	30	50	40
New Orleans	50	10	60	50
Kansas City	30	50	30	30

Solution:

	Oklahoma City	Baton Rouge	Denver	Tulsa	
Houston	(20) / 4	(10) / 2	+2 / 6	+1 / 5	30
Dallas	−2 / 3	(20) / 3	(10) / 5	−1 / 4	30
New Orleans	−1 / 5	−3 / 1	(40) / 6	−1 / 5	40
Kansas City	0 / 3	+4 / 5	(0) / 3	(50) / 3	50
	20	30	50	50	

Chart 4.11

All of the information is brought together in chart 4.11. Note again that the last column is the supply, the last row is the demand, and the upper right-hand entries are the shipping costs expressed in $10 units for simplicity. The northwest corner method gives us an initial solution whose total shipping cost is $6000. The cost factors have been entered, and it is left as an exercise for the reader to justify these figures.

(20) / 4	(10) / 2	−1 / 6	−2 / 5
+1 / 3	+3 / 3	(30) / 5	−1 / 4
+2 / 5	(20) / 1	(20) / 6	−1 / 5
+3 / 3	+7 / 5	(0) / 3	(50) / 3

Chart 4.12

Chart 4.12 is the result of a pattern change on (3,2), the square with the −3 cost factor. We are able to move 20 units around that loop, which is a savings of $600. The product (20)(−3) gives 60 ten-dollar bills, a $600 savings.

[4] (20)	[2] +2	[6] +1	[5] (10)
[3] −1	[3] +3	[5] (30)	[4] −1
[5] 0	[1] (30)	[6] (10)	[5] −1
[3] +1	[5] +7	[3] (10)	[3] (40)

Chart 4.13

The total shipping cost has now been reduced to $5400, as expected, and by making a pattern change on (1,4), we can reduce our cost by another $200. (Note that the smallest value in a subtract square for this loop is 10, and 10 times the cost factor of −2 expressed in 10's equals $200.) See chart 4.13.

[4] (20)	[2]	[6]	[5] (10)
[3]	[3]	[5]	[4] (30)
[5]	[1] (30)	[6] (10)	[5] −1
[3]	[5]	[3] (40)	[3] (10)

Chart 4.14

Chart 4.14 is the result of a pattern change involving (2,4) for a savings of $300. The only cost factor computed in chart 4.14 was for square (3,4). It should be observed that as we progress through our problem, the negative cost factors all change in a positive direction if they change at all. And since (−1) was the "most negative" cost factor in chart 4.13, chart 4.14 can do no better.

Chart 4.15 is the result of the pattern change on (3,4). There are no remaining negative cost factors; thus we have arrived at an optimal solution. A second optimal solution is found by performing a pattern change on (2,1) with its zero cost factor (chart 4.16). The cheapest shipping cost is $4800 and the two shipping schedules follow.

[4] (20)	[2] +1	[6] +1	[5] (10)
[3] 0	[3] +3	[5] +1	[4] (30)
[5] +1	[1] (30)	[6] +1	[5] (10)
[3] +1	[5] +6	[3] (50)	[3] (0)

Chart 4.15

[4] 0	[2] +1	[6] +1	[5] (30)
[3] (20)	[3] +3	[5] +1	[4] (10)
[5] +1	[1] (30)	[6] +1	[5] (10)
[3] +1	[5] +6	[3] (50)	[3] (0)

Chart 4.16

From	**To:**	*Okla. City*	*Baton Rouge*	*Denver*	*Tulsa*
Houston		20	—	—	10
Dallas		—	—	—	30
New Orleans		—	30	—	10
Kansas City		—	—	50	—

From	**To:**	*Okla. City*	*Baton Rouge*	*Denver*	*Tulsa*
Houston		—	—	—	30
Dallas		20	—	—	10
New Orleans		—	30	—	10
Kansas City		—	—	50	—

■

Exercise Set 4.3

1. Solve the transportation problem defined by the tables below.

Demand	
A	100
B	120
C	340
D	140

Supply	
I	120
II	80
III	100
IV	130
V	270

Shipping Cost

To	**From**	*I*	*II*	*III*	*IV*	*V*
A		3	2	4	4	2
B		4	5	6	4	3
C		3	3	5	3	2
D		2	2	4	3	2

2. Solve the following transportation problems.

a.

1	2	3	1	100
2	3	4	2	120
2	2	3	4	130
2	3	4	1	250
200	200	100	100	

b.

3	4	4	2	60
1	1	1	2	28
0	2	1	3	18
4	5	5	3	14
30	20	60	10	

c.

[3] (5)	[1] (10)	[3] (5)	[2]	[2]
[2]	[1]	[4] (2)	[1] (6)	[1] (7)
[2]	[1]	[3]	[2]	[1] (6)
[2]	[2]	[3]	[1]	[2] (8)

d.

[9] (5)	[8] (4)	[8] (11)	[7] (3)	[9]
[8]	[7]	[8]	[6] (9)	[6] (8)
[6]	[6]	[7]	[5]	[4] (10)

3. Solve the transportation problem defined by the following tables.

Supply	
I	80
II	60
III	30
IV	30
V	20

Demand	
A	40
B	50
C	60
D	70

Shipping Cost

From \ To	A	B	C	D
I	4	2	1	3
II	1	3	2	2
III	3	2	1	3
IV	2	2	1	2
V	2	2	2	3

4.4 Alternate Initial Solutions

In most cases it denies common sense to use the northwest corner method as an initial solution to a transportation problem. Any **proper initial solution** will serve, and if we use common sense in our selection, we can reduce the number of pattern changes necessary in finding an optimal solution.

A *proper initial solution* must satisfy the following conditions:

1. It contains exactly $m + n - 1$ entries.
2. Each blank identifies exactly one loop.
3. There are no paths without a blank.

If the first two conditions are satisfied, the third will follow. The third condition is included as a checking device, i.e., the best way to spot an improper initial solution is to locate a path without a blank.

Using a common-sense approach, we first locate the single best square in the chart, or at least the one we think is the best, and enter the maximum amount possible. This entry has satisfied either a row or a column. The next entry will be taken as the square in the unfulfilled row or column that has the smallest shipping cost. This procedure is repeated until an initial solution is obtained, i.e., when every row and column is satisfied.

If some entry satisfies both a row and a column simultaneously, a circled zero must be included. A step may be saved if the circled zero is entered in the blank with the next lowest shipping cost. Of course, we must select the blank in the same row or the same column as the blank we last filled. It is also necessary that this is a square in a new, empty row or column. The following example will illustrate the process.

■ Example 10. Make a **common-sense initial solution** to the following transportation problem.

Solution: In making our common-sense initial solution, the squares have been lettered alphabetically in the order of their selection (see chart 4.17). Once an initial square is chosen, a kind of domino effect takes over as our example shows.

[3]	[2]	[4]	[1]	[2]	100
[1]	[3]	[1]	[4]	[5]	100
[3]	[5]	[4]	[2]	[1]	100
[1]	[3]	[4]	[4]	[5]	50
120	30	50	80	70	

Sample

[3]	[2] (20)	[4]	[1] (80)	[2]
[1] (80)	[3]	[1] (20)	[4]	[5]
[3]	[5]	[4] (30)	[2] −3	[1] (70)
[1] (40)	[3] (10)	[4]	[4]	[5]

Chart 4.17

a: the square with the smallest shipping cost that allows the largest entry. Other good starting squares are (1,4), (2,3), (3,5), and (4,1). Enter 100, check off row 2.

b: must complete column 1. Check off column 1.

c: must complete row 4. Check off row 4 and column 2.

d: Since the entry in *c* satisfied both row 4 and column 2, a circled zero is necessary. We choose square (1,2) because it has the smallest shipping cost of all possible unfilled squares in column 2.

e: the best remaining choice. Check off column 4.

f: with only two columns to choose from, we continue to pick "best square." Check off row 1.

g: the only choice left in column 5. Check off column 5.

h: the final entry. Our initial solution is complete. Note that we have $m + n - 1$, or $(5 + 4 - 1 = 8)$, entries. ■

■ Example 11. Complete the transportation problem in example 10.

[3]	d [2] (0)	[4]	e [1] (80)	f [2] (20)	100 √
a [1] (100)	[3]	[1] −5	[4]	[5]	100 √
[3]	[5]	h [4] (50)	[2]	g [1] (50)	100
b [1] (20)	c [3] (30)	[4]	[4]	[5]	50 √
120 √	30 √	50	80 √	70 √	

Chart 4.18

Once we have made our *initial common-sense solution*, each blank must be checked to determine if the solution is optimal. All blank squares must be examined for possible negative cost factors. However, examine first the squares with small shipping cost and perform a pattern change on the blank with the first negative cost factor you uncover.

Solution: In chart 4.17, blank (2,3) has the smallest shipping cost. We examine the loop and find the cost factor to be −5. A pattern change on this loop produces chart 4.18. Now the blanks (in chart 4.18) with the smallest unused shipping cost are (3,4) and (1,5) with cost factors of −3 and +5, respectively. A pattern change filling (3,4) gives chart 4.19, in which the cost factor in blank (3,1) is −1. A final pattern change results in chart 4.20, which contains only positive cost factors.

[3]	[2] (30)	[4]	[1] (70)	[2]
[1] (70)	[3]	[1] (30)	[4]	[5]
[3] −1	[5]	[4] (20)	[2] (10)	[1] (70)
[1] (50)	[3]	[4]	[4]	[5]

Chart 4.19

[3] +1	[2] (30)	[4] +2	[1] (70)	[2] +2
[1] (50)	[3] +2	[1] (50)	[4] +4	[5] +6
[3] (20)	[5] +2	[4] +1	[2] (10)	[1] (70)
[1] (50)	[3] +2	[4] +3	[4] +4	[5] +6

Chart 4.20 ■

It is instructive to examine (3,3) in charts 4.17, 4.18, 4.19, and 4.20. In our initial solution, chart 4.17, this entry was obviously undesirable (too large an entry in a square with too high a shipping cost). It took three pattern changes to empty this square. Had we been foresighted (lucky) and started by filling blank (2,3) first, the immediate solution would have been our final result, with no pattern changes needed. We would still have to check all the cost factors, of course, to be sure of our result. This is necessary in any solution.

When problems are large, it may be to your advantage to try several initial solutions and pick the one with the smallest total shipping cost as a starting place. Being a little foresighted helps also.

The best starting square is often found by looking at the respective differences in shipping costs in each row and column. In the preceding example, column 3 stands out as critical, since the difference in the shipping cost in this column jumps from 1 to 4 if square (2,3) is not selected.

In the preceding examples we minimized the total shipping cost. When required to maximize the total shipping cost, we would make pattern changes for positive cost factors only, thus increasing the total shipping cost. When no positive cost factors remain, the maximization is complete.

Exercise Set 4.4

1. Complete the following transportation problems to minimize the total shipping cost and find one alternate optimal solution if it exists.

a.

3	1	2	2	50
2	2	3	1	40
4	4	3	4	40
1	1	2	1	20
40	60	10	40	

b.

1	2	1	3	40
2	4	2	3	50
5	6	4	5	60
3	2	2	5	50
30	90	30	50	

2. Complete the following transportation problems to maximize the total shipping cost and find one alternate optimal solution if it exists.

a.

2	3	1	4	15
4	4	3	4	25
2	2	1	3	10
4	2	1	4	20
3	2	2	3	30
40	20	20	20	

b.

5	4	4	4	3	100
3	2	1	1	1	200
4	3	4	3	2	300
4	5	3	3	4	400
300	250	200	150	100	

3. Find the cheapest shipping cost and produce a shipping schedule for the transportation problem defined by the following tables:

Supply

Factory	Amount
A	110
B	50
C	30
D	40
E	180

Demand

Store	Order Size
1	50
2	60
3	70
4	100
5	80
6	50

Shipping Cost

From \ To:	1	2	3	4	5	6
A	1	3	2	1	2	3
B	2	1	1	4	1	2
C	3	4	2	2	1	2
D	3	2	1	1	1	2
E	1	1	2	1	3	1

4. Solve the following transportation problem.

5	6	4	7	5	5	15
2	4	2	0	4	4	20
4	6	5	4	6	6	20
7	8	6	6	7	7	40
2	2	4	5	6	6	38
4	5	6	4	6	6	67
20	30	20	40	10	80	

5. Owing to an unexpected wildcat strike, the Ace Canning Plant needs to replace 120 workers on the floor at once. Everyone is pressed into service: 5 vice-presidents, 10 junior executives, 10 foremen and middle management personnel, 20 salespeople, and 18 office workers. Mrs. Handy, the president, also authorized the temporary hiring of 57 day workers. She knows more are needed but does not wish to replace the 120 plant workers with more than 120 for fear the union will claim their people are being overworked. The recruited personnel replace 20 sorters, 30 cutters, 20 quality controllers, and 50 packers and stackers. The main problem, and this has happened before, is that not all of the new workers can produce 8 hours' work in an 8-hour day. The following chart is the expected output for each category of worker in an 8-hour day. To which jobs should each of the new workers be assigned?

Personnel	Sorter	Cutter	*Quality Controller*	*Packer*
Senior executives	7	6	8	5
Junior executives	8	8	8	8
Middle management	8	7	7	7
Salespeople	5	4	6	5
Office workers	7	7	6	7
Day workers	7	6	5	7

6. Five upperclassmen in a private boarding school agree to tutor some of the freshmen. The following tables give the number of hours that each upperclassman will devote to tutoring, the hours needed for each subject, the letter grade each of the upperclassmen made in the subject he or she may tutor. Make the best assignment.

Upperclassman	*Number of Hours of Help*
1	20
2	30
3	30
4	30
5	30

Subject	*Number of Hours Needed*
Math	30
English	40
History	40
Science	30

Grade

Subject	*1*	*2*	*3*	*4*	*5*
Math	D	C	B	B	C
English	B	A	A	B	C
History	C	C	B	B	B
Science	B	C	C	D	C

4.5 Uneven Supply and Demand

When the supply is greater than the demand, the excess must be dealt with. Extra columns will need to be added to make the supply equal to the demand. These extra columns reflect the disbursement of the over-supply, and the shipping cost reflects the cost incurred. The most common extra columns are warehouse, salvage, discount sale, and promotion, but there are others.

If the demand exceeds the supply, extra rows will need to be added. Again, the supply is made equal to the demand. These new rows are usually overtime production or paper factories, which produce nothing but imaginary products. In any case, it is important to remember that the shipping cost may be any cost incurred.

■ **Example 12.** The Windswift Bicycle Company was caught short by a gas shortage. The total production at their three plants was 5000, 5000, and 6000 units respectively. Windswift's five distribution centers sent in orders for 3000, 4000, 4000, 3000, and 3000 bicycles respectively. The shipping costs are given in the following table.

Shipping Cost

	A	*B*	*C*	*D*	*E*
I	1	2	3	4	3
II	4	2	5	4	2
III	4	2	4	3	3

The transportation problem is set up in the following chart and solved in the usual way.

	A	*B*	*C*	*D*	*E*	
I	[1] 3000	[2]	[3] 2000	[4]	[3]	5000
II	[4]	[2] 2000	[5]	[4]	[2] 3000	5000
III	[4]	[2] 2000	[4] 1000	[3] 3000	[3]	6000
IV	[0]	[0]	[0] 1000	[0]	[0]	1000
	3000	4000	4000	3000	3000	

Note that distribution center *C* will be short 1000 bicycles, since IV is a paper factory. ■

■ **Example 13.** The management of the company in example 12 decides to go into overtime to produce the 1000 extra bicycles. The per item overtime cost is given in the table below, left. In addition, shipments from factory III to stores *B* and *C* have proved unsafe, and an extra insurance cost of $2 per unit must be included.

Per Item Overtime Cost

Factory	*Per Item*
I	8
II	7
III	6

Shipping + Overtime + Insurance

Factory	*A*	*B*	*C*	*D*	*E*
I	9	10	11	12	11
II	11	9	12	11	9
III	10	10	12	9	9

The combined cost arrived at by adding the shipping cost, the extra overtime cost, and the insurance fee are given in the table above, right. The lowest cost in each column has been circled, and these values are used as the "shipping cost" for the new row, overtime production. If store *A* or *C* has to receive any bicycles made on overtime, they will be made at factory I; the cheapest source for store *B* will be factory II; and store *D* will get their bicycles from factory III. Store *E* will have a choice. The solution to this transportation problem changes the shipping schedule. The problem is set up in the following table. The solution is left as an exercise.

	A	B	C	D	E	
Factory I	1	2	3	4	3	5000
Factory II	4	2	5	4	2	5000
Factory III	4	2	4	3	3	6000
Overtime	9	9	11	9	9	1000
	3000	4000	4000	3000	3000	

■ Example 14. The production at the three main plants has exceeded demand. A discount firm has agreed to buy 100 items at a $9 an item loss to the company. The remainder must be warehoused. All the necessary information is given in the following tables. Find the most effective solution to this transportation problem.

Plants	*Production*
A	150
B	200
C	170

Stores	*Orders*
I	100
II	50
III	50
IV	100

Shipping Cost

	Stores			
Plants	I	II	III	IV
A	3	2	3	3
B	2	4	1	3
C	4	2	4	2

Warehousing Cost

Plant	*Cost*
A	1
B	1
C	2

Solution: The chart below leads to three possible optimal shipping schedules, one of which is shown.

	I	II	III	IV	Ware-house	Disc. Sales	
A	3	2	3	3	1	9	150
B	2	4	1	3	1	9	200
C	4	2	4	2	2	9	170
	100	50	50	100	120	100	

	I	II	III	IV	Ware-house	Disc. Sales
A	—	50	—	—	70	30
B	100	—	50	—	50	—
C	—	—	—	100	—	70 ■

Exercise Set 4.5

1. Solve the transportation problem in example 13.
2. Find the alternate shipping schedules in example 14.
3. Business is good for a small fishing operation off the Keys in Florida. Their boats always catch enough fish. A portion of their catch is prepared by hand for fresh delivery to selected restaurants in the area, and the rest is sold to the big cannery farther up the bay.

   The profit per pound is used as the shipping cost. Restaurants III and V will not serve frozen fish, so a zero is used as their shipping cost. Restaurant VI deep fries everything and will not pay as much for fresh fish as the rest. Restaurant V is willing to pay extra to make sure they get first choice. Find the day's gross profits for this part of the operation by maximizing the chart below.

	I	II	III	IV	V	VI	Supply
Regular	3	3	3	3	4	2	100
Overtime	2	2	2	2	2	1	40
Frozen	1	1	0	2	0	1	Whatever amount is necessary.
Orders	50	50	20	30	30	20	

4. The production at four factories exceeds the demand. One hundred of the excess units are to be sold at $5 per unit discount to a chain store, 50 of the excess units are donated to Boy Scouts as a promotion, and the rest warehoused. Construct an optimal shipping schedule. The tables below contain all the necessary information.

Factory	*Production*
I	300
II	200
III	400
IV	200

Outlets	*Orders*
A	100
B	200
C	150
D	50
E	70

Shipping Cost

From	*A*	*B*	*C*	*D*	*E*
I	3	2	2	1	2
II	2	3	4	2	5
III	2	3	3	3	4
IV	1	3	2	1	3

From	*Actual Cost to Donate to Boy Scouts*
I	40
II	40
III	35
IV	42

Per Unit Warehousing Cost

From	*Warehouse*
I	2
II	2
III	3
IV	1

Shipping Cost to Discount Store

From	*Cost*
I	1
II	2
III	1
IV	3

4.6 Assignment Problems

An **assignment problem** is a special case of the transportation problem. In the assignment problem each supplier produces one unit and each requirement is one unit. The result is a square matrix with exactly one assignment in every row and exactly one assignment in every column. If the matrix is not square, dummy rows or columns are added as necessary.

■ Example 15. Three people apply for two jobs and are tested. The results of the test are given in the following table.

	Job I	*Job II*	*No Job*
A	85	71	0
B	82	92	0
C	80	93	0

The highest score for job I was made by A and the highest score for job II was made by C and these are the logical, optimal, choices. Using our transportation model with the northwest corner method as our initial solution, our problem would look like chart 4.21 below. Making pattern changes about any positive cost factors, since we wish to maximize our result, we arrive at the final optimal solution in chart 4.22.

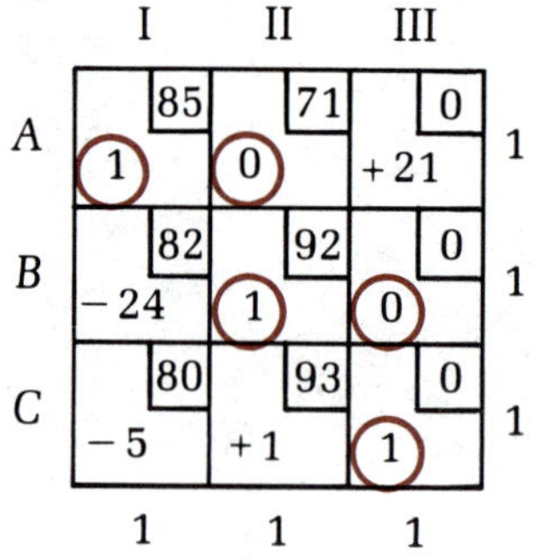

Chart 4.21

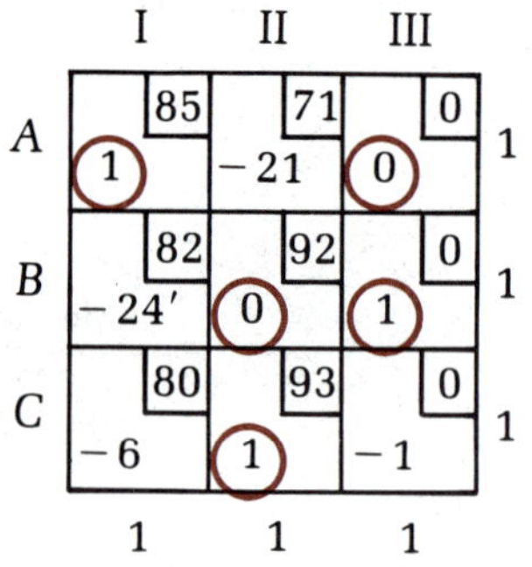

Chart 4.22

Solving assignment problems in this way becomes cumbersome as problems become larger. There is a clever technique that reduces the data to a form in which optimal assignments can be easily made. This method, known as the **Hungarian Method,** is outlined in the following diagram.

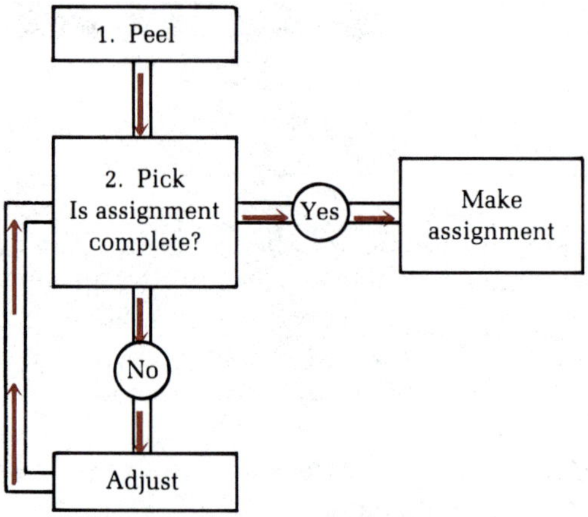

In using the Hungarian Method, the "best" values are equated with zeros, and assignments are then made from these zeros. **Peeling** is the process used to locate zeros initially, and produce a **zero pattern.**

> **1. Peeling to produce a zero pattern.**
>
> **a.** Peel the columns first by replacing each value by the non-negative difference between the value and the "best pick." If you are making maximum assignments, the largest number in each column is the best pick for that column. If you are making minimum assignments, the smallest number in each column is the best pick for that column.
>
> **b.** Regardless of which, maximum or minimum, assignment you seek, after the columns have been peeled, subtract the smallest entry in each row from every entry of that row.

NOTE: If a row contains a zero, the row remains unchanged, so only rows without zeros will be affected.

■ Example 16. Peel the following matrix in order to make a maximum assignment.

Solution:

a.
$$\begin{bmatrix} \textcircled{8} & 7 & 6 & \textcircled{5} \\ 4 & 5 & 3 & 2 \\ 6 & \textcircled{8} & \textcircled{8} & 5 \\ 3 & 2 & 4 & 2 \end{bmatrix}$$

The best picks in each column have been circled for emphasis.

b.
$$\begin{bmatrix} 0 & 1 & 2 & 0 \\ 4 & 3 & 5 & 3 \\ 2 & 0 & 0 & 0 \\ 5 & 6 & 4 & 3 \end{bmatrix}$$

The peeling for columns, part (a), has been completed. In columns 1, 2, and 3, each entry was subtracted from 8. In column 4, each entry was subtracted from 5.

c.
$$\begin{bmatrix} 0 & 1 & 2 & 0 \\ 1 & 0 & 2 & 0 \\ 2 & 0 & 0 & 0 \\ 2 & 3 & 1 & 0 \end{bmatrix}$$

The peeling for rows, part (b), has been completed. Only rows 2 and 4 were changed, since they were the only rows that did not already contain at least one zero. ■

■ Example 17. Peel the following matrix in order to make a minimum assignment. This is the same matrix that appears in example 16, where a maximum assignment was to be made. Note the difference.

a.
$$\begin{bmatrix} 8 & 7 & 6 & 5 \\ 4 & 5 & \textcircled{3} & 2 \\ 6 & 8 & 8 & 5 \\ \textcircled{3} & \textcircled{2} & 4 & \textcircled{2} \end{bmatrix}$$

The best picks in each column have been circled for emphasis.

b. $\begin{bmatrix} 5 & 5 & 3 & 3 \\ 1 & 3 & 0 & 0 \\ 3 & 6 & 5 & 3 \\ 0 & 0 & 1 & 0 \end{bmatrix}$

The peeling for columns, part (a), has been completed. In columns 1 and 3, 3 was subtracted from each entry. In columns 2 and 4, each entry was reduced by 2.

c. $\begin{bmatrix} 2 & 2 & 0 & 0 \\ 1 & 3 & 0 & 0 \\ 0 & 3 & 2 & 0 \\ 0 & 0 & 1 & 0 \end{bmatrix}$

The peeling for rows, part (b), has been completed. Only rows 1 and 3 were adjusted since they were the only rows that did not contain at least one zero. ■

✓ **NOTE:** The only difference in making a maximum or a minimum assignment occurs in the peeling of the columns, part (a). The remainder of the process is exactly the same for both maximum and minimum problems! Once the columns have been peeled, all of the problems become minimum problems. Since "best values" are converted to zeros, high values in maximum problems become low values.

2. **Picking:** Once columns and rows have been peeled, we are ready to attempt an assignment based on zeros. We begin by taking any "must picks." "Must picks" occur when there is only one zero in a row or a column. As we make a pick, we place an "X" over that zero to indicate that it is part of the assignment. When a zero is picked, we place a "√" over all other zeros in the same row and column as our picked zero. These are no longer eligible choices. After taking all "must picks," we continue with other zeros until every zero has either a "X" or a "√." To have an optimal solution, there must be exactly one "X" in each row and each column. If, after all "must picks" are taken, any row or column contains two or more zeros, alternate optimal solutions can be found. If it is not possible to complete an assignment with this zero pattern, then an adjustment is necessary.

■ **Example 18.** Make picks based on the following zero patterns.

a. $\begin{bmatrix} 0 & 1 & 2 & 0 \\ 1 & 0 & 2 & 0 \\ 2 & 0 & 0 & 0 \\ 0 & 3 & 1 & 0 \end{bmatrix}$ **b.** $\begin{bmatrix} 2 & 2 & 0 & 0 \\ 1 & 3 & 0 & 0 \\ 0 & 3 & 2 & 0 \\ 0 & 0 & 1 & 0 \end{bmatrix}$ **c.** $\begin{bmatrix} 0 & 0 & 2 & 1 \\ 1 & 3 & 0 & 1 \\ 2 & 2 & 0 & 0 \\ 2 & 6 & 1 & 0 \end{bmatrix}$

Solution:

a. $$\begin{bmatrix} \overset{\times}{0} & 1 & 2 & \overset{\checkmark}{0} \\ 1 & \overset{\times}{0} & 2 & \overset{\checkmark}{0} \\ 2 & \overset{\checkmark}{0} & \overset{\times}{0} & \overset{\checkmark}{0} \\ \overset{\checkmark}{0} & 3 & 1 & \overset{\times}{0} \end{bmatrix}$$

The (3,3) entry is a must pick and after it is chosen the (2,2) entry is a must pick. After these choices, we have zeros at the four corners and we can choose (1,1) and (4,4) or (1,4) and (4,1). The assignment is complete since there is exactly one X in every row and column.

b. $$\begin{bmatrix} 2 & 2 & \overset{\checkmark}{0} & \overset{\times}{0} \\ 1 & 3 & \overset{\times}{0} & \overset{\checkmark}{0} \\ \overset{\times}{0} & 3 & 2 & \overset{\checkmark}{0} \\ \overset{\checkmark}{0} & \overset{\times}{0} & 1 & \overset{\checkmark}{0} \end{bmatrix}$$

The (4,2) entry is a must pick and after it is chosen, the (3,1) entry is a must pick. After these choices there are four zeros in the upper right corner. We can choose (2,3) and (1,4) or (1,3) and (2,4). In either case the assignment is complete.

c. $$\begin{bmatrix} \overset{\times}{0} & \overset{\checkmark}{0} & 2 & 1 \\ 1 & 3 & \overset{\times}{0} & 1 \\ 2 & 2 & \overset{\checkmark}{0} & \overset{\checkmark}{0} \\ 2 & 6 & 1 & \overset{\times}{0} \end{bmatrix}$$

The (1,1), (2,3), and (4,4) entries are must picks. So is (1,2), but after (1,1) is chosen (1,2) cannot be. Since neither row 3 nor column 2 has an X, the assignment is not complete, and an adjustment is necessary.

---

3. Adjusting: Once a matrix is peeled and an optimal solution cannot be picked from the combination of zeros present, an adjustment must be made. We adjust by first covering all zeros with horizontal and/or vertical lines. The procedure requires that we use fewer lines than the number of assignments we are trying to make. If the zeros cannot be covered with fewer lines than the number of assignments, then no adjustment is necessary and an optimal solution can be picked.

After covering all zeros with $k < n$ lines, find the smallest uncovered number and **1.** subtract this number from every uncovered number, **2.** add this number to every double covered number, and **3.** do nothing to the numbers with one cover.

NOTE: In some large problems, it may not be obvious how to draw the lines so as to cover the zeros with fewer than n lines. The following procedure is frequently helpful. Cover first zeros with $\surd$ marks, beginning with a row or column having the largest number of these. After all such zeros have been covered, draw any additional lines needed to cover zeros with "Xs." ■

■ **Example 19.** Solve example 15 by the Hungarian Method.

Solution:

$$\begin{bmatrix} 85 & 71 & 0 \\ 82 & 92 & 0 \\ 80 & 93 & 0 \end{bmatrix} \longrightarrow \begin{bmatrix} 0 & 22 & 0 \\ 3 & 1 & 0 \\ 5 & 0 & 0 \end{bmatrix} \longrightarrow \begin{bmatrix} \not{0} & 22 & 0^{\checkmark} \\ 3 & 1 & \not{0} \\ 5 & \not{0} & 0^{\checkmark} \end{bmatrix}$$

NOTE: Once the matrix is peeled, column 1, column 2, and row 2 each have "must picks." The picks were made with "X" marks over the zeros and "√" marks over zeros that are no longer subject to choice; e.g., in row 1, once the zero at (1,1) is picked, the zero at (1,3) must be checked off, since only one pick can be made in each row. ■

■ **Example 20.** Find the minimum assignment in the following matrix.

$$\begin{bmatrix} 3 & 2 & 5 & 4 \\ 4 & 5 & 3 & 4 \\ 8 & 7 & 6 & 6 \\ 5 & 8 & 4 & 3 \end{bmatrix} \xrightarrow[\text{columns}]{\text{Peeling}} \begin{bmatrix} 0 & 0 & 2 & 1 \\ 1 & 3 & 0 & 1 \\ 5 & 5 & 3 & 3 \\ 2 & 6 & 1 & 0 \end{bmatrix} \xrightarrow[\text{row 3}]{\text{Peeling}} \begin{bmatrix} 0 & 0 & 2 & 1 \\ 1 & 3 & 0 & 1 \\ 2 & 2 & 0 & 0 \\ 2 & 6 & 1 & 0 \end{bmatrix}$$

Solution: We now try making an assignment. A complete assignment is not possible here because we will not be able to make a selection in both of the first two columns. We can cover all zeros with three lines ($3 < 4$), so an adjustment is necessary. The smallest uncovered value is 1, so we subtract 1 from all uncovered values, add 1 to all double-covered (intersection of lines) values, and leave single-covered values as they were.

$$\begin{bmatrix} \not{0} & 0^{\checkmark} & 2 & 1 \\ 1 & 3 & \not{0} & 1 \\ 2 & 2 & 0 & \not{0} \\ 2 & 6 & 0 & 0^{\checkmark} \end{bmatrix} \xrightarrow[\text{and picking}]{\text{Adjusting}} \begin{bmatrix} 0^{\checkmark} & \not{0} & 3 & 2 \\ \not{0} & 2 & 0^{\checkmark} & 1 \\ 1 & 1 & \not{0} & 0^{\checkmark} \\ 1 & 5 & 1 & \not{0} \end{bmatrix} \xrightarrow[\text{circled}]{\text{Assignments}} \begin{bmatrix} 3 & \textcircled{2} & 5 & 4 \\ \textcircled{4} & 5 & 3 & 4 \\ 8 & 7 & \textcircled{6} & 6 \\ 5 & 8 & 4 & \textcircled{3} \end{bmatrix}$$

The assignments have been circled in the original matrix for identification.

We have included more steps than would normally be included. The actual sequence would look like the following.

$$\begin{bmatrix} 3 & 2 & 5 & 4 \\ 4 & 5 & 3 & 4 \\ 8 & 7 & 6 & 6 \\ 5 & 8 & 4 & 3 \end{bmatrix} \xrightarrow[\text{columns}]{\text{Peeling}} \begin{bmatrix} 0 & 0 & 2 & 1 \\ 1 & 3 & 0 & 1 \\ 5 & 5 & 3 & 3 \\ 2 & 6 & 1 & 0 \end{bmatrix} \xrightarrow[\text{row 3}]{\text{Peeling}} \begin{bmatrix} 0 & 0 & 2 & 1 \\ 1 & 3 & 0 & 1 \\ 2 & 2 & 0 & 0 \\ 2 & 6 & 1 & 0 \end{bmatrix} \xrightarrow[\text{picking}]{\text{Adjusting}} \begin{bmatrix} 0^{\checkmark} & \not{0} & 3 & 2 \\ \not{0} & 2 & 0^{\checkmark} & 1 \\ 1 & 1 & \not{0} & 0^{\checkmark} \\ 1 & 5 & 1 & \not{0} \end{bmatrix}$$ ■

■ **Example 21.** Find the maximum assignment in the matrix in example 20.

$$\begin{bmatrix} 3 & 2 & 5 & 4 \\ 4 & 5 & 3 & 4 \\ 8 & 7 & 6 & 6 \\ 5 & 8 & 4 & 3 \end{bmatrix} \xrightarrow{\text{Peeling}} \begin{bmatrix} 5 & 6 & 1 & 2 \\ 4 & 3 & 3 & 2 \\ 0 & 1 & 0 & 0 \\ 3 & 0 & 2 & 3 \end{bmatrix} \xrightarrow{\text{Peeling}} \begin{bmatrix} 4 & 5 & 0 & 1 \\ 2 & 1 & 1 & 0 \\ 0 & 1 & 0 & 0 \\ 3 & 0 & 2 & 3 \end{bmatrix}$$

$$\xrightarrow{\text{Picking}} \begin{bmatrix} 4 & 5 & 0 & 1 \\ 2 & 1 & 1 & 0 \\ 0 & 1 & 0 & 0 \\ 3 & 0 & 2 & 3 \end{bmatrix} \xrightarrow[\text{Circled}]{\text{Assignments}} \begin{bmatrix} 3 & 2 & \textcircled{5} & 4 \\ 4 & 5 & 3 & \textcircled{4} \\ \textcircled{8} & 7 & 6 & 6 \\ 5 & \textcircled{8} & 4 & 3 \end{bmatrix}$$

■ **Example 22.** Find a minimum assignment in the matrix below.

$$\begin{bmatrix} 7 & 3 & 7 & 6 & 8 & 9 & 3 \\ 8 & 4 & 5 & 7 & 9 & 6 & 2 \\ 7 & 2 & 8 & 8 & 6 & 9 & 3 \\ 7 & 4 & 6 & 6 & 6 & 8 & 1 \\ 6 & 5 & 4 & 7 & 9 & 9 & 3 \\ 7 & 8 & 8 & 7 & 8 & 8 & 3 \\ 7 & 6 & 9 & 6 & 8 & 8 & 4 \end{bmatrix}$$

Solution:

$$\begin{bmatrix} 1 & 1 & 3 & 0 & 2 & 3 & 2 \\ 2 & 2 & 1 & 1 & 3 & 0 & 1 \\ 1 & 0 & 4 & 1 & 0 & 3 & 2 \\ 1 & 2 & 2 & 0 & 0 & 2 & 0 \\ 0 & 3 & 0 & 1 & 3 & 3 & 2 \\ 1 & 6 & 4 & 1 & 2 & 2 & 2 \\ 1 & 4 & 5 & 0 & 2 & 2 & 3 \end{bmatrix} \longrightarrow \begin{bmatrix} 1 & 1 & 3 & 0 & 2 & 3 & 2 \\ 2 & 2 & 1 & 1 & 3 & 0 & 1 \\ 1 & 0 & 4 & 2 & 0 & 3 & 2 \\ 1 & 2 & 2 & 0 & 0 & 2 & 0 \\ 0 & 3 & 0 & 1 & 3 & 3 & 2 \\ 0 & 5 & 3 & 0 & 1 & 1 & 1 \\ 1 & 4 & 5 & 0 & 2 & 2 & 3 \end{bmatrix} \longrightarrow \begin{bmatrix} 1 & 0 & 3 & 0 & 1 & 3 & 1 \\ 2 & 1 & 1 & 1 & 2 & 0 & 1 \\ 2 & 0 & 5 & 3 & 0 & 4 & 2 \\ 2 & 2 & 3 & 1 & 0 & 3 & 0 \\ 0 & 2 & 0 & 1 & 2 & 3 & 1 \\ 0 & 4 & 3 & 0 & 0 & 1 & 0 \\ 1 & 3 & 5 & 0 & 1 & 2 & 3 \end{bmatrix}$$

$$\longrightarrow \begin{bmatrix} 7 & \textcircled{3} & 7 & 6 & 8 & 9 & 3 \\ 8 & 4 & 5 & 7 & 9 & \textcircled{6} & 2 \\ 7 & 2 & 8 & 7 & \textcircled{6} & 9 & 3 \\ 7 & 4 & 6 & 6 & 6 & 8 & \textcircled{1} \\ 6 & 5 & \textcircled{4} & 7 & 9 & 9 & 3 \\ \textcircled{7} & 8 & 8 & 7 & 8 & 8 & 3 \\ 7 & 6 & 9 & \textcircled{6} & 8 & 8 & 4 \end{bmatrix}$$

In certain problems some assignments should never be made. We insert a **non-selector symbol** ∞ when a particular assignment is to be excluded. Peeling and adjusting leaves this symbol unchanged.

■ Example 23 Better Buy retail chain plans to open a new outlet store in southern Illinois. Five key people from within the company will set up the operation. Their efficiency ratings in each area are given in the following table. The symbol ∞ is used when an assignment cannot be made for either lack of training or past performance. Find an optimal assignment.

Employee	*Credit Manager*	*Stockroom Manager*	*Personnel Manager*	*Public Relations Manager*	*Advertising and Promotion Manager*
Adams	∞	8	7	6	8
Brown	8	∞	∞	9	7
Crain	7	10	∞	8	∞
Davis	9	8	8	9	10
Evans	9	8	7	∞	∞

Solution:

$$\begin{bmatrix} \infty & 2 & 1 & 3 & 2 \\ 1 & \infty & \infty & 0 & 3 \\ 2 & 0 & \infty & 1 & \infty \\ 0 & 2 & 0 & 0 & 0 \\ 0 & 2 & 1 & \infty & \infty \end{bmatrix} \longrightarrow \begin{bmatrix} \infty & 1 & 0 & 2 & 1 \\ 1 & \infty & \infty & 0 & 3 \\ 2 & 0 & \infty & 1 & \infty \\ 0 & 2 & 0 & 0 & 0 \\ 0 & 2 & 1 & \infty & \infty \end{bmatrix}$$

Adams becomes personnel manager.
Brown becomes public relations manager.
Crain becomes stockroom manager.
Davis becomes advertising and promotion manager.
Evans becomes credit manager. ■

Exercise Set 4.6

1. Make a minimum assignment in the following matrix.

$$\begin{bmatrix} 3 & 3 & 4 & 4 \\ 4 & 3 & 3 & 4 \\ 5 & 4 & 6 & 4 \\ 6 & 5 & 4 & 5 \end{bmatrix}$$

2. Find one optimal maximum assignment to the matrix in problem 1.
3. Complete problem 2 by finding all alternate optimal maximum assignments.
4. Better Buy retail stores plan to open a new outlet store in southern Illinois next month. Five key people from within the company will

set up the operation. The efficiency rating for each person in each respective area is given below. The symbol ∞ is used when an assignment cannot be made for either lack of training or past performance. Find an optimal assignment.

Employee	Credit Manager	Stockroom Manager	Personnel Manager	Public Relations Manager	Advertising and Promotion Manager
Adams	8	6	7	∞	8
Brown	7	9	6	8	8
Crain	8	8	7	7	7
Davis	∞	∞	7	6	9
Evans	9	10	8	8	∞

5. Five college students—Angela, Bob, Carl, Dana, and Ellen—have applied for summer jobs in a local company. Although company policy encourages employees' children to apply, they do not allow the students to work in the same department as their parent. Angela's father is the supervisor of inspection, Bob's mother is in charge of advertising, and Carl's father is in charge of shipping. The company assigns jobs on the basis of efficiency ratings. The efficiency ratings are given in the following table.

Student	Public Relations	Advertising	Inspection	Packaging	Shipping
Angela	9	9	10	6	3
Bob	9	10	5	5	7
Carl	9	5	3	2	10
Dana	10	9	10	9	10
Ellen	8	2	1	7	3

Find an optimal assignment and determine the average efficiency rating for this assignment.

6. Find a maximum assignment to the matrix in example 22.

7. From the following data

	1	2	3	4	5	6
A	\$3	\$3	\$4	\$5	\$6	\$3
B	\$5	\$9	\$5	\$5	\$5	\$5
C	\$7	\$5	\$5	\$7	\$7	\$7
D	\$3	\$7	\$3	\$4	\$7	\$9
E	\$8	\$9	\$7	\$8	\$4	\$9
F	\$10	\$5	\$6	\$7	\$9	\$9

a. Make a minimum assignment.
b. Is there an alternate minimum assignment? If so, produce one.
c. Make a maximum assignment.
d. Find an alternate maximum assignment if one exists.

8. Find a minimum assignment.

$$\begin{bmatrix} 91 & 85 & 88 & 92 & 74 & 105 & 88 \\ 94 & 85 & 87 & 90 & 75 & 106 & 91 \\ 90 & 89 & 88 & 92 & 73 & 100 & 94 \\ 98 & 89 & 85 & 94 & 70 & 103 & 93 \\ 90 & 92 & 89 & 90 & 75 & 104 & 92 \\ 92 & 87 & 93 & 90 & 73 & 107 & 90 \\ 94 & 86 & 91 & 95 & 70 & 106 & 92 \end{bmatrix}$$

4.7 Chapter Review

Important Terms and Symbols in Chapter 4.

transportation problems (introduction)
shipping schedule (section 4.1)
matrix schedule (section 4.1)
northwest corner method (section 4.1)
loops (section 4.2)
proper initial assignment (section 4.2)
pattern change (section 4.2)
cost factor (section 4.3)
proper initial solution (section 4.4)
common-sense initial solution (section 4.4)
assignment problems (section 4.6)
Hungarian Method (section 4.6)
peeling (section 4.6)
zero pattern (section 4.6)
picking (section 4.6)
adjusting (section 4.6)
non-selector symbol (∞) (section 4.6)

Review Exercises

1. Use the northwest corner method to find a shipping schedule subject to the information in the following charts.

Factory	*Output*
A	120
B	340
C	280
D	260

Outlets	*Orders*
1	284
2	166
3	342
4	108
5	100

2. With reference to the following chart, complete the letter sequence that defines the loop necessary to fill the blank square identified by the letter (1) *c*, (2) *e*, (3) *s*, and (4) *m*.

a 20	*b* 5	*c*	*d*	*e*	*f*	25
g	*h* 10	*i* 30	*j* 5	*k*	*l*	45
m	*n*	*o*	*p* 10	*q* 10	*r*	20
s	*t*	*u*	*v*	*w* 5	*x* 10	15
20	15	30	15	15	10	

3. With reference to the following chart, complete the letter sequence necessary to fill the blank square identified by the letter (1) *a*, (2) *d*, (3) *q*, (4) *l*, and (5) *w*.

a	*b*	*c* 10	*d*	*e* 15	*f*	25
g 20	*h* 5	*i* 20	*j*	*k*	*l*	45
m	*n* 10	*o*	*p* 10	*q*	*r*	20
s	*t*	*u*	*v* 5	*w*	*x* 10	15
20	15	30	15	15	10	

4. Consider the following chart. The number of entries is too large. The correct number of entries is $m + n - 1 = 10$. Remove two of the entries to achieve a proper number of entries. Remember that the sum of the entries will remain constant at 250.

		(20)			20
(20)			(10)		30
	(30)			(10)	40
(10)	(20)		(20)		50
(20)				(40)	60
		(10)		(40)	50
50	50	30	30	90	

5.

[2] a	[1] (10)	[1] (20)	[2]	[3] b	30
[2]	[1] c	[2]	[1] (40)	[1]	40
[3]	[2] (20)	[2]	[0] d	[1] (0)	20
[2] (10)	[1] e	[3]	[2] (40)	[1]	50
[2]	[2]	[3] (10)	[1]	[2] (140)	150
[1] (40)	[2] (60)	[2] f	[3]	[2] g	100
50	90	30	80	140	

a. Compute the cost factors for squares indicated by letters *a*, *b*, *c*, *d*, *e*, *f*, and *g*.

b. What is the maximum amount that can be used to fill each of the blank squares lettered *a*, *b*, *c*, *d*, *e*, *f*, and *g*?

c. What is the resulting total shipping cost change induced by these pattern changes if each was the only pattern change made?

6. Compute the cost factors for the blank squares labeled *A*, *B*, *C*, *D*, *E*, *F*, *G*, *H*, *I*, and *J* in the following chart.

[1] (50)	[3] (60)	[2] (0)	[1]	[2] *A*	[3]	110
[2] *B*	[1]	[1] (50)	[4] *C*	[1]	[2] *D*	50
[3]	[4]	[2] (20)	[2] E	[1] (10)	[2]	30
[3] *F*	[2] *G*	[1] *H*	[1] (40)	[1]	[2]	40
[1]	[1] *I*	[2] *J*	[1] (60)	[3] (70)	[1] (50)	180
50	60	70	100	80	50	

7. With reference to the chart in problem 6, what is the maximum number that can be moved to fill each of the blanks labeled *A* through *J*?
8. What is the net change in total shipping cost for each pattern change in problems **6** and **7** if each was the only pattern change made?
9. Solve the following transportation problem and give the minimum shipping cost.

Supply

Factory	Number
A	30
B	30
C	40
D	50

Demand

Store	Number
1	20
2	30
3	50
4	50

Shipping Costs

From	**To:** 1	2	3	4
A	4	2	6	5
B	3	3	5	4
C	5	1	6	5
D	3	5	3	3

10. Solve the following transportation problem to minimize the total shipping cost.

[2]	[3]	[2]	[5]	[4]	50
[6]	[4]	[3]	[5]	[4]	60
[4]	[3]	[3]	[4]	[4]	140
[2]	[1]	[1]	[2]	[4]	150
60	60	80	100	100	

11. Find the cheapest shipping cost and produce a shipping schedule for the transportation problem using the following data.

Supply

Producer	Amount
1	20
2	30
3	25
4	45

Demand

Consumer	Amount Consumed
A	30
B	40
C	50

Shipping Cost

From To:	A	B	C
1	5	8	7
2	6	7	9
3	8	8	8
4	4	7	6

12. Solve the transportation problem defined by the following tables to obtain the minimum total shipping cost.

Supply

Factory	Number
I	150
II	200
III	200

Demand

Store	Number
1	150
2	150
3	75
4	175

Shipping Costs

From To:	1	2	3	4
I	5	7	5	3
II	6	4	3	7
III	5	6	6	4

13. Solve the following transportation problem to find a minimum cost shipping schedule and give the total cost.

Supply

Factory	Number
A	30
B	15
C	15

Demand

Store	Number
1	20
2	15
3	25

Shipping Costs

From To:	1	2	3
A	3	2	2
B	2	2	2
C	3	1	3

14. Solve the following transportation problem where supply and demand are not equal and give the minimum shipping cost.

Supply

Factory	Number
1	100
2	100
3	100
4	50

Demand

Store	Number
A	120
B	30
C	50
D	80

Shipping Costs

From To:	A	B	C	D
1	3	2	4	1
2	1	3	1	4
3	3	5	4	2
4	1	3	4	4

Warehousing Costs

Factory	1	2	3	4
Cost	2	5	1	5

15. A company has decided to use overtime production, as necessary, to compensate for a supply shortage. Find a minimum cost shipping schedule and give the minimum cost, based on the following information.

Supply

Factory	*Number*
A	200
B	300
C	150

Demand

Store	*Number*
1	300
2	250
3	250

Shipping Costs

From	**To:** *1*	*2*	*3*
A	2	5	4
B	2	4	3
C	8	2	3

Overtime Costs

Factory	*A*	*B*	*C*
Cost	6	5	6

16. A poultry processing company has discovered that it has produced too many turkeys for Christmas sales at its plants. The company has produced 10,000 turkeys, but the regional distributors have ordered only 7,000. The company has storage space in Jackson for 2,000 of the extra turkeys. A cannery in Houston will buy 900 of the remaining 1,000 turkeys, but the company will lose $3.50 per turkey sold to the cannery. The company president decides that, rather than let the other 100 spoil, he will give them to his alma mater in Minneapolis. But this will result in a $6.00 loss per turkey. Solve the transportation problem based on the following tables, where all costs are given in 10¢ units; i.e., a cost of 8 means 80¢, a cost of 12 means $1.20, and so on. *Hint:* Don't forget the losses for the gift turkeys and those sold to the cannery.

Supply

Plant	*Number*
Jackson	4000
Little Rock	3000
Shreveport	3000

Demand

Distributor	*Number*
Philadelphia	1500
Atlanta	1000
Denver	2000
Los Angeles	2500

Shipping Costs

From	**To:** *Philadelphia*	*Atlanta*	*Denver*	*Los Angeles*	*Minneapolis*	*Houston*	*Storage*
Jackson	8	5	10	12	10	5	2
Little Rock	8	6	9	11	8	5	4
Shreveport	9	6	9	10	9	3	5

17. Make a maximum assignment based on the following:

	1	2	3	4	5
A	7	3	7	6	8
B	8	4	5	7	9
C	7	2	8	8	6
D	7	4	6	6	6
E	6	5	4	7	9

18. Make a minimum assignment based on the table in problem 17.

19. Make a maximum assignment based on the following table.

	1	2	3	4
A	52	95	38	94
B	75	23	55	75
C	80	75	76	64
D	32	90	88	55

20. Make a minimum assignment based on the table in problem 19.

21. Doris, Joy, Judy, and Sue have been hired by a company to fill openings in advertising, sales, public relations, and personnel. They take a series of tests and have interviews with the heads of each department. They are then assigned an efficiency rating between 0 and 10, where 10 is the maximum possible. How should the women be assigned, based on the efficiency ratings given in the following table?

	Assignment			
Applicant	*Advertising*	*Sales*	*Public Relations*	*Personnel*
Doris	9	8	7	6
Joy	8	6	4	6
Judy	7	8	8	8
Sue	7	6	8	6

22. James, Jerry, Terry, and Vern have agreed to test some new electronics equipment. Each person will test one of the new products for a period of time. The following table gives the time, in weeks, each individual would require to test each item before making an evaluation. How should the company assign workers to products to minimize time, since the workers will be paid for their time?

	Item			
Worker	I	II	III	IV
James	6	4	5	7
Jerry	7	5	6	5
Terry	7	5	6	6
Vern	6	6	6	6

23. Six men—Tom, Ron, J.D., Fred, Joe, and Dick—have agreed to chair committees for the local scout exhibition. They are asked to rate their interest in each of the six committees on a scale of 1 to 10 (10 is highest). How should they be assigned to committees on the basis of the following ratings?

	Committee					
Person	*Advertising*	*Set-up*	*Donations*	*Refreshments*	*Tickets*	*Clean-up*
Tom	3	7	6	8	9	3
Ron	4	5	7	9	6	2
J.D.	2	8	8	6	9	3
Fred	4	6	6	6	8	1
Joe	5	4	7	9	9	3
Dick	8	8	7	8	8	3

24. In order to facilitate the collection of county taxes, the county collector's office has been given permission to establish collection offices in seven communities outside the county seat. Seven residents—Mary, Judi, Pat, Connie, Vicki, Marilyn, and Candy—have agreed to staff the offices for ten days as a civic project. They will not charge the county for anything but mileage. The table below gives the mileage for each volunteer to each community. How should the volunteers be assigned to minimize mileage and, hence, cost?

	Assignment						
Volunteer	*1*	*2*	*3*	*4*	*5*	*6*	*7*
Mary	14	16	6	18	14	6	12
Judi	16	18	8	12	10	4	14
Pat	14	12	4	18	16	6	16
Connie	14	12	8	16	12	2	12
Vicki	12	18	10	18	8	6	14
Marilyn	14	16	16	16	16	6	14
Candy	14	16	12	16	18	8	12

25. As part of a class project on careers, eight children are to visit adults and learn about their occupations. The children will give a report to the class. Each child is asked to rank each of the eight jobs in order of their preference (1 is most desirable). The table gives the results of the ranking.

	Occupation							
Student	*Construction Worker*	*Doctor*	*Dentist*	*Merchant*	*Nurse*	*Pharmacist*	*Lawyer*	*Teacher*
Brent	3	1	2	4	7	5	8	6
Erik	1	2	3	4	8	5	7	6
Gina	8	3	4	7	1	6	5	2
Kirk	4	2	3	1	8	5	6	7
Nickie	6	2	1	7	8	3	5	4
Ronnie	4	3	1	2	8	5	7	6
Timmy	6	1	2	4	3	5	7	8
Wendy	8	7	6	5	1	4	2	3

After looking over the rankings, the teacher decides that since Erik's father is in construction, Gina's mother is a teacher, and Timmy's mother is a nurse, these children should not be allowed those particular assignments. How should the assignments be made, subject to the stated restrictions, so that there is a minimum of dissatisfaction?

5 Sets and Counting

Introduction

One of the most difficult concepts to master is the concept of counting. We first learn to count by using a one-to-one matching technique with the counting numbers 1, 2, 3, 4, . . . , and so on. Counting by 2s, by 5s, by 10s, and, finally, addition and multiplication are logical extensions of counting. There is a segment of the business community that counts by the dozen and prices items by the gross. These people know, without contemplation, that an increase in the wholesale cost of \$7.60 to \$9.40 per dozen is a \$.15 per item increase.

Conversion tables help us to count the number of inches in $3\frac{1}{3}$ yards, to calculate the speed of a ship in nautical miles per hour, or to determine the metric equivalent to a 12-gallon tank. Counting the number of possible poker hands, or the number of ways ten books can be arranged on a shelf, or finding the number of committees of four people that can be formed from a group of eight men and six women when not all committee members can be of the same sex involves more complex counting techniques.

It is possible to state many different kinds of counting problems. Easy, straightforward problems often sound the same as difficult problems. The subtle differences between problems are understood only after repeated practice. The student is well advised to analyze and solve a large number of problems.

In this chapter, several counting techniques will be examined. These techniques will be useful in the next chapter when we consider probability.

One of the concepts that provides assistance in counting, as well as convenient notation and terminology for studying probability, is the concept of a set. In the first part of this chapter, some of the notation, terminology, and definitions of set theory will be developed as a preparation for the study of counting and probability.

5.1 Sets

A **set** is a collection of objects. The objects in a set are called **members** or **elements.** When sets are written out to specify the elements, braces { } are used to indicate "set." For example, $\{a,e,i,o,u\}$ is "the set of vowels in the alphabet." The letter "a" is a member, or element, of this set, and we write $a \in \{a,e,i,o,u\}$. The symbol $\in$ is read "is a member of" or "is an element of." The symbol $\notin$ is used for "is not a member of" or "is not an element of." For example, $b \notin \{a,e,i,o,u\}$. Sometimes, sets contain so many elements that it is not practical, or possible, to write them all. The set of counting numbers less than 500 contains all of the numbers 1 to 499. It is not practical to write them all. Instead, we abbreviate by writing $\{1,2,3,\ldots,499\}$, where the three dots represent the missing elements. The infinite set of all counting numbers can be written $\{1,2,3,\ldots\}$ without a last element.

When working with infinite sets, or sets whose elements are difficult to list, we often use a slightly different notation, called a **set-builder notation.** The idea is to include a description of the set within the braces to identify exactly what elements are present.

■ Example 1. The set of real numbers (R) between 0 and 1, including 0 and 1, can be written $\{x \in R | 0 \leq x \leq 1\}$ or $\{x | x \in R \text{ and } 0 \leq x \leq 1\}$. The vertical bar is read "such that." The two preceding sets could be read "the set of all x in R such that x is greater than or equal to 0 but less than or equal to 1." This notation is often shortened to $\{x | 0 \leq x \leq 1\}$, where the general restriction on x (x is a real number) is understood. ■

Capital letters, such as A, B, S, T, and so forth, are commonly used to represent sets, whereas small letters are used for elements of sets. If we let $V = \{a,e,i,o,u\}$, we say $a \in V$, $e \in V$, $r \notin V$, and so on. We use the symbol $\emptyset$ or an empty brace ({ }) to represent a set with no elements. Such a set, and it is a set, is called the **empty set** or the **null set.**

Now that some set terminology has been developed, we can begin to develop some of the formal ideas of set theory. We begin by considering ways of comparing two sets, called **set relations.**

> **Definition 5.1** If A and B are two sets, then A is equal to B, written $A = B$, providing A and B are identical sets, that is, A and B contain exactly the same elements.

Later in this chapter, when we consider properties of sets, we will sometimes refer to sets as being **equal sets.** Definition 5.1 states precisely what we mean by equal sets.

Another important set relation is the **subset** relation, given in definition 5.2.

> **Definition 5.2** If A and B are two sets, then A is a **subset** of B, written $A \subset B$, providing every element of A is also an element of B.

To determine if a set, C, is a subset of another set, B, we check to make sure that each element of C is an element of B. If there is at least one element of C that is not an element of B, then C is not a subset of B, written $C \not\subset B$.

■ Example 2. Given $A = \{0,1,2\}$, $B = \{0,1,2,3,4\}$, and $C = \{1,3,5\}$, then $A \subset B$ because each element of A (0,1, and 2) is also an element of B. On the other hand, $C \not\subset B$ because $5 \in C$ and $5 \notin B$ (read "5 is an element of C and 5 is not an element of B"). ■

If $A \not\subset B$, then there is at least one element of A that is not an element of B. Then $\emptyset \not\subset A$ would mean there must be at least one element in $\emptyset$ which is not in A. This is not possible since $\emptyset$ has no elements. We conclude, therefore, that $\emptyset \subset A$ for every set A. In addition, every element of S is an element of S, so $S \subset S$ for every set S.

■ Example 3. Given $B = \{1,2,3\}$, find all subsets of B.

Solution: According to the preceding remarks, $\emptyset \subset B$ and $B \subset B$. In addition, the three single element sets $\{1\}$, $\{2\}$, and $\{3\}$ are subsets of B.

There are also three subsets with two elements each: $\{1,2\}$, $\{1,3\}$, and $\{2,3\}$. Note here that $\{1,2\} = \{2,1\}$, so we count this only as one subset.

Therefore, there are eight subsets of the set $B = \{1,2,3\}$. They are as follows:

$\emptyset$	$\{2\}$	$\{1,2\}$	$\{2,3\}$
$\{1\}$	$\{3\}$	$\{1,3\}$	$\{1,2,3\}$

■

It can be shown that if A has n elements, then there are 2^n subsets of A. The set $V = \{a,e,i,o,u\}$ has 5 elements, so there are $2^5 = 32$ subsets of V.

We have considered two ways of comparing sets, called set relations. We turn our attention now to ways of combining sets, called **set operations.** We will consider three set operations.

Definition 5.3 Let A and B be sets. The **intersection** of A and B, written $A \cap B$, is the single set consisting of all elements common to both A and B. Written in set notation, $A \cap B = \{x|x \in A \text{ and } x \in B\}$.

Definition 5.4 Let A and B be sets. The **union** of A and B, written $A \cup B$, is the set consisting of all elements that are in A or B or both. The union is, then, the set of elements in at least one of the two sets. Written in set notation,

$$A \cup B = \{x|x \in A \text{ or } x \in B\}.$$

It should be emphasized that if an element appears in both A and B, it appears only once in the union of A and B.

■ Example 4. Given $A = \{0,1,2,3,4\}$, $B = \{0,2,4,6\}$, and $C = \{1,3,5,7\}$, find **1.** $A \cap B$, **2.** $A \cap C$, **3.** $B \cap C$, **4.** $A \cup B$, **5.** $A \cup C$, **6.** $B \cup C$, **7.** $A \cap \emptyset$, and **8.** $B \cup \emptyset$.

Solutions:

1. $A \cap B = \{0,2,4\}$
2. $A \cap C = \{1,3\}$
3. $B \cap C = \emptyset$ since there are no common elements. When the intersection of two sets is the empty set, the sets are said to be **disjoint sets.**
4. $A \cup B = \{0,1,2,3,4,6\}$
5. $A \cup C = \{0,1,2,3,4,5,7\}$
6. $B \cup C = \{0,1,2,3,4,5,6,7\} = \{0,2,4,6,1,3,5,7\}$. The order in which the elements are written doesn't matter.
7. $A \cap \emptyset = \emptyset$
8. $B \cup \emptyset = B$ ■

In most problems there is a stated or implied restriction on the elements in our sets. The set of acceptable values is called the **universal set,** or simply the universe, and is denoted by U. The third set operation relies on the universe and can now be defined.

Definition 5.5 Let U be the universe and let $A \subset U$. Then the **complement** of A, written A', is the set consisting of all elements of the universe that are not in set A. Written with set notation, $A' = \{x|x \in U \text{ and } x \notin A\}$ or simply $\{x|x \notin A\}$.

■ Example 5. Let $U = \{0,1,2,3,4,5\}$, $A = \{0,1,2\}$, $B = \{0,2,4\}$, and $C = \{1,3,5\}$. Find **1.** A', **2.** B', **3.** C', **4.** $A \cap C'$, **5.** $\emptyset'$, and **6.** U'.

Solutions:

1. $A' = \{3,4,5\}$
2. $B' = \{1,3,5\} = C$
3. $C' = \{0,2,4\} = B$
4. $A \cap C' = \{0,1,2\} \cap \{0,2,4\} = \{0,2\}$
5. $\emptyset' = U$
6. $U' = \emptyset$ ■

Exercise Set 5.1

1. Given $A = \{0,1,2,3,4,5\}$, $B = \{1,3,5\}$, $C = \{0,2,4\}$, and $D = \{0,3\}$, determine whether the following are true or false.
   - **a.** $1 \in B$
   - **b.** $2 \subset C$
   - **c.** $B = C$
   - **d.** $\{3\} \subset D$
   - **e.** $\emptyset \subset C$
   - **f.** $3 \in C$
   - **g.** $5 \in B$
   - **h.** $7 \in A$
   - **i.** $B \subset A$
   - **j.** $D \subset B$
   - **k.** $\emptyset \in B$
   - **l.** $1 \in (B \cap C)$
   - **m.** $0 \in (C \cap D)$
   - **n.** $\emptyset = B \cap C$
   - **o.** $C \subset A$
   - **p.** $\{3\} \in B$
   - **q.** $5 \subset A$
   - **r.** $1 \notin D$
   - **s.** $A \subset B$
   - **t.** $B \subset B$
2. Use set notation to write each of the following:
   - **a.** The set of numbers whose square is 4.
   - **b.** The set of letters in your last name.
   - **c.** The set of letters in the word "Mississippi."
   - **d.** The set of letters in the word "Illinois."
   - **e.** The set of women who have been president of the United States.
   - **f.** The set of United States senators from your state.
   - **g.** The set of counting numbers greater than 20.
   - **h.** The set of counting numbers that satisfy $x + 1 = 9$.
   - **i.** The set of real numbers that satisfy $x^2 + 1 = 10$.
   - **j.** The set of real numbers that satisfy $x^2 + 4 = 0$.
   - **k.** The set of real numbers that satisfy $x + 3 > 5$.
   - **l.** The set of counting numbers that satisfy $x^2 + 1 > 17$.

3. Given the universal set $\{0,1,2,3,4,5,6,7\}$
   a. Write the set of even numbers.
   b. Write the set of numbers less than 6.
   c. Write the set $\{x|3 < x < 7\}$.
   d. Write the set of numbers where $x < 8$.
4. Use set-builder notation to write the following sets.
   a. $\{2,4,6,8, \ldots ,24\}$
   b. {Kennedy, Johnson, Nixon, Eisenhower, Ford, Carter, Reagan}
   c. $\{1,4,9,16, \ldots\}$
   d. {I, me, you, us, we, our, them, they, those, . . .}
5. Given $S = \{2,3,\{2\},\{2,4\}\}$, which of the following statements are true?
   a. $2 \in S$
   b. $\{2\} \subset S$
   c. $\{2\} \in S$
   d. $2 \subset S$
   e. $\emptyset \in S$
   f. $\{2,4\} \subset S$
   g. $\{2,4\} \in S$
   h. $\{3,4\} \in S$
   i. $\{3,4\} \subset S$
   j. $\emptyset \subset S$
6. Find $A = \{x^2|x \in X\}$, given $X = \{1,2,3,4,5\}$.
7. Given $U = \{0,1,2,3,4,5,6,7,8,9\}$, $A = \{0,3,6,9\}$, $B = \{0,2,4,6,8\}$, $C = \{1,3,5,7,9\}$, $D = \{1,4,7\}$, and $E = \{0,5\}$, find
   a. $A \cup B$.
   b. $A \cap C$.
   c. $C \cup D$.
   d. $A \cup E$.
   e. $A \cap (B \cup C)$.
   f. $A' \cap C$.
   g. $B \cap C'$.
   h. $(D \cup E)'$.
   i. $B \cap C$.
   j. $D \cup E$.
   k. C'.
   l. $A \cap B$.
   m. $B \cup C'$.
   n. $A' \cup C'$.
   o. $(A \cap B) \cup (A \cap C)$.
   p. $(D \cap E)'$.
   q. $C \cap D$.
   r. $A \cap E$.
   s. $B \cap D$.
   t. $B \cup C$.
   u. $B' \cup C$.
   v. $D' \cup E'$.
   w. $(A \cap C)'$.
   x. $(A \cap B) \cap C$.
8. Find all of the subsets of $B = \{a,b,c,d\}$.
9. If $A \cup B = A$, what can be said about A and B?
10. If $A \cap B = A$, what can be said about A and B?
11. Suppose $A \cup B = \emptyset$. What must be true about A and B?
12. Suppose $A \cup B = U$ and $A \cap B = \emptyset$. What is the relationship between A and B?

5.2 Venn Diagrams and Properties of Sets

It is frequently helpful, when mathematical concepts are being worked with, to draw pictures to represent the given information. *Venn Diagrams* are pictures used to represent sets or properties of sets. Venn Diagrams can take a variety of forms, but we will use only a simple form in which the universe is represented by points in a rectangle and specific subsets of the universe are represented by points in circles. It is understood that the elements of a particular set are represented by points inside the circle and those not in the set are represented by points outside the circle, whereas everything (the universe) is represented by points inside the rectangle. Figure 5.1 illustrates a universe and a set A that is a subset of the universe. The set A' is also identified.

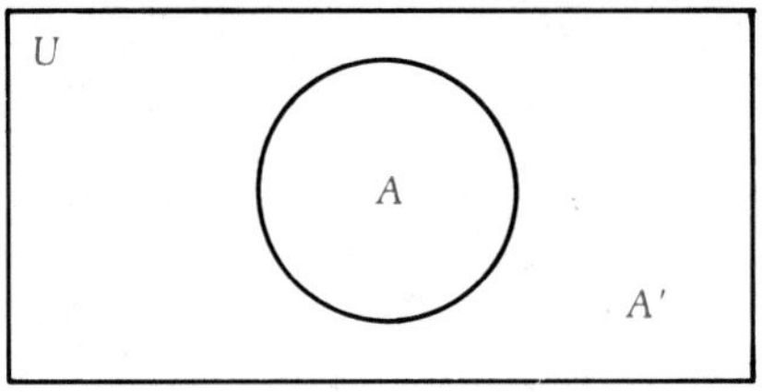

Figure 5.1

Figure 5.2 shows Venn Diagrams illustrating the three ways in which two sets can be related.

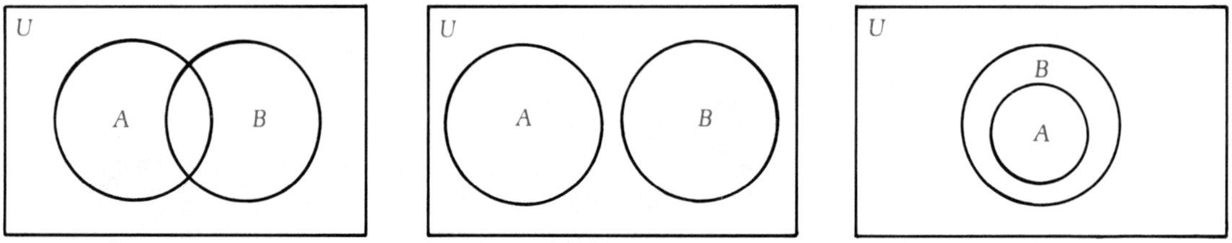

a. Overlapping sets **b.** Disjoint sets **c.** One set as a subset of the other

Figure 5.2

When working with two general sets, we will normally use figure 5.2(a). In figure 5.2(a), the circles overlap and divide the universe into four disjoint regions, as shown in figure 5.3.

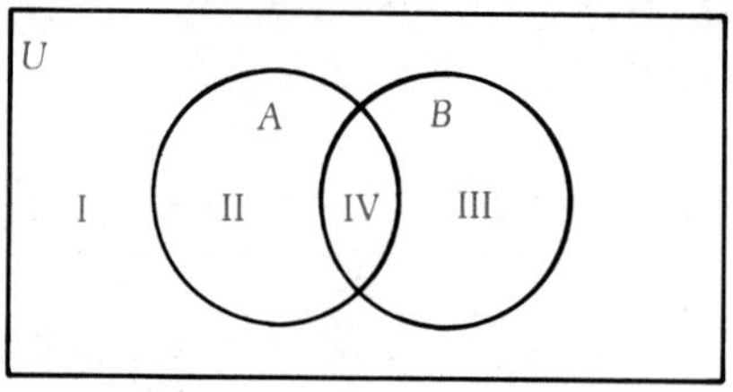

Region I is the set $(A \cup B)'$.
Region II is the set $A \cap B'$.
Region III is the set $A' \cap B$.
Region IV is the set $A \cap B$.

Figure 5.3

The set $A \cup B$ consists of the regions II, III, and IV. Any of these four regions could contain no elements with the following results.

1. If region I, $(A \cup B)'$, is empty, then $A \cup B = U$.
2. If region II, $A \cap B'$, is empty, then all of the elements of A are in region IV and, hence, in B. Therefore, $A \subset B$.
3. If region III, $A' \cap B$, is empty, then $B \subset A$.
4. If region IV, $A \cap B$, is empty, then A and B are disjoint.

Statements 2 and 3 are examples of figure 5.2(c), whereas statement 4 is an example of figure 5.2(b).

The idea is that figure 5.3 is general enough to be used for any situation when only two sets are involved. Figures 5.2(b) and 5.2(c) are not that general.

Venn Diagrams can also be used to illustrate set operations. The technique is to shade the desired regions. In figure 5.4 the Venn Diagrams for $A \cup B$ and $A \cap B$ have been drawn.

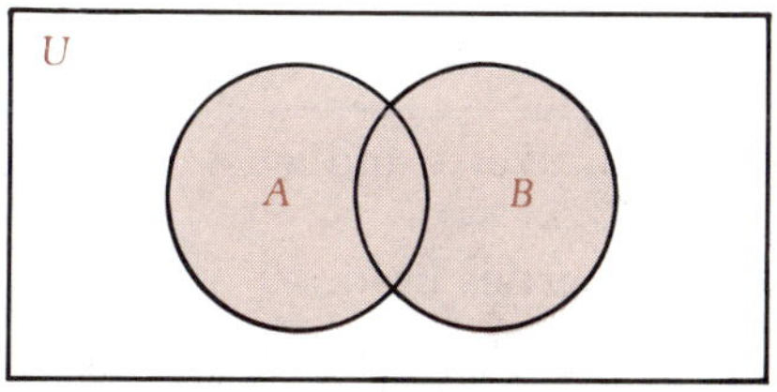

a. $A \cup B$

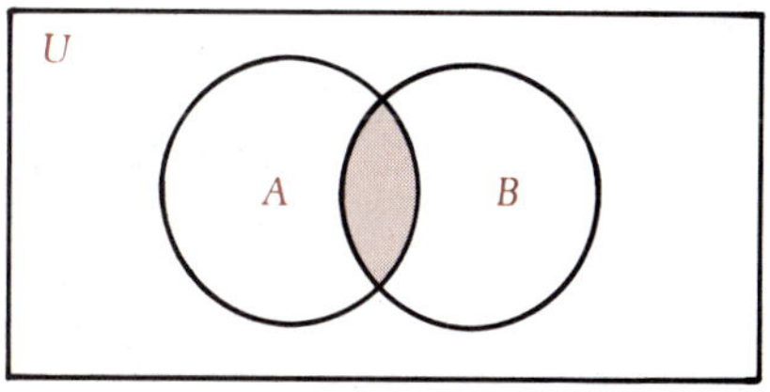

b. $A \cap B$

Figure 5.4

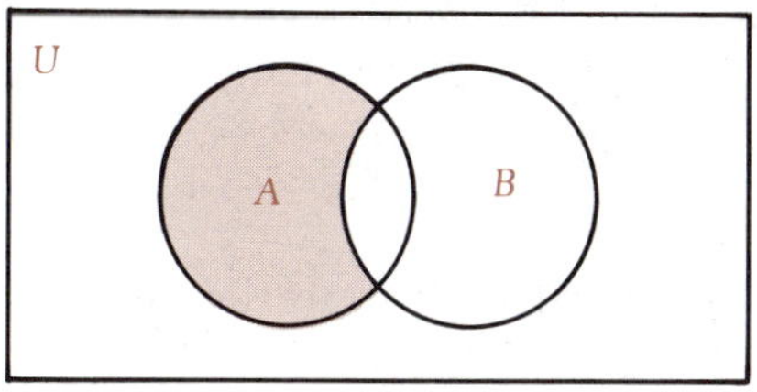

Figure 5.5 is the Venn Diagram for $A \cap B'$, the area inside A and outside B.

Figure 5.5

The use of Venn Diagrams can be extended to include more than two sets. Figure 5.6 illustrates the union and intersection of three sets in the most general case. When three sets are being considered, there are numerous possibilities. It is frequently easier to work with the general case than it is to identify all of the possibilities. There are eight distinct, disjoint regions that are identified. Any combination of these regions can be empty; thus, one figure can be used to encompass all of the possibilities.

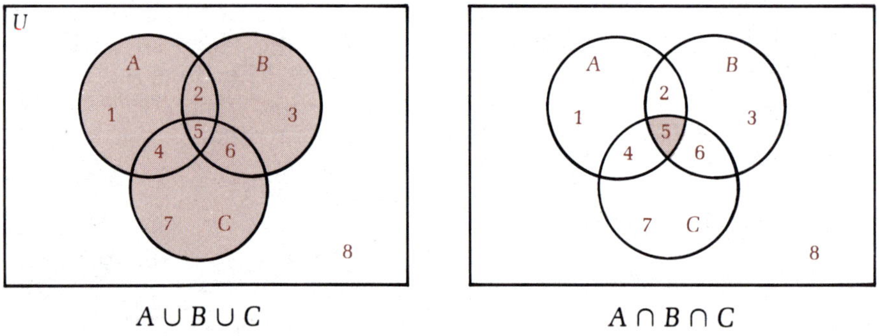

Figure 5.6

Before saying more about the eight regions and what they represent, we state several properties of set operations.

Property 5.1 Let U be a universe and let A, B, and C be subsets of U. Then

1. The **commutative properties** hold.
   a. $A \cup B = B \cup A$
   b. $A \cap B = B \cap A$
2. The **associative properties** hold.
   a. $(A \cup B) \cup C = A \cup (B \cup C)$
   b. $(A \cap B) \cap C = A \cap (B \cap C)$
3. The **distributive properties** hold.
   a. $A \cup (B \cap C) = (A \cup B) \cap (A \cup C)$
   b. $A \cap (B \cup C) = (A \cap B) \cup (A \cap C)$

Venn Diagrams can be used to illustrate the validity of these properties. Note that such an illustration does not constitute a mathematical proof. The Venn Diagrams serve only as an intuitive verification.

■ Example 6. Use Venn Diagrams to illustrate

$$A \cap (B \cup C) = (A \cap B) \cup (A \cap C).$$

Solution: We first find the region for the left side by shading the set $B \cup C$ (figure 5.7) and then intersecting this region with set A (figure 5.8).

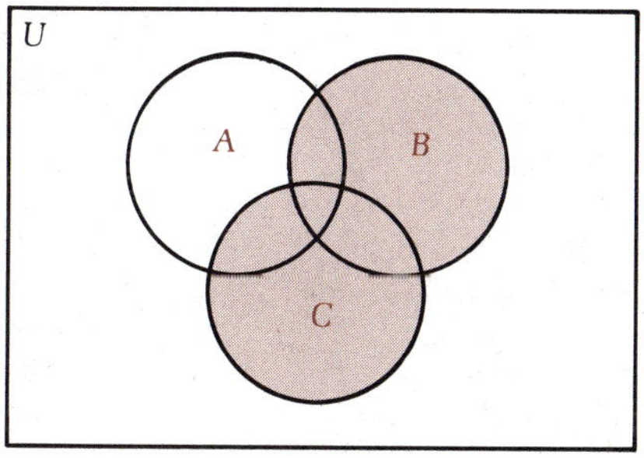

Figure 5.7

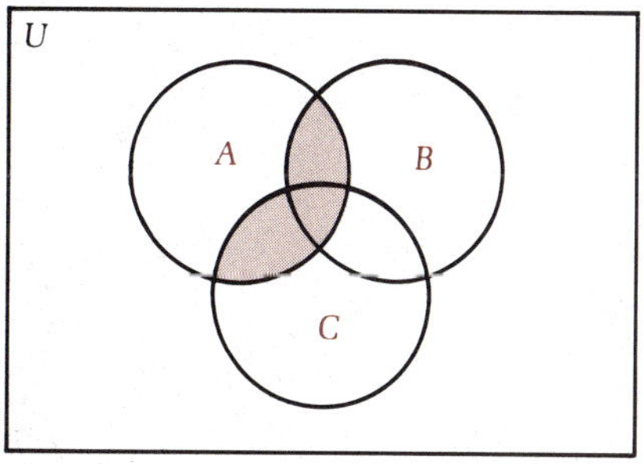

Figure 5.8

$A \cap (B \cup C)$ is the portion of $B \cup C$ that intersects A. For $(A \cap B) \cup (A \cap C)$, we first shade $A \cap B$ (figure 5.9) and then $A \cap C$ (figure 5.10).

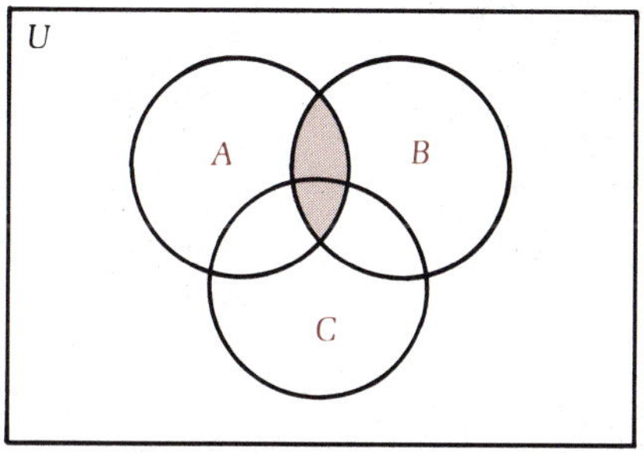

Figure 5.9

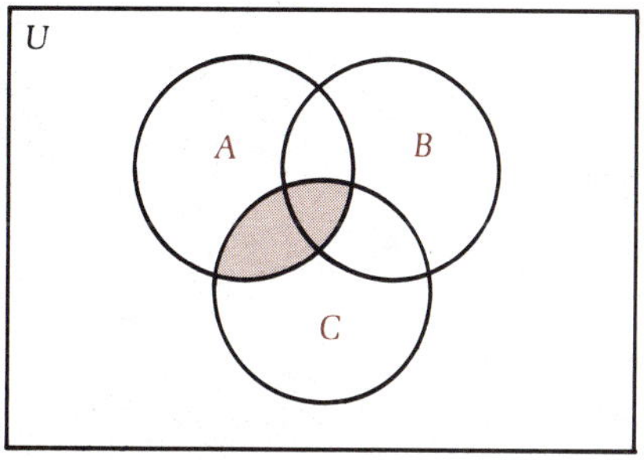

Figure 5.10

$(A \cap B) \cup (A \cap C)$ is that portion that is shaded in at least one of the two figures. The union of these two shaded regions (figure 5.11) is exactly the same as figure 5.8.

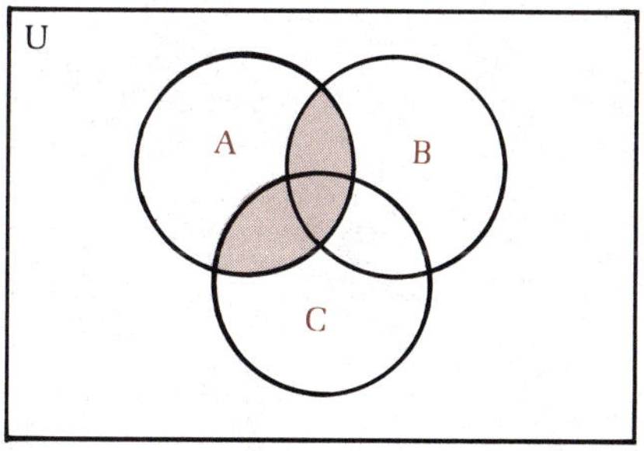

$(A \cap B) \cup (A \cap C)$

Figure 5.11

Hence, we conclude that $A \cap (B \cup C) = (A \cap B) \cup (A \cap C)$. ■

The following properties are also useful when working with sets.

Property 5.2 If U is a universe and A and B are subsets of U, then

1. $A \subset (A \cup B)$.
2. $B \subset (A \cup B)$.
3. $(A \cap B) \subset A$.
4. $(A \cap B) \subset B$.
5. $A \cup \emptyset = A$.
6. $A \cap \emptyset = \emptyset$.
7. $A \cup U = U$.
8. $A \cap U = A$.
9. $(A')' = A$.
10. $A \cup A' = U$.
11. $A \cap A' = \emptyset$.

Two other important set properties will be useful to us. The first is that the complement of the union of two sets is the intersection of the complements of the sets. The other is that the complement of the intersection of two sets is the union of the complements of the sets. These two statements are known as **DeMorgan's Laws.**

Property 5.3 *DeMorgan's Laws*—if A and B are subsets of a universe U, then

1. $(A \cup B)' = A' \cap B'$.
2. $(A \cap B)' = A' \cup B'$.

■ Example 7. Use Venn Diagrams to illustrate $(A \cup B)' = A' \cap B'$.

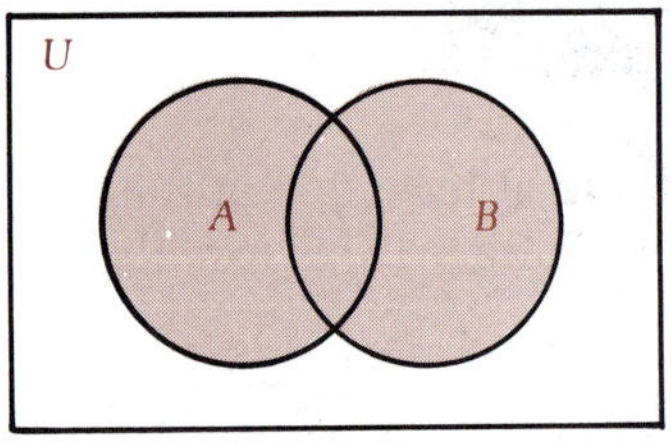

$A \cup B$

Figure 5.12

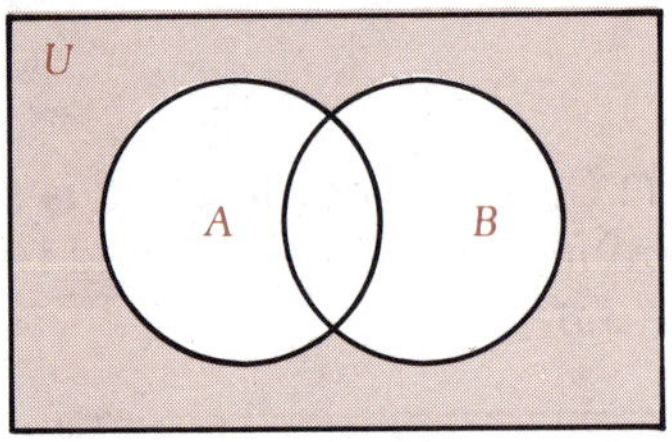

$(A \cup B)'$

Figure 5.13

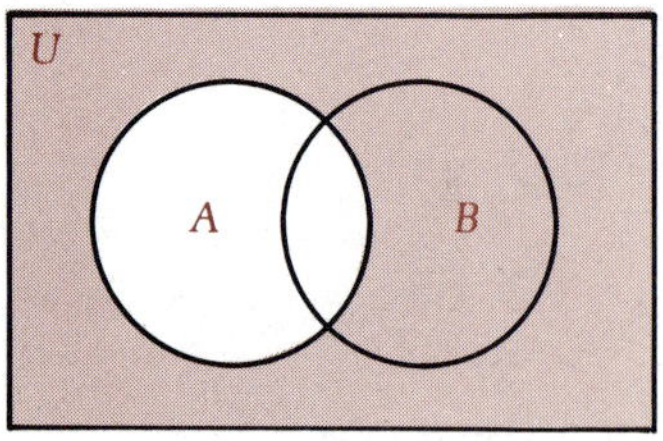

A'

Figure 5.14

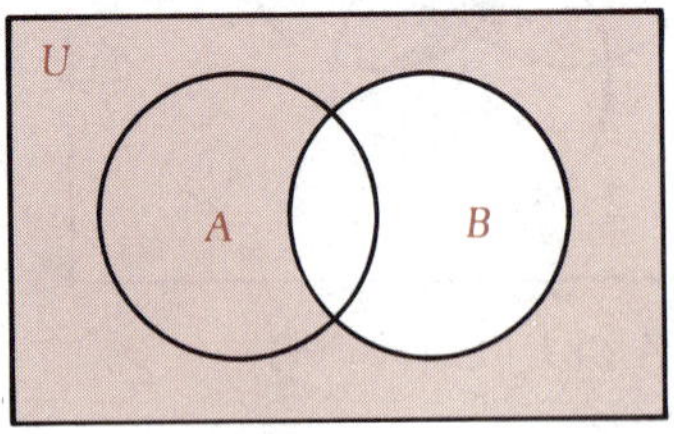

B'

Figure 5.15

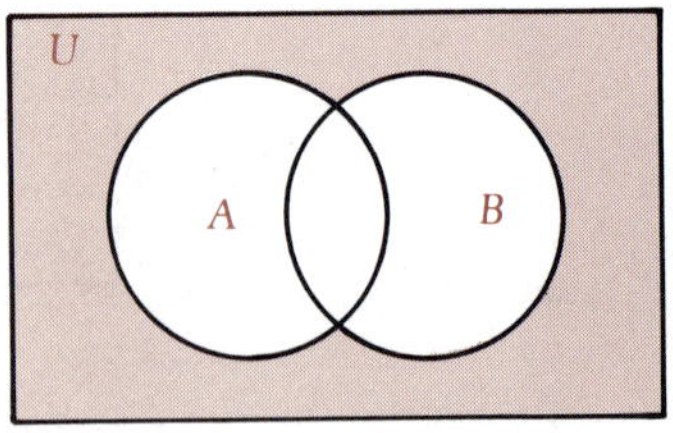

$A' \cap B'$

Figure 5.16

Since figures 5.13 and 5.16 are the same, we conclude $(A \cup B') = A' \cap B'$. ■

Certain other properties of sets will be useful to us when we work with probability. The notation $n(A)$ is used to represent the number of elements (n) in set A.

■ Example 8. Given $A = \{0,1,2,4,7\}$, $S = \{2,4,\{2,4\}\}$, $R = \{A,B,C,D,E,F\}$, and $T = \{X|X$ is a letter of the alphabet$\}$, find

a. $n(A)$ b. $n(S)$ c. $n(R)$ d. $n(T)$ e. $n(\emptyset)$

Solutions: a. $n(A) = 5$ b. $n(S) = 3$ c. $n(R) = 6$ d. $n(T) = 26$ e. $n(\emptyset) = 0$ ■

Using this new notation, we are now in a position to state our first counting principle, called the **inclusion-exclusion principle.**

> **Property 5.4** ***The Inclusion-Exclusion Principle:*** For any two sets A and B, $n(A \cup B) = n(A) + n(B) - n(A \cap B)$.

Since $n(A)$ counts all of the elements in A, including those in $A \cap B$, and $n(B)$ counts all elements in B, the elements in $A \cap B$ are counted twice. Then $n(A \cup B)$ is computed by adding $n(A)$ to $n(B)$ and subtracting the number of elements that have been counted twice, namely $n(A \cap B)$. If A and B are disjoint sets, then $n(A \cap B) = 0$ and $n(A \cup B) = n(A) + n(B)$.

■ Example 9. Given $A = \{1,2,3,4\}$ and $B = \{3,4,5,6\}$, use the inclusion-exclusion principle to find $n(A \cup B)$.

Solution: Solving, we find $n(A) = 4$, $n(B) = 4$, and $n(A \cap B) = n\{3,4\} = 2$. Hence, $n(A \cup B) = 4 + 4 - 2 = 6$. Checking, since $A \cup B = \{1,2,3,4,5,6\}$, $n(A \cup B) = 6$, by a straight-forward count. ■

Using the inclusion-exclusion principle and some arithmetic, we obtain a second formula:

$$n(A \cap B) = n(A) + n(B) - n(A \cup B).$$

■ Example 10. Given $n(A \cup B) = 62$, $n(A) = 41$, and $n(B) = 37$, find $n(A \cap B)$.

Solution: $n(A \cap B) = n(A) + n(B) - n(A \cup B) = 41 + 37 - 62 = 16.$ ■

■ Example 11. Given $n(U) = 20$, $n(A) = 8$, $n(B) = 14$, and $n(A \cap B) = 4$, find $n(A' \cap B')$.

Solution: Solving, we find $n[(A \cap B)'] = n(U) - n(A \cap B) = 20 - 4 = 16$. By DeMorgan's Law, $(A \cap B)' = A' \cup B'$. Hence, $n(A' \cup B') = n[(A \cap B)'] = 16$. By the inclusion-exclusion principle, $n(A' \cap B') = n(A') + n(B') - n(A' \cup B') = 12 + 6 - 16 = 2$. ■

Exercise Set 5.2

1. Use Venn Diagrams to illustrate that $A \cup (B \cap C) = (A \cup B) \cap (A \cup C)$.
2. Use Venn Diagrams to illustrate that $(A \cap B)' = A' \cup B'$.
3. Draw a Venn Diagram and shade the region represented by each of the following:
   a. $A \cup B'$
   b. $A' \cap (B \cup C)$
   c. $(A \cap B) \cup (A \cap B')$
   d. $A \cap (B \cup C)'$
   e. $(A \cup B) \cup C'$
   f. $(A \cup B)' \cap C$
   g. $(A \cap B) \cup (A \cap C) \cup (B \cap C)$
   h. $(A \cup B) \cap (A \cup C) \cap (B \cup C)$
4. Use the sets given and $\cap$, $\cup$, $'$, to write an expression for the shaded region in each of the following Venn Diagrams.

a.

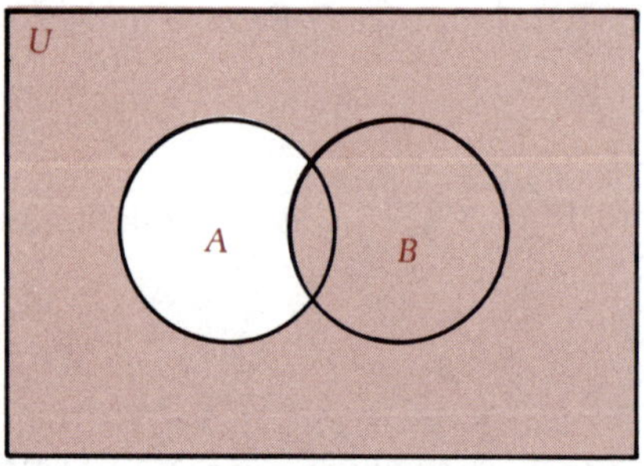

c.

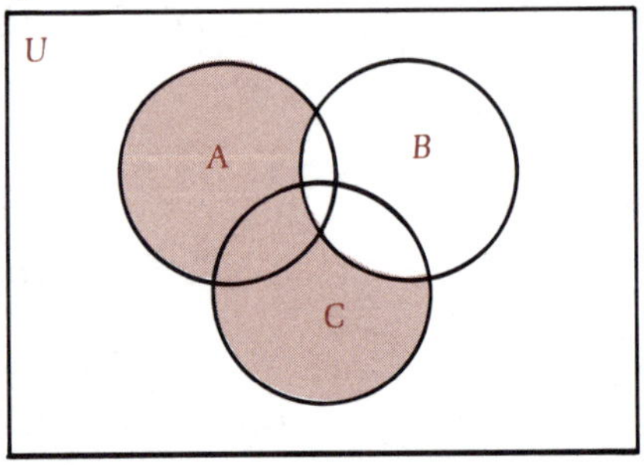

b.

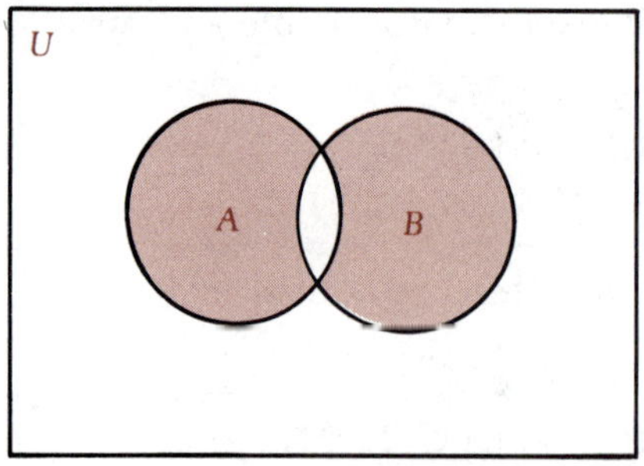

d.

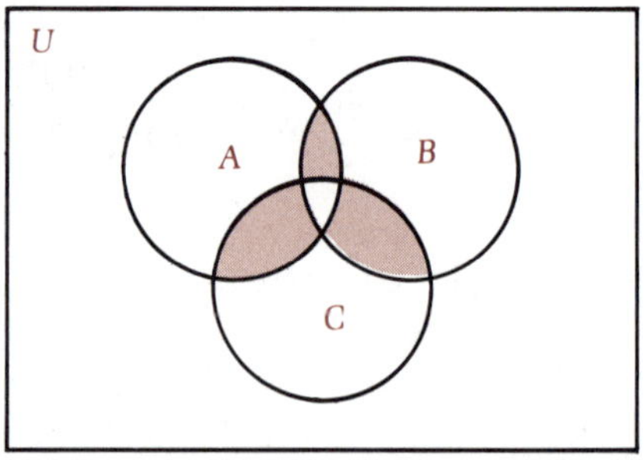

e.

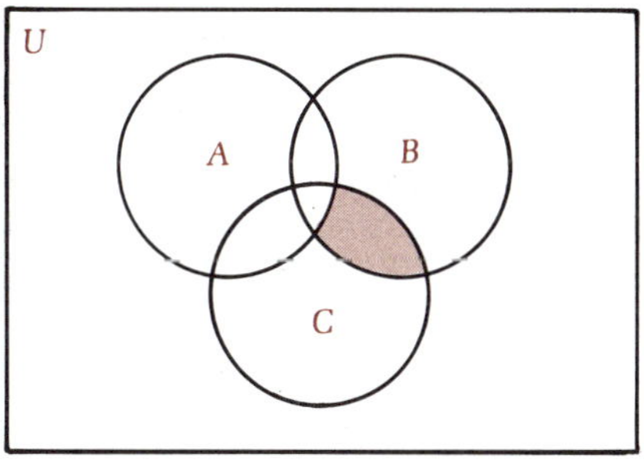

g.

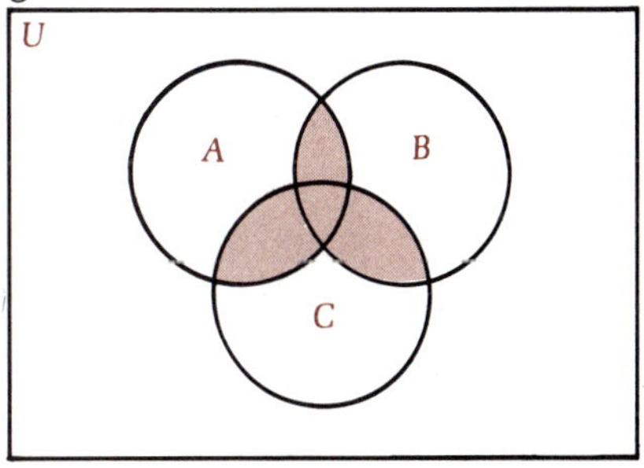

f.

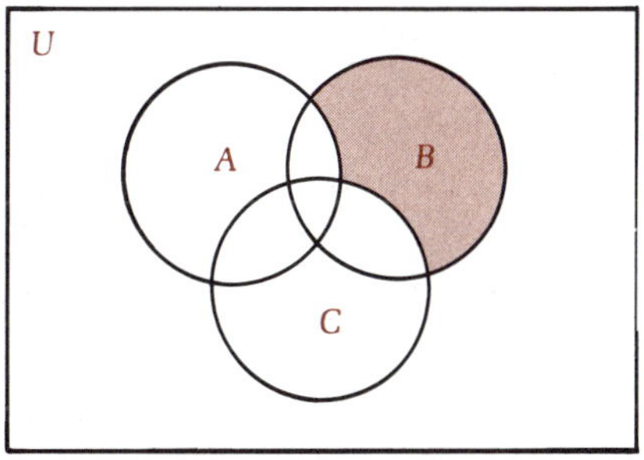

h.

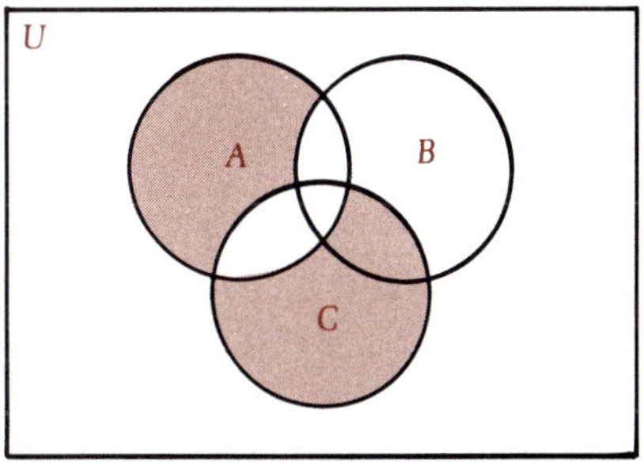

5. Given $n(A) = 25$, $n(B) = 35$, and $n(A \cap B) = 15$, find $n(A \cup B)$.

6. Find $n(A \cup B)$ if $n(A) = 27$, $n(B) = 34$, and $n(A \cap B) = 18$.

7. Given $n(A \cup B) = 150$, $n(A) = 85$, and $n(B) = 85$, find $n(A \cap B)$.

8. Find $n(A \cap B)$ if $n(A) = 50$, $n(B) = 60$, and $n(A \cup B) = 70$.

9. If $n(A) = 75$, $n(A \cup B) = 125$, and $n(A \cap B) = 35$, find $n(B)$.

10. If $n(A \cup B) = 150$, $n(A \cap B) = 30$, and $n(B) = 100$, find $n(A)$.

11. Find $n[(A \cap B)']$ if $n(U) = 200$, $n(A) = 120$, $n(B) = 50$, and $n(A \cup B) = 130$.

12. Find $n[(A \cup B)']$ if $n(U) = 80$, $n(A) = 50$, $n(B) = 30$, and $n(A \cap B) = 20$.

13. If $n(U) = 150$, $n(A) = 100$, $n(B) = 80$, and $n(A \cap B') = 60$, find $n(A' \cap B)$.

14. Given $n(U) = 500$, $n(A \cup B) = 450$, $n(A \cap B) = 75$, and $n(B) = 325$, find $n(A')$.

15. A local supermarket carries three brands of canned green beans—a generic brand, the store brand, and a national brand. The manager asks 50 customers which brands they like. If 20 indicate that they like the generic brand, 25 say that they like the store brand, and 15 say that they like both of these two, how many like neither of these two?

16. The manager of a local theater is under pressure from the owners to show more movies rated "R" or "X." The manager decides to sample the community before deciding. She randomly questions 5,000 people and finds that 4,850 would not consider seeing an X-rated movie and 3,800 would not see an R-rated movie. Only 150 indicated that they would see both kinds. How many people would attend at least one of the two kinds of movies?

17. A total of 5,651 people voted in a local election. If 3,211 voted against the mayor's sales tax proposal, 3,341 voted for the mayor, and 2,015 voted for the tax and for the mayor, how many voted against the tax and against the mayor? Assume everyone voted on the tax proposal and in the mayoral race.
18. A manufacturer reports that in a sample of 25,000 women, 17,000 use hand cream X, 15,000 use face cream Y, 13,000 use both X and Y, and only 1,000 do not use either of the products. What is wrong with this report?
19. Use the properties of this section to show that if A and B are subsets of a universe U, then
    a. $A = (A \cap B) \cup (A \cap B')$
    b. $(A \cap B)$ and $(A \cap B')$ are disjoint sets, and
    c. $n(A) = n(A \cap B) + n(A \cap B')$.
20. Using problem 19(c) and the properties of this section, show that $n(A \cap B') = n(A \cup B) - n(B)$.
21. If 350 people use product A and/or product B and 200 use product B, how many use A but not B? (*Hint:* Use the result in problem 20.)
22. A marketing research firm estimates that in a community of 20,000 people, 60% subscribe to the local newspaper, 45% subscribe to a metropolitan paper, and 30% subscribe to both. What percent of the people do not subscribe to either?
23. Using the information in problem 22, what percent of the people subscribe to the metropolitan paper but not the local paper?
24. A college fraternity wants to sponsor an event to raise money for a charity. The two types of events with the most success in the past have been raffles and softball games involving local celebrities and city officials. It is known that 5,000 people would buy ball game tickets but would not buy raffle tickets, 5,000 people would buy raffle tickets only, and 2,000 people would support either event. If each raffle ticket raises $1.00 for charity and each ball game ticket raises $.75 for charity, which event will raise the most money, and how much will it raise?

5.3 Sorting

Sets and Venn Diagrams can be used in other situations as well. Suppose a high school is planning a field trip to a local business, where the students will be provided with lunch. The business wants to know how many students must be fed. Two classes are to be combined for the trip. The typing class has 30 students in it; the shorthand class has 20 students. How many students should the business anticipate? On the surface, it may appear that 50 students will be involved. However, if some of the students are in both classes, the number will actually be less than 50.

Suppose we know that 10 students are enrolled in both classes in the preceding example. How many students will be making the trip? Let's use a Venn Diagram to help "sort out" the information. Let T be the set of students in the typing class and S the set of students in the shorthand class.

We know $n(T) = 30$, $n(S) = 20$, and $n(T \cap S) = 10$. Using the inclusion-exclusion principle,

$$\begin{aligned} n(T \cup S) &= n(T) + n(S) - n(T \cap S) \\ &= 30 + 20 - 10 \\ &= 40 \end{aligned}$$

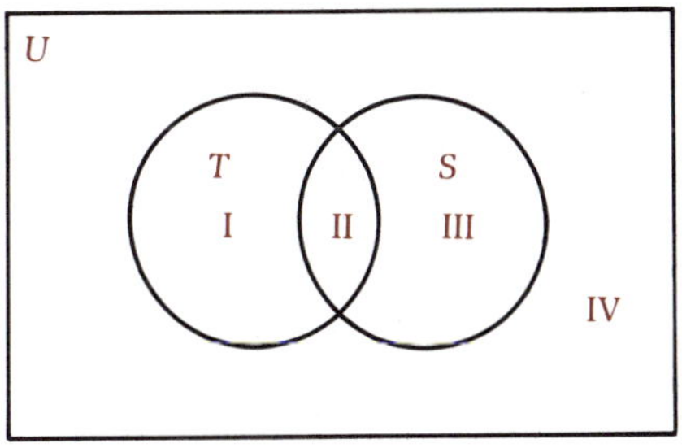

Figure 5.17

Using a Venn Diagram we have the picture in figure 5.17. Since $n(T)$ counts regions I and II and $n(S)$ counts regions II and III, region II has been counted twice.

We start the sorting process by filling in the overlapping regions first and then work our way out through the remaining regions, one at a time.

Step 1

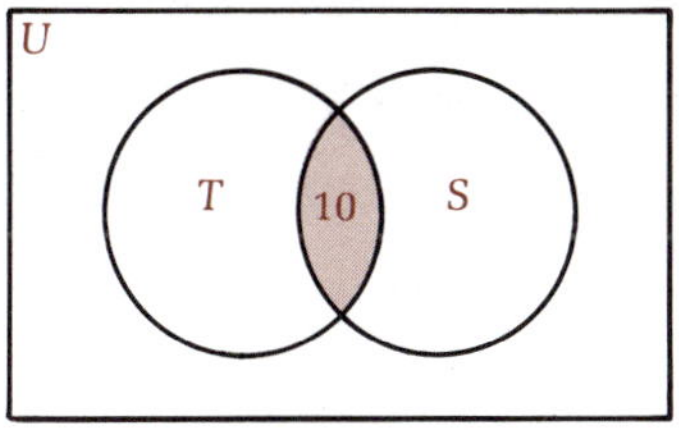

$n(T \cap S) = 10$

Step 2

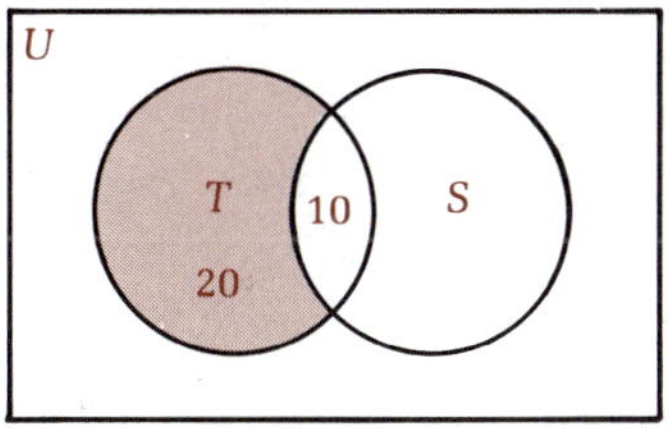

$$\begin{aligned} n(T \cap S') &= n(T) - n(T \cap S) \\ &= 30 - 10 \\ &= 20 \end{aligned}$$

Step 3

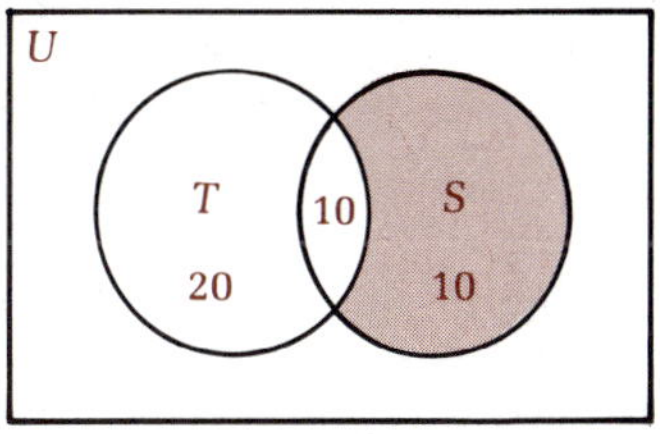

$$\begin{aligned} n(S \cap T') &= n(S) - n(S \cap T) \\ &= 20 - 10 \\ &= 10 \end{aligned}$$

Then $n(S \cup T) = 20 + 10 + 10 = 40$ is the number of people who will eat lunch. If we were given $n(U)$, the number of people in school, we could compute the number of students in region IV.

When only two sets are involved, the inclusion-exclusion principle is simpler than Venn Diagrams. However, as the number of sets increases, the situation will change somewhat. The Venn Diagram approach, particularly when working with three sets, is an efficient method for solving problems where overlapping information is given. The following example will help illustrate the technique.

■ Example 12. The owner of a restaurant is interested in customer dining habits so he surveys 150 of his customers. From his survey he obtains the following results:

60	Ordered salad
65	Ordered dessert
50	Ordered soup
20	Ordered salad and dessert
15	Ordered salad and soup
25	Ordered dessert and soup
10	Ordered all three

Use a Venn Diagram to sort out this information and determine how many of the 150 people ordered none of the three items.

Solution: Let S represent those ordering salad.
Let D represent those ordering dessert.
Let P represent those ordering soup.

The information in the survey is summarized as follows:

$n(S) = 60$
$n(D) = 65$
$n(P) = 50$
$n(S \cap D) = 20$
$n(S \cap P) = 15$
$n(D \cap P) = 25$
$n(S \cap D \cap P) = 10$

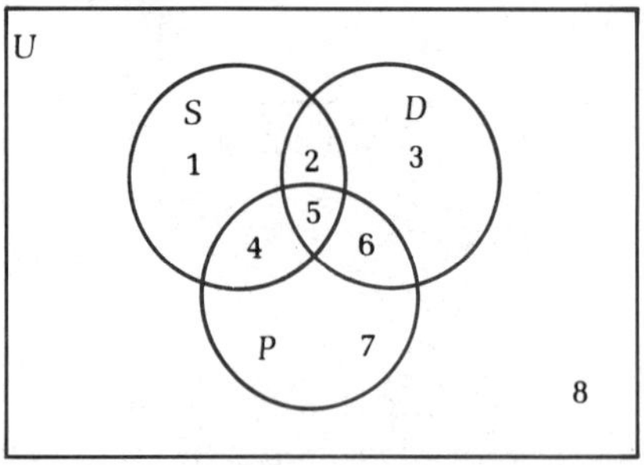

The figure on the left identifies the eight regions. Region 1 is $S \cap D' \cap P'$, Region 2 is $S \cap D \cap P'$, etc.

We fill in the number in each region, circled in the following diagram for clarity, starting with the innermost region and working our way outward.

The symbol for the number in the ith region will be $n(i)$.

$$n(5) = n(S \cap D \cap P) = 10$$
$$n(2) - n(S \cap D) - n(5) = 20 - 10 = 10$$
$$n(6) = n(D \cap P) - n(5) = 25 - 10 = 15$$
$$n(4) = n(S \cap P) - n(5) = 15 - 10 = 5$$
$$n(1) = n(S) - n(2) - n(5) - n(4) = 60 - 10 - 10 - 5 = 35$$
$$n(3) = n(D) - n(2) - n(5) - n(6) = 65 - 10 - 10 - 15 = 30$$
$$n(7) = n(P) - n(4) - n(5) - n(6) = 50 - 5 - 10 - 15 = 20$$
$$n(8) = n(U) - n(1) - n(2) - n(3) - n(4) - n(5) - n(6) - n(7)$$
$$= 150 - 35 - 10 - 30 - 5 - 10 - 15 - 20 = 25$$

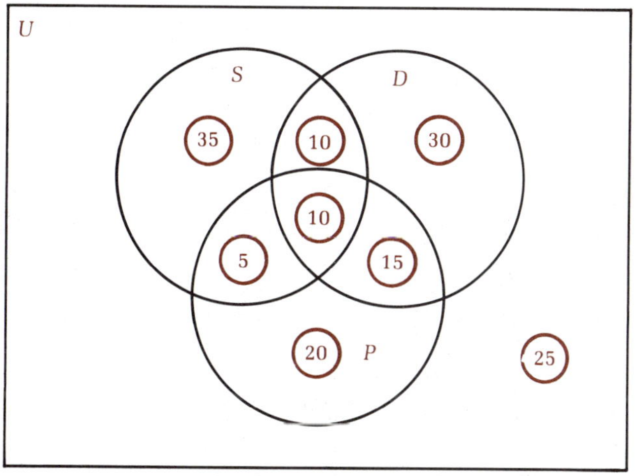

Therefore $n(8) = 25$ is the answer to the original question.

Once the information is sorted in this way, we are able to answer many questions. A few examples follow.

1. How many ordered only dessert? $n(3) = 30$
2. How many ordered exactly one of the three items?
   $n(1) + n(3) + n(7) = 35 + 30 + 20 = 85$
3. How many ordered salad and soup, but not dessert? $n(4) = 5$

The original information included overlap. That is, people were counted more than once. Sorting the information into disjoint categories where each unit of data is counted exactly once makes for a complete analysis. ■

Although Venn Diagrams are useful they are not a complete tool. When more than three sets are involved it becomes difficult, if not impossible, to draw and identify all possible disjoint regions.

For 2 sets, there are $2^2 = 4$ disjoint regions.
For 3 sets, there are $2^3 = 8$ disjoint regions.
For 4 sets, there are $2^4 = 16$ disjoint regions, and
for n sets, there are 2^n disjoint regions.

When the number of sets increases, it is often useful to expand the inclusion-exclusion principle.

$$\begin{aligned} n(A \cup B \cup C) &= n[(A \cup (B \cup C)] = n(A) + n(B \cup C) \\ &\quad - n[A \cap (B \cup C)] \\ &= n(A) + [n(B) + n(C) - n(B \cap C)] \\ &\quad - n[(A \cap B) \cup (A \cap C)] \\ &= n(A) + n(B) + n(C) - n(B \cap C) \\ &\quad - \{n(A \cap B) + n(A \cap C) \\ &\quad - n[(A \cap B) \cap (A \cap C)]\} \\ &= [n(A) + n(B) + n(C)] - [n(A \cap B) + n(A \cap C) \\ &\quad + n(B \cap C)] + [n(A \cap B \cap C)] \end{aligned}$$

In a similar way: $n(A \cup B \cup C \cup D) =$ [sum in each single set] $-$ [sum in all possible pairs of intersecting sets] $+$ [sum of the intersections of all possible 3 set combinations] $- n(A \cap B \cap C \cap D)$.

The inclusion-exclusion principle can be extended to any number of sets. Note that in the case of two, three, and four sets, a pattern is developing. We start with the sum of the numbers in the individual sets and subtract the number of elements in each two-set intersection. We then add numbers in three-set intersections and subtract numbers in four-set intersections. The $+$ and $-$ signs alternate, where $+$ signs are used for an odd number of sets and minus signs for an even number of sets. We always end with the number of elements in the intersection of all of the sets.

■ **Example 13.** Sort the following data into disjoint categories given:

1. $n(A \cup B \cup C) = 100$
2. $n(A) = 42$
3. $n(B) = 37$
4. $n(C) = 54$
5. $n(U) = 120$
6. $n(A \cap B) = 15$
7. $n(B \cap C) = 12$
8. $n(A \cap C) = 15$.

Solution: We begin by using the inclusion-exclusion principle for three sets to find $n(A \cap B \cap C)$. This will allow us to fill in the center region of the Venn Diagram to begin sorting the data.

$$n(A \cup B \cup C) = n(A) + n(B) + n(C) - n(A \cap B) - n(B \cap C) - n(A \cap C) + n(A \cap B \cap C)$$

$$100 = 42 + 37 + 54 - 15 - 12 - 15 + n(A \cap B \cap C).$$

Hence $n(A \cap B \cap C) = 9$ and we proceed exactly as in example 12. We begin with region 5 and work our way out, ending with the following diagram:

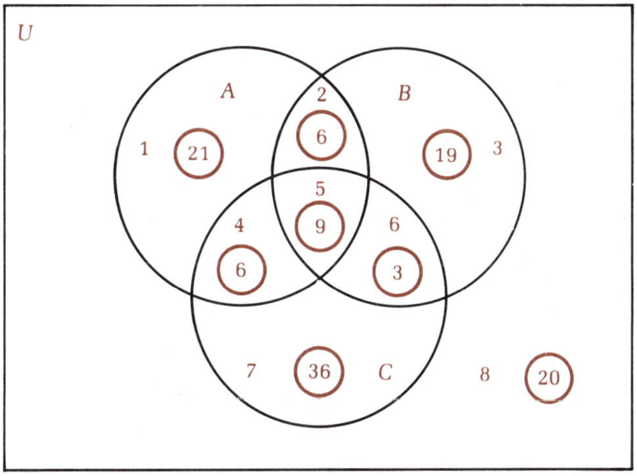

■ Example 14. In a survey of 100 business executives in a major city, it was found that 49 like the ballet, but 27 of these do not like opera. The survey also revealed that 35 enjoy symphony concerts and 19 of these also like ballet, 21 enjoy opera but not the symphony or ballet, 16 like opera and the symphony, and 7 of these also like ballet. How many of the 100 executives

1. like none of the three?,
2. like opera?, and
3. like at least two of the activities?

Solution: Let B, O, and S represent ballet, opera, and symphony respectively. Then we are given:

	Regions		Regions
$n(B) = 49$	1,2,4,5	$n(O \cap B' \cap S') = 21$	3
$n(B \cap O') = 27$	1,4	$n(O \cap S) = 16$	5,6
$n(S) = 35$	4,5,6,7	$n(O \cap S \cap B) = 7$	5
$n(S \cap B) = 19$	4,5	$n(U) = 100$	all 8

The regions of the diagram represented by each of these sets have been included beside the number of elements in the set. The information is sorted in the diagram below. We start with region 5, $(O \cap S \cap B)$, and work backwards through the information.

Region	*Number*
5	7
6	16 − 7 = 9
3	21
4	19 − 7 = 12
7	35 − 12 − 9 − 7 = 7
1	27 − 12 = 15
2	49 − 15 − 12 − 7 = 15
8	100 − 15 − 15 − 21 − 12 − 7 − 9 − 7 = 14

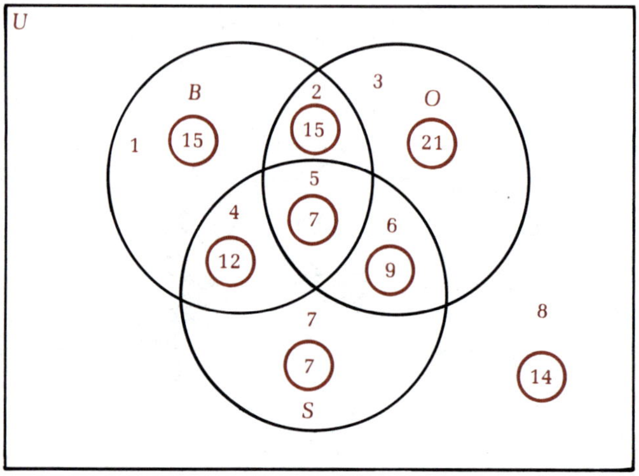

1. How many like none of the three? Answer: 14 (region 8).
2. How many like opera? Answer: 52 (regions 2, 3, 5, and 6).
3. How many like at least two? Answer: 43 (regions 2, 4, 5, and 6). ■

Exercise Set 5.3

1. Find $n(A \cap B')$ given $n(A) = 18$, $n(A \cap B) = 10$.
2. Given $n(U) = 120$, $n(A) = 75$, $n(B) = 60$, and $A \cup B = U$, find:
   a. $n(A \cap B')$
   b. $n(A \cap B)$
   c. $n(A' \cap B)$
3. The results of a survey show that 80 stores are open on Sunday and 175 are open late on Friday evenings. If 25 stores are open both

Friday evenings and Sundays, how many stores were in the survey if 15 stores were closed both Friday evenings and Sundays?

4. If $n(R) = 58$, $n(S) = 42$, $n(R \cap S) = 20$ and $n(U) = 90$, then is it true that $R \cup S = U$?

5. In a group of 50 students, 30 are taking English, 25 are taking math, and 10 are taking both. How many are not taking either subject?

6. In a high school current events class, all 50 students are required to read *Time* or *Newsweek*. If 20 students read *Time* but not *Newsweek* and 15 read *Newsweek* but not *Time*, how many read both magazines?

7. Given $n(U) = 200$, $n(A) = 90$, $n(B) = 70$, $n(C) = 90$, $n(A \cap B) = 25$, $n(A \cap C) = 40$, $n(B \cap C) = 35$, and $n(A \cap B \cap C) = 15$, find

a. $n(A \cup B \cup C)$
b. $n[(A \cup B \cup C)']$
c. $n(A \cap B')$
d. $n(B \cap C \cap A')$
e. $n[(A \cup C) \cap B']$
f. $n(A \cap B) \cup (A \cap C)$

8. Given $n(A) = 30$, $n(A \cap B') = 15$, $n(C) = 20$, $n(A \cap C) = 10$, $n[B \cap (A \cup C)'] = 5$, $n(B \cap C) = 10$, $n(A \cap B \cap C) = 5$, and $n(A \cup B \cup C)' = 5$, find

a. $n(U)$
b. $n(A \cup B \cup C)$
c. $n(B)$
d. $n(A \cap B)$
e. $n[C \cap (A \cup B)]$
f. $n\{[A \cap (B \cup C)]'\}$

9. Routine physical examinations of 500 pre-school children revealed that 40 had dental problems, 45 had vision problems, 55 had hearing problems, 15 had dental and vision problems, 15 had dental and hearing problems, 20 had vision and hearing problems, and 10 had dental, vision, and hearing problems. How many of the children had none of the three kinds of problems?

10. A local merchant uses television, radio, and newspaper advertising. To determine the effectiveness of advertising, he questions 200 customers during a special after-hours sale to see how they knew about the sale. He found that 115 had seen television ads, 75 had heard radio ads, and 125 had read newspaper ads. He also found that 30 received information from television and radio, 70 from television and newspapers, 25 from radio and newspapers, and 10 from all three. If everyone else said they heard it from a friend, how many heard it from a friend?

11. In order to prepare a report on agricultural prospects for his county, the county farm advisor questions 100 farmers about their crop plans for the following year. He finds that 75 intend to plant corn, 55 will plant soybeans, 35 will plant wheat, 35 will plant corn and soybeans, 25 will plant corn and wheat, 15 will plant soybeans and wheat, and 10 will plant all three. How many of the farmers will plant only one crop? How many will plant at least two crops?

12. In a group of 150 primary students, 100 watch "Sesame Street," 55 watch "Electric Company," and 65 watch "Mr. Rogers' Neighborhood." If, in addition, 35 watch "Sesame Street" and "Electric Company," 45 watch "Sesame Street" and "Mr. Rogers," 30 watch "Electric Company" and "Mr. Rogers," and 20 watch all three, how many watch none of the three? How many watch only "Sesame Street"?

13. An auto dealer wants to increase the sales of new cars by offering a free option package. Each package will include one major option, either an AM/FM radio with tape player (electronics package), automatic transmission (driving package), or reclining seats (comfort package). The dealer wants to use the package with the most popular major option, hoping this will increase customer interest. Records on the last 200 cars sold reveal that 105 had radios but 45 of these did not have automatic transmissions, 50 had reclining seats, and 35 of these also had radios, 50 had automatic transmissions but not the other two options, and 30 had automatic transmissions and reclining seats. All three options are on 20 of the cars. How many of the cars had none of the three options? Which option was most popular, and which package should the dealer try?

14. The production department of a local television station has a 30-minute early evening time period to fill. One staff member feels reruns of old game shows should be used since there would be less work to do for the staff. Other staff members disagree and feel a program of local interest should be developed. For the latter option, the two suggestions are an interview show with a different local political figure each week or a weekly documentary dealing with local problems and community projects. The program director is interested in community opinion, so she sends out 10,000 questionnaires. Of the 5,000 responses received, 200 indicate they are interested in none of the three, 700 would be interested in reruns but not an interview show, 1,200 are interested in the interviews but not documentaries, and 500 are interested in documentaries and reruns. The responses also revealed that 300 are interested in interviews and reruns, 1,500 are interested in interviews and documentaries but not reruns, and 200 are interested in interviews and reruns but not documentaries.

a. How many are interested in reruns?
b. How many are interested in documentaries?
c. How many are interested in interviews?
d. Which type of program should the director choose?

15. Human blood type is determined by the presence (or absence) of three antigens—A, B, and Rh. If the Rh antigen is present, this is indicated by a "+." If it is not present a "−" is used. If neither the A nor the B antigen is present, the blood type is O. Therefore, there are eight possible blood types denoted: AB+, AB−, A+, A−, B+, B−, O+, O−. For example, A+ blood has the A and Rh antigens but does not have the B antigen whereas O− blood has none of the three. The following data comes from the American National Red Cross. Of a random group of 1,000 people approximately 480 have antigen A and 440 of these do not have antigen B; 855 have the Rh antigen and 415 of these also have the A antigen; 20 have only the B antigen and not the other two; 115 have both B and Rh antigens and 35 of these also have the A antigen.

a. Use a Venn Diagram to sort the data and determine the approximate percentage of people who have each of the eight distinct blood types.

b. What percentage of people have the B antigen?

c. What percentage have type O blood?

d. Which specific blood type is the most common? Least common?

16. Because of the energy crisis, the owner of a large apartment complex has decided to provide bus service to one of the three large shopping centers nearby. The owner conducted a survey of the 2,000 units in the complex with the following results: 800 shopped at A but 560 of these did not shop at B, 820 shopped at C and 280 of these also shopped at A, 420 shopped only at B, 360 shopped at B and C and 100 of these also shopped at A.

a. How many shopped at none of the three?

b. How many shopped at exactly one of the three?

c. How many shopped at exactly two of the three?

d. How many in all shopped at B?

e. Based on this survey, which of the three should the owner provide bus service to?

5.4 Basic Counting Tools

Suppose a man has two shirts and three ties. How many combinations of one shirt and one tie can he choose if either shirt can be worn with any tie?

Below is a **tree diagram** which will help illustrate the basic counting principle used to solve this problem.

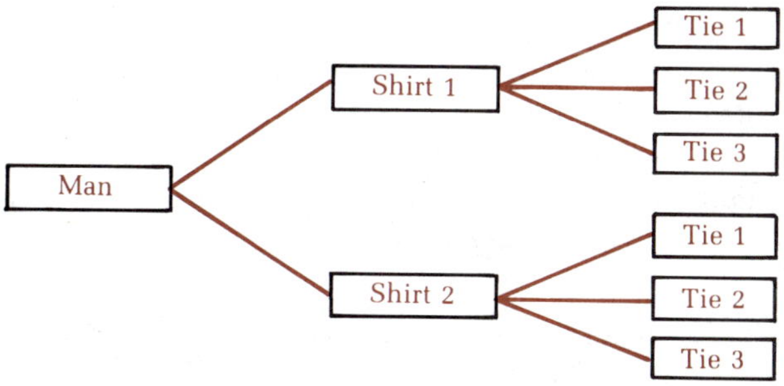

The shirt can be selected in two ways, and once the shirt is selected, a tie is selected from the three choices. Counting the ends of the "tree branches" yields a total of six combinations. Selecting the tie first and the shirt second would give us a different looking tree diagram but the ends of the respective branches would still identify the same six possible combinations.

The number of combinations = (number of ties) × (number of shirts); i.e., $6 = 3 \times 2$.

If two selections are to be made and the first can be made in n_1 ways and the second in n_2 ways, then the number of possible selections is given by $n_1 \cdot n_2$, known as the **multiplication principle.**

Property 5.5 ***The Multiplication Principle:*** If a task consisting of two choices is to be completed and the first choice can be made in n_1 ways and the second in n_2 ways, then the number of ways that the task can be completed is given by $N = n_1 \cdot n_2$.

Tree diagrams are useful for listing all possible choices, or for tracing specific choices where order is important. However, for counting purposes, the multiplication principle is usually quicker.

It is possible to generalize the multiplication principle to include situations that involve more than two selections in sequence.

The number of choices possible in a selection process consisting of a sequence of t consecutive choices can be counted and is equal to the product of the number of ways each of these t choices can be made. If the first selection can be made in n_1 ways, the second in n_2 ways, and so on until the last can be made in n_t ways, then the total number of possible outcomes is given by $N = n_1 \cdot n_2 \cdot n_3 \ldots \cdot n_t$. This property is called the **fundamental counting principle (F.C.P.).**

Property 5.6 ***The Fundamental Counting Principle:*** If a task consists of a sequence of t selections and there are n_1 choices for the first selection, n_2 choices for the second selection, and so on until there are n_t choices for the last selection, then the number of ways of completing the task is given by $N = n_1 \cdot n_2 \cdot n_3 \cdot \ldots \cdot n_t$.

■ Example 15. A fast food chain offers 6 kinds of sandwiches and 4 kinds of drinks in 3 sizes: large, medium, and small. As a sales promotion, a wheel-of-fortune is set up where all possible combinations of one sandwich and one drink of a specific size are the outcomes. A customer, after placing his order, spins the wheel, and if the result matches his order, he does not have to pay. How many outcomes are listed on the wheel?

Solution:

Let n_1 = the number of kinds of sandwiches = 6,

n_2 = the number of kinds of drinks = 4,

and n_3 = the number of different size drinks = 3.

Then $N = n_1 \cdot n_2 \cdot n_3 = 6 \cdot 4 \cdot 3 = 72$. ■

■ Example 16. A history quiz consists of 10 true-false questions. In how many different ways can the quiz be completed?

Solution: We consider the quiz to consist of 10 tasks, each of which can be performed in 2 ways. There are, then,

$\underbrace{2 \cdot 2 \cdot 2 \cdot \ldots \cdot 2}_{10 \text{ factors}} = 2^{10} = 1024$ ways to complete the quiz. ■

Exercise Set 5.4

1. If a woman has 5 skirts, 8 blouses, and 4 pairs of shoes, how many different outfits does she have?
2. John has 10 pairs of socks scattered loosely in his drawer. Each sock has only one mate. If he reaches in the drawer and picks 2 socks, in how many ways can he get 2 socks that do not match?
3. In how many ways can first, second, and third place occur in a horse race with 10 entries?
4. Sue wants to travel from Kansas City to New York by way of Chicago. To get from Kansas City to Chicago, she has a choice of 5 planes, 2 trains, or 4 buses. From Chicago to New York she can choose from 8 planes, 3 trains, or 6 buses. In how many different ways can she make the trip?
5. Sharon has just been informed that she has won a dream trip. She can choose any one of 9 cities, any one of 6 dates, and any one of 3 hotels in each city. How many possibilities does Sharon have to choose from?
6. Mary wants to redecorate her kitchen. She wants wallpaper on one wall, paint on the other walls, and vinyl covering on the floor. She has a choice of 5 vinyl patterns, 4 paint colors, and 12 wallpaper patterns. Assuming any of the combinations would match, how many combinations does Mary have to choose from?
7. How many license plates are possible if each is to consist of 2 letters followed by 4 digits?
8. In some states, license plates use a county designation. Suppose a state uses the county designation and then 6 digits.
   a. How many plates are possible for each county?
   b. If there are 105 counties in the state, how many plates are possible for the state?
9. In a certain state, all license plates consist of 2 letters followed by 4 digits, 3 letters followed by 3 digits, or 5 digits. How many different license plates are possible?
10. How many different license plates can be made if the first 2 symbols are letters and the last 3 digits are numbers, the first of which cannot be zero?

11. On a true-false test consisting of 20 questions, how many ways can the test be completed?
12. How large a population is necessary to assure that at least 2 people have exactly the same 3 initials?
13. On a Saturday morning Roberto is allowed to watch cartoons for 3 hours. During the first hour he has a choice of 3 cartoon shows, the second hour only 2 cartoon shows, but the third hour he has a choice of 4 cartoon shows.
    a. In how many different ways can he watch for 3 hours?
    b. In how many ways can he spend his time if not watching a cartoon show is counted? Assume he either watches or does not watch television for a full hour at a time.
14. If Pierre took a 10-question, multiple-choice exam in which each question had 5 possible answers, in how many ways could he answer the questions?

5.5 Permutations and Combinations

Suppose we are given the set of letters $A = \{a,b,c,d\}$. We define a two-letter "word" to be any sequence of two letters, whether it makes sense or not. How many two-letter words can be found using the letters from A?

We need to decide whether or not a letter can be repeated. If repeats are allowed, we can choose the first letter in 4 ways and the second letter in 4 ways. Hence, by the multiplication principle, there are $4 \cdot 4 = 16$ "words" possible.

Our interest here, however, is the situation in which a letter cannot be repeated. Again, using the multiplication principle, since the first letter can be chosen in 4 ways and the second can be chosen in 3 ways, there are $4 \cdot 3 = 12$ "words" possible. These 12 words are given as follows.

ab	*ba*	*ca*	*da*
ac	*bc*	*cb*	*db*
ad	*bd*	*cd*	*dc*

Basically, we are being asked to find the number of different arrangements or orderings of 4 things (letters) taken 2 at a time, where repeats are not allowed. These arrangements are called **permutations.**

Suppose a group of four people—Anita, Bob, Carol, and Dana—needs to choose 2 people to buy refreshments for a party. How many different pairs are possible? We can proceed as in the previous situation, obtaining

12 pairs. However, we see now that the pair (Anita, Bob) and the pair (Bob, Anita) both appear in the 12 possibilities, even though they are not actually different pairs. The only difference is in the order in which they were chosen, and order of selection is not critical here. Further investigation reveals that each of the pairs appears two times if we are not concerned with order of selection. There are, then, $12 \div 2 = 6$ different pairs to buy the refreshments.

Here we are asked to find the number of unordered arrangements of 4 things (people) taken 2 at a time. These *unordered arrangements* are called **combinations.**

The difference, then, between permutations and combinations is strictly a matter of order. When order is involved, we are after permutations. If order is unimportant, we are looking for combinations. The following examples will help illustrate the distinction between permutations and combinations.

■ Example 17. In how many ways can two marbles be selected from a collection of 5 marbles? The marbles are colored red (R), white (W), blue (B), green (G), and yellow (Y).

Solution: We could, using F.C.P., select the first marble in 5 ways and select the second marble from the remaining 4 so $N = 5 \cdot 4 = 20$. Listed below are these 20 possibilities:

(R,W)	(W,R)	(G,R)	(B,R)	(Y,R)
(R,G)	(W,G)	(G,W)	(B,W)	(Y,W)
(R,B)	(W,B)	(G,B)	(B,G)	(Y,G)
(R,Y)	(W,Y)	(G,Y)	(B,Y)	(Y,B)

We see that a blue marble first followed by a green marble second (B,G) is counted as different from green first, blue second (G,B). In fact, each pair of marbles selected is counted twice in deference to order. If we are not interested in the order of selection, then the number of combinations is 10, in this case half the number of permutations. The correct answer, 10 or 20, depends entirely upon our interpretation regarding the importance of order in the selection. ■

■ Example 18. In how many ways can 5 men be selected for a bowling team from a roster of 6 players?

Solutions: Case 1: *Combinations.*
We can select the one player who does not bowl in 6 ways. Therefore, those who do bowl are selected in the same 6 ways. The 5 men left to bowl have not been ordered and are considered as one unit; hence, there are 6 possible combinations.

Case 2: *Permutations.*
If we are concerned with not only who bowls but also in what order they bowl, we can select our line-up as follows:

n_1 = number of ways we can pick first bowler = 6
n_2 = number of ways we can pick second bowler = 5
n_3 = number of ways we can pick third bowler = 4
n_4 = number of ways we can pick fourth bowler = 3
n_5 = number of ways we can pick fifth bowler = 2

Hence, $N = 6 \cdot 5 \cdot 4 \cdot 3 \cdot 2 = 720$.

Which is the correct answer, 6 or 720? The answer depends on how you interpret the question. There are 6 combinations and 720 permutations. ■

By agreement, when no reference to order is stated or implied, we will assume the question asks for combinations.

The number of permutations of n things taken r at a time is written symbolically **P(n,r)** and is equal to $n(n-1)(n-2)\ldots(n-r+1)$ by direct application of the fundamental counting principle.

Permutation Formula
$P(n,r) = n(n-1)(n-2)\ldots(n-r+1)$

■ Example 19. Find **a.** $P(5,3)$ **b.** $P(10,3)$

Solution:

a. $P(5,3) = 5 \cdot 4 \cdot 3 = 60$
Note that $n = 5$, $r = 3$, and the last term
$(n - r + 1) = 5 - 3 + 1 = 3$.

b. $P(10,3) = 10 \cdot 9 \cdot 8 = 720$
Note that $n = 10$, $r = 3$, and the last term
$(n - r + 1) = 10 - 3 + 1 = 8$. ■

There is a convenient mathematical concept that can be used here called **n factorial.** This concept and its notation help simplify computational formulas and make them more compact.

Definition 5.6 If n is a counting number, then the product of the numbers from n down to 1 is called **n factorial** and is denoted $n!$. That is, $n! = n(n-1)(n-2)\ldots(3)(2)(1)$. By special convention (we need it to make our formulas compute) we agree that $0! = 1$.

■ **Example 20.**

$$0! = 1$$
$$1! = 1$$
$$2! = 2 \cdot 1 = 2$$
$$3! = 3 \cdot 2 \cdot 1 = 6$$
$$4! = 4 \cdot 3 \cdot 2 \cdot 1 = 24$$
$$5! = 5 \cdot 4 \cdot 3 \cdot 2 \cdot 1 = 120$$
$$6! = 6 \cdot 5 \cdot 4 \cdot 3 \cdot 2 \cdot 1 = 6(120) = 720$$
$$7! = 7(6!) = 7(720) = 5{,}040$$
$$8! = 8(7!) = 8(5{,}040) = 40{,}320$$
$$n! = n(n-1)!$$
■

This last expression, $n! = n(n-1)!$, and a generalized form related to it, are sometimes useful when we are working with factorials. If $k < n$, then $n! = n(n-1) \cdot (n-2) \cdot \ldots \cdot (n - [n-k-1]) \cdot k!$. Note here that $n - [n-k-1] = k+1$. Hence $n! = n(n-1) \cdot (n-2) \cdot \ldots \cdot (k+1) \cdot k!$.

■ **Example 21.** Compute $\frac{7!}{5!}$.

Solution:

$$\frac{7!}{5!} = \frac{7 \cdot 6 \cdot (5!)}{5!} = 7 \cdot 6 = 42$$

Here $n = 7$ and $k = 5$, so we write $7! = 7 \cdot 6 \cdot 5!$. It would be pointless and self-defeating to solve the problem in the following way: $\frac{7!}{5!} = \frac{7 \cdot 6 \cdot 5 \cdot 4 \cdot 3 \cdot 2 \cdot 1}{5 \cdot 4 \cdot 3 \cdot 2 \cdot 1} = \frac{5{,}040}{120} = 42.$ ■

■ **Example 22.** Compute $\frac{8!}{3!5!} \div \frac{10!}{3!7!}$.

Solution:

$$\frac{8!}{3!5!} \div \frac{10!}{3!7!} = \frac{8!}{3!5!} \cdot \frac{3!7!}{10!} = \frac{7 \cdot 6 \cdot (5!)(8!)}{(5!)(10)(9)(8!)} = \frac{7 \cdot 6}{10 \cdot 9} = \frac{7}{15}$$

We have used the identities $10! = (10)(9)(8!)$ and $7! = (7)(6)(5!)$. ■

Factorial notation can be used to express formulas for finding the number of permutations and combinations of n things taken r at a time. The number of permutations of n things taken r at a time is given by

$$P(n,r) = n(n-1)(n-2) \cdot \ldots \cdot (n-r+1).$$

If we multiply and then divide on the right by $(n-r)!$ we obtain

$$P(n,r) = n(n-1)(n-2) \cdot \ldots \cdot (n-r+1)$$
$$= \frac{n(n-1)(n-2) \cdot \ldots \cdot (n-r+1)(n-r)!}{(n-r)!} = \frac{n!}{(n-r)!}.$$

Since multiplying and dividing an expression by the same nonzero quantity does not change the value of the quantity, we see that there is no difference between the two formulas for $P(n,r)$. We will see that the second formula, $P(n,r) = \frac{n!}{(n-r)!}$, is convenient when we want to find a formula for combinations.

The expression $P(n,n)$ is of special interest. By the preceding formula, $P(n,n) = \frac{n!}{(n-n)!} = \frac{n!}{0!} = n!$. The number of different orderings of n objects is, then, $n!$. For example, if five books are to be lined up on a shelf, there are $P(5,5) = 5! = 120$ ways of lining up the books.

The number of combinations of n things taken r at a time will be denoted $C(n,r)$ or $\binom{n}{r}$. Either of these notations is acceptable. The symbol $\binom{n}{r}$ is frequently used in probability. Remember, order is not important in combinations. To determine the number of combinations, start with the number of permutations, $P(n,r) = \frac{n!}{(n-r)!}$. It is necessary now to determine how many times each set of r objects appears. But, the number of times each set of r objects appears is the number of permutations of r objects taken r at a time, $P(r,r)$.

From our above discussion, $P(r,r) = r!$. The number of combinations of n things taken r at a time is, then,

$$C(n,r) = \frac{P(n,r)}{P(r,r)}$$
$$= \frac{P(n,r)}{r!} = \frac{n!}{r!(n-r)!}$$

Permutation Formula	**Combination Formula**
$P(n,r) = \frac{n!}{(n-r)!}$	$C(n,r) = \frac{n!}{r!(n-r)!}$

These two formulas, $P(n,r) = \frac{n!}{(n-r)!}$ and $C(n,r) = \frac{n!}{r!(n-r)!}$, can be used to solve a variety of counting problems. The relationship between permutations and combinations is the factor $r!$. That is, $P(n,r) = r!C(n,r)$ or $C(n,r) = \frac{P(n,r)}{r!}$.

In example 17, where 2 marbles were selected, the number of permutations (20) is $2! = 2$ times the number of combinations. In example

18, where we were selecting 5 bowlers, the number of combinations (6) equals the number of permutations (720) divided by 5! = 120, or the number of permutations is 5! times the number of combinations.

■ Example 23. In how many ways can a 3-member committee be selected from a group of 28 students?

Solution: Since the order of selection to the committee is not important, we are asked to find the number of combinations where $n = 28$ and $r = 3$. Then

$$C(28,3) = \frac{28!}{3!25!} = \frac{(28)(27)(26)(25!)}{2 \cdot 3 \cdot (25!)}$$
$$= \frac{(28)(27)(26)}{2 \cdot 3} = (28)(9)(13) = 3{,}276.$$

This is the number of combinations of 28 things taken 3 at a time and is the number of 3-member committees possible. ■

■ Example 24. In how many ways can we select a president, a vice-president, and a secretary from a class of 28 students?

Solution: In this example, order of selection (the specific office) is important, so we want the number of permutations.

$$P(28,3) = \frac{28!}{25!} = (28)(27)(26) = 19{,}656.$$

This is the number of permutations of 28 students taken 3 at a time. ■

When the F.C.P. is used to count the total number of possible combinations where more than one type of selection is involved, the sequence of events must **1.** Be defined in such a way to allow for all possible outcomes, and **2.** not allow for duplications. A general guideline is to make each task in the sequence a combination and to check if there are possible duplications.

■ Example 25. In how many ways can we select 3 aces and a pair of matching cards from a regular deck of 52 cards?

Solution: We define three tasks as follows: T_1 is selection of 3 aces, T_2 is selection of a non-ace denomination, and T_3 is selection of a pair.

$n(T_1) = n_1 =$ the number of combinations of 3 aces $= C(4,3) = 4$.
$n(T_2) = n_2 =$ the number of ways to pick a denomination of a non-ace $= C(12,1) = 12$.
$n(T_3) = n_3 =$ the number of combinations of pairs of a specific denomination $= C(4,2) = 6$.
$N = 4 \cdot 12 \cdot 6 = 288$. ■

Exercise Set 5.5

1. Evaluate each of the following.

a. 10!	**e.** $C(12,8)$
b. $8! - 7!$	**f.** $P(8,4) \div P(10,4)$
c. $P(8,3)$	**g.** $C(9,6) \div C(12,5)$
d. $C(9,6)$	**h.** $C(9,2) \div C(12,2)$

2. How many 5-card poker hands can be dealt from an ordinary deck of 52 cards?
3. In how many ways can 3 black cards be drawn from an ordinary deck of 52 cards?
   a. If the order is important?
   b. If the order is unimportant?
4. In how many ways can 3 heads appear if 5 coins are tossed?
5. Show $C(12,4) = C(12,8)$.
6. Show that in general $C(n,r) = C(n,n-r)$.
7. An urn contains 12 black and 8 white balls.
   a. How many different combinations of 2 balls can we choose?
   b. How many different combinations of 2 black balls can be selected?
   c. How many different combinations of 2 white balls can be selected?
   d. How many different combinations of 1 white and 1 black ball can we select?
8. A dozen eggs contains 4 rotten and 8 good eggs. If two eggs are selected at random,
   a. How many ways can 2 eggs be selected?
   b. How many of these ways will have both good eggs?
   c. How many of these ways will have at least 1 bad egg?
   d. Answer questions **a**, **b**, and **c** as they relate to a 3-egg selection.
   e. Answer questions **a**, **b**, and **c** as they relate to a 4-egg selection.
9. A coin is tossed 5 times and the sequence of heads and tails is recorded.
   a. In how many different ways can 3 heads occur?
   b. In how many different ways can 2 tails occur?
   c. Why is question **a** the same as question **b**?
10. A gross of lightbulbs are to be examined by taking a random sample of 3 bulbs and testing them. If 10% of the lightbulbs are defective, in how many ways can the test pass inspection? What percent of the time will at least one defective lightbulb be picked?
11. How many poker hands consist of 5 spades?
12. Without back tracking, how many routes lead from point X to point Y on the map below? (Hint: All paths from X to Y are 9 blocks long, 4 down and 5 across, not necessarily in that order.)

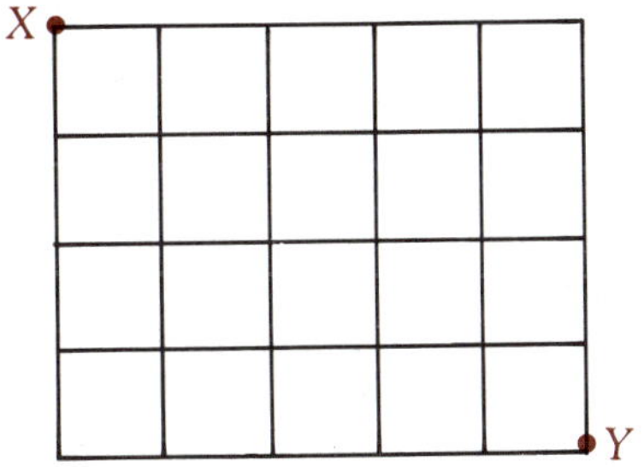

13. Relate problem 12 with a coin tossing problem.
14. In the map below, a car must be driven from point A to point B to point C to point D. The shortest route is 15 blocks long. In how many different ways can the trip be made?

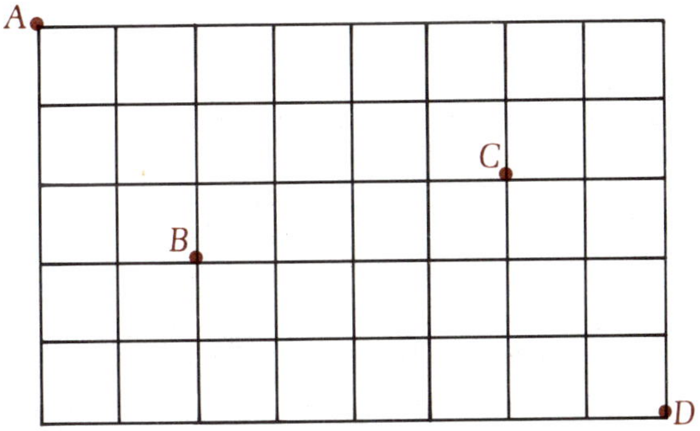

15. A club whose membership consists of 6 women and 5 men wish to send a delegation of 2 women and 2 men to Washington. In how many ways can they be chosen?
16. In December it was nice 11 days. It rained 5 days and snowed 15 days. In how many ways could this have happened?
17. A rating service wishes to rate 12 hotels as 1-star, 2-star, and 3-star. In how many ways can this be done if each rating is used the same number of times?
18. There are 100 people in a contest. The awards are one first place finish, three second place finishes, and ten third place finishes. In how many ways can the winners be picked?
19. Which is larger, $C(100,4)$ or $C(100,93)$?
20. Three cards are drawn at random from a deck of 52 playing cards. Will an ace, a deuce, or a jack be drawn at least half the time?
    a. Without replacement?
    b. With replacement?
21. A television director has 10 time slots to show 3 different commercials, one 4 times and the others 3 times each. In how many ways can she schedule these commercials?

22. If the commercials in problem 21 are not allowed to follow themselves, how does this affect the number of choices open to the director?
23. A family has three children, two boys and a girl. In how many different orders could this occur?
24. A sample of size N is taken from a population of 20 things, half of which are defective. How large should N be in order that the sample should contain a defective at least half the time?
25. In problem 24, how large should we make N in order to select a sample with a defective at least 90% of the time?
26. A five card poker hand is called a full house if it contains 3 cards of one denomination and 2 of another. How many hands consisting of a full house are possible from a deck of 52 cards?
27. How many poker hands consisting of a flush (5 cards of the same suit) are possible?
28. How many different 13-card bridge hands are possible from a deck of 52 cards?
29. Eight horses are entered in a race. A man wishes to pick the top three horses. How many choices does he have if
    a. he wants to pick the exact order of finish?
    b. he doesn't care about the order of finish?
30. An auto race has 30 cars entered and prize money is awarded for the first five finishers.
    a. How many different orders of finish of the 30 cars are possible?
    b. How many different possibilities for the top 5 finishers, in order, are possible?
    c. How many different sets of prize winners are possible?

5.6 Partitions and the Binomial Theorem

To **partition** a set is to divide the set into two or more disjoint subsets. If we view the subsets as different, we have an **ordered partition.** If the various subsets are not ordered, i.e., not numbered or named or in any way identified, we have an **unordered partition** or, simply, a partition.

For example, if we wish to select 3 balls from an urn containing 5 balls, we can do this in $C(5,3) = 10$ ways. If we wish to partition 5 balls into two groups, one containing 3 balls and the other containing 2 balls, this can be done in the same $C(5,3) = 10$ ways.

Since taking 3 balls in one group is different from taking 2 balls, the subsets are different, and we have an ordered partition. Consider selecting 3 balls from an urn that contains 6 balls. This can be done in $C(6,3) = 20$ ways. However, each time we select a combination of 3 balls, we leave another combination of 3 balls behind. When the 3 remaining balls are selected, the first 3 balls picked will now be the remaining balls. We would count each of these as a different selection, even though there is only one division. For example, from {a,b,c,d,e,f} we can select C(6,3) = 20 combinations of 3 letters, of which {a,b,c} and {d,e,f} are two possible outcomes, whereas the division of {a,b,c,d,e,f} into the two sets {a,b,c} and {d,e,f} is a single unordered partition.

The number of ways of arranging n objects into t cells where n_1 things go into cell 1, n_2 things go into cell 2, and so on until n_t things are placed in the last cell and where $n_1 + n_2 + \cdots + n_t = n$, is computed by the formula $\binom{n}{n_1,n_2,\ldots,n_t} = \dfrac{n!}{n_1!\,n_2!\ldots n_t!}$. Note that the cells are numbered, making this an ordered partition.

■ Example 26. Calculate $\binom{9}{2,3,4}$.

Solution: $\binom{9}{2,3,4} = \dfrac{9!}{2!\,3!\,4!} = \dfrac{9\cdot 8\cdot 7\cdot 6\cdot 5\cdot 4!}{1\cdot 2\cdot 1\cdot 2\cdot 3\cdot 4!} = 9\cdot 4\cdot 7\cdot 5 = 1260.$

We have already studied a special case $C(n,r) = \binom{n}{r,n-r}$ where n things are divided into 2 ordered subsets. The r things we select, and the $n - r$ things we leave unselected. The symbol $\binom{n}{r,n-r}$ is frequently abbreviated $\binom{n}{r}$ and both are defined to be $\dfrac{n!}{r!(n-r)!}$.

If unordered partitions are required, the preferred method is to compute the number of ordered partitions and then compensate by dividing by $m!$ where m is the number of subsets in a partition.

Of special interest is the case where we divide a set into m subsets of equal size. The number of ordered partitions is $\binom{n}{r,r,r,\ldots r}$ where r is the size of each subset. Then $\binom{n}{r,r,r,\ldots,r} = \dfrac{n!}{r!\,r!\ldots r!} = \dfrac{n!}{(r!)^m}$ and $m = \dfrac{n}{r}$ is the number of subsets in a partition. It follows, then, that the number of unordered partitions (divisions) is equal to $\dfrac{n!}{(r!)^m} \div m! = \dfrac{n!}{m!\,(r!)^m}$. ■

■ **Example 27.** In how many ways can twelve members of a professional basketball team be

1. Assigned roommates, two men to a room?
2. Assigned rooms numbered 101 to 106, two men to a room?

Solution:

1. This is an unordered partition, and the number of ways is then $\dfrac{12!}{6!(2!)^6} = \dfrac{12 \cdot 11 \cdot 10 \cdot 9 \cdot 8 \cdot 7}{2 \cdot 2 \cdot 2 \cdot 2 \cdot 2 \cdot 2} = 10{,}395.$
2. This is an ordered partition; hence, the number of ways is $\dfrac{12!}{(2!)^6} = 7{,}484{,}400.$ This could also be solved by F.C.P. $N = \binom{12}{2}\binom{10}{2}\binom{8}{2}\binom{6}{2}\binom{4}{2}\binom{2}{2}$ by filling rooms 101 to 106 in order.

   The arithmetic produces the same answer. ■

A slight variation of the preceding idea is sometimes useful. Suppose we want to partition n objects into m cells, where each of k cells contain s objects and each of the remaining $m - k$ cells contain t objects. Note that

$$\underbrace{s + s + s + \dots + s}_{k \text{ terms}} + \underbrace{t + t + \dots + t}_{m - k \text{ terms}} = n.$$

That is, $ks + (m - k)t = n$. How many unordered subsets are there? There are $\binom{n}{s,s,\dots,s,t,t,\dots,t} = \dfrac{n!}{s! \cdot s! \cdot \dots \cdot s! \cdot t! \cdot t! \cdot \dots \cdot t!} = \dfrac{n!}{(s!)^k(t!)^{m-k}}$ ordered subsets.

To find the number of unordered subsets, the number of ordered subsets must be divided by the number of repetitions. The number of repetitions in the k cells of size s is $k!$, and the number of repetitions in the $m - k$ cells of size t is $(m - k)!$. The number of unordered partitions is

$$\frac{n!}{(s!)^k(t!)^{m-k}} \div k!(m - k)! = \frac{n!}{k!(m - k)!(s!)^k(t!)^{m-k}}.$$

■ **Example 28.** If the number of ordered subsets is given by $\binom{n}{t,t,t,s,s,s,s} = \dfrac{n!}{(t!)^3(s!)^4}$, find the number of unordered partitions.

Solution: The answer is $\dfrac{n!}{3!4!(t!)^3(s!)^4}$ where we divide by *3!* and *4!* to exclude the repeated combinations. ■

■ **Example 29.** Twelve sailors are to be assigned to man two boats that require 3-man crews and three boats that require 2-man crews. In how many ways can the crews be made up?

Solution: $\frac{12!}{2!3!(3!)^2(2!)^3} = 660.$ ■

The symbol $\binom{n}{r}$ plays an interesting role in the expansion of $(a + b)^n$. The numerical coefficient of each of the $n + 1$ terms in the expansion of $(a + b)^n$ is expressible in the form $\binom{n}{r}$ where r ranges from 0 to n.

Recall that $\binom{n}{r} = C(n,r)$. Hence, the symbol $\binom{n}{r}$ is referred to as either "the combination of n things taken r at a time" or "the **binomial coefficient.**"

Property 5.7 ***Binomial Theorem:*** The binomial theorem states that

$$(a + b)^n = \binom{n}{0}a^n + \binom{n}{1}a^{n-1}b + \binom{n}{2}a^{n-2}b^2 + \ldots + \binom{n}{n-1}ab^{n-1} + \binom{n}{n}b^n.$$

For those who understand summation notation, this can be abbreviated

$$(a + b)^n = \sum_{r=0}^{n} \binom{n}{r} a^{n-r}b^r.$$

A device known as **Pascal's Triangle** (Blaise Pascal, 1623–1662) has been used for years to determine these binomial coefficients when they are relatively small. Observe:

$$
\begin{aligned}
(a + b)^0 &= 1 \\
(a + b)^1 &= 1a + 1b \\
(a + b)^2 &= 1a^2 + 2ab + 1b^2 \\
(a + b)^3 &= 1a^3 + 3a^2b + 3ab^2 + 1b^3 \\
(a + b)^4 &= 1a^4 + 4a^3b + 6a^2b^2 + 4ab^3 + 1b^4 \\
&\vdots
\end{aligned}
$$

Looking only at the coefficients, we have

$n = 0$,	$(a + b)^0$,	1					
$n = 1$,	$(a + b)^1$,	1	1				
$n = 2$,	$(a + b)^2$,	1	2	1			
$n = 3$,	$(a + b)^3$,	1	3	3	1		
$n = 4$,	$(a + b)^4$,	1	4	6	4	1	
$n = 5$,	$(a + b)^5$,	1	5	10	10	5	1
$\vdots$		$\vdots$	$\vdots$	$\vdots$			

The row where $n = 6$, i.e., the coefficients of $(a + b)^6$, is obtainable directly from the preceding row where $n = 5$.

$n = 5,$	$(a + b)^5$	1	5	10	10	5	1	
$n = 6,$	$(a + b)^6$	1	r	s	t	u	v	w

Figure 5.18

The missing numbers can be determined by adding the number directly above the missing number and the preceding number. If $\begin{matrix} x & y \\ & z \end{matrix}$ appears, then $z = y + x$. In figure 5.18, $s = 10 + 5$ and in fact

$$\begin{aligned} r &= 5 + 1 = 6 \\ s &= 10 + 5 = 15 \\ t &= 10 + 10 = 20 \\ u &= 5 + 10 = 15 \\ v &= 1 + 5 = 6 \end{aligned}$$

and the final entry $w = 0 + 1 = 1$.

■ Example 30. Expand $(A + B)^5$.

Solution: $(A + B)^5 = A^5 + 5A^4B + 10A^3B^2 + 10A^2B^3 + 5AB^4 + B^5$.

NOTE: The powers of A decrease from 5 to 0 while the powers of B increase from 0 to 5, and the sum of the powers, in each term, adds to 5. ■

■ Example 31. Four coins are tossed. Analyze the results.

Solution: Each of the four coins can come up heads or tails. Any combination is possible and there are $2 \cdot 2 \cdot 2 \cdot 2 = 16$ possible results.

If we expand $(H + T)^4$, the terms will partition these 16 possible outcomes into 5 distinct sets and the binomial coefficient will count the number of elements in each set.

$$(H + T)^4 = H^4 + 4H^3T + 6H^2T^2 + 4HT^3 + T^4$$

term 1	H^4	(H,H,H,H)	Exactly 4 heads	$\binom{4}{0} = 1$
term 2	$4H^3T$	(H,H,H,T) (H,H,T,H) (H,T,H,H) (T,H,H,H)	Exactly 3 heads and one tail	$\binom{4}{1} = 4$

term 3	$6H^2T^2$	(H,H,T,T) (H,T,H,T) (H,T,T,H) (T,H,H,T) (T,H,T,H) (T,T,H,H)	Exactly 2 heads and 2 tails	$\binom{4}{2} = 6$
term 4	$4HT^3$	(H,T,T,T) (T,H,T,T) (T,T,H,T) (T,T,T,H)	Exactly 1 head and 3 tails	$\binom{4}{3} = 4$
term 5	T^4	(T,T,T,T)	All tails	$\binom{4}{4} = 1$

Notice that the sum of the binomial coefficients is

$$\binom{4}{0} + \binom{4}{1} + \binom{4}{2} + \binom{4}{3} + \binom{4}{4} = 1 + 4 + 6 + 4 + 1 = 16 = 2^4.$$ ■

Exercise Set 5.6

1. Compute each of the following:

   a. $\binom{6}{2,2,2}$ e. $\binom{6}{2,1,3}$

   b. $\binom{8}{5,2,1}$ f. $\binom{10}{2,2,3,3}$

   c. $\binom{9}{4,3,2}$ g. $\binom{12}{2,2,2,4,2}$

   d. $\binom{10}{4,3,3}$ h. $\binom{16}{10,2,4}$

2. For each part of exercise 1, compute the number of unordered partitions.
3. In how many ways can a set of 8 things be divided into two sets, 5 in one set and 3 in the other?
4. In how many ways can a set of 8 things be divided into two equal sets?
5. In how many ways can a set of 8 things be partitioned into four equal sets?
6. Twenty teams are to be assigned times for a bowling tournament. Six of the teams are to bowl at 1:00 p.m., four at 3:00 p.m., and ten at 5:00 p.m. In how many ways can the assignment be made?

7. The college housing office has 30 late applications for dormitory rooms. If 22 spaces are available in dorm A, 6 spaces in dorm B, and 2 spaces in dorm C, how many ways can the assignments be made?
8. Eleven members of a college sorority have been chosen to plan a party. One woman is to serve as the general chairperson and the others are to be assigned to a subcommittee. Three are to be on an entertainment subcommittee, four on a decorations subcommittee, and three on a refreshment subcommittee. In how many ways can the committee be subdivided?
9. Find the coefficient of each of the following terms in the expansion of $(a + b)^7$.
   a. a^2b^5 **b.** a^3b^4 **c.** a^6b
10. Find the coefficient of each of the following terms in the expansion of $(p + q)^8$.
    a. pq^7 **b.** p^6q^2 **c.** p^5q^3 **d.** p^4q^4
11. Find the coefficient of each of the following terms in the expansion of $(a + 2b)^6$.
    a. a^5b **b.** a^3b^3 **c.** a^2b^4 **d.** ab^5
12. In the expansion of $(2a + 3b)^5$, what is the coefficient of each of the following terms?
    a. a^3b^2 **b.** a^2b^3 **c.** ab^4 **d.** a^4b
13. In the expansion of $(p + .2)^5$, find the entire term that contains the factor p^4.
14. The Cubs and Cardinals have a three game series scheduled. Use the expansion of $(W + L)^3$ to analyze the won-loss performance of the Cubs in the series.

5.7 Chapter Review

Important Terms and Symbols in Chapter 5.

set (section 5.1)
member, element (section 5.1)
$\{\quad\}$ (section 5.1)
set-builder notation (section 5.1)
empty set, null set (section 5.1)
set relations (section 5.1)
equal sets (section 5.1)
subset (section 5.1)
set operations (section 5.1)
intersection (section 5.1)
union (section 5.1)
disjoint sets (section 5.1)
universal set (section 5.1)
complement (section 5.1)
Venn Diagram (section 5.2)
commutative properties (section 5.2)
associative properties (section 5.2)
distributive properties (section 5.2)
DeMorgan's Laws (section 5.2)
inclusion-exclusion principle (section 5.2)
tree diagram (section 5.4)
multiplication principle (section 5.4)

fundamental counting principle (F.C.P.) (section 5.4)
permutations (section 5.5)
combinations (section 5.5)
permutation formula, or $P(n,r)$ (section 5.5)
n factorial ($n!$) (section 5.5)
combination formula, or $C(n,r)$ $\binom{n}{r}$ (section 5.5)
partition (section 5.6)
ordered partition (section 5.6)
unordered partition (section 5.6)
binomial coefficient (section 5.6)
binomial theorem (section 5.6)
Pascal's Triangle (section 5.6)

Review Exercises

1. Write each of the following using set notation.
a. The set of letters in the word "Pennsylvania."
b. The set of even numbers greater than 10 and less than 20.
c. The set of counting numbers whose square is less than 50.
d. The set of United States presidents whose last name was Smith.

2. Given $U = \{0,1,2,3,4,5,6,7\}$, $A = \{2,4,6\}$, $B = \{0,3,6\}$, $C = \{1,3,5,7\}$, $D = \{0,2,4,6\}$, and $E = \{1,4,7\}$, find
a. $A \cap B$.
b. $A \cup C$.
c. $B \cap D$.
d. $C \cap E$.
e. $C \cup D$.
f. D'.
g. $(D \cup E)'$.
h. $A \cap (D \cup E)$.
i. $(A \cap D)'$.
j. $B \cup (C \cap D)$.
k. $A \cap (B \cup C)'$.
l. $(A \cap E)'$.
m. $C \cup (D \cap E)$.
n. $C \cap D'$.
o. $C' \cap D'$.
p. $A' \cup D'$.

3. Draw a Venn Diagram and shade the region represented by each of the following.
a. $A \cap (B \cap C)'$
b. $A \cup (B \cap C)$
c. $(A \cap B) \cup (A \cap C)$
d. $(A \cap B) \cap C'$
e. $(A \cup B) \cap C'$
f. $(A \cup B) \cap C$

4. Use the given sets and ∪, ∩, and ′ to write an expression for the shaded region in each of the following:

a.

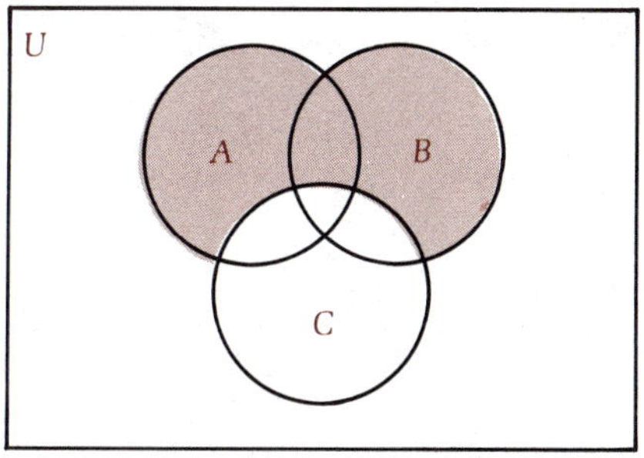

b.

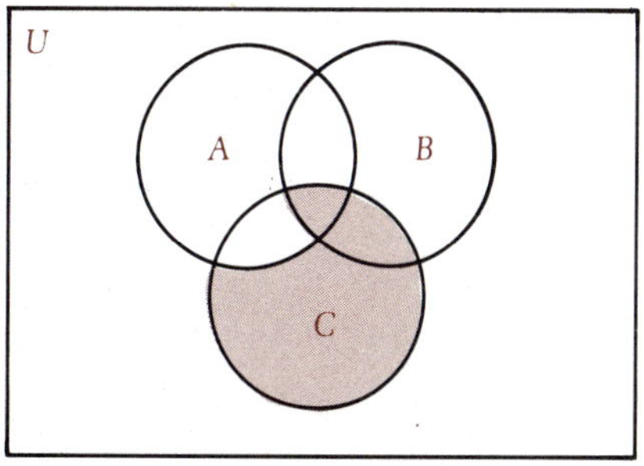

5. Given $n(A) = 60$, $n(B) = 75$, and $n(A \cap B) = 30$, find $n(A \cup B)$.
6. Given $n(A) = 40$, $n(B) = 60$, and $n(A \cup B) = 90$, find $n(A \cap B)$.
7. Given $n(A) = 5000$, $n(B) = 4000$, and $n(A \cap B) = 2000$, find $n(A \cup B)$.
8. Given $n(A) = 300$, $n(B) = 200$, and $n(A \cup B) = 500$, find $n(A \cap B)$.
9. Find $n(A)$, given $n(A \cup B) = 500$, $n(A \cap B) = 100$, and $n(B) = 200$.
10. Find $n(B)$, given $n(A \cup B) = 5000$, $n(A \cap B) = 4000$, and $n(A) = 4000$.
11. Given $n(U) = 100$, $n(A) = 35$, $n(B) = 45$, and $n(A \cap B) = 20$, find $n[(A \cup B)']$.
12. Given $n(U) = 500$, $n(A \cup B) = 290$, $n(A) = 190$, and $n(B) = 210$, find $n[(A \cap B)']$.
13. Find $n(A \cap B')$, given $n(U) = 100$, $n(A) = 55$, $n(B) = 45$, and $n(A' \cap B) = 30$.
14. Find $n(A \cup B')$, given $n(U) = 250$, $n(A) = 110$, $n(B) = 120$, and $n(A' \cup B) = 155$.
15. Evaluate each of the following.
    a. $P(8,5)$ **b.** $P(15,9)$ **c.** $P(9,6)$ **d.** $P(12,4)$
    e. $P(7,1)$ **f.** $P(11,11)$ **g.** $C(8,5)$ **h.** $C(15,9)$
    i. $C(9,6)$ **j.** $C(12,4)$ **k.** $C(7,1)$ **l.** $C(11,11)$
16. Evaluate each of the following.

    a. $\binom{12}{3,4,2,3}$ **b.** $\binom{8}{4,2,2}$ **c.** $\binom{10}{1,2,3,4}$

    d. $\binom{12}{2,2,2,3,3}$ **e.** $\binom{9}{3,3,3}$ **f.** $\binom{16}{8,5,3}$

17. Find the number of unordered partitions for each part of problem 16.
18. Find the coefficient of each of the following terms in the expansion of $(a + b)^9$.
    a. a^5b^4 **b.** a^2b^7 **c.** a^6b^3 **d.** ab^8
19. If $(2a + b)^6$ is expanded, what is the coefficient of each of the following terms?
    a. a^3b^3 **b.** a^2b^4 **c.** a^4b^2 **d.** a^5b
20. In the expansion of $(a + 2)^6$, find the entire term containing each of the following factors:
    a. a^4 **b.** a^3 **c.** a^2 **d.** a^5
21. If a state license plate consists of 2 letters followed by 3 digits followed by 2 letters, how many different license plates are possible?
22. In problem 21, how many different license plates are possible if
    a. No letter and no digit can be used more than once?
    b. Letters and digits can be used more than once, but the first two letters cannot be the same, the last two letters cannot be the same, and consecutive digits must be different?
23. In a college calculus class, 25 students are business majors, 15 students are political science minors, and 5 business majors are political science minors.

a. How many students are business majors or political science minors?

b. If there are 40 students in the class, how many are neither business majors nor political science minors?

24. In the college elections, 7,621 people voted for Dawn or Lori, the two winners for the student senate. If Dawn received 5,231 votes and Lori received 3,692 votes, how many people

a. voted for Dawn and Lori?

b. voted for Dawn but not Lori?

25. Given $n(U) = 1000$, $n(A) = 550$, $n(C) = 400$, $n(A \cap B') = 300$, $n(A \cap C) = 150$, $n(B \cap C) = 200$, $n[B \cap (A \cup C)'] = 175$, and $n(A \cap B \cap C) = 75$, find

a. $n[(A \cup B \cup C)']$.

b. $n[A \cap (B \cup C)']$.

c. $n(A \cap B)$.

d. $n[(A \cap B) \cap C']$.

26. In a small high school, 25 seniors are enrolled in math, 18 are enrolled in chemistry, 15 are enrolled in physics, 12 are enrolled in math and chemistry, 8 are enrolled in math and physics, 5 are enrolled in chemistry and physics, and 3 are enrolled in all three. If the classes are to be combined for a weekend field trip, accommodations will have to be made for how many students?

27. Given $n(U) = 100$, $n(A) = 40$, $n(B) = 30$, $n(C) = 20$, $n(A \cap B) = 10$, $n(A \cap C) = 10$, $n(B \cap C) = 15$, and $n(A \cap B \cap C) = 5$, find

a. $n[(A \cup B \cup C)]$.

b. $n[(A \cup B \cup C)']$.

c. $n[C \cap (A \cup B)']$.

d. $n[(A \cap B) \cap C']$.

e. $n[(A \cap B) \cup (A \cap C) \cup (B \cap C)]$.

f. $n[A \cap (B \cup C)']$.

28. There are 1,000 students living in a college dormitory. The cafeteria manager polls the students to find out how many plan to eat lunch and dinner on Friday and breakfast on Saturday before the dorm closes for spring break. The survey results show that 830 plan to eat lunch on Friday, 550 plan to eat dinner on Friday, 310 plan to eat breakfast on Saturday, 450 plan to eat lunch and dinner, 250 plan to eat lunch and breakfast, 200 plan to eat dinner and breakfast, and 150 plan to eat all three. Assume all students responded.

a. How many plan to eat at least one meal?

b. How many plan to eat none of the three meals?

c. How many plan to eat at least two meals?

d. How many plan to eat exactly one of the three meals?

29. A survey was conducted among 500 senior citizens to determine what activities are the most popular. Of the 500 senior citizens, 300 like bingo but 200 of these do not like cards; 290 enjoy movies and 150 of these also like bingo; 40 like only cards; and 110 like movies and cards, with 50 of these also enjoying bingo.

a. How many of the senior citizens like at least one of the three?

b. How many like none of the three?

c. How many in all like cards?

d. Which activity seems to be the most popular?

30. In a horse race with 12 entries,
   a. How many orders of finish for the 12 horses are possible?
   b. How many orders of finish are possible for the top 3 places, in order?
   c. How many orders of finish are possible for the top 3 horses in any order?

31. To win the "daily double," you must pick the winning horse in each of the first two races. How many choices do you have if
   a. There are 12 horses in the first race and 11 in the second?
   b. There are 12 horses in each race?

32. A horse race has 10 horses entered, and you wish to pick the top 3 horses.
   a. How many possibilities are there if you must pick the correct order of finish?
   b. How many choices do you have if the order of finish doesn't matter?

33. In a football conference with 10 teams, how many orders of finish are possible for the top 3 places?

34. A college fraternity with 50 members needs to select 5 officers. In how many ways can the officers be selected?

35. The fraternity in problem 34 is having a party and needs a central planning committee. If the committee is to consist of 5 members, how many committees are possible?

36. In order to prepare a contract proposal, the 20-member union negotiating team is to be broken into four subcommittees to work on specific areas. The group working on salary is to have 6 members, the fringe benefit group is to have 5 members, the group looking into working conditions is to have 4 members, and the group working on grievance procedures is to have 5 members. In how many ways can the negotiating team be subdivided?

37. In how many ways can 20 children be subdivided into 4 teams of 5 students each?

38. There are 18 candidates for the school basketball team. The coach is going to divide the players into three teams—red, white, and blue—in that order.
   a. In how many ways can this be done?
   b. If the order is not important, how many possibilities are there?

39. In the class election, Jeff received 15 votes for president, Dana received 12 votes for vice-president, and Sara received 16 votes for secretary. There were 9 students who voted for Jeff and Dana, 9 who voted for Jeff and Sara, 9 who voted for Dana and Sara, and 7 who voted for all three. If there were 23 students in the class,
   a. How many voted for none of the 3?
   b. How many voted for Dana and Sara but not Jeff?
   c. How many voted for exactly 2 of the 3?

40. In a recent city election, 19,675 people voted for Brown for mayor, but 14,112 of these did not vote for his son for city treasurer. Of the 14,190 who voted for Brown's wife for city clerk, 6,612 also voted for Brown. There were 5,894 people who voted for Brown's son but not Brown or his wife, and 5,086 people voted for Brown's son and wife, with 2,351 of these also voting for Brown.

a. If the second-place finisher in the treasurer's race had 13,150 votes, did Brown's son win or lose and by how many votes?

b. How many people voted for Brown but not his wife or son?

c. Which candidate—Brown, his son, or his wife—had the lowest vote total and by how much?

6 Probability

INTRODUCTION

Probability and statistics are playing an increasingly important role in modern society. People in all walks of life are finding that some knowledge of probability and statistics is useful, for both personal and occupational reasons. We constantly encounter statements about the probability of rain, the median family income, the average corn yield, the Dow Jones average, or the EPA (Environmental Protection Agency) average fuel economy for a new car. These are just a few examples of how probability and statistics are involved in our daily lives.

A rigorous study of the theory of probability requires mathematical concepts not included in this book. For students interested in a more formal treatment of probability, we recommend a course in probability and statistics.

In this chapter some elementary concepts of probability will be developed. Although probability is sometimes associated with games of chance, its use is more widespread than this. In fact, modern statistical theory is based on the theory of probability.

The development of probability in this chapter is somewhat more intuitive than formal. Some properties will be proved when it seems appropriate. These proofs will be based, for the most part, on the set and counting properties of chapter 5.

6.1 Sample Spaces and Probabilities

In probability, an **experiment** is any process that results in an observation or a measurement. When an experiment is designed, the result of any one trial is referred to as an **outcome**. The set of all possible outcomes is called the **sample space**. In set language, the sample space is the universal set, and the outcomes are elements. Subsets of the sample space are referred

to as **events**. The subsets consisting of single elements are called **simple events**. Subsets with more than one outcome are sometimes called **composite events**.

■ Example 1. Two six-sided dice, (1) red and (2) blue, are tossed. Define the sample space.

Solution: By the multiplication principle, if we let w_1 be the number of ways the first die can turn up and w_2 be the number of ways the second die can turn up, the size of our sample space is $w_1 \cdot w_2 = 36$. Below is a listing of this sample space, where the first number in the ordered pair is for die 1 and the second is for die 2.

(1,1)	(1,2)	(1,3)	(1,4)	(1,5)	(1,6)
(2,1)	(2,2)	(2,3)	(2,4)	(2,5)	(2,6)
(3,1)	(3,2)	(3,3)	(3,4)	(3,5)	(3,6)
(4,1)	(4,2)	(4,3)	(4,4)	(4,5)	(4,6)
(5,1)	(5,2)	(5,3)	(5,4)	(5,5)	(5,6)
(6,1)	(6,2)	(6,3)	(6,4)	(6,5)	(6,6)

Figure 6.1. Sample Space ■

■ Example 2. Find the number of outcomes in the following events. (Refer to the sample space of example 1.)

1. $n(E)$, where E is the event that one of the dice shows a 6. Since $E = \{(1,6),(2,6),(3,6),(4,6),(5,6),(6,6),(6,5),(6,4),(6,3),(6,2),(6,1)\}$, $n(E) = 11$.
2. $n(F)$, where F is the event that both dice show a 6. Thus, $F = \{(6,6)\}$ and $n(F) = 1$.
3. $n(G)$, where G is the event that the sum of the numbers shown is 8. $G = \{(2,6),(6,2),(5,3),(3,5),(4,4)\}$; therefore, $n(G) = 5$.
4. $n(H)$, where H is the event that the sum is 3. Here $H = \{(1,2),(2,1)\}$ and $n(H) = 2$. ■

In some experiments, more than one sample space may be possible. The choice of a sample space is somewhat dependent on whatever aspect of the experiment is of interest to us at the time. We choose the sample space that best fits our needs.

■ Example 3. Consider again the experiment of tossing two dice.

a. If we are interested in the number showing on the face of each die, the appropriate sample space would be $S = \{(x,y) | x$ is the number on the first die, and y is the number on the second die$\}$. This is the 36-element set of ordered pairs in figure 6.1.

b. If our interest is in the sum of the numbers showing on the faces of the dice, we might consider a different sample space. Since the only sums we can get are 2,3,4,5, . . . ,12, we could let the sample space be $T = \{2,3,4, \ldots ,12\}$. This sample space contains 11 elements instead of 36.

c. Suppose our only concern is whether the sum of the two faces is an even number or an odd number. We could then use as a sample space the two-element set $U = \{\text{sum is even, sum is odd}\} = \{E,O\}$.

d. Suppose we are interested in whether each face shows an even or an odd number. Using ordered pair notation, (E,O) would represent an even number on the first die and an odd number on the second die. Our sample space would now be the four-element set $V = \{(E,E),(E,O),(O,E),(O,O)\}$. ■

■ Example 4. Three coins—a penny, a nickel, and a dime—are flipped. Construct a sample space for this experiment if

a. We are interested in the face showing on each coin.

b. We are interested in the number of heads showing.

Solutions:

a. We consider an outcome to be an ordered triple (P,N,D), where P is the face showing on the penny, N is the face showing on the nickel, and D is the face showing on the dime. The sample space consists of eight possible outcomes, since each coin has two ways to appear, H (heads) or T (tails), and $2 \cdot 2 \cdot 2 = 8$ by F.C.P. Then the sample space is given by $S = \{(T,T,T),(T,T,H), (T,H,T),(H,T,T), (T,H,H),(H,T,H), (H,H,T),(H,H,H)\}$.

b. Since the number of heads will be 0,1,2, or 3, the sample space would be the set $H = \{0,1,2,3\}$, which consists of four elements. ■

We begin our discussion of the **probability** of an event by considering the probability of simple events, or outcomes. The probability of an outcome is a real number that expresses the chance of that outcome occurring. Probabilities can be related to percents. If a weather forecaster says there is a 100% chance of rain, she is saying that she is certain it will rain.

Since 100% expressed as a decimal is 1.00, the forecaster is saying there is a 1 in 1 chance of rain. This corresponds to a probability of 1. If the forecaster says there is a 40% chance of rain, she is indicating that there is a 4 in 10 chance of rain (40% = 0.40 = 0.4). The corresponding probability is $\frac{4}{10}$ or $\frac{2}{5}$.

If a particular outcome must occur (100% chance), the probability of occurrence is 1. If, on the other hand, an outcome cannot occur (0% chance), the corresponding probability is 0. Intuitively, then, the probability of an outcome should be a number from 0 through 1 (the 0 and 1 are included).

How are probabilities actually assigned to the outcomes in the sample space? There is no single way of making these assignments. The assignments may be based on personal opinion, on past experience, or on the basis of other factors.

If, for example, John says that the probability the Cardinals will win a particular baseball game is $\frac{3}{5}$, this represents his personal opinion. This opinion may be based on who the pitchers are, where the game is being played, or other factors. However, this opinion may also be based on his personal prejudice.

When a weather forecaster says the probability of rain today is 0.4, the forecaster may be relying on past information. That is, it has rained about 4 out of every 10 days when similar weather conditions were observed.

Suppose an organization sells raffle tickets where the prize is a new car. They will sell only 100 tickets. The tickets are placed in a container and thoroughly mixed, then one ticket is drawn. What is the probability that you win if you purchased one ticket? Since you purchased 1 out of 100 tickets, your chance of winning is 1 in 100. Hence, it would be reasonable to assign a probability of $\frac{1}{100}$ or 0.01 to the outcome "your ticket is drawn."

Many factors may be involved in the assignment of probabilities to the outcomes of an experiment. The nature of the experiment and the type of outcome that is of interest determine the sample space under consideration. The sample space, in turn, influences the assignment of probabilities to the outcomes.

For example, if a pair of dice are tossed, we could use as a sample space the 36-element set of ordered pairs or the 11-element set of integers from 2 through 12, depending on what aspect of the experiment is of interest. The assignment of probabilities to outcomes would not be the same for both sample spaces.

One method of assigning probabilities that works well for certain kinds of experiments is the use of **relative frequency.** That is, if the experiment is performed a large number of times, what proportion of the times would we obtain, or expect to obtain, a particular outcome. If a coin is flipped 5,000 times, what proportion of the times would we expect to have a head showing? If we assume the coin is fair and there is no reason to anticipate one outcome more often than the other, we would expect to obtain about 2,500 heads.

The relative frequency is the ratio of the number of occurrences, or expected occurrences, to the number of times the experiment is performed. For the coin experiment, the relative frequency of heads is given by $\frac{2500}{5000} = \frac{1}{2}$. Hence, we assign a probability of $\frac{1}{2}$ to the outcome "heads."

Suppose a weather forecaster says the probability of rain today is 0.4 because past information indicates it rains about 40% of the time when similar conditions exist. The forecaster is using this relative frequency method for assigning probabilities.

Once probabilities have been assigned to outcomes, or simple events, these probabilities can be used to find the probabilities of composite events. Before stating a definition, we will introduce some notation.

If $S = \{e_1, e_2, \ldots, e_n\}$ is the sample space for an experiment, we will write $P(e_1)$, $P(e_2)$, . . . , $P(e_n)$ for the probabilities of the individual outcomes. We could also write $P(\{e_1\})$, $P(\{e_2\})$, . . . , $P(\{e_n\})$ for the probabilities of the simple events. Remember, an event is a subset of the sample space so the set notation is appropriate. However, the choice between outcome and simple event is not critical for determining probabilities. If E is any event, the probability of E will be denoted $P(E)$.

The following definition can be used for the probability of an event. It should be noted that it does depend on knowing the probabilities of the outcomes of the experiment.

Definition 6.1. If E is any event from a sample space S, the **probability** of E, denoted $P(E)$, is defined to be the sum of the probabilities of the outcomes in E.

■ Example 5. Suppose $S = \{e_1, e_2, e_3, e_4, e_5, e_6\}$ and $P(e_1) = 0.10$, $P(e_2) = 0.15$, $P(e_3) = 0.20$, $P(e_4) = 0.25$, $P(e_5) = 0.15$, and $P(e_6) = 0.15$.

a. Find $P(E)$ where $E = \{e_1, e_3\}$.
b. Find $P(F)$ where $F = \{e_1, e_3, e_5\}$.
c. Find $P(G)$ where $G = \{e_1, e_2, e_3, e_4\}$.

Solutions:

a. Since E contains the outcomes e_1 and e_3,
$P(E) = P(e_1) + P(e_3) = 0.10 + 0.20 = 0.30.$

b. Since F contains the outcomes e_1, e_3, and e_5,
$P(F) = P(e_1) + P(e_3) + P(e_5) = 0.10 + 0.20 + 0.15 = 0.45$

c.
$$\begin{aligned} P(G) &= P(e_1) + P(e_2) + P(e_3) + P(e_4) \\ &= 0.10 + 0.15 + 0.20 + 0.25 \\ &= 0.70 \end{aligned}$$

■

A special case of the relative frequency assignment of probabilities can be considered when the outcomes are **equally likely.** Outcomes are equally likely if there is no reason for one outcome to occur more often than any other outcome when the experiment is repeated a large number of times. That is, each outcome has the same chance of occurring.

If a fair coin is flipped, the two possible outcomes, heads or tails, are equally likely. Similarly, the six possible outcomes when a fair die is tossed are equally likely. If the names of the 100 people who purchased raffle tickets on a new car are placed in a box and mixed thoroughly, the 100 possible outcomes (name drawn) are equally likely.

All sample spaces are not made up of equally likely outcomes. Suppose a pair of fair dice are tossed. If we use the sample space consisting of the 36 ordered pairs, each of the outcomes has the same chance of occurring so the outcomes are equally likely. However, if we use the sample space consisting of the integers from 2 through 12, the outcomes are not equally likely. In this sample space the outcome 5 is more likely to occur than the outcome 2 since there are 4 ways for the sum to be 5 but there is only one way for the sum to be 2. The probability that "the sum is 5" is, then, four times the probability that "the sum is 2."

Consider the experiment of tossing a fair die a large number of times. Since there are six equally likely possible outcomes, we would expect to get each outcome about $\frac{1}{6}$ of the time. Hence, we would assign a probability of $\frac{1}{6}$ to each outcome, or simple event. When a fair coin is flipped, we would expect to get each of the two possible outcomes about $\frac{1}{2}$ of the time. Therefore, we assign each outcome a probability of $\frac{1}{2}$.

Suppose a sample space consists of N equally likely outcomes ($n(S) = N$). Then the probability assigned to each outcome is $\frac{1}{N}$. This can be extended to composite events from a sample space of equally likely outcomes. If an event E consists of r outcomes from a sample space of N equally likely outcomes, by definition 6.1 and the preceding discussion,

$$P(E) = \overbrace{\frac{1}{N} + \frac{1}{N} + \ldots + \frac{1}{N}}^{r \text{ terms}}$$

$$= \frac{\overbrace{1 + 1 + \ldots + 1}^{r \text{ terms}}}{N}$$

$$= \frac{r}{N}$$

That is, the probability of E is the number of outcomes in E (the number of favorable outcomes) divided by the number of outcomes in S. This is stated more formally as definition 6.2.

Definition 6.2. If a sample space S consists of N equally likely outcomes and E is an event consisting of r of these outcomes, the probability of E is given by

$$P(E) = \frac{\text{number of favorable outcomes}}{\text{number of outcomes in } S} = \frac{n(E)}{n(S)} = \frac{r}{N}.$$

NOTE: We are considering only finite sample spaces here. Modifications of this discussion are required for infinite sample spaces.

If the elements of a sample space (outcomes) are not equally likely to occur, special treatment is necessary. In most problems, to avoid this unnecessary difficulty, we model our sample spaces to contain only equally likely outcomes.

We return now to the experiment of tossing three coins in example 4. Two different sample spaces, S and H, were constructed. In the sample space S, each of the eight outcomes is equally likely. Each of the simple events would have a probability of $\frac{1}{8}$. If E is the event "the penny shows a tail," then

$$E = \{(T,T,T),(T,T,H),(T,H,T),\ (T,H,H)\} \text{ and } P(E) = \frac{n(E)}{n(S)} = \frac{4}{8} = \frac{1}{2}.$$

If F is the event "either the penny or the dime or both show tails," then

$$F = \{(T,T,T),(T,T,H),(T,H,T),(T,H,H),(H,T,T),(H,H,T)\} \text{ and}$$

$$P(F) = \frac{n(F)}{n(S)} = \frac{6}{8} = \frac{3}{4}.$$

The sample space $H = \{0,1,2,3\}$ consists of four elements indicating the number of heads. In order to talk about probabilities, further analysis is necessary. We begin by looking at the four simple events and how they come about in terms of the larger, eight-element sample space.

1. $\{0\}$ is obtained only from the outcome (T,T,T).
2. $\{1\}$ is obtained from (T,T,H), (T,H,T), or (H,T,T).
3. $\{2\}$ is obtained from (T,H,H), (H,T,H), or (H,H,T).
4. $\{3\}$ is obtained from (H,H,H) only.

We can now determine probabilities based on the relative frequency of occurrence in S.

1. $P(\{0\}) = \dfrac{\text{number of cases}}{n(S)} = \dfrac{1}{8}$
2. $P(\{1\}) = \dfrac{\text{number of cases}}{n(S)} = \dfrac{3}{8}$
3. $P(\{2\}) = \dfrac{\text{number of cases}}{n(S)} = \dfrac{3}{8}$
4. $P(\{3\}) = \dfrac{\text{number of cases}}{n(S)} = \dfrac{1}{8}$

The outcomes in this sample space are not equally likely. That is, $P(\{0\}) \neq \dfrac{n(\{0\})}{n(S)} = \dfrac{1}{4}$.

Remember special care must be taken when dealing with sample spaces where the outcomes are not equally likely. It may be necessary to refer to a larger sample space, with equally likely outcomes, in order to find probabilities.

■ Example 6. Suppose two dice are tossed and an outcome is the sum of the faces showing. Find the probability that

a. the sum is equal to 7.
b. the sum is less than 6.
c. the sum is 7 or 11.

Solutions: Since we are interested in the sum of the faces, we let the sample space be $S = \{2,3,4,5,6,7,8,9,10,11,12\}$. However, these outcomes are not equally likely since some of them come from more combinations than others. For example, the only way to get a 2 is to have a 1 on both dice, whereas a 3 can be obtained in two ways, (1,2) or (2,1).

To find the probabilities in question, we refer back to the 36-element sample space of example 1.

a. If E is the event "sum of 7," then $E = \{(1,6),(2,5),(3,4),(4,3),(5,2),(6,1)\}$ and $P(E) = \frac{n(E)}{36} = \frac{6}{36} = \frac{1}{6}$.

b. Let F be the event "sum is less than 6."
Then $F = \{(1,1),(1,2),(1,3),(1,4),(2,1),(2,2),(2,3),(3,1),(3,2),(4,1)\}$ and
$P(F) = \frac{n(F)}{36} = \frac{10}{36} = \frac{5}{18}$.

c. If G is the event "sum is 7 or 11,"
then $G = \{(1,6),(2,5),(3,4),(4,3),(5,2),(6,1),(5,6),(6,5)\}$ and
$P(G) = \frac{n(G)}{36} = \frac{8}{36} = \frac{2}{9}$. ■

Remember: To compute **probability as a ratio** for events from finite sample spaces with equally likely outcomes, the probability of an event E, written $P(E)$, is computed as the ratio of $n(E)$ to $n(S)$; i.e.,

$$P(E) = \frac{n(E)}{n(S)} = \frac{\text{number of favorable outcomes}}{\text{total possible number of outcomes}}.$$

■ Example 7. A dozen eggs contains 2 eggs with double yolks. If 3 eggs are selected at random, what is the probability that no eggs with double yolks are selected?

Solution: Since there are 12 eggs to choose from, the total number of possible outcomes is $\binom{12}{3}$. Since there are 10 eggs with single yolks, the number of favorable outcomes is $\binom{10}{3}$. Hence

$$P(\text{3 eggs with single yolks}) = \frac{\binom{10}{3}}{\binom{12}{3}} = \frac{10!}{3!7!} \cdot \frac{3!9!}{12!} = \frac{8 \cdot 9 \cdot 10}{10 \cdot 11 \cdot 12} = \frac{6}{11}.$$ ■

■ Example 8. From a class of 15 girls and 10 boys, a committee of 4 people, 2 girls and 2 boys, is selected at random for hall duty.

a. What is the probability that Ann is selected?

b. What is the probability that Ann is chosen if any combination of boys and girls may be selected as long as the selection process is random?

Solutions: **a.** $P(\text{Ann is selected}) = \dfrac{14}{\binom{15}{2}} = \dfrac{14 \cdot 2!13!}{15!} = \dfrac{2}{15}$ where there are 14 favorable combinations, Ann with any of the other 14 girls, and the total number of possible outcomes is $\binom{15}{2}$. More directly, there are 2 girls selected from 15. Hence, the probability is $\frac{2}{15}$.

b. The favorable outcomes are Ann with any other 3 people. The total number of favorable outcomes is $\binom{24}{3}$. The total number of outcomes is, of course, $\binom{25}{4}$.

Then $P(\text{Ann is selected}) = \dfrac{\binom{24}{3}}{\binom{25}{4}} = \dfrac{24!}{3!21!} \cdot \dfrac{4!21!}{25!} = \dfrac{4}{25}$.

Ann's chances have improved slightly. ■

If E is an event from a finite sample space S, then $E \subset S$ and $n(E) \leq n(S)$. The following properties are true for $P(E)$.

Property 6.1 Let S be a sample space and let $E \subset S$ be an event. Then,

1. $0 \leq P(E) \leq 1$. The $P(E)$ is the ratio of two non-negative numbers where the numerator is never larger than the denominator.
2. $P(E) = 0$ implies the event E cannot occur.
3. $P(E) = 1$ implies the event E is bound to occur.

CAUTION: The interpretations in statements 2 and 3 refer only to finite sample spaces. If, in an infinite sample space, an event occurs only a finite number of times, then $P(E)$ may be zero or approach zero as a limit. The study of infinite sample spaces that cannot be reduced to finite models will not be considered here. In a later chapter we will consider one particular situation related to infinite sample spaces.

If E is an event from a sample space S, then $E \subset S$. The set E' is also a subset of S and, hence, is also an event. The events E and E' are referred to as **complementary events.** Recall here that for any set E, $E \subset U$, where U is the universe, and $E \cup E' = U$.

For any sample space S, ∅ and S are complementary events. The probabilities of these two events, ∅ and S, are given in the following property.

Property 6.2 Let S be a sample space. Then
1. $P(S) = 1$ and
2. $P(\emptyset) = 0$.

Proof (finite sample space, equally likely outcomes):

1. $P(S) = \dfrac{n(S)}{n(S)} = 1.$
2. $P(\emptyset) = \dfrac{n(\emptyset)}{n(S)} = \dfrac{0}{n(S)} = 0.$

It is possible to state a more general property relating the probability of a set and its complement. This is done in property 6.3.

Property 6.3 If S is a sample space and E is an event from this sample space ($E \subset S$), then $P(E) = 1 - P(E')$. It is sometimes easier to find $P(E')$ than $P(E)$ by direct computation. This property will then give us a way to find $P(E)$.

Proof:

Since $S = E \cup E'$

$n(S) = n(E \cup E')$	
$n(S) = n(E) + n(E') - n(E \cap E')$	Inclusion-exclusion principle
$n(S) = n(E) + n(E')$	Since $E \cap E' = \emptyset$ and $n(\emptyset) = 0$.
$\dfrac{n(S)}{n(S)} = \dfrac{n(E)}{n(S)} + \dfrac{n(E')}{n(S)}$	Divide by $n(S)$.
$1 = P(E) + P(E')$	Definition of $P(E)$ and $P(E')$
$P(E) = 1 - P(E')$	Algebra
$P(E') = 1 - P(E)$	Algebra

■ **Example 9.** Two dice are tossed, and the sum of the faces is observed. Find the probability that the sum

a. does not equal 7.
b. is greater than or equal to 6.

Solutions:

a. If E is the event "sum is 7," then E′ is the event "sum is not 7." In example 6(a), we found

$$P(E) = \frac{1}{6}, \text{ so } P(E') = 1 - P(E) = 1 - \frac{1}{6} = \frac{5}{6}.$$

b. If F is the event "sum is greater than or equal to 6," then F' is the event "sum is less than 6." We have

$$P(F) = 1 - P(F')$$

$$= 1 - \frac{5}{18} \qquad \text{from example 6(b), } P(F') = \frac{5}{18}$$

$$= \frac{13}{18}.$$

■

Exercise Set 6.1

1. Two dice are tossed, and the sum of the faces is observed. Use the sample space of example 1 to find the probability of each simple event in the sample space {2,3,4,5, . . . ,12}.
2. Use the sample space in example 1 to calculate the probabilities of the following events.
   - **a.** The face of at least one die shows a 4.
   - **b.** Neither face shows an odd number.
   - **c.** The faces of both dice show either a 1 or a 2.
3. If 5 coins are tossed, what is the probability that there are exactly 3 heads?
4. A dozen eggs contains 3 rotten eggs. If 2 eggs are selected at random, what is the probability that
   - **a.** no bad eggs were selected?
   - **b.** exactly 1 bad egg was selected?
   - **c.** both eggs are bad?
   - **d.** at least 1 egg is bad?
5. Five cards are selected at random from a deck of 52 cards. What is the probability that
   - **a.** 5 hearts are selected?
   - **b.** 5 cards of the same suit are selected?
   - **c.** 4 cards of the same suit are selected, the fifth card being of another suit?
6. Playing draw poker (5 cards are dealt each player), you are dealt 4 hearts and 1 spade.
   - **a.** What is the probability that you are dealt a heart on your one-card draw?
   - **b.** What effect does the number of players have on this probability?
7. You buy a ticket on a 3-digit number in the Illinois State Lottery. You pick three distinct digits and you win if the digits come up in any order. What is the probability that you pick a winner at random?

8. What is the probability that you will pass through point B on one of your short routes from A to C if your route is random?

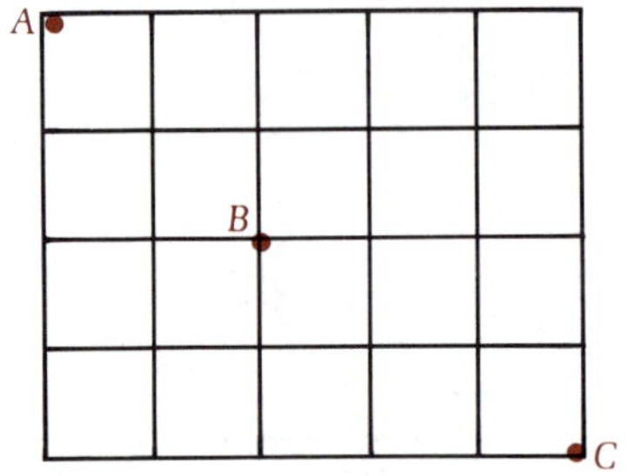

9. If $P(A) = \frac{2}{3}$ and $P(B) = \frac{1}{5}$, compute $P(A') \cdot P(B')$.
10. In a class of 14 girls and 26 boys, the instructor picks, each day, 4 students to respond to questions. What is the probability that at least 1 boy is asked to respond?
11. A drawer contains 10 pairs of socks. Five of the pairs are black and interchangeable. The other 5 pairs have only 1 mate each. It's 3 a.m., and the baby is about due. The proud father grabs a couple of socks from the drawer and takes his wife to the hospital. What is the probability that the socks match?
12. Your 3 pet deer go wild and now run with a herd of 50 deer which needs to be reduced to 40 so the range can support them. What is the probability that all of your former pets will escape?
13. A coin is tossed 5 times. What is the probability of obtaining more heads than tails?
14. A man has 20 golf balls in a bag. Five of the balls are new, and the other 15 are well used. The man selects 5 balls at random to practice with. What is the probability that exactly 2 of these practice balls are new?
15. There are 20 employees at "Texas Oil" who share offices, 4 employees to an office. If 15 of the employees like country music, and the other 5 really prefer the Boston Pops Orchestra, what is the probability that in any one office, where the employees have been randomly assigned, at least one non-country music lover will exist?
16. A 3-digit number is drawn at random from 000 to 999. What is the probability that three distinct digits are drawn?
17. *Playboy* magazine has 180 pages. A page may have writing on both sides. What is the probability that, if the magazine is opened at random, the centerfold will appear? *Hint:* Use the sample space {(1,2),(2,3),(3,4), . . . ,(179,180)}, where (8,9) indicates the magazine is open between pages 8 and 9.
18. Find the size of a sample space of equally likely events to find the outcomes of the experiment of tossing 2 dice and flipping 2 coins.

19. In problem 18, what is the probability of getting the faces of the dice to match? What is the probability that the faces on the dice will match and the coins will also match?

20. Given a sample space U of equally likely events, where $S \subset U$ and $n(U) = 100$ while $n(S) = 47$. Find $P(T)$ if $S \cup T = U$ and $S \cap T = \emptyset$.

6.2 Compound Events

In the preceding section, probabilities of single events have been considered. Many probability problems involve more than one event. This section will discuss probabilities of combinations of events, referred to as **compound events.** In particular, if E and F are events from a sample space S, what is the probability that E or F will occur? What is the probability that E and F will occur?

Some of the set ideas from chapter 5 will be used here. By $E \cup F$, we mean the event that E or F (or both) occurs. Then $P(E \cup F)$ (read "the probability of E or F") is the probability that at least one of the events occurs.

The event $E \cap F$ is the event that both E and F occur. The probability that both events occur is written $P(E \cap F)$ and is read "the probability of E and F."

The set operation symbols $\cup$ and $\cap$ are used intentionally. If E and F are events, then E and F will occur precisely when the outcome is in both sets; i.e., the outcome is in $E \cap F$. Similarly, E or F will occur when the outcome is in at least one of E or F; i.e., the outcome is in $E \cup F$. The following example should help illustrate.

■ Example 10. Three coins—a penny, nickel, and dime—are flipped. Let E_1, E_2, and E_3 be the events as follows:

E_1 is event "H on the penny."
E_2 is event "H on the dime."
E_3 is event "T on the dime."
Find the following events:

1. $E_1 \cup E_2$. **2.** $E_1 \cup E_3$. **3.** $E_2 \cup E_3$.
4. $E_1 \cap E_2$. **5.** $E_1 \cap E_3$. **6.** $E_2 \cap E_3$.

Solutions: The sample space, example 4(a), is

$$S = \{(T,T,T),(T,T,H),(T,H,T),(H,T,T),(T,H,H),(H,T,H),(H,H,T),(H,H,H)\}$$

and

$$E_1 = \{(H,T,T),(H,T,H),(H,H,T),(H,H,H)\}.$$
$$E_2 = \{(T,T,H),(T,H,H),(H,T,H),(H,H,H)\}.$$
$$E_3 = \{(T,T,T),(T,H,T),(H,T,T),(H,H,T)\}.$$

1. $E_1 \cup E_2 = \{(H,T,T),(H,T,H),(H,H,T),(H,H,H),$
$(T,T,H),(T,H,H)\}$
2. $E_1 \cup E_3 = \{(H,T,T),(H,T,H),(H,H,T),(H,H,H),$
$(T,T,T),(T,H,T)\}$
3. $E_2 \cup E_3 = \{(T,T,H),(T,H,H),(H,T,H),(H,H,H),$
$(T,T,T),(T,H,T),(H,T,T),(H,H,T)\} = S$
4. $E_1 \cap E_2 = \{(H,T,H),(H,H,H)\}$
5. $E_1 \cap E_3 = \{(H,T,T),(H,H,T)\}$
6. $E_2 \cap E_3 = \emptyset$

Note that E_2 and E_3 are complementary events. That is, $E_2 = E_3'$ or $E_3 = E_2'$. ■

In situations where it is possible to write out the sets $E \cup F$ and $E \cap F$, probabilities can be determined using the method of section 6.1. For the preceding example, since $n(S) = 8$,

$$P(E_1 \cup E_2) = \frac{n(E_1 \cup E_2)}{n(S)} = \frac{6}{8} = \frac{3}{4}$$

$$\text{and } P(E_1 \cap E_2) = \frac{n(E_1 \cap E_2)}{n(S)} = \frac{2}{8} = \frac{1}{4}.$$

Other probabilities could be computed in a similar manner.

However, it is sometimes not possible to compute the probabilities in this way. The following property, which could be called the inclusion-exclusion principle for probability, is useful for some of these situations. The proof is given for the case where all outcomes are equally likely. The property is not restricted to this case, however. This property is true for any sample space and any events.

Property 6.4 If E and F are events from a sample space S, then $P(E \cup F) = P(E) + P(F) - P(E \cap F)$.

Proof (assuming all outcomes are equally likely):
From the inclusion-exclusion principle stated in chapter 5, $n(E \cup F) = n(E) + n(F) - n(E \cap F)$. Dividing through by $n(S)$ gives

$$\frac{n(E \cup F)}{n(S)} = \frac{n(E)}{n(S)} + \frac{n(F)}{n(S)} - \frac{n(E \cap F)}{n(S)}.$$

Then $P(E \cup F) = P(E) + P(F) - P(E \cap F)$, by the definition of probability.

Consider the events E_1 and E_2 of example 10. On the basis of the sets given, $P(E_1) = \frac{4}{8} = \frac{1}{2}$, $P(E_2) = \frac{4}{8} = \frac{1}{2}$, and $P(E_1 \cap E_2) = \frac{2}{8} = \frac{1}{4}$.

We then could find $P(E_1 \cup E_2)$ using property 6.4.

$$P(E_1 \cup E_2) = P(E_1) + P(E_2) - P(E_1 \cap E_2) = \frac{1}{2} + \frac{1}{2} - \frac{1}{4} = \frac{3}{4}.$$

It is also possible to use property 6.4 to find $P(E \cap F)$ when $P(E \cup F)$ is known. Using a little elementary algebra, we solve for $P(E \cap F)$ and obtain $P(E \cap F) = P(E) + P(F) - P(E \cup F)$.

■ **Example 11.** If $P(E) = 0.7$, $P(F) = 0.5$, and $P(E \cap F) = 0.3$, what is $P(E \cup F)$?

Solution:

$$\begin{aligned} P(E \cup F) &= P(E) + P(F) - P(E \cap F) \\ &= 0.7 + 0.5 - 0.3 \\ &= 1.2 - 0.3 = 0.9 \end{aligned}$$

■

■ **Example 12.** If $P(E) = 0.5$, $P(F) = 0.4$, and $P(E \cup F) = 0.7$, find $P(E \cap F)$.

Solution:

$$\begin{aligned} P(E \cap F) &= P(E) + P(F) - P(E \cup F) \\ &= 0.5 + 0.4 - 0.7 \\ &= 0.9 - 0.7 = 0.2 \end{aligned}$$

■

■ **Example 13.** In a college sorority with 50 members, 21 are business majors, 12 are math minors, and 5 are majoring in business and minoring in math. What is the probability that a member selected at random is a business major or math minor?

Solution: Let B be the event "is a business major" and M be the event "is a math minor."

Then $P(B) = \frac{21}{50} = 0.42$, $P(M) = \frac{12}{50} = 0.24$, and $P(B \cap M) = \frac{5}{50} = 0.10$.

$$\begin{aligned} P(B \cup M) &= 0.42 + 0.24 - 0.10 \\ &= 0.66 - 0.10 \\ &= 0.56 \end{aligned}$$

Alternate solution using Venn Diagram:
The Venn Diagram in figure 6.2 has the information sorted, as was done in chapter 5.

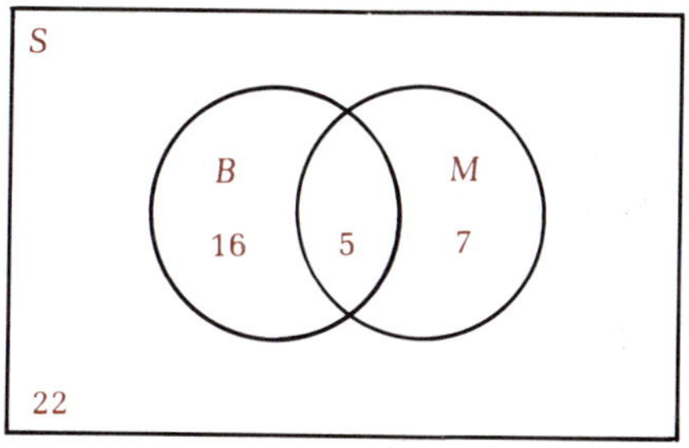

$$n(B \cup M) = 16 + 5 + 7 = 28$$

$$P(B \cup M) = \frac{n(B \cup M)}{n(S)} = \frac{28}{50} = 0.56$$

Figure 6.2

■

Example 14. A 5-card hand is drawn from a standard deck of 52 cards. Find the probability of getting

1. A straight (5 consecutive denominations where an ace can be counted as high or low),
2. a flush (5 cards of the same suit), and
3. a straight or a flush.

Solutions: Let E be the event "draw a straight" and F the event "draw a flush."

1. $P(E) = \frac{n(E)}{n(S)}$. S is the sample space consisting of all 5-card hands, so $n(S) = \binom{52}{5} = 2{,}598{,}960$. To find $n(E)$, reason as follows: A straight consists of 5 cards of consecutive denominations. The high card in a straight can be in any one of the 4 suits, the second high card can be in any one of the 4 suits, and so on. Then there are $4 \cdot 4 \cdot 4 \cdot 4 \cdot 4 = 4^5$ possible straights with any particular card as the high card. The high card can be any of the 10 cards A,K,Q,J, . . . ,5,(5,4,3,2,A counts). Then there are $10 \cdot 4^5$ possible straights.

$$P(E) = \frac{n(E)}{n(S)} = \frac{10.4^5}{\binom{52}{5}} = \frac{10{,}240}{2{,}598{,}960} \text{ which is about } 0.003940.$$

2. $P(F) = \frac{n(F)}{n(S)} = \frac{n(F)}{2{,}598{,}960}$

   To find $n(F)$, consider a sequence of two events. First pick a suit, and then choose 5 cards from this suit. The number of ways of obtaining a flush is

$$n(F) = 4 \cdot \binom{13}{5} = \frac{4 \cdot 13 \cdot 12 \cdot 11 \cdot 10 \cdot 9 \cdot 8!}{5!8!}$$
$$= \frac{4 \cdot 13 \cdot 12 \cdot 11 \cdot 10 \cdot 9}{5 \cdot 4 \cdot 3 \cdot 2 \cdot 1}$$
$$= 5148$$

   Then $P(F) = \frac{5148}{2{,}598{,}960}$, or about 0.001981.

3. $P(E \cup F) = \frac{n(E \cup F)}{n(S)}$ or $P(E \cup F) = P(E) + P(F) - P(E \cap F)$ where $P(E \cap F) = \frac{n(E \cap F)}{n(S)}$.

   It is easier in this case to compute $n(E \cap F)$ than it is to find $n(E \cup F)$. To find $n(E \cap F)$, we consider a sequence of two events where we first select one of the 4 suits and then select a straight in that suit. There are, as before, 10 cards that could be the highest card in the straight. However, once this highest card has been chosen, there is only one choice for a second card, one for a third card, and so on. *Remember*: the suit has been selected first.

Then $n(E \cap F) = 4 \cdot 10 \cdot 1 \cdot 1 \cdot 1 \cdot 1$, so there are 40 straight flushes.

$$P(E \cap F) = \frac{40}{2{,}598{,}960}, \text{ or about } 0.000015.$$

$$\begin{aligned} P(E \cup F) &= P(E) + P(F) - P(E \cap F) \\ &= 0.003940 + 0.001981 - 0.000015 \\ &= 0.005906 \end{aligned}$$

■

Suppose E and F are events from a sample space S and suppose, further, that, as sets, E and F are disjoint. Then $n(E \cap F) = 0$ and $P(E \cap F) = \dfrac{n(E \cap F)}{n(S)} = \dfrac{0}{n(S)} = 0$. Such events are called **mutually exclusive.**

Definition 6.3 Two events, E and F, are called *mutually exclusive*, providing, as sets, $E \cap F = \emptyset$.

For finite sample spaces, if $P(E \cap F) = 0$, then $E \cap F = \emptyset$ and E and F are mutually exclusive. This provides a way of determining when two events are mutually exclusive in some problems.

It follows from property 6.4, that if E and F are mutually exclusive, then

$$P(E \cup F) = P(E) + P(F).$$

In general, if $E_1, E_2, E_3, \ldots, E_n$ are pairwise mutually exclusive events (each pair of them being mutually exclusive), then

$$P(E_1 \cup E_2 \cup E_3 \cup \ldots \cup E_n) = P(E_1) + P(E_2) + P(E_3) + \ldots + P(E_n).$$

In section 6.1, we showed that for any event E, $P(E) = 1 - P(E')$. This result could also be shown by using property 6.4. Since $E \cap E' = \emptyset$ and $E \cup E' = S$, we have, from property 6.4, that

$$P(E \cup E') = P(E) + P(E') - P(E \cap E').$$

$1 = P(E) + P(E') - 0$	$E \cup E' = S$ so $P(E \cup E') = 1$
	$E \cap E' = \emptyset$ so $P(E \cap E') = 0$
$P(E) = 1 - P(E')$	Algebra

■ Example 15. Suppose $P(E) = 0.4$, $P(F) = 0.3$, and $P(E \cup F) = 0.7$. Find $P(E \cap F)$. What can be said about E and F?

Solution:

$$\begin{aligned} P(E \cap F) &= P(E) + P(F) - P(E \cup F) \\ &= 0.4 + 0.3 - 0.7 \\ &= 0 \end{aligned}$$

Hence E and F are mutually exclusive. ■

■ **Example 16.** If $P(E) = 0.7$, $P(F) = 0.5$, and $P(E \cap F) = 0.4$, find $P(E' \cap F')$.

Solution: We are asked to find the probability that E does not occur and that F does not occur.

Using one of DeMorgan's Laws (see chapter 5), we see that

$$E' \cap F' = (E \cup F)', \text{ so}$$

$$\begin{aligned} P(E' \cap F') &= P[(E \cup F)'] \\ &= 1 - P(E \cup F). \end{aligned}$$

Start by using property 6.4 to find $P(E \cup F)$ and then use the result obtained previously.

$$\begin{aligned} P(E \cup F) &= P(E) + P(F) - P(E \cap F) \\ &= 0.7 + 0.5 - 0.4 \\ &= 0.8 \end{aligned}$$

$$\begin{aligned} \text{Then } P(E' \cap F') &= P[(E \cup F)'] \\ &= 1 - P(E \cup F) \\ &= 1 - 0.8 = 0.2 \end{aligned}$$

■

CAUTION: It would be incorrect to say that

$$\begin{aligned} P(E' \cap F') &= 1 - P(E \cap F) \\ &= 1 - 0.4 \\ &= 0.6. \end{aligned}$$

This is *not* true, since $(E' \cap F') \neq (E \cap F)'$. Care must be taken when working with complements to ensure that the properties are used correctly.

In some situations, it becomes useful to be able to find mutually exclusive events with certain properties. If two events are mutually exclusive, then we know that the probability of their union is the sum of their probabilities.

One important example of this observation begins with a consideration of any two events, E and F. It doesn't matter if E and F are mutually exclusive or not. Consider the two compound events $(E \cap F)$ and $(E \cap F')$. Property 6.5 makes two claims about these events.

Property 6.5 If E and F are any two events from a sample space S, then

1. $(E \cap F)$ and $(E \cap F')$ are mutually exclusive and
2. $[(E \cap F) \cup (E \cap F')] = E$.

Proof:
Recall that $F \cap F' = \emptyset$ and $F \cup F' = S$.

1. $(E \cap F) \cap (E \cap F') = E \cap [F \cap (E \cap F')]$ Associative Property
$= E \cap [(F \cap E) \cap F']$ Associative Property
$= E \cap [(E \cap F) \cap F']$ Commutative Property
$= E \cap [E \cap (F \cap F')]$ Associative Property
$= E \cap (E \cap \emptyset)$ $F \cap F' = \emptyset$
$= E \cap \emptyset$ $E \cap \emptyset = \emptyset$ for any set E
$= \emptyset$

By definition, if $(E \cap F) \cap (E \cap F') = \emptyset$, then $(E \cap F)$ and $(E \cap F')$ are mutually exclusive.

2. $(E \cap F) \cup (E \cap F') = E \cap (F \cup F')$ Distributive Property
$= E \cap S$ $F \cup F' = S$
$= E$ $E \cap S = E$ for any set E

Hence, $E \cap F$ and $E \cap F'$ are mutually exclusive events whose union is E. Then $P(E) = P(E \cap F) + P(E \cap F')$ and $P(E \cap F') = P(E) - P(E \cap F)$.

■ **Example 17.** Suppose, in a fraternity of 100 members, 25 are agriculture majors, 35 are business minors, and 5 agriculture majors are business minors. What is the probability that a randomly selected fraternity member is majoring in agriculture but not minoring in business?

Solution: Let A be the event "is an agriculture major" and B be the event "is a business minor." We are asked to find $P(A \cap B')$.

Since $P(A) = P(A \cap B) + P(A \cap B')$,

$$P(A \cap B') = P(A) - P(A \cap B) = \frac{25}{100} - \frac{5}{100} = \frac{20}{100} = 0.2.$$ ■

■ **Example 18.** In a survey of 45 kindergarten children, it is found that 40 watch "Sesame Street" or "Electric Company," 30 watch "Sesame Street," and 25 watch "Electric Company." What is the probability that a randomly selected student from this group watches "Electric Company" but not "Sesame Street"?

Solution: If E is the event "watches 'Electric Company'" and F is the event "watches 'Sesame Street,'" then $P(E) = \frac{25}{45} = \frac{5}{9}$, $P(F) = \frac{30}{45} = \frac{6}{9}$, and $P(E \cup F) = \frac{40}{45} = \frac{8}{9}$.

We need to find $P(E \cap F')$. By property 6.4,

$$P(E \cap F) = P(E) + P(F) - P(E \cup F) = \frac{5}{9} + \frac{6}{9} - \frac{8}{9} = \frac{3}{9}.$$

Then

$$P(E \cap F') = P(E) - P(E \cap F) = \frac{5}{9} - \frac{3}{9} = \frac{2}{9}.$$

■

Property 6.5 can, under certain conditions, be extended. This extension is stated, without proof, as property 6.6.

Property 6.6 Let $E_1, E_2, E_3, \ldots, E_n$ be **pairwise mutually exclusive** (each pair E_i, E_j with $i \neq j$ satisfies $E_i \cap E_j = \emptyset$) events from a sample space S with $E_1 \cup E_2 \cup \ldots \cup E_n = S$. Then, for any event $F \subset S$,

1. $(F \cap E_1), (F \cap E_2), \ldots, (F \cap E_n)$ are pairwise mutually exclusive.
2. $(F \cap E_1) \cup (F \cap E_2) \cup \ldots \cup (F \cap E_n) = F$, and
3. $P(F) = P(F \cap E_1) + P(F \cap E_2) + \ldots + P(F \cap E_n)$.

The following example will illustrate the use of property 6.6. You may see other ways of working this example, but it is intended to serve only as an illustration.

■ **Example 19.** Human blood is classified in terms of four blood groups designated A, B, AB, or O. When the Rh factor, either + or −, is added to each of the basic blood groups, eight distinct blood types are obtained. According to data from the American National Red Cross, approximately 38% of the people have A+ blood, 8% have B+ blood, 3% have AB+ blood, and 36% have O+ blood. What is the probability that a randomly selected person has Rh− blood?

Solution: Let E_1, E_2, E_3, and E_4 represent the events that "a person is in blood groups A, B, AB, and O, respectively," and let F represent the event that "a person has Rh+ blood." We seek $P(F') = 1 - P(F)$. Converting percents to probabilities, we have

$$P(F \cap E_1) = 0.38$$

$$P(F \cap E_2) = 0.08$$

$$P(F \cap E_3) = 0.03$$

$$P(F \cap E_4) = 0.36.$$

The four events E_1, E_2, E_3, E_4 satisfy the conditions of property 6.6. That is, they are pairwise mutually exclusive and $E_1 \cup E_2 \cup E_3 \cup E_4 = S$. By property 6.6,

$$\begin{aligned} P(F) &= P(F \cap E_1) + P(F \cap E_2) + P(F \cap E_3) + P(F \cap E_4) \\ &= 0.38 + 0.08 + 0.03 + 0.36 \\ &= 0.85 \end{aligned}$$

Hence $P(F') = 1 - 0.85$
$= 0.15$

About 15% of all people have Rh− blood. ■

In the statement of property 6.6, it is very important that we require the events to be pairwise mutually exclusive. The three sets $A = \{0,1,2,3\}$, $B = \{0,2,4\}$, and $C = \{1,3,4\}$ satisfy $A \cap B \cap C = \emptyset$, so we could say they are mutually exclusive. However, they are not pairwise mutually exclusive, since $A \cap B = \{0,2\}$, $A \cap C = \{1,3\}$, and $B \cap C = \{4\}$, and none of these intersections is empty.

The next example will indicate why the events are required to be pairwise mutually exclusive.

■ **Example 20.** Consider the experiment of tossing a fair die. Then $S = \{1,2,3,4,5,6\}$, and let $E_1 = \{2,4,6\}$, $E_2 = \{1,3,5\}$, $E_3 = \{3,6\}$, and $F = \{1,2,3\}$. Find $P(F)$.

Solution: Since each outcome in S is equally likely, we have

$$P(F) = \frac{n(F)}{n(S)} = \frac{3}{6} = \frac{1}{2}.$$

However, if we try to use property 6.6, we encounter a problem. It is true that $E_1 \cup E_2 \cup E_3 = S$ and $E_1 \cap E_2 \cap E_3 = \emptyset$. Note, however, that $E_1 \cap E_3 \neq \emptyset$ and $E_2 \cap E_3 \neq \emptyset$. To use property 6.6, we need

$$P(F \cap E_1) = P(\{2\}) = \frac{1}{6},$$

$$P(F \cap E_2) = P(\{1,3\}) = \frac{2}{6}, \text{ and}$$

$$P(F \cap E_3) = P(\{3\}) = \frac{1}{6}.$$

Then $P(F) = \frac{1}{6} + \frac{2}{6} + \frac{1}{6} = \frac{4}{6} \neq \frac{1}{2}$. ■

Be sure, before using property 6.6 (or any other property for that matter), that all of the conditions are satisfied. Although it is true that if $E_1, E_2, \ldots, E_n$ are pairwise mutually exclusive, then $E_1 \cap E_2 \cap \ldots \cap E_n = \emptyset$, the converse is not necessarily true. That is, as seen in example 20,

$E_1 \cap E_2 \cap \ldots \cap E_n = \emptyset$ does not imply that $E_1, E_2, \ldots, E_n$ are pairwise mutually exclusive.

■ **Example 21.** Use a Venn Diagram to construct a model sample space where $P(A) = 0.29$, $P(B) = 0.41$, and $P(A \cap B) = 0.12$. Use the model to calculate

a. $P(A \cap B')$.
b. $P(A' \cap B)$.
c. $P(A' \cap B')$.
d. $P(A \cup B)$.

Solution: Since the probabilities given are correct to two decimal places, we choose $n(S) = 100$, and our Venn Diagram is:

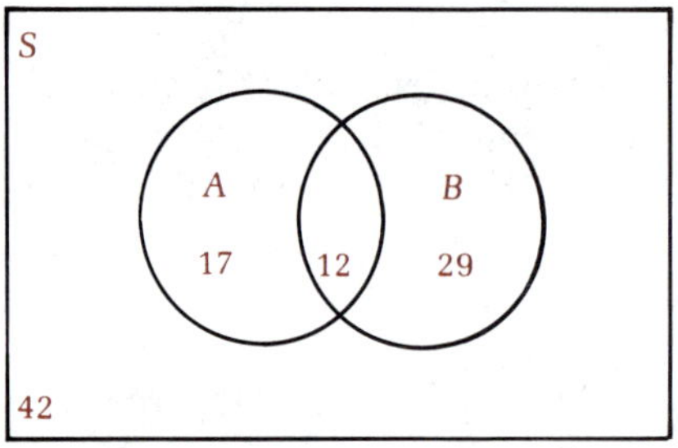

Since $P(A \cap B) = \dfrac{n(A \cap B)}{n(S)}$, then it follows that $0.12 = \dfrac{n(A \cap B)}{100}$ and $n(A \cap B) = 12$. Similarly, $n(A) = 29$, and $n(B) = 41$, and the usual sorting techniques are employed.

a. $P(A \cap B') = \dfrac{17}{100} = 0.17$

b. $P(A' \cap B) = \dfrac{29}{100} = 0.29$

c. $P(A' \cap B') = \dfrac{42}{100} = 0.42$

d. $P(A \cup B) = \dfrac{17 + 12 + 29}{100} = \dfrac{58}{100} = 0.58.$ ■

Exercise Set 6.2

1. Given $P(A) = 0.43$, $P(B) = 0.59$, and $P(A \cup B) = 0.72$,
a. Find $P(A \cap B)$.
b. Find $P(A \cap B')$.
c. Find $P(A' \cap B)$.
d. Find $P(A' \cap B')$.

2. Given $P(A) = 0.20$, $P(B) = 0.70$, and $P(A \cap B') = 0.10$,
a. Find $P(A \cup B')$.
b. Find $P(A \cup B)$.
c. Find $P(A \cap B)$.
d. Find $P(A' \cap B)$.

3. Given $P(A \cup B) = 0.85$, $P(A \cap B) = 0.35$, and $P(A) = P(B)$,
   a. Find $P(A)$.
   b. Find $P(A \cap B')$.
   c. Find $P(A' \cap B')$.
   d. Find $P(A \cup B')$.
4. If A and B are mutually exclusive events where $P(A) = 0.40$ and $P(B) = 0.30$,
   a. Find $P(A \cap B)$.
   b. Find $P(A \cup B)$.
   c. Find $P(A \cap B')$.
   d. Find $P(A' \cap B')$.
   e. Find $P(A' \cup B)$.
5. On the toss of 2 coins, $P(H,H) = \frac{1}{4}$ and $P(T,T) = \frac{1}{4}$. Are the events $\{(H,H)\}$ and $\{(T,T)\}$ mutually exclusive?
6. If $P(E) = 0.62$ and $P(F) = 0.51$, show that the events E and F cannot be mutually exclusive.
7. Given $P(A) = 0.20$, $P(B) = 0.30$, and $P(C) = 0.40$, where A, B, and C are pairwise mutually exclusive events, find $P[(A \cup B \cup C)']$.
8. Given $A \cup B \cup C = D$ and A, B, and C are mutually disjoint sets where $P(D \cap A) = 0.05$, $P(D \cap B) = 0.08$, and $P(D \cap C) = 0.09$, find
   a. $P(D)$ b. $P[(A \cup B) \cap D)]$ c. $P(A \cap B \cap D)$
9. Given $P(A \cap B \cap C) = 0.04$, $P(A \cap B) = 0.07$, $P(A \cap C) = 0.06$, $P(B \cap C) = 0.05$, $P(A) = 0.14$, $P(B) = 0.18$, and $P(C) = 0.18$, use a Venn Diagram to find
   a. $P(A' \cap B' \cap C')$.
   b. $P(A \cap B' \cap C')$.
   c. $P(A' \cap B \cap C)$.
   d. $P[A \cap (B \cup C)]$.
   e. $P[A \cup (B \cap C)]$.
   f. $P[(A' \cap B' \cap C') \cup (A \cap B \cap C)]$.
10. If $P(A \cup B) = 1$, are events A and B mutually exclusive? If not, give an example.
11. Without using Venn Diagrams, find $P(E' \cap F')$, given $P(E) = 0.8$, $P(F) = 0.4$, and $P(E \cap F) = 0.3$.
12. Use Venn Diagrams to construct a model to solve problem 11.
13. Two dice are tossed, and the sum of the faces is observed. Find the probabilities of each of the following events. The results of exercise 1 from section 6.1 will help.
    a. The sum is either 6 or 10.
    b. The sum is 5, 7, or 11.
    c. The sum is greater than 8.
    d. The sum is less than 6.
    e. The sum is less than or equal to 7.
    f. The sum is between 5 and 10, not including 5 or 10.
14. A coin is flipped 6 times. What is the probability of getting
    a. An odd number of heads?
    b. An even number of heads?

15. In a college fraternity, 60% of the members are interested in athletics and 45% are interested in theatre. If 25% of the members are interested in athletics but not theatre, what is the probability that a randomly selected member is not interested in either athletics or theatre?

16. If the probability that the Cubs win the pennant is 0.10 and the probability that the Cardinals win the pennant is 0.15, what is the probability that neither of these teams wins the pennant?

6.3 Tree Diagrams and Probability

When the multiplication principle was discussed in chapter 5, it was introduced through the use of a tree diagram. Tree diagrams can also be used to find probabilities. In certain situations, particular outcomes can be traced along the branches of the tree to determine a desired probability.

■ Example 22. A single fair coin is flipped twice. Find a sample space and use a tree diagram to find the probability of each outcome.

Solution: A convenient sample space, S, is a set of ordered pairs to represent the result on the two flips. $S = \{(H,H),(H,T),(T,H),(T,T)\}$.

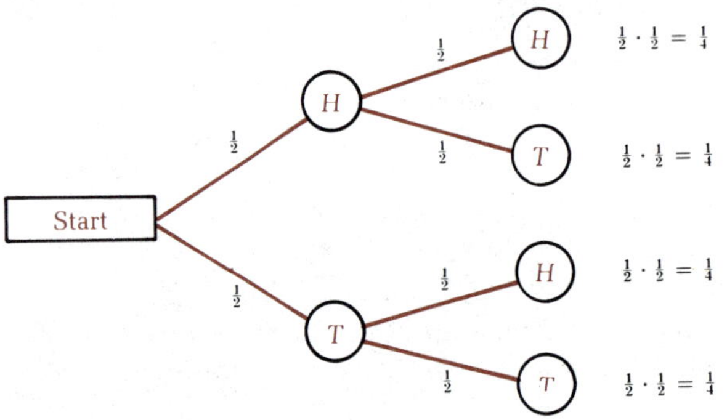

The tree diagram to find probabilities will include a probability on each branch. For this example, each initial branch has a probability of $\frac{1}{2}$ and the second stage branches also have probability of $\frac{1}{2}$.

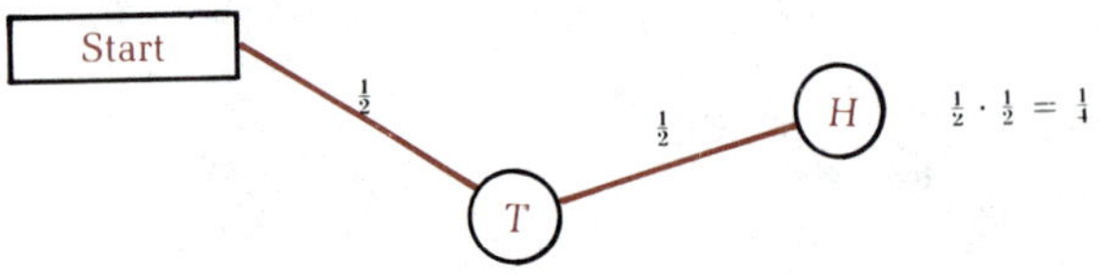

The probability of a particular outcome is determined by taking the product of the branch probabilities used to obtain the outcome. For the outcome (T,H) we follow the branches to obtain

$$P(\{(T,H)\}) = \frac{1}{2} \cdot \frac{1}{2} = \frac{1}{4}.$$ ■

Tree diagrams are particularly appropriate when an experiment consists of a sequence of experiments to be performed in order. Each path from the start to the end of a sequence of branches represents an outcome of the experiment. Hence, the sum of the probabilities for the ends of the completed paths will be 1, since every possible outcome of the experiment is represented by a final branch of the tree.

If we consider the branches beginning at any common point in the tree to be a cluster, then the sum of the probabilities for each cluster of branches will also be 1. In the above example, there are three clusters of branches. One cluster begins at "Start," one begins at H, and one begins at T. Note that the sum of the probabilities for the branches in each cluster is 1.

■ **Example 23.** An urn contains 3 red balls, 3 yellow balls, and 4 blue balls. A ball is drawn, replaced, and then a second ball is drawn.

a. Draw a tree diagram to represent the probabilities of the outcomes of the experiment.

b. Find the probability of the outcome "first ball is yellow and second is blue."

c. Find the probability that at least one of the balls is red.

Solutions: **a.** The tree diagram will be

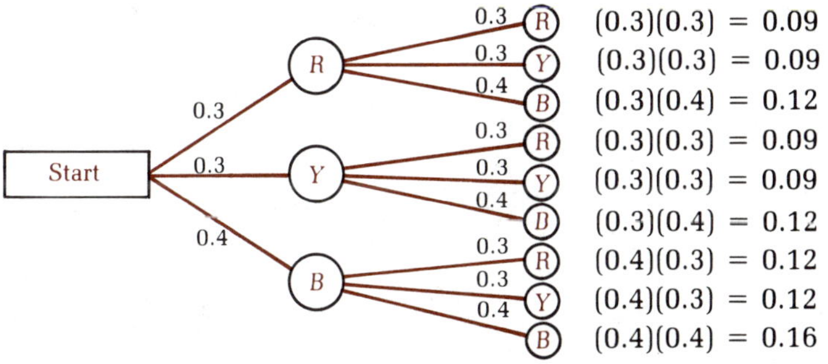

For each of the four clusters, the sum of the probabilities is $0.3 + 0.3 + 0.4 = 1$. The sum of the probabilities for all outcomes is $0.09 + 0.09 + 0.12 + 0.09 + 0.09 + 0.12 + 0.12 + 0.12 + 0.12 + 0.16 = 1$.

b. The probability of "first ball is yellow" and "second is blue" is determined from the path

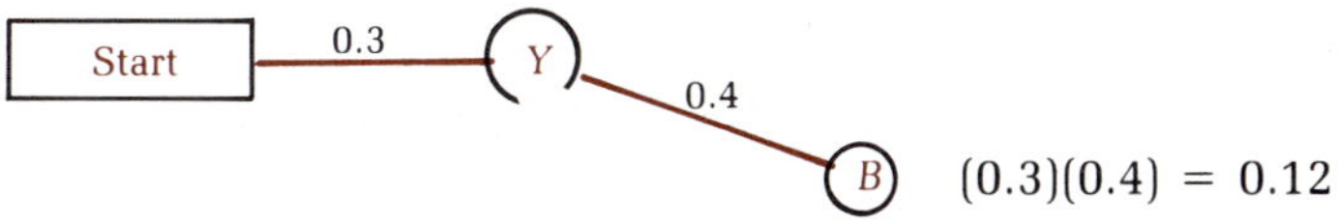

and is equal to 0.12.

c. To find the probability of at least one red ball, we add the probabilities of all paths with at least one red ball. These paths are

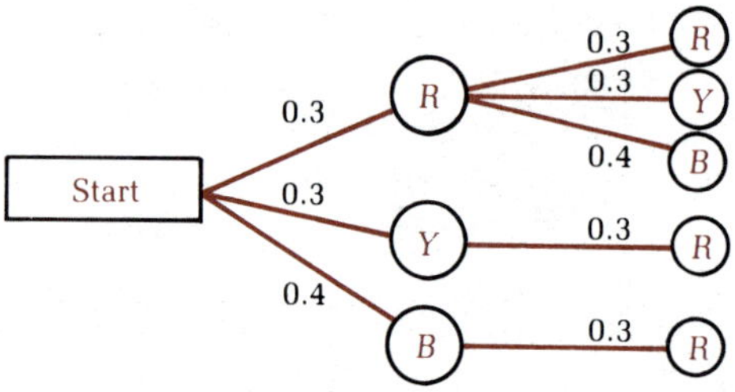

$(0.3)(0.3) = 0.09$

$(0.3)(0.3) = 0.09$

$(0.3)(0.4) = 0.12$

$(0.3)(0.3) = 0.09$

$(0.4)(0.3) = 0.12$

So P (at least one red ball) $= 0.09 + 0.09 + 0.12 + 0.09 + 0.12 = 0.51.$ ■

■ **Example 24.** The same experiment is performed as in example 23 except the first ball is not replaced before the second ball is drawn. Construct the tree diagram and then

a. find the probability of a yellow ball on the first draw and a blue ball on the second draw, and

b. find the probability of at least one red ball.

Solution: The tree diagram is drawn as in example 23. However, since the first ball is not replaced, the probabilities on the second stage branches will be determined by the ratio of the number of balls of that color remaining to the total number of balls remaining (9).

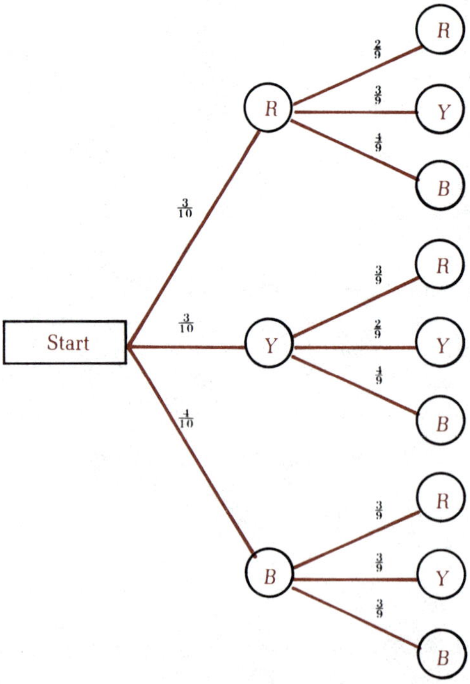

$\frac{3}{10} \cdot \frac{2}{9} = \frac{6}{90}$

$\frac{3}{10} \cdot \frac{3}{9} = \frac{9}{90}$

$\frac{3}{10} \cdot \frac{4}{9} = \frac{12}{90}$

$\frac{3}{10} \cdot \frac{3}{9} = \frac{9}{90}$

$\frac{3}{10} \cdot \frac{2}{9} = \frac{6}{90}$

$\frac{3}{10} \cdot \frac{4}{9} = \frac{12}{90}$

$\frac{4}{10} \cdot \frac{3}{9} = \frac{12}{90}$

$\frac{4}{10} \cdot \frac{3}{9} = \frac{12}{90}$

$\frac{4}{10} \cdot \frac{3}{9} = \frac{12}{90}$

a. The probability of yellow on first draw and blue on second is $\frac{3}{10} \cdot \frac{4}{9} = \frac{12}{90}$ or $\frac{2}{15}$.

b. The probability of at least one red (from the diagram) is $\frac{6}{90} + \frac{9}{90} + \frac{12}{90} + \frac{9}{90} + \frac{12}{90} = \frac{48}{90}$ or $\frac{8}{15}$. ■

Tree diagrams are applicable to a variety of situations. Even though the diagrams begin to get large rather quickly, they provide ready reference for tracing probabilities of specific outcomes.

We close this section with one additional example. This example illustrates a slightly different use of a tree diagram.

■ Example 25. A store has three brands of light bulbs. They have three cartons of brand A, two cartons of brand B, and five cartons of brand C. On the average, each brand A carton contains 96% good bulbs, each brand B carton contains 92% good bulbs, and each brand C carton contains 98% good bulbs. A carton is selected at random and from it a bulb is selected at random. What is the probability that the selected bulb is bad?

Solution: We begin by drawing a tree diagram and converting the percents to probabilities.

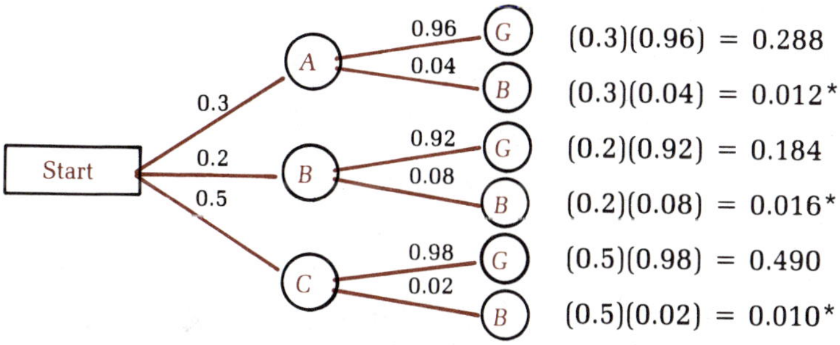

To find the probability that the bulb is bad, add the probabilities of the paths ending at a B (bad bulb).

So $P(\text{bulb is bad}) = 0.012 + 0.016 + 0.010$
$= 0.038.$ ■

Exercise Set 6.3

Use tree diagrams to answer the following questions.

1. Three machines, A, B, and C, are known to produce defective nails at rates of 2%, 3%, and 4%, respectively. Given A produces 30% of the nails, B produces 50% of the nails, and C the rest, what is the percent of defective nails produced?

2. Urn I contains 5 red, 3 white, and 2 blue marbles, and urn II contains 4 red, 3 white, and 1 blue marble. If an urn is selected at random and a marble is drawn,
   a. Find the probability that a red marble is selected.
   b. Find the probability that a white marble is selected.
   c. Find the probability that a blue marble is selected.
3. Three urns colored red, white, and blue contain different combinations of red, white, and blue marbles according to the following table.

	Marbles		
	R	*W*	*B*
Red urn	3	2	5
White urn	2	2	6
Blue urn	6	3	1

   An event is defined as selecting an urn at random and picking a colored marble, replacing the marble, and selecting at random a second marble from the urn whose color matches the first marble. Find the probability that
   a. A red marble is selected last.
   b. A white marble is selected last.
   c. Both marbles are red.
   d. Both marbles are the same color.
   e. If the second marble is returned and a third is selected from the urn matching the color of the second marble, then what is the probability that the marbles are red, white, and blue, in that order?
4. In the upcoming election, the employees of a certain plant indicate they plan to vote according to the following table:

	Percent Voting			
	Democrat	*Republican*	*Do not plan to vote*	*Total employees*
Hourly employees	60	30	10	100
Salary employees	40	40	20	100
Piece-work employees	30	40	30	100

   a. Draw a tree diagram to indicate how a person, randomly selected, will vote.
   b. What is the probability that an employee, selected at random, will vote Democrat?
   c. What percent of the employees plan to vote?
5. The principal of an elementary school is to select randomly 1 kindergarten student to present flowers to the teacher. There are two kindergarten classes. The morning class has 13 boys and 12 girls. The afternoon class has 6 boys and 19 girls. The principal is going to flip a coin to choose a class (heads for the morning and tails for the

afternoon) and then select a student from this class. Use a tree diagram to find the probability that the selected student is a boy.

6. Six urns, numbered 1 to 6, contain red and green balls. A die is to be tossed and a ball is to be drawn from the urn whose number matches the face of the die. The distribution of balls in the urns is given in the following table:

Urn	*Red*	*Green*
1	10	15
2	3	7
3	5	20
4	7	3
5	8	17
6	9	1

a. What is the probability that a red ball is drawn?

b. If all of the balls are dumped into one urn and a ball is drawn, what is the probability the ball drawn is red?

7. A game consists of a sequence of two tasks. First, a ball is drawn from an urn containing 18 red balls, 6 green balls, and one blue ball. After the ball is drawn, a card is selected from a box whose color matches the ball drawn. The red box contains 3 cards marked W (win) and 22 cards marked L (lose). The green box has 5 W cards and 10 L cards, and the blue box has 8 W cards and 2 L cards. If you play the game, what is the probability you win (draw a card marked W)?

8. Of the 25 employees of a local firm, 8 are married men, 11 are married women, 2 are single men, and 4 are single women. Two people are to be chosen to represent the company at a charity dinner. The names are put in a box and a name is drawn. A second name is then drawn from those remaining. What is the probability that

a. both are women?

b. both are single?

c. one is married and the other is single?

d. both single men are chosen?

e. at least one of the two is a single woman?

f. one man and one woman are chosen?

6.4 Conditional Probability and Independent Events

Two dice are tossed. One shows a 5, and the other rolls under the chair. Does this change the probabilities of the outcomes?

Before the dice were tossed, the probability that the sum would equal 7 was $\frac{6}{36}$. Now the sum is 7 only if the die under the chair shows a 2, and that probability is $\frac{1}{6}$. The resulting probability this time is unchanged. Before the dice were tossed, the probability that the sum would be 9 was $\frac{4}{36}$; now the probability is $\frac{1}{6}$.

When the results of one or more of the simple events are known, then the probabilities of compound events may change. It could be argued that the probability of an event can be computed only before the experiment is executed, since once the experiment has been performed, the event either happened or did not happen. That is, the probability is 1 or 0.

From a practical standpoint, additional information about the occurrence of an event simply reduces the set under consideration to a subset of the original sample space. When one of two dice shows a 5 and the other is unknown, we have a new sample space,

$$S = \{(5,1),(5,2),(5,3),(5,4),(5,5),(5,6)\},$$

and the changes in the probabilities of specific events are obvious.

Because the probability of an event can change when certain conditions are met, a new notation is necessary to reflect this change. The commonly used notation is $P(A|B)$, referred to as the *conditional probability* of A, given B. We read this as the probability that A will occur, given that B has occurred.

Since it is known that B has occurred, B now becomes the universal set, or sample space. We are no longer concerned about anything outside B.

■ Example 26. In a group of 30 students, 20 students like math, 10 students receive an A, and half of the A students like math. Find the probability that a student who likes math got an A.

Solution: Let L be the event "likes math" and A the event "received an A." We want $P(A|L)$, the probability that a student got an A, given he or she likes math.

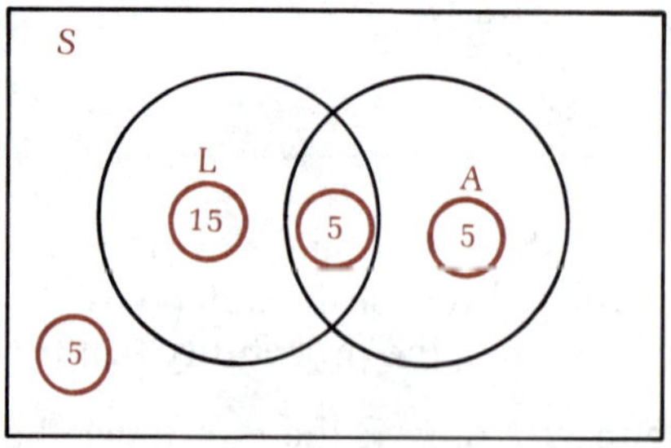

Since we know the selected student likes math, we have restricted ourselves to a consideration of the set L.

Looking at 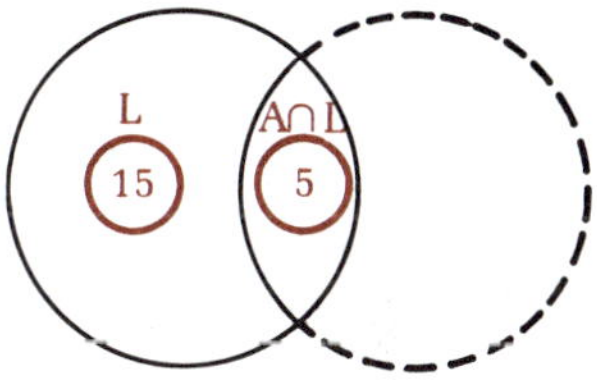

we compute $P(A|L) = \frac{5}{20} = \frac{1}{4}$. ■

$P(L|A)$ is read "the probability that a student likes math if it is known that he or she received an A." Our new sample space is A and $P(L|A) = \frac{5}{10} = \frac{1}{2}$.

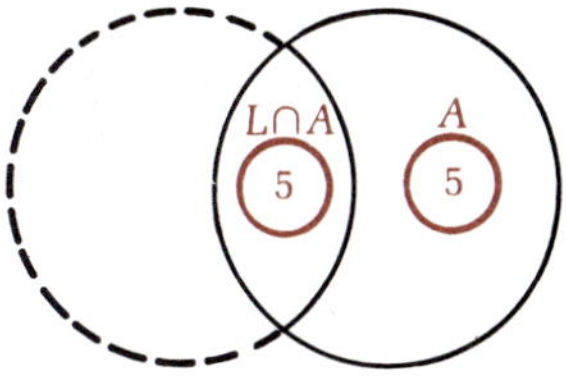

When the outcomes in a sample space S are equally likely, the probability of an event E is given by $P(E) = \frac{n(E)}{n(S)}$. If E and F are events, then $P(E|F)$ can be found in a similar manner. The set under consideration is now the set F, and the favorable outcomes, outcomes in the set E, will be the outcomes in E that are also in F. Hence, the favorable outcomes are those in $E \cap F$. Then $P(E|F) = \frac{n(E \cap F)}{n(F)}$. Note that $P(E|F)$ implies that F has occurred, so $F \neq \emptyset$ and hence $n(F) \neq 0$.

Dividing both numerator and denominator of this expression by $n(S)$ yields

$$P(E|F) = \frac{\frac{n(E \cap F)}{n(S)}}{\frac{n(F)}{n(S)}} = \frac{P(E \cap F)}{P(F)}.$$

The conditional probability of E, given F, $P(E|F)$, can be found in terms of $P(E \cap F)$ and $P(F)$.

NOTE: This formula is valid even when the outcomes are not equally likely.

■ Example 27. In a finite math class of 40 students, 25 are boys and 35 are business majors. If 22 of the boys are business majors, what is the probability that a randomly selected student is a business major, given the student is a girl?

Solution: Let G be the event "student is a girl" and B be the event "student is a business major."

Since there are 15 girls $(40 - 25)$, $P(G) = \frac{15}{40}$.

There are 22 boys who are business majors, so $35 - 22 = 13$ girls are business majors, and $P(B \cap G) = \frac{13}{40}$.

Then $P(B|G) = \frac{P(B \cap G)}{P(G)} = \frac{\frac{13}{40}}{\frac{15}{40}} = \frac{13}{15}$. ■

■ **Example 28.** If 38% of all people have A+ blood and 85% of all people have Rh+ blood, what is the probability that a person is type A if the blood is Rh+?

Solution: We know $P(Rh+) = 0.85$ and $P(A+) = P(A \cap Rh+) = 0.38$, so $P(A|Rh+) = \frac{0.38}{0.85} = 0.45$. In other words, about 45% of all the people who are Rh+ are actually A+. ■

If both sides of the conditional probability formula, $P(E|F) = \frac{P(E \cap F)}{P(F)}$ are multiplied by $P(F)$, a formula for $P(E \cap F)$ is obtained. That is,

$$P(E \cap F) = P(F) \cdot P(E|F).$$

This formula can be used to find the probability of the joint occurrence of two events, $P(E \cap F)$, if $P(F)$ and $P(E|F)$ are known or can be found.

■ **Example 29.** An urn contains 10 red balls, 7 yellow balls, and 8 green balls. Two balls are drawn without replacement. What is the probability that both balls are red?

Solution: Let R_1 and R_2 be the events of drawing the red ball on the first and second draws, respectively.

Then $P(R_1) = \frac{10}{25} = \frac{2}{5}$. If the first ball drawn is red, there are 9 red balls left of 24, so $P(R_2|R_1) = \frac{9}{24} = \frac{3}{8}$.

$$\text{Then } P(R_2 \cap R_1) = P(R_1) \cdot P(R_2|R_1) = \frac{2}{5} \cdot \frac{3}{8} = \frac{6}{40} = \frac{3}{20}.$$ ■

This problem could also be worked using other techniques. Since there are $\binom{25}{2}$ combinations of 2 balls and $\binom{10}{2}$ combinations of 2 red balls,

$P(2\text{red}) = \dfrac{\binom{10}{2}}{\binom{25}{2}} = \dfrac{3}{20}$. A tree diagram could also be used. The choice of a method is not important. Various methods are presented here for your use. It is not always possible to use more than one, but when it is, use the one you are the most comfortable with.

■ **Example 30.** Find the probability that a person has blood type B+ if it is known that 85% of the population has Rh+ blood and 9% of people with Rh+ blood have type B.

Solution: $P(B \text{ and } Rh+) = P(Rh+) \cdot P(B|Rh+) = (0.85) \cdot (0.09) = 0.0765.$
About 8% of the population have B+ blood. ■

We have been considering situations involving conditional probability, $P(E|F)$. Is there any useful relationship between $P(E|F)$ and $P(E'|F)$? Recall that any event F can be written as

$$\begin{aligned} F &= (F \cap E) \cup (F \cap E') \\ &= (E \cap F) \cup (E' \cap F) \end{aligned}$$

where $E \cap F$ and $E' \cap F$ are mutually exclusive. Then

$$P(F) = P(E \cap F) + P(E' \cap F).$$

Dividing by $P(F)$ (we assume $P(F) \neq 0$) yields

$$\begin{aligned} \frac{P(F)}{P(F)} &= \frac{P(E \cap F)}{P(F)} + \frac{P(E' \cap F)}{P(F)} \\ 1 &= \frac{P(E \cap F)}{P(F)} + \frac{P(E' \cap F)}{P(F)} \\ 1 &= P(E|F) + P(E'|F). \end{aligned}$$

Then solving for $P(E'|F)$ we have

$$P(E'|F) = 1 - P(E|F).$$

In example 28, we found that $P(A|Rh+)$ was 0.45. Then $P(A'|Rh+) = 1 - 0.45 = 0.55$. That is, the probability a person's blood is in one of the groups B, AB, or O given he or she is Rh+, is 0.55.

■ **Example 31.** Find the probability that a person is Rh− if it is known that the person is type O, where 36% of all people are type O+ and 42% are in blood group O.

Solution: Note that an answer of 0.06 = 0.42 − 0.36 is not correct, even though it is the probability that a randomly selected person has O− blood.

The correct method of solution is

$$
\begin{aligned}
P(\text{Rh}-|\text{O}) &= 1 - P(\text{Rh}+|\text{O}) \\
&= 1 - \frac{P(\text{Rh}+ \text{ and O})}{P(\text{O})} \\
&= 1 - \frac{P(\text{O}+)}{P(\text{O})} \\
&= 1 - \frac{0.36}{0.42} = 1 - 0.86 = 0.14.
\end{aligned}
$$

About 14% of all people in blood group O have type O− blood. ■

Suppose an urn contains 8 red balls and 2 green balls. A ball is drawn, the color is recorded, and the ball is returned to the urn. A second ball is then drawn. What is the probability that both balls are red? If E is the event "red chosen on first draw" and F is the event "red chosen on second draw," the desired probability is $P(E \cap F)$. Since 8 of the 10 balls are red, $P(E) = 0.8$. The first ball drawn is replaced, so 8 of 10 balls are still red and $P(F) = 0.8$. But $P(F|E) = 0.8$, since the first drawing has no effect on the second. Then

$$
\begin{aligned}
P(E \cap F) &= P(E) \cdot P(F|E) \\
&= (0.8) \cdot (0.8) \\
&= 0.64.
\end{aligned}
$$

When the occurrence of one event has no effect on the probability of the occurrence of another event, as in the situation described in the preceding paragraph, the events are said to be **independent.** The following definition provides a more formal statement about independence.

Definition 6.4 Two events, E and F are said to be independent provided $P(E|F) = P(E)$ and $P(F|E) = P(F)$.

When E and F are independent events,

$$
\begin{aligned}
P(E \cap F) &= P(E) \cdot P(F|E) \\
&= P(E) \cdot P(F).
\end{aligned}
$$

There are two ways to check for independence. Two events are independent if $P(F|E) = P(F)$ and $P(E|F) = P(E)$ or if $P(E \cap F) = P(E) \cdot P(F)$.

Definition 6.4 states that E and F are independent, providing $P(F|E) = P(F)$ and $P(E|F) = P(E)$. However, it is not necessary to verify both of these to show that E and F are independent. It is only necessary to show that one of them is satisfied.

Property 6.7 If $P(F|E) = P(F)$, $P(E) \neq 0$, and $P(F) \neq 0$, then $P(E|F) = P(E)$.

Proof

$$P(F) = P(F|E) = \frac{P(E \cap F)}{P(E)}$$

$$P(E) \cdot P(F) = P(E \cap F) \qquad \text{Multiplying by } P(E)$$

$$P(E) = \frac{P(E \cap F)}{P(F)} \qquad \text{Dividing by } P(F),\ P(F) \neq 0$$

$$P(E) = P(E|F) \qquad \text{By definition.}$$

CAUTION: Mutually exclusive events and independent events are not equivalent ideas. Two events can be mutually exclusive and not independent or independent and not mutually exclusive.

■ **Example 32.** Suppose two coins are flipped. The events $E = \{(H,H)\}$ and $F = \{(T,T)\}$ are mutually exclusive. $P(E) = \frac{1}{4}$ and $P(F) = \frac{1}{4}$. However, $P(E|F) = 0 \neq P(E)$. Hence, the two events are not independent. ■

■ **Example 33.** A die is tossed twice. Let E be the event "first toss gives a 5" and F be the event "second toss gives a 3." Then

$$E = \{(5,1),(5,2),(5,3),(5,4),(5,5),(5,6)\},$$
$$F = \{(1,3),(2,3),(3,3),(4,3),(5,3),(6,3)\}, \text{ and}$$
$$E \cap F = \{(5,3)\}.$$

$P(E) = \frac{1}{6}$, $P(F) = \frac{1}{6}$, $P(E|F) = \frac{1}{6} = P(E)$, and $P(F|E) = \frac{1}{6} = P(F)$. E and F are independent, but $E \cap F = \{(5,3)\} \neq \emptyset$, so E and F are not mutually exclusive. Remember, mutually exclusive events are disjoint sets. ■

Consider $P(E \cap F) = P(E) \cdot P(F)$ as a test for the independence of events. For E and F to be mutually exclusive, $P(E \cap F) = 0$. Then the only way mutually exclusive events can be independent is for at least one of $P(E)$ or $P(F)$ to be zero.

The formula $P(E \cap F) = P(E) \cdot P(F|E)$ is still valid for mutually exclusive events. If E and F are mutually exclusive and E has occurred, then F cannot occur and $P(F|E) = 0$. Then

$$\begin{aligned} 0 &= P(E \cap F) \\ &= P(E) \cdot P(E|F) \\ &= P(E) \cdot 0 \\ &= 0. \end{aligned}$$

■ **Example 34.** Two cards are drawn from a deck of 52 cards. What is the probability of getting 2 spades if

a. The drawing is without replacement?
b. The drawing is with replacement?

Solutions: Let E be the event "first card is a spade" and F be the event "second card is a spade."

$$P(E \cap F) = P(E) \cdot P(F|E)$$

a. $P(E) = \frac{13}{52} = \frac{1}{4}$. To find $P(F|E)$, note that there will be 12 spades remaining and 51 cards remaining. Then $P(F|E) = \frac{12}{51} = \frac{4}{17}$.

$$P(E \cap F) = \frac{1}{4} \cdot \frac{4}{17} = \frac{1}{17}$$

These events are not independent, since the probability of occurrence of F is affected by the occurrence of E. That is,

$$P(F) = \frac{13}{52} \neq P(F|E) = \frac{4}{17}.$$

b. Since the first card is replaced before the second is drawn (we assume the cards are reshuffled), $P(F|E) = \frac{13}{52} = \frac{1}{4}$. There are still 52 cards and 13 spades. Then

$$\begin{aligned} P(E \cap F) &= P(E) \cdot P(F|E) \\ &= P(E) \cdot P(F) \\ &= \frac{1}{4} \cdot \frac{1}{4} = \frac{1}{16}. \end{aligned}$$

Here the two events are independent. ■

In general, when objects are drawn from a collection, if the drawing is done with replacement, the events are independent. If the drawing is done without replacement, the events are not independent.

Suppose we want to make more than two drawings from a collection and are interested in the joint occurrence of certain events. Or, more generally, if $E_1, E_2, \ldots, E_n$ are events, what is $P(E_1 \cap E_2 \cap \ldots \cap E_n)$? The following property is stated in terms of three events. However, it can be generalized to more than three. No proof is given for this property.

Property 6.8 If E_1, E_2, and E_3 are events from a sample space S, then

$$P(E_1 \cap E_2 \cap E_3) = P(E_1) \cdot P(E_2|E_1) \cdot P(E_3|E_1 \cap E_2).$$

The events are considered one at a time, conditioning on the occurrence of preceding events. The following examples will help illustrate.

■ Example 35. Three cards are to be drawn from a deck of 52 cards without replacement. What is the probability of getting

a. Three aces?
b. An ace, king, queen, in that order?
c. An ace, king, queen in any order?

Solutions: **a.** Let A_1, A_2, and A_3 be the events "ace chosen on first, second, third draws, respectively." Then

$$\begin{aligned} P(A_1 \cap A_2 \cap A_3) &= P(A_1) \cdot P(A_2|A_1) \cdot P(A_3|A_1 \cap A_2) \\ &= \frac{4}{52} \cdot \frac{3}{51} \cdot \frac{2}{50} \\ &= \frac{1}{13} \cdot \frac{1}{17} \cdot \frac{1}{25} \\ &= \frac{1}{5525}. \end{aligned}$$

b. Let A, K, and Q be the events "ace chosen on first draw, king on second draw, and queen on third draw, respectively." Then

$$\begin{aligned} P(A \cap K \cap Q) &= P(A) \cdot P(K|A) \cdot P(Q|A \cap K) \\ &= \frac{4}{52} \cdot \frac{4}{51} \cdot \frac{4}{50} \\ &= \frac{1}{13} \cdot \frac{4}{51} \cdot \frac{2}{25} \\ &= \frac{8}{16{,}575}. \end{aligned}$$

c. Since the order of selection is not important here, any one of the 4 aces, 4 kings, or 4 queens is acceptable on the first draw. Any of the

8 cards from an unselected denomination is acceptable on the second draw, and any of 4 cards from the remaining denomination is acceptable on the third draw. Let E_1 be the event "first draw results in one of the 12 acceptable cards," E_2 be the event "second draw gives one of 8 remaining acceptable cards," and E_3 be the event "third draw gives one of the 4 remaining acceptable cards." Then

$$\begin{aligned} P(E_1 \cap E_2 \cap E_3) &= P(E_1) \cdot P(E_2|E_1) \cdot P(E_3|E_2 \cap E_1) \\ &= \frac{12}{52} \cdot \frac{8}{51} \cdot \frac{4}{50} \\ &= \frac{3}{13} \cdot \frac{8}{51} \cdot \frac{2}{25} \\ &= \frac{16}{5525}. \end{aligned}$$

Note that $\frac{16}{5525} = \frac{48}{16{,}575} = 6 \cdot \frac{8}{16{,}575} = 6 \cdot P(A \cap K \cap Q)$ in the specified order. Using our answer from part **b**, we could have reasoned as follows. There are $3! = 6$ permutations of the 3 cards A, K, and Q. So there are 6 times as many ways of getting A, K, and Q in some order. Hence, the probability is 6 times as great. In part **b**, only 1 of the 6 permutations was acceptable. ■

■ Example 36. If 5 cards are drawn without replacement from a deck of 52 cards, what is the probability of getting a royal flush (A, K, Q, J, and 10 of the same suit)?

Solution: The drawing can be considered a sequence of 5 events E_1, E_2, E_3, E_4, and E_5, where E_1 represents A, K, Q, J, or 10 of one of the suits, E_2 represents one of the 4 remaining acceptable cards, and so on.

Then, using an extension of property 6.8, we have

$$\begin{aligned} P(E_1 \cap E_2 \cap E_3 \cap E_4 \cap E_5) &= P(E_1) \cdot P(E_2|E_1) \cdot P(E_3|E_1 \cap E_2) \\ &\quad \cdot P(E_4|E_1 \cap E_2 \cap E_3) \cdot \\ &\quad P(E_5|E_1 \cap E_2 \cap E_3 \cap E_4) \\ &= \frac{20}{52} \cdot \frac{4}{51} \cdot \frac{3}{50} \cdot \frac{2}{49} \cdot \frac{1}{48} \\ &= \frac{5}{13} \cdot \frac{4}{51} \cdot \frac{3}{50} \cdot \frac{2}{49} \cdot \frac{1}{48} \\ &= \frac{1}{12 \cdot 17 \cdot 10 \cdot 49 \cdot 6} \\ &= \frac{1}{649{,}740}. \end{aligned}$$

We could solve this problem in a slightly different way, without using conditional probability. If we let E be the event "get a royal flush," then

$P(E) = \frac{n(E)}{n(S)}$, where the sample space S is the set of all 5 card hands. Then $n(E) = 4$, one for each suit, and $n(S) = \binom{52}{5} = \frac{52!}{47!5!}$, the number of combinations of 52 cards, taken 5 at a time.

$$P(E) = \frac{4}{\binom{52}{5}} = \frac{4}{2,598,960}$$

$$= \frac{1}{649,740}$$

■

We conclude this section with one final comment about conditional probability and the $P(E \cap F)$. By our definition of conditional probability,

$$P(E|F) = \frac{P(E \cap F)}{P(F)} \text{ or } P(E \cap F) = P(F) \cdot P(E|F).$$

$$P(F|E) = \frac{P(F \cap E)}{P(E)} = \frac{P(E \cap F)}{P(E)} \text{ or } P(E \cap F) = P(E) \cdot P(F|E).$$

Since $P(E \cap F) = P(F) \cdot P(E|F)$ and $P(E \cap F) = P(E) \cdot P(F|E)$, it follows that $P(E) \cdot P(F|E) = P(F) \cdot P(E|F)$. This does not imply that $P(E) = P(F)$ and $P(F|E) = P(E|F)$. It only implies that the products are equal.

■ **Example 37.** In a high school class of 40 students, 25 take English, 15 take French, and 10 take both. Find the probability that a student takes both French and English.

Solutions: We will solve this problem in three different ways. Let E be the event "takes English" and F be the event "takes French."

1. By direct computation, since 10 of the 40 take both,
   $$P(E \cap F) = \frac{10}{40} = \frac{1}{4}.$$
2. $$P(E \cap F) = P(E) \cdot P(F|E)$$
   $$= \frac{25}{40} \cdot \frac{10}{25} \quad \text{(10 of the 25 in English are also in French.)}$$
   $$= \frac{10}{40} = \frac{1}{4}$$
3. $$P(E \cap F) = P(F) \cdot P(E|F)$$
   $$= \frac{15}{40} \cdot \frac{10}{15} \quad \text{(10 of the 15 in French are also in English.)}$$
   $$= \frac{10}{40} = \frac{1}{4}$$

■

Exercise Set 6.4

In all problems that involve percent, convert the percents to probabilities, find appropriate probabilities, and then convert the probabilities back to percents.

1. Two eggs are selected from a dozen eggs, of which 9 are good and 3 are rotten.
   a. What is the probability that both eggs are good?
   b. If the eggs are selected one at a time and examined, what is the probability that the second egg is good given that the first egg was good?
   c. If the eggs are selected one at a time and the first egg is rotten, what is the probability that the second egg is good?
2. If 14% of those receiving a direct mail survey send back the coupon for the free gift and 30% of those who get the free gift buy the product, what percent of the people contacted by direct mail buy the product?
3. In a study of 300 students, the following data on first date kissing were obtained.

Frequency for Kissing on First Date

	Always	*Sometimes*	*Rarely*	*Never*	*Total*
Male	42	56	12	20	130
Female	38	62	28	42	170
Total	80	118	40	62	300

Suppose M, F, A, S, R, N are defined as follows:

M - subject is male.

F - subject is female.

A - subject always kisses on first date.

S - subject sometimes kisses on first date.

R - subject rarely kisses on first date.

N - subject never kisses on first date.

Compute each of the following.

a. $P(M)$
b. $P(M|A)$
c. $P(F|S)$
d. $P(S|F)$
e. $P(R|M)$
f. $P[F|A \cup S)]$

4. In a box of 25 light bulbs, 5 are defective. Two bulbs are drawn from the box. What is the probability that both are good if the drawing was
   a. without replacement? **b.** with replacement?

5. Given $P(E) = 0.56$, $P(F) = 0.42$, and $P(E \cap F) = 0.21$, find
a. $P(E|F)$.
b. $P(F|E)$.
c. $P(E|F')$.
d. $P(F'|E)$.

6. If $P(E) = 0.45$, $P(F) = 0.63$, and $P(E \cap F) = 0.36$, find
a. $P(E|F)$.
b. $P(F|E)$.
c. $P(E'|F)$.
d. $P(F|E')$.

7. Suppose $P(E) = 0.72$, $P(F) = 0.45$, and $P(E \cup F) = 0.90$. Find
a. $P(E|F)$.
b. $P(F|E)$.
c. $P(E'|F)$.
d. $P(F'|E)$.
e. $P(E'|F')$.
f. $P(F'|E')$.

8. Given $P(E) = 0.46$, $P(F) = 0.69$, and $P(E \cup F) = 0.92$, find
a. $P(E|F)$.
b. $P(F|E)$.
c. $P(E'|F)$.
d. $P(F'|E)$.
e. $P(E'|F')$.
f. $P(F'|E')$.

9. Based on a random survey, a pollster estimates that 40% of all adults enjoy sports on TV, 60% enjoy situation comedies, and 24% enjoy both.
a. What is the probability that a person likes sports, given he or she likes situation comedies?
b. What is the probability that a person likes situation comedies, given he or she likes sports?
c. What can be said about the events "likes sports" and "likes situation comedies"?

10. If 30% of all teenagers have used marijuana, 60% have used alcohol, and 25% have used both, what is the probability a teenager has used
a. alcohol, given he or she has used marijuana?
b. marijuana, given he or she has used alcohol?

11. In a small community, 80% of the businesses advertise in the local newspaper and 40% of these also advertise on the radio. What percent use both newspapers and radio?

12. In a survey of legislators, it was found that 60% favor tax cuts and 25% of these favor increased defense spending. What percent of the legislators favor both?

13. If 85% of all people have Rh+ blood and 42% of all people with Rh+ blood are type O, what percent are O+?

14. If 10% of all people have the B antigen in their blood and 40% of these also have the A antigen, what percent have both antigens?

15. A new movie is shown in a "sneak preview" and people are asked to give their opinion of the movie. The results are tabulated in the following table:

	Liked	*Disliked*	*No Opinion*
Adult males	100	40	60
Adult females	120	160	20
Minor males	80	100	20
Minor females	100	100	100

a. What is the probability that a person who liked the movie was an adult male?
b. What is the probability that a person who liked the movie was a minor?
c. What is the probability that a person who disliked the movie was an adult female?
d. What is the probability that a person who disliked the movie was a female?

16. In an opinion poll on satisfaction with the president's performance in office, the following data were obtained

Party	*Approve*	*Disapprove*	*No Opinion*
Democrat	100	125	25
Republican	150	75	25
Independent	100	100	50

a. What is the probability that a person who approved of the president's performance was a Democrat?
b. What is the probability that a person who disapproved of the president's performance was a Republican?

6.5 Bayes' Theorem

Now that we have considered conditional probability, we are in a position to use this concept to develop **Bayes' Theorem.** We will begin with an illustration of the type of problem to be solved.

A college accounting professor is teaching two sections of beginning accounting, Accounting 100, and a graduate accounting course, Accounting 581. The department chairperson has heard of some problems the professor is having and wants to select one student at random to discuss the situation with in an attempt to determine the seriousness of the problem. The professor's grade distribution is given in the following table.

Course	*A*	*B*	*C*	*D*	*F*	*Total*
Accounting 100, section 1	0	3	12	10	5	30
Accounting 100, section 2	2	2	15	6	5	30
Accounting 581	4	4	5	2	5	20

If the chairperson selects a course at random and a student at random from this course, what is the probability that the student selected is failing the course? If it is known that the selected student is failing, what is the probability that he or she is in the graduate course?

Before trying to answer these questions, we will establish two formulas to use for the computation. In our development of these formulas, we will work with a specific situation. The formulas can, however, be stated in more general terms.

Suppose we are given a universe U and three sets A, B, $C \subset U$, where $A \cap B = \emptyset$, $A \cap C = \emptyset$, and $B \cap C = \emptyset$. In addition, suppose $A \cup B \cup C = U$. These three sets partition U. In figure 6.3 we have drawn a Venn Diagram to illustrate.

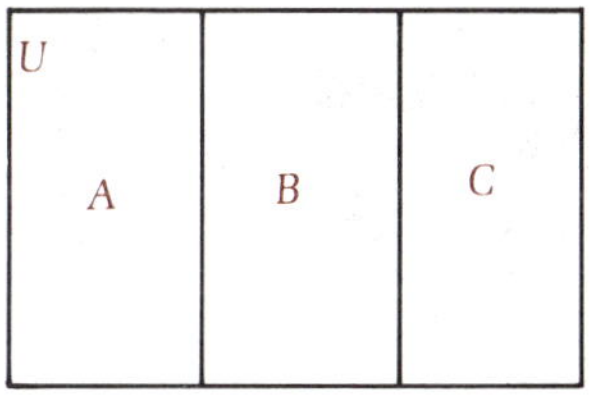

Figure 6.3

Note that circles are not used here for the sets A, B, and C. Since the sets are pairwise disjoint (each pair is disjoint) and they include all of U, figure 6.3 is easier to work with.

Consider now a fourth set, $X \subset U$. In figure 6.4 this set has been included in the Venn Diagram.

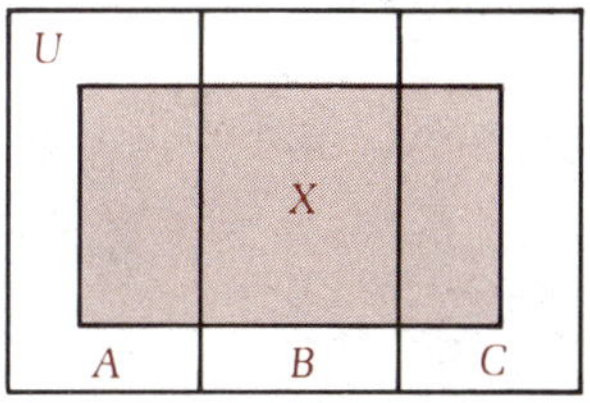

Figure 6.4

NOTE: In the problem used to introduce this section, U is the set of all students in the three classes, A, B, and C are the three classes, and X is the set of students receiving F's. It is $P(X)$ that we seek as an answer for the first question and $P(C|X)$ for the second question.

We look first at $P(X)$. The set X can be written as

$$X = (A \cap X) \cup (B \cap X) \cup (C \cap X).$$

Since A, B, and C are pairwise disjoint, so are $(A \cap X)$, $(B \cap X)$, and $(C \cap X)$, and, treated as events, they are mutually exclusive. Using the inclusion-exclusion principle yields

$$\begin{aligned} P(X) &= P[(A \cap X) \cup (B \cap X) \cup (C \cap X)] \\ &= P(A \cap X) + P(B \cap X) + P(C \cap X) \\ &= P(A) \cdot P(X|A) + P(B) \cdot P(X|B) + P(C) \cdot P(X|C). \end{aligned}$$

We now generalize to state the following property in terms of events and the sample space.

Property 6.9 If $E_1, E_2, \ldots, E_k, F$ are events from a sample space S where $E_1, E_2, \ldots, E_k$ are pairwise mutually exclusive, with $E_1 \cup E_2 \cup \ldots \cup E_k = S$, $P(E_i) \neq 0$, and $P(F) \neq 0$, then $P(F) = P(E_1) \cdot P(F|E_1) + P(E_2) \cdot P(F|E_2) + \ldots + P(E_k) \cdot P(F|E_k)$.

This formula is sometimes referred to as the *total probability formula.*

We will not prove this property in this text. However, the property can be illustrated by using a tree diagram (see section 6.3). In fact, example 25 of section 6.3 is an example of a problem that fits the conditions of this property. The solution was determined using a tree diagram.

To construct a tree diagram, we consider a two stage process where we first select one of the pairwise mutually exclusive events $E_1, E_2, \ldots, E_k$. We then select an outcome in this event. This outcome is either in F or in F', giving two branches from each of the sets E_i. The probabilities for the first stage branches are $P(E_1), P(E_2), \ldots, P(E_k)$. The probabilities for the second stage branches are $P(F|E_i)$ and $P(F'|E_i)$ for the branch ends F and F' respectively.

The tree diagram will have the following appearance.

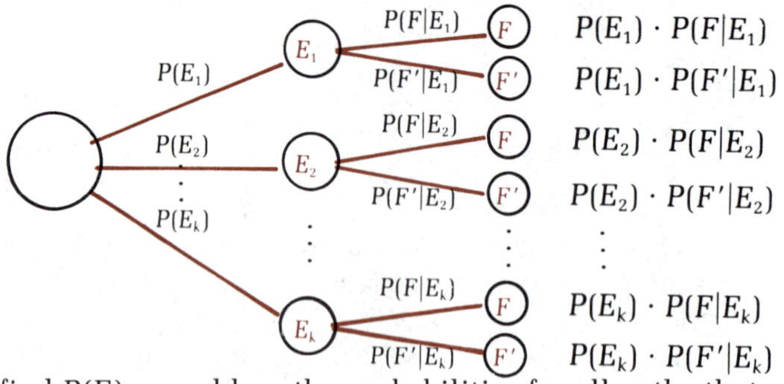

To find $P(F)$, we add up the probabilities for all paths that end at F. That is,

$$P(F) = P(E_1) \cdot P(F|E_1) + P(E_2) \cdot P(F|E_2) + \ldots + P(E_k) \cdot P(F|E_k)$$

This is precisely what property 6.9 gives for $P(F)$.

Example 38. Find the probability that the student selected by the department chairperson in the illustration given at the beginning of this section is failing.

Solution: Let E_1 be event of selecting Accounting 100, section 1.
Let E_2 be event of selecting Accounting 100, section 2.
Let E_3 be event of selecting Accounting 581.
Since the class is randomly chosen first, we have $P(E_1) = P(E_2) = P(E_3) = \frac{1}{3}$. Letting F be the event that the student chosen is failing, we have $P(F|E_1) = \frac{5}{30} = \frac{1}{6}$, $P(F|E_2) = \frac{5}{30} = \frac{1}{6}$, and $P(F|E_3) = \frac{5}{20} = \frac{1}{4}$. By the total probability formula,

$$\begin{aligned} P(F) &= P(E_1)P(F|E_1) + P(E_2)P(F|E_2) + P(E_3)P(F|E_3) \\ &= \frac{1}{3}\cdot\frac{1}{6} + \frac{1}{3}\cdot\frac{1}{6} + \frac{1}{3}\cdot\frac{1}{4} \\ &= \frac{1}{18} + \frac{1}{18} + \frac{1}{12} = \frac{2}{36} + \frac{2}{36} + \frac{3}{36} = \frac{7}{36}. \end{aligned}$$

We could also use a tree diagram to find the solution.

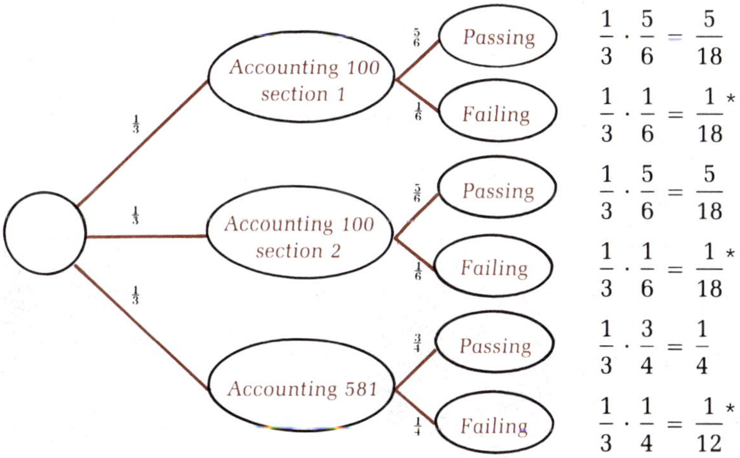

To find the probability that the selected student is failing, we add up the probabilities of all paths that end at failing, indicated by the *. Then

$$P(F) = \frac{1}{18} + \frac{1}{18} + \frac{1}{12} = \frac{2}{36} + \frac{2}{36} + \frac{3}{36} = \frac{7}{36}.$$

■

If the 80 students had been lumped together and 1 student selected at random, we would have $P(F) = \frac{15}{80} = \frac{3}{16} \neq \frac{7}{36}$. The difference occurs because the probability of an F is not the same for each of the three classes. Had we not chosen a class first, $\frac{3}{16}$ would have been the correct answer.

To handle the second question posed at the beginning of this section, we recall from section 6.4 that

$$P(A|B) = \frac{P(A \cap B)}{P(B)} \qquad [P(B) > 0]$$

and $$P(B|A) = \frac{P(A \cap B)}{P(A)} \qquad [P(A) > 0].$$

Then $P(A \cap B) = P(B) \cdot P(A|B) = P(A) \cdot P(B|A)$.

Therefore $P(A|B) = \dfrac{P(A) \cdot P(B|A)}{P(B)}$, $P(B) > 0$.

If A, B, C, and X are subsets of U with A, B, and C pairwise mutually exclusive and $A \cup B \cup C = U$, then

$$P(A|X) = \frac{P(A) \cdot P(X|A)}{P(X)}.$$

By the total probability formula,

$$P(X) = P(A)P(X|A) + P(B)P(X|B) + P(C)P(X|C).$$

Then $$P(C|X) = \frac{P(C)P(X|C)}{P(A)P(X|A) + P(B)P(X|B) + P(C)P(X|C)}.$$

This formula can be generalized to a result known as *Bayes' Theorem*.

Property 6.10 ***Bayes' Theorem*** If $E_1, E_2, \ldots, E_k$, and F are events from a sample space S, where $E_1, E_2, \ldots, E_k$ are pairwise mutually exclusive, with $E_1 \cup E_2 \cup \ldots \cup E_k = S$, $P(E_i) > 0$, and $P(F) > 0$, then

$$P(E_i|F) = \frac{P(E_i) \cdot P(F|E_i)}{P(F)}$$

$$= \frac{P(E_i) \cdot P(F|E_i)}{P(E_1) \cdot P(F|E_1) + P(E_2) \cdot P(F|E_2) + \ldots + P(E_k) \cdot P(F|E_k)}.$$

Bayes' Theorem, like the total probability formula, can be illustrated using a tree diagram. We are trying to find $P(E_i|F)$. But

$$P(E_i|F) = \frac{P(E_i \cap F)}{P(F)}.$$

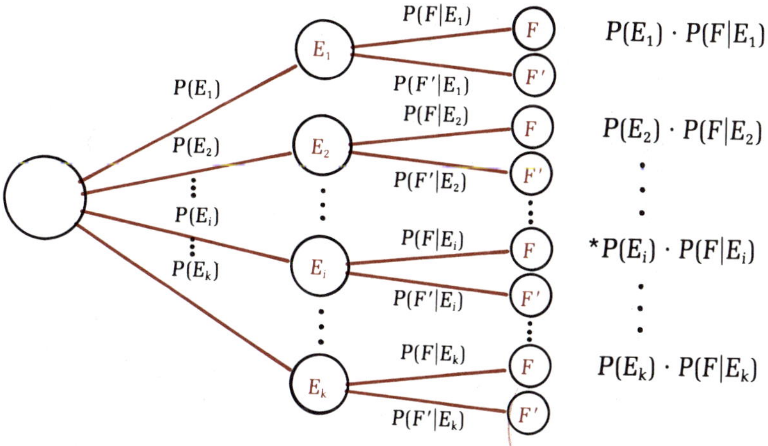

On the tree diagram, $P(F)$ is the sum of the probabilities for the paths ending with F. To find $P(E_i \cap F)$, we trace the path through E_i to F and $P(E_i \cap F)$ is the probability for this path. You should notice that this path is one of the paths used in finding $P(F)$.

$$P(E_i|F) = \frac{P(E_i \cap F)}{P(F)} = \frac{P(E_i) \cdot P(F|E_i)}{P(F)}$$

■ Example 39. Find the probability that the student selected (in the preceding example) was in the graduate course, given he or she was receiving an F.

Solution: To use Bayes' Theorem, we need the following: $P(E_1) = P(E_2) = P(E_3) = \frac{1}{3}$, $P(F|E_1) = \frac{5}{30} = \frac{1}{6}$, $P(F|E_2) = \frac{5}{30} = \frac{1}{6}$, $P(F|E_3) = \frac{5}{20} = \frac{1}{4}$. By Bayes' Theorem,

$$P(E_3|F) = \frac{P(E_3)P(F|E_3)}{P(E_1)P(F|E_1) + P(E_2)P(F|E_2) + P(E_3)P(F|E_3)}$$

$$= \frac{\frac{1}{3} \cdot \frac{1}{4}}{\frac{1}{3} \cdot \frac{1}{6} + \frac{1}{3} \cdot \frac{1}{6} + \frac{1}{3} \cdot \frac{1}{4}} = \frac{\frac{1}{12}}{\frac{7}{36}} = \frac{1}{12} \cdot \frac{36}{7} = \frac{3}{7}.$$

We could also compute $P(E_1|F)$ and $P(E_2|F)$ as

$$P(E_1|F) = \frac{P(E_1)P(F|E_1)}{P(F)} = \frac{\frac{1}{3} \cdot \frac{1}{6}}{\frac{7}{36}} = \frac{\frac{1}{18}}{\frac{7}{36}} = \frac{1}{18} \cdot \frac{36}{7} = \frac{2}{7} \text{ and}$$

$$P(E_2|F) = \frac{P(E_2)P(F|E_2)}{P(F)} = \frac{\frac{1}{18}}{\frac{7}{36}} = \frac{2}{7}.$$

Note that $P(E_1|F) + P(E_2|F) + P(E_3|F) = \frac{2}{7} + \frac{2}{7} + \frac{3}{7} = 1$. This will always happen, since this is really $P(S|F)$.

This problem could also be solved using a tree diagram.

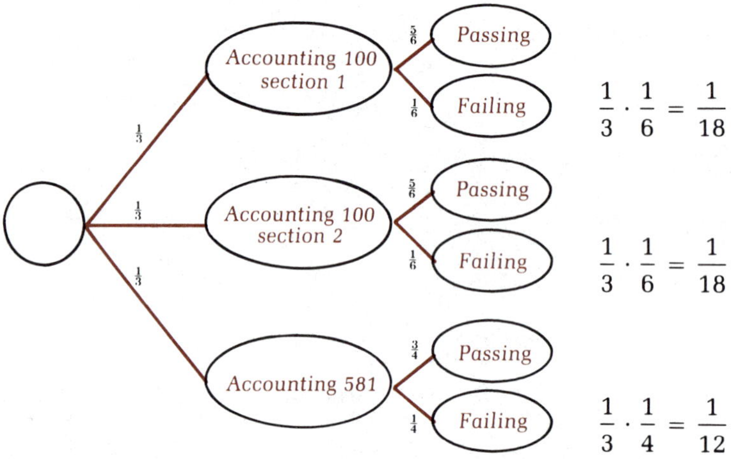

$$
\begin{aligned}
P(\text{Accounting 581}|\text{Failing}) &= \frac{P(\text{Accounting 581 and Failing})}{P(\text{Failing})} \\
&= \frac{\frac{1}{12}}{\frac{1}{18} + \frac{1}{18} + \frac{1}{12}} \\
&= \frac{\frac{1}{12}}{\frac{7}{36}} = \frac{1}{12} \cdot \frac{36}{7} = \frac{3}{7}
\end{aligned}
$$

■

■ **Example 40.** Suppose you have 3 urns. Urn I contains 25 green balls, urn II contains 25 green balls, and urn III contains 50 red balls.

a. An urn is selected at random, and then a ball is selected from the urn. Find the probability that the ball is green.
b. The same selection process is done as in part a. Find the probability that the ball selected is red.
c. Compare the results in parts a and b with the following: All of the balls are mixed in a single urn, and a ball is drawn. What is the probability that the ball is green? Red?
d. A ball is drawn as in part a and found to be green. What is the probability it came from urn I? Urn II? Urn III?

Solutions: For parts **a, b,** and **d,** let E_1, E_2, and E_3 represent drawing from urns I, II, and III, respectively. G represents a green ball, and R represents a red ball. Then $P(E_1) = P(E_2) = P(E_3) = \frac{1}{3}$.

$$P(G|E_1) = \frac{25}{25} = 1 \qquad P(R|E_1) = \frac{0}{25} = 0$$

$$P(G|E_2) = \frac{25}{25} = 1 \qquad P(R|E_2) = \frac{0}{25} = 0$$

$$P(G|E_3) = \frac{0}{50} = 0 \qquad P(R|E_3) = \frac{50}{50} = 1$$

a. By the total probability formula,

$$\begin{aligned} P(G) &= P(E_1)P(G|E_1) + P(E_2)P(G|E_2) + P(E_3)P(G|E_3) \\ &= \frac{1}{3}\cdot 1 + \frac{1}{3}\cdot 1 + \frac{1}{3}\cdot 0 = \frac{2}{3}. \end{aligned}$$

b. $P(R) = \frac{1}{3}\cdot 0 + \frac{1}{3}\cdot 0 + \frac{1}{3}\cdot 1 = \frac{1}{3}$.

c. If the balls are mixed in 1 urn, there will be 100 balls, 50 green and 50 red. Then $P(G) = \frac{50}{100} = \frac{1}{2}$ and $P(R) = \frac{50}{100} = \frac{1}{2}$. The two-stage process of first selecting the urn and then the ball affects the probability, since the probability of selecting an urn containing all green balls, $\frac{2}{3}$, is twice the probability of selecting the urn containing all red balls.

d. By Bayes' Theorem,

$$P(E_1|G) = \frac{P(E_1)P(G|E_1)}{P(G)} = \frac{\frac{1}{3}\cdot 1}{\frac{2}{3}} = \frac{1}{2}$$

$$P(E_2|G) = \frac{P(E_2)P(G|E_2)}{P(G)} = \frac{\frac{1}{3}\cdot 1}{\frac{2}{3}} = \frac{1}{2}$$

$$P(E_3|G) = \frac{P(E_3)P(G|E_3)}{P(G)} = \frac{\frac{1}{3}\cdot 0}{\frac{2}{3}} = 0.$$

We could also solve parts **a**, **b**, and **d** using a tree diagram.

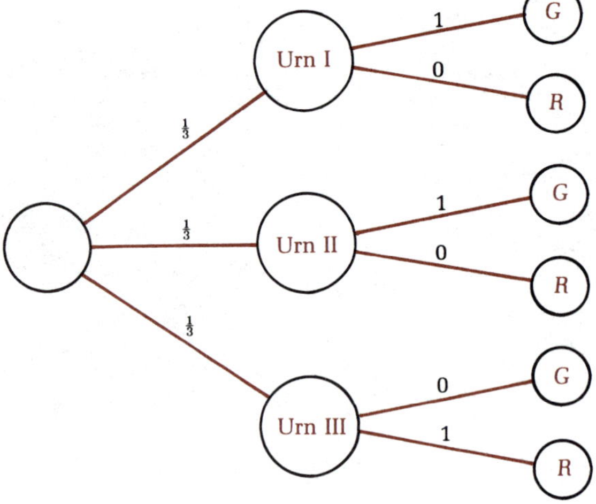

$$\frac{1}{3}\cdot 1 = \frac{1}{3}$$
$$\frac{1}{3}\cdot 0 = 0$$
$$\frac{1}{3}\cdot 1 = \frac{1}{3}$$
$$\frac{1}{3}\cdot 0 = 0$$
$$\frac{1}{3}\cdot 0 = 0$$
$$\frac{1}{3}\cdot 1 = \frac{1}{3}$$

a. $P(G) = \frac{1}{3} + \frac{1}{3} + 0 = \frac{2}{3}$

b. $P(R) = 0 + 0 + \frac{1}{3} = \frac{1}{3}$

d. $P(E_1|G) = \frac{\frac{1}{3}}{\frac{2}{3}} = \frac{1}{2}$

$P(E_2|G) = \frac{\frac{1}{3}}{\frac{2}{3}} = \frac{1}{2}$

$P(E_3|G) = \frac{0}{\frac{2}{3}} = 0.$ ■

■ **Example 41.** A store has received a shipment of four cartons of size D batteries, each containing 100 batteries. During normal production, it is expected that 2 batteries of each 100 will be bad. The manager is notified that because of a problem in production, one carton has been shipped that is estimated to have 40 bad batteries based on testing done by the manufacturer. The store manager decides to select one of the cartons at random and from this carton select 1 battery to test.

a. What is the probability that this battery is bad?
b. If the battery selected is found to be bad, what is the probability it came from the carton with 40 bad batteries?

Solution: Since the cartons are not distinguishable, we will let them be A_1, A_2, A_3, and X for the bad carton. We let B represent a bad battery.

Then $P(A_1) = P(A_2) = P(A_3) = P(X) = \frac{1}{4}$,

$P(B|A_1) = P(B|A_2) = P(B|A_3) = \frac{2}{100} = \frac{1}{50}$, and $P(B|X) = \frac{40}{100} = \frac{2}{5}$.

a.
$$\begin{aligned} P(B) &= P(A_1)P(B|A_1) + P(A_2)P(B|A_2) \\ &\quad + P(A_3)P(B|A_3) + P(X)P(B|X) \\ &= \frac{1}{4}\cdot\frac{1}{50} + \frac{1}{4}\cdot\frac{1}{50} + \frac{1}{4}\cdot\frac{1}{50} + \frac{1}{4}\cdot\frac{2}{5} \\ &= \frac{1}{200} + \frac{1}{200} + \frac{1}{200} + \frac{2}{20} \\ &= \frac{1}{200} + \frac{1}{200} + \frac{1}{200} + \frac{20}{200} = \frac{23}{200} \end{aligned}$$

b.
$$\begin{aligned} P(X|B) &= \frac{P(X)\cdot P(B|X)}{P(B)} \\ &= \frac{\frac{1}{4}\cdot\frac{2}{5}}{\frac{23}{200}} = \frac{\frac{2}{20}}{\frac{23}{200}} = \frac{2}{20}\cdot\frac{200}{23} = \frac{20}{23} \end{aligned}$$

The tree diagram solution would be as follows.

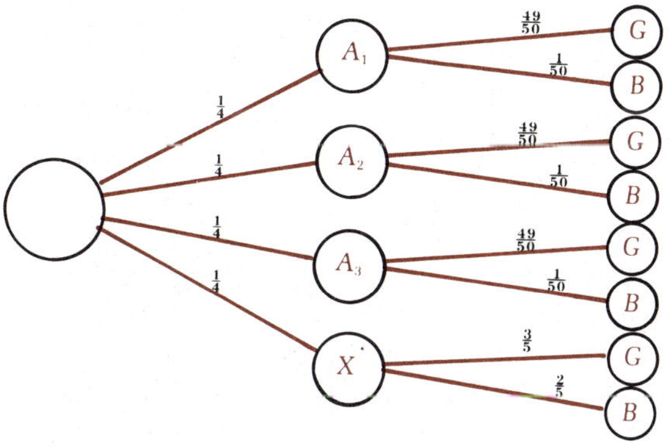

$$\frac{1}{4}\cdot\frac{1}{50} = \frac{1}{200}$$

$$\frac{1}{4}\cdot\frac{1}{50} = \frac{1}{200}$$

$$\frac{1}{4}\cdot\frac{1}{50} = \frac{1}{200}$$

$$\frac{1}{4}\cdot\frac{2}{5} = \frac{1}{10}$$

$$P(B) = \frac{1}{200} + \frac{1}{200} + \frac{1}{200} + \frac{1}{10} = \frac{23}{200}.$$

$$P(X|B) = \frac{\frac{1}{10}}{\frac{23}{200}} = \frac{1}{10}\cdot\frac{200}{23} = \frac{20}{23}.$$

Since this probability is relatively high, the store manager may want to test 2 or 3 other batteries from this carton to see if it is, in fact, the bad carton.

We could use our sorting techniques from chapter 5 to investigate part **b**, since each carton contains the same number of batteries. In figure 6.5 we have partitioned the universe using A_1, A_2, A_3, and X and have included B. We have also filled in the numbers in the regions.

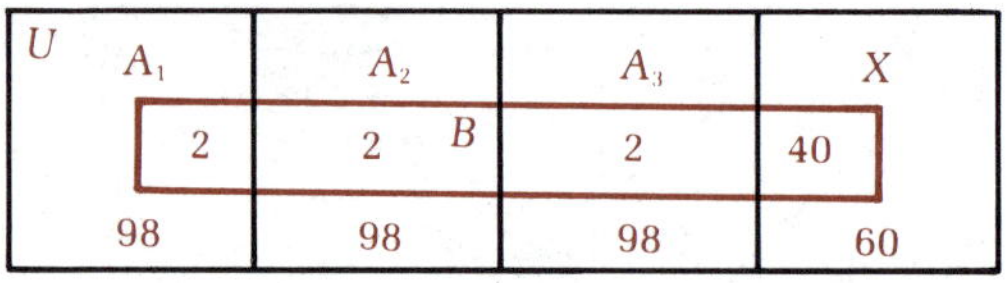

Figure 6.5

Since we are conditioning on B, this becomes our reduced sample space. Figure 6.6 shows this reduced sample space with the appropriate numbers filled in.

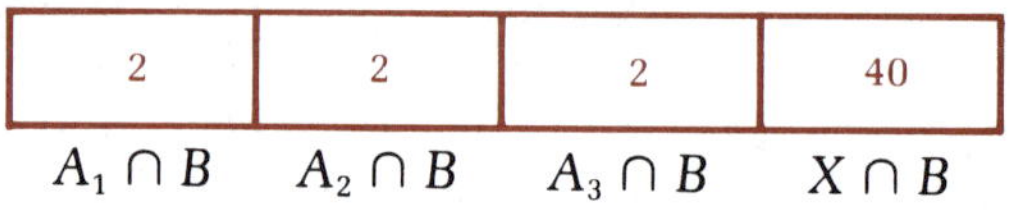

Figure 6.6

Since $n(B) = 46$ and $n(X \cap B) = 40$, $P(X|B) = \dfrac{40}{46} = \dfrac{20}{23}$. ■

NOTE: In part **a** of this problem, $P(B) = \dfrac{23}{200}$. If all the batteries had been dumped into one big container and mixed up, the same results would have been obtained. That is, there are 400 batteries, and 46 of them are bad. $P(B) = \dfrac{46}{400} = \dfrac{23}{200}$. In examples 38 and 39, this would not be true. This result will be true when all of the partitioning sets or events have the same number of elements.

■ Example 42. Five different professors are teaching sections of a certain course. The following table gives the percent of total enrollment each professor has and the percent of each instructor's students who are failing. If a student is selected at random and found to be failing, what is the probability he has professor D?

Professor	Percent of enrollment	Professor (P)	Percent of F's	$P(F\|X)$	$P(X) \cdot P(F\|X)$
A	15	0.15	3	0.03	0.0045
B	20	0.20	5	0.05	0.0100
C	10	0.10	2	0.02	0.0020
D	30	0.30	6	0.06	0.0180*
E	25	0.25	1	0.01	0.0025
				Total	0.0370

*The fourth row (professor D) is the row of interest.

Solution: We want to find $P(D|F)$, where F is the event "student is failing" and D is the event "student has professor D." Columns 2 and 4 give the percentages as probabilities. The fifth column of the table gives, in decimal form, $P(\text{a professor}) \cdot P(F|\text{this professor})$.

By Bayes' Theorem, $P(D|F) = \dfrac{P(D) \cdot P(F|D)}{P(F)}$.

$$\begin{aligned} P(F) &= P(A) \cdot P(F|A) + P(B) \cdot P(F|B) + P(C) \cdot P(F|C) \\ &\quad + P(D) \cdot P(F|D) + P(E) \cdot P(F|E) \\ &= (0.15)(0.03) + (0.20)(0.05) + (0.10)(0.02) \\ &\quad + (0.30)(0.06) + (0.25)(0.01) \\ &= 0.0370 \end{aligned}$$

$$\begin{aligned} P(D|F) &= \frac{P(D) \cdot P(F|D)}{0.037} = \frac{(0.30)(0.06)}{0.037} \\ &= \frac{0.0180}{0.0370} = \frac{18}{37} = 0.4865, \text{ or about } 49\%. \end{aligned}$$

This problem can also be solved using a tree diagram.

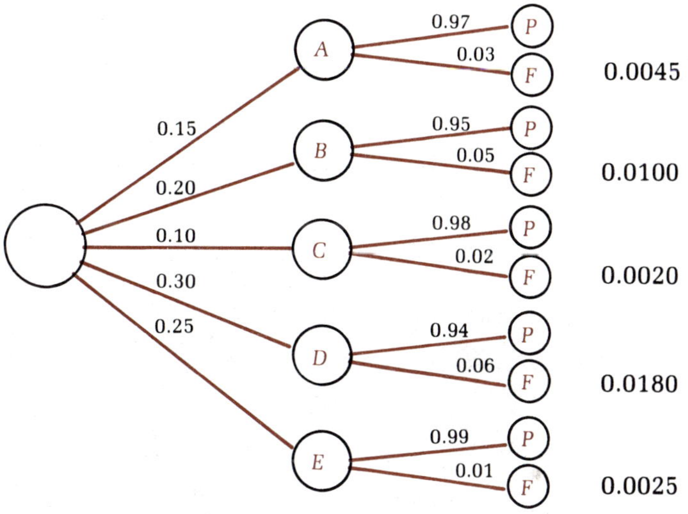

$$\begin{aligned} P(D|F) &= \frac{0.0180}{0.0045 + 0.0100 + 0.0020 + 0.0180 + 0.0025} \\ &= \frac{0.0180}{0.0370} = \frac{18}{37} = 0.4865. \end{aligned}$$

■

■ Example 43. Referring back to example 42, find

1. $P(A|F)$. **2.** $P(B|F)$. **3.** $P(C|F)$. **4.** $P(E|F)$.

Solutions: From example 42 we have $P(F) = 0.0370$. Using this and Bayes' Theorem, we compute

1. $$P(A|F) = \frac{P(A) \cdot P(F|A)}{P(F)} = \frac{(0.15)(0.03)}{0.0370} = 0.1216$$
2. $$P(B|F) = \frac{P(B) \cdot P(F|B)}{P(F)} = \frac{(0.20)(0.05)}{0.0370} = 0.2703$$
3. $$P(C|F) = \frac{P(C) \cdot P(F|C)}{P(F)} = \frac{(0.10)(0.02)}{0.0370} = 0.0541$$
4. $$P(E|F) = \frac{P(E) \cdot P(F|E)}{P(F)} = \frac{(0.25)(0.01)}{0.0370} = 0.0676$$

Note that $P(A|F) + P(B|F) + P(C|F) + P(D|F) + P(E|F) = 1.0001$ and not 1. This is due to round-off error. Using fractions instead of decimals would yield

$$\frac{0.0045}{0.0370} + \frac{0.0100}{0.0370} + \frac{0.0020}{0.0370} + \frac{0.0180}{0.0370} + \frac{0.0025}{0.0370}$$
$$= \frac{45}{370} + \frac{100}{370} + \frac{20}{370} + \frac{180}{370} + \frac{25}{370}$$
$$= \frac{370}{370} = 1.$$

■

■ **Example 44.**

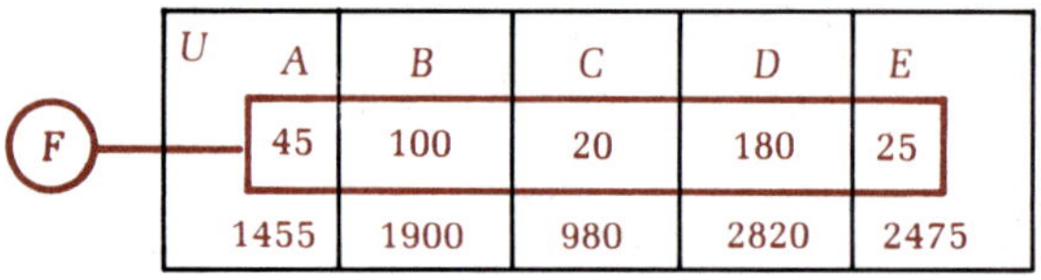

Example 42 could also be solved by sorting and choosing a model space large enough so that fractions are not involved.

Solution: If we take $n(U) = 10{,}000$, then this figure will result. Now with $n(F) = 370$ and $n(D \cap F) = 180$, $P(D|F) = \frac{180}{370} = \frac{18}{37} = 0.4865$. ■

The solution to example 43 is also straightforward.

$$P(A|F) = \frac{45}{370} \qquad P(B|F) = \frac{100}{370} \qquad P(C|F) = \frac{20}{370} \qquad P(E|F) = \frac{25}{370}$$

Exercise Set 6.5

1. An instructor is teaching three classes. Class I has 5 girls and 15 boys, class II has 10 girls and 20 boys, and class III has 15 girls and 25 boys. A class is chosen at random and then a student is randomly chosen from this class. What is the probability

a. the selected student is a girl?
b. the selected student is in class I, given the student is a girl?
c. the student is a girl, if all of the students' names are put on cards, mixed in a box, and a card drawn?

2. Three instructors are teaching a particular course. Instructor A has 5 girls and 25 boys, B has 10 girls and 20 boys, and C has 15 girls and 15 boys. An instructor is chosen at random and a student is picked from his class. What is the probability the selected student
   a. is a boy?
   b. is in Instructor A's class, given the student is a boy?
3. A store has three brands of light bulbs. They have 1 carton of Brand A, which averages 1% bad, 3 cartons of Brand B, which averages 3% bad, and 7 cartons of Brand C, which average 10% bad. A bulb is selected at random and tested. What is the probability that
   a. the bulb is bad?
   b. the bulb is Brand C, given it is bad?
4. a. Rework problem 5 in exercise set 6.3 using the results of this section.
   b. What is the probability that the coin showed "tails," given a boy was selected?
5. a. Rework problem 6(a) in exercise set 6.3 using the results of this section.
   b. What is the probability that the die showed 6, given the ball drawn was red?
   c. What is the probability that the die showed 3, given the ball drawn was red?
6. a. Rework exercise 7 in exercise set 6.3 using results of this section.
   b. What is the probability the ball drawn was green, given you won?
   c. What is the probability the ball drawn was blue, given you won?
7. Twenty percent of the people at a rock concert were high school students, 70% were college students, and 10% were adults. Suppose 10% of the high school students, 10% of the college students, and 60% of the adults did not like the concert.
   a. What percent of those in attendance did not like the concert?
   b. If a person did not like the concert, what is the probability that that person is an adult?
8. A television network wants to preview a new show for possible inclusion in its fall schedule. The network decides to test the show in four cities—New York, Chicago, Dallas, and Los Angeles. They sample people in the four cities to get their opinion, counting only people who watched the entire show. Of the people whose opinions were counted, 35% were from New York, 20% were from Chicago, 15% were from Dallas, and 30% were from Los Angeles. Sixty percent of the people from New York, 75% of the people from Chicago, 90% of the people from Dallas, and 65% of the people from Los Angeles liked the show.
   a. What percent of the people in the poll liked the show?

b. If a randomly selected person liked the show, what is the probability that the person is from Dallas? From New York?

9. A store sells batteries of 4 different brands to give people a choice of prices. The following table gives the percent of the total number sold for each brand and the estimated percent bad for the brand.

Brand	*Percent of sales*	*Percent bad*
A	10%	1%
B	30%	2%
C	40%	2%
D	20%	4%

a. What is the probability that a randomly chosen battery is bad?
b. If a customer calls and complains that he purchased a bad battery, what is the probability that it is brand D?

10. A farm advisor has prepared a report on spring planting for his six-county area. The table below gives the proportion of total cropland for each county and the proportion planted.

County	*Proportion of land*	*Proportion planted*
I	.12	.40
II	.22	.40
III	.34	.30
IV	.08	.60
V	.14	.20
VI	.10	.70

a. What percent of the total six-county acreage is planted?
b. What percent of the planted land is in county III? county VI?

11. In a certain high school, 23% of the students are freshmen, 30% are sophomores, 27% are juniors, and 20% are seniors. In addition, 60% of the freshmen, 50% of the sophomores, 30% of the juniors, and 10% of the seniors are enrolled in a math class.
a. What is the probability that a randomly selected student is enrolled in a math class?
b. What is the probability that Mary Anne is a senior, given she is enrolled in a math class?
c. What is the probability that Mark is a freshman, given that he is enrolled in math?

12. At a certain college, 30% of the students are freshmen, 25% are sophomores, 25% are juniors, 15% are seniors, and 5% are graduate students. Of these, 90% of the freshmen, 85% of the sophomores, 60% of the juniors, 30% of the seniors, and 80% of the graduate students live in college housing.

a. What percent of the students live in college housing?
b. What percent of the students in college housing are graduate students?
c. What percent of the students in college housing are freshmen?
d. What percent of the students in college housing are underclassmen (freshmen or sophomores)?

6.6 Odds and Expected Value

Have you ever seen a horse race? If you have, either in person or on television, you have probably seen the *odds* quoted on each horse in the race. What is meant by odds of 3 to 1, 5 to 2, 2 to 5, or 10 to 1?

Odds are just another way of making a probability statement. Suppose there are 10 horses in a race with the odds on each given in the following table.

Horse	*Odds*	*Horse*	*Odds*
A	8 to 1	F	30 to 1
B	10 to 1	G	50 to 1
C	9 to 2	H	2 to 1
D	6 to 1	I	12 to 1
E	7 to 2	J	14 to 1

Which horse is the favorite? Which horse is least likely to win?

For any event, the odds can be stated as the odds in favor of the event or the odds against the event. The odds, as is common in racing, are the odds against each horse. The odds of 50 to 1 against horse G would certainly make G an unlikely winner. Horse H, on the other hand, is the favorite, with odds of 2 to 1 against H (1 to 2 for H).

In order to better evaluate situations involving odds, the odds are converted to probabilities. Odds can be converted to probabilities using the following relationships.

If the odds in favor of an event E are a to b, then $P(E) = \dfrac{a}{a+b}$. If the odds against an event E are a to b, then $P(E) = \dfrac{b}{a+b}$.

The connection between odds in favor of an event and odds against the event is a simple one. Consider the events E and E'. If the odds against

E are a to b, then the odds in favor of E' (odds against E) are a to b. We have already seen that $P(E) = 1 - P(E')$. Since the odds in favor of E' are a to b, then

$$P(E') = \frac{a}{a+b},$$

so $P(E) = 1 - P(E')$

$$= 1 - \frac{a}{a+b}$$

$$= \frac{a+b}{a+b} - \frac{a}{a+b} = \frac{b}{a+b}.$$

We return to the table and fill in the probability of each horse winning (according to the quoted odds). Remember, the odds quoted are odds against the horse. Then for odds of a to b, the probability of a win is $\dfrac{b}{a+b}$.

Horse	*Odds*	*Probability of winning*	*Horse*	*Odds*	*Probability of winning*
A	8 to 1	$\frac{1}{8+1} = \frac{1}{9}$	F	30 to 1	$\frac{1}{30+1} = \frac{1}{31}$
B	10 to 1	$\frac{1}{10+1} = \frac{1}{11}$	G	50 to 1	$\frac{1}{50+1} = \frac{1}{51}$
C	9 to 2	$\frac{2}{9+2} = \frac{2}{11}$	H	2 to 1	$\frac{1}{2+1} = \frac{1}{3}$
D	6 to 1	$\frac{1}{6+1} = \frac{1}{7}$	I	12 to 1	$\frac{1}{12+1} = \frac{1}{13}$
E	7 to 2	$\frac{2}{7+2} = \frac{2}{9}$	J	14 to 1	$\frac{1}{14+1} = \frac{1}{15}$

On the basis of these probabilities, horse H is the clear-cut favorite, whereas horse G is an overwhelming underdog.

NOTE: The sum of these probabilities is greater than 1. This assures the track of a profit. The odds are affected by the money bet on the horses. The more money bet on a horse, the lower the odds against that horse. In other words, the people placing bets determine the final odds and hence the probabilities. The sum of the probabilities will always be greater than 1 to assure the track of a profit.

■ Example 45. According to a poll, the odds are 4 to 3 in favor of Congress passing the president's tax cut proposal. What is the probability that the bill will pass?

Solution: Since the odds are 4 to 3 in favor of passage, the probability that the bill will pass is $\frac{4}{4+3} = \frac{4}{7}$. ■

Suppose we know the probability of an event E. Can we apply this principle in reverse and convert this probability to odds? The odds in favor of E are given by the ratio $\frac{P(E)}{P(E')} = \frac{a}{b}$. The odds against E are given by $\frac{P(E')}{P(E)} = \frac{b}{a}$. Here a and b are integers and the odds are quoted a to b and b to a, respectively.

■ Example 46. The probability of a Republican candidate's winning a particular election is given by a pollster as 0.45. What are the odds in favor of, and against, the candidate?

Solution: Since $P(E) = 0.45$, $P(E') = 0.55$. Then the odds in favor of the candidate winning are given by $\frac{P(E)}{P(E')} = \frac{0.45}{0.55} = \frac{45}{55} = \frac{9}{11}$, or 9 to 11. The odds against will be 11 to 9. ■

■ Example 47. Five cards are drawn from a standard 52-card deck. What are the odds against drawing a full house (3 cards of one denomination and 2 of another)?

Solution: Let F be the event "draw a full house." The odds against F are given by $\frac{P(F')}{P(F)}$. We begin by finding $P(F)$.

Consider the event F to be composed of the following four tasks.

T_1: Select a denomination from the 13 possible for the three of a kind. $n_1 = \binom{13}{1}$

T_2: Select the 3 cards from this denomination. $n_2 = \binom{4}{3}$

T_3: Select a denomination from the remaining 12. $n_3 = \binom{12}{1}$

T_4: Select 2 cards from this denomination. $n_4 = \binom{4}{2}$

$$\begin{aligned} \text{Then } n(F) &= n_1 \cdot n_2 \cdot n_3 \cdot n_4 \\ &= \binom{13}{1} \cdot \binom{4}{3} \cdot \binom{12}{1} \cdot \binom{4}{2} \\ &= \frac{13!}{12!1!} \cdot \frac{4!}{3!1!} \cdot \frac{12!}{11!1!} \cdot \frac{4!}{2!2!} \\ &= 13 \cdot 4 \cdot 12 \cdot 6 = 3744 \end{aligned}$$

$$P(F) = \frac{n(F)}{n(S)} = \frac{3744}{\binom{52}{5}} = \frac{3744}{2{,}598{,}960}$$

$$\begin{aligned} P(F') &= 1 - P(F) \\ &= 1 - \frac{3744}{2{,}598{,}960} \\ &= \frac{2{,}595{,}216}{2{,}598{,}960} \end{aligned}$$

$$\text{The odds against } P = \frac{P(E')}{P(F)} = \frac{\frac{2{,}595{,}216}{2{,}598{,}960}}{\frac{3744}{2{,}598{,}960}} = \frac{2{,}595{,}216}{3744}.$$

The odds against F are 2,595,216 to 3744. The final fraction given could be reduced to $\frac{4159}{6}$. The odds, then, are given by 4159 to 6. Some books approximate this by quoting the odds as 693 to 1. ■

We close this section with one final idea related to probability. Suppose a coin is flipped 100 times. How many heads would you "expect" to get? If we assume that the coin is fair, we would expect about half of the tosses to result in a head, so we would expect about 50 heads—that is, the number of flips times the probability of a head on each flip, or $(100)\left(\frac{1}{2}\right) = 50$.

The concept of **expected value,** sometimes called average value, for an experiment can be used in a variety of situations. The expected value of an experiment is based on the outcomes of the experiment and the probability of each outcome. Before the concept is stated more formally, the following example will be considered.

■ Example 48. Suppose a coin is tossed 2 times. How many heads would you expect?

Solution: Intuitively, assuming a fair coin, we would expect 1 head. Consider a more formal approach to the problem. The possible outcomes for the experiment are 0, 1, or 2 heads. From earlier work, these outcomes are associated with the sets $\{(T,T), (H,T)\}$, $\{(T,H)\}$, and $\{(H,H)\}$, respectively. Then $P(0 \text{ heads}) = \frac{1}{4}$, $P(1 \text{ head}) = \frac{2}{4}$, and $P(2 \text{ heads}) = \frac{1}{4}$.

Then $0 \cdot P(0 \text{ heads}) + 1 \cdot P(1 \text{ head}) + 2 \cdot P(2 \text{ heads})$

$= 0 \cdot \frac{1}{4} + 1 \cdot \frac{2}{4} + 2 \cdot \frac{1}{4}$

$= 0 + \frac{2}{4} + \frac{2}{4} = 1$, which we obtained as our intuitive answer. ■

In words, the expected value is the sum of the values obtained by taking each outcome times the probability of that outcome. For this to have meaning, the outcomes must be numerical values. Hence the expected value is a numerical value.

> **Definition 6.5** Let an experiment consist of k possible numerical outcomes $n_1, n_2, n_3, \ldots, n_k$, with respective probabilities $p_1, p_2, p_3, \ldots, p_k$. Then the expected value of the experiment is given by $n_1 \cdot p_1 + n_2 \cdot p_2 + n_3 \cdot p_3 + \cdots + n_k \cdot p_k$.

■ Example 49. A fair coin is flipped 5 times. What is the expected number of heads?

Solution: The possible outcomes are 0, 1, 2, 3, 4, or 5. There are $2^5 = 32$ possible sequences of 5 heads and tails. The probability of getting k heads is

$$\frac{\text{number of sequences with } k \text{ heads}}{32}. \text{ Then}$$

$$P(0) = \frac{\binom{5}{0}}{32} = \frac{1}{32} \qquad P(3) = \frac{\binom{5}{3}}{32} = \frac{10}{32}$$

$$P(1) = \frac{\binom{5}{1}}{32} = \frac{5}{32} \qquad P(4) = \frac{\binom{5}{4}}{32} = \frac{5}{32}$$

$$P(2) = \frac{\binom{5}{2}}{32} = \frac{10}{32} \qquad P(5) = \frac{\binom{5}{5}}{32} = \frac{1}{32}$$

So Expected Value

$$= 0 \cdot \frac{1}{32} + 1 \cdot \frac{5}{32} + 2 \cdot \frac{10}{32} + 3 \cdot \frac{10}{32} + 4 \cdot \frac{5}{32} + 5 \cdot \frac{1}{32}$$

$$= \frac{5 + 20 + 30 + 20 + 5}{32}$$

$$= \frac{80}{32} = 2.5.$$ ■

NOTE: The expected value here is not one of the possible outcomes.

■ Example 50. John is one of 2,000,000 people who purchase a $1 lottery ticket. There is 1 grand prize of $1,000,000 and 1 prize of $100,000, and there are 5 winners of $50,000 each. What is John's expected value?

Solution: There are 4 possible outcomes: receive $1,000,000 (or win $999,999), receive $100,000, receive $50,000, or lose $1 (win −$1.00).

$$EV = 999{,}999 \cdot P(1{,}000{,}000) + 99{,}999 \cdot P(100{,}000) + 49{,}999 \cdot P(50{,}000) + (-1.00) \cdot P(-1.00)$$

$$= [999{,}999]\left(\frac{1}{2{,}000{,}000}\right) + [99{,}999]\left(\frac{1}{2{,}000{,}000}\right) + [49{,}999]\left(\frac{5}{2{,}000{,}000}\right) + (-1.00)\left(\frac{1{,}999{,}993}{2{,}000{,}000}\right)$$

$$= \frac{999{,}999 + 99{,}999 + 249{,}995 - 1{,}999{,}993}{2{,}000{,}000}$$

$$= \frac{-650{,}000}{2{,}000{,}000}$$

$$= -.325$$

John's expected value is to lose 32.5¢.

Alternate Solution 1.

If John bought all the tickets, he would pay out $2,000,000 and receive all prize money in return. Since the sum of the prize money is $1,350,000 this is a net loss of $650,000 or $32\frac{1}{2}$¢ on each of the 2,000,000 tickets.

The expected value has been computed taking the average of the total cost minus the total gain. This method is the reason why the expected value is often referred to as average value.

Alternate Solution 2.

The expected value can also be computed as the difference in the *expected return* and the initial cost. The resulting formula is:

$$EV = [\text{expected return}] - [\text{cost}]$$

In our problem

$$EV = \left[\frac{1{,}000{,}000}{2{,}000{,}000} + \frac{100{,}000}{2{,}000{,}000} + \frac{5(50{,}000)}{2{,}000{,}000}\right] - [1]$$

$$= \left[\frac{1}{2} + \frac{1}{20} + \frac{1}{16}\right] - 1$$

$$= -\frac{31}{80} = -.325 = -32\tfrac{1}{2}¢.$$

■

■ Example 51. Bill and Fred agree to play a game with a die. If the die shows a 1 or a 6, Fred pays Bill \$3.00. If the die shows any other number, Bill pays Fred \$1.50. What is Bill's expected value?

Solution: There are two possible outcomes for Bill. Either he wins \$3.00 or he loses \$1.50 (wins −\$1.50).

$$EV = \frac{2}{6}(3.00) + \frac{4}{6}(-1.50) = 1.00 - 1.00 = 0$$

Fred's expected value is $\frac{4}{6}(1.50) + \frac{2}{6}(-3.00) = 0$ also.

This game is considered to be *fair*. A game is fair if it has expected value equal to 0. ■

■ Example 52. A company has a committee of 10 employees, 6 men and 4 women, that is charged with investigating personnel matters. When a grievance is filed, a subcommittee of 3 people is randomly chosen to make an initial investigation and recommendation to the full committee. What is the expected number of women on such a subcommittee?

Solution: The total number of 3-member committees that can be formed is given by $\binom{10}{3} = 120.$

To find the probabilities of 0, 1, 2, or 3 women, we need to count the number of ways each of these events can occur and divide by 120. The number of ways that 1 woman can be on the committee, for example, is formed by taking $\binom{4}{1}\binom{6}{2}$. That is, any combination of 1 woman from the 4 can occur with any combination of 2 men from 6. Hence the total number of ways of having 1 woman on the subcommittee is $\binom{4}{1}\binom{6}{2}$. The probabilities, then, are

$$P(0) = \frac{\binom{4}{0}\binom{6}{3}}{120} = \frac{(1)(20)}{120} = \frac{20}{120}$$

$$P(1) = \frac{\binom{4}{1}\binom{6}{2}}{120} = \frac{(4)(15)}{120} = \frac{60}{120}$$

$$P(2) = \frac{\binom{4}{2}\binom{6}{1}}{120} = \frac{(6)(6)}{120} = \frac{36}{120}$$

$$P(3) = \frac{\binom{4}{3}\binom{6}{0}}{120} = \frac{(4)(1)}{120} = \frac{4}{120}$$

The expected number of women is given by

$$\begin{aligned} EV &= 0 \cdot \frac{20}{120} + 1 \cdot \frac{60}{120} + 2 \cdot \frac{36}{120} + 3 \cdot \frac{4}{120} \\ &= \frac{60 + 72 + 12}{120} = \frac{144}{120} = \frac{6}{5} \end{aligned}$$

The expected number of women is 1.2. ■

Exercise Set 6.6

1. In each of the following, the odds for or against an event E are given. Find P(E) and P(E′).
   - **a.** Odds against E are 7 to 1.
   - **b.** Odds against E are 5 to 2.
   - **c.** Odds against E are 6 to 5.
   - **d.** Odds against E are 30 to 1.
   - **e.** Odds against E are 1 to 1.
   - **f.** Odds in favor of E are 3 to 2.
   - **g.** Odds in favor of E are 5 to 1.
   - **h.** Odds in favor of E are 7 to 3.
   - **i.** Odds in favor of E are 2 to 7.
   - **j.** Odds in favor of E are 11 to 3.
2. In each of the following, P(E) or P(E′) is given. Find the odds in favor of E and the odds against E.
   - **a.** $P(E) = 0.35$
   - **b.** $P(E) = \frac{5}{7}$
   - **c.** $P(E) = \frac{3}{5}$
   - **d.** $P(E) = 0.55$
   - **e.** $P(E) = \frac{1}{5}$
   - **f.** $P(E) = \frac{2}{3}$
   - **g.** $P(E') = \frac{4}{5}$
   - **h.** $P(E') = \frac{4}{7}$

i. $P(E') = 0.25$ **k.** $P(E') = 0.80$
j. $P(E') = 0.56$ **l.** $P(E') = 0.30$

3. The weather forecaster says there is a 30% chance of rain tonight. What are the odds against rain?

4. If the weather forecast includes a 75% chance of snow, what are the odds in favor of snow?

5. You are playing a game that involves tossing a pair of dice and observing the sum of the faces. What are the odds against you winning if, to win, you must get a sum
a. equal to 7? **c.** less than 6?
b. equal to 11? **d.** at least as large as 9?

6. If five coins are flipped simultaneously, what are the odds against
a. exactly three heads? **c.** no heads?
b. exactly one head? **d.** at least 3 heads?

7. Suppose 10 envelopes of money are placed in front of you. Five of the envelopes contain $1.00 each, 3 contain $5.00 each, 1 contains $10.00, and 1 contains $50.00. If you randomly select one envelope, what is your expected value?

8. An urn contains 50 balls—30 red, 15 white, and 5 blue. You pay $1.00 to draw a ball. If you draw a blue ball, you win $10.00 and if you draw a white ball you win $2.00. You lose if you draw a red ball. What is the expected value of the game if your $1.00 is returned if you win? What is the expected value if the $1.00 is not returned?

9. In a state lottery, 5,000,000 tickets are sold at $1.00 each. The top prize is $1,000,000, second prize is $500,000, and there are to be two prizes of $250,000 each. An additional 1000 prizes of $100 each and 200,000 prizes of $2.00 each are to be awarded. What is the expected value of a ticket in the lottery?

10. A horse owner has estimated that the probability his horse would win a particular race if he entered is 0.05. The probability his horse would finish second is estimated to be 0.15 and the probability his horse would finish third is 0.20. The purse is $120,000 for first, $90,000 for second, and $60,000 for third, with an entry fee of $20,000. What is the owner's expected value if he enters his horse?

11. If you pay $2.00 to play a game in which the odds against you winning are 4 to 1, how much should you be paid for winning so that the game is fair?

12. An urn contains 100 balls—5 blue, 20 red, and 75 white. If you draw a red or blue ball, you win. The payoff for a blue ball is 4 times the payoff for a red ball. If the cost of drawing a ball is $8.00, what should the respective payoffs be so that the game is fair?

6.7 Chapter Review

Important Terms and Formulas in Chapter 6.

experiment (section 6.1)
outcomes (section 6.1)
sample space (section 6.1)
events (simple event) (section 6.1)
composite events (section 6.1)
probability (definition) (section 6.1)
relative frequency (section 6.1)
favorable outcomes (section 6.1)
equally likely (section 6.1)
probability as a ratio (section 6.1)
$P(E) = \dfrac{n(E)}{n(S)}$ (section 6.1)
$0 \leq P(E) \leq 1$ (section 6.1)
complementary events (section 6.1)
$P(E) = 1 - P(E')$ (section 6.1)
compound events (section 6.2)
$P(E \cup F) = P(E) + P(F) - P(E \cap F)$ (section 6.2)
mutually exclusive events (section 6.2)
pairwise mutually exclusive (section 6.2)
DeMorgan's Laws (section 6.2)
tree diagram (section 6.3)
conditional probability (section 6.4)
$P(E|F) = \dfrac{P(E \cap F)}{P(F)}$ (section 6.4)
$P(E \cap F) = P(F) \cdot P(E|F)$ (section 6.4)
independent events (section 6.4)
Bayes' Theorem (section 6.5)
total probability formula (section 6.5)
odds (section 6.6)
$\dfrac{P(E')}{P(E)} = \dfrac{b}{a}$ and $\dfrac{P(E)}{P(E')} = \dfrac{a}{b}$ (section 6.6)
expected value (section 6.6)
fair game (section 6.6)

Review Exercises

1. In a population of size 30, 5 items are known to be defective. Compute the probability that a sample will contain at least 1 defective if the sample is of size
   a. 1. **b.** 2. **c.** 3. **d.** 10.
2. You draw 5 cards from a standard deck of 52 playing cards. Four of the cards would be part of a straight (consecutive denominations). But the fifth card doesn't fit, so you discard it and draw another card.
   a. What is the probability of filling an inside straight (the missing card is in the middle of the straight)?
   b. What is the probability of filling the straight if it is open at both ends (your 4 cards are consecutive and you need 1 at either end)?
   c. What is the probability of completing a straight flush (all cards are the same suit) if your 4 cards are the same suit and you need an inside card? (The fifth card was a different suit.)
   d. What is the probability that you fill a straight flush that is open on both ends?
   e. What is the probability of completing a flush that is not a straight if the straight must fill inside (part c)?

f. What is the probability of completing a flush that is not a straight if the straight is open on both ends (part d)?

3. Given $P(A') = .42$, $P(B) = .38$, and $P(A \cap B) = .21$,
   a. Find $P(A)$
   b. Find $P(A \cup B)$.
   c. Find $P(A' \cup B')$.
   d. Find $P(A \cap B')$.
4. Three dice are tossed, and the sum of the faces is observed. Find the probability of each of the following events.
   a. The sum equals 10.
   b. The sum is greater than 9.
   c. The sum is an even number.
5. A Cub Scout pack with 9 members must select 3 members to act as color guard in the opening ceremony at the next council meeting. The 3 will be selected by drawing names out of a hat. John and Jim Morley are both members of this pack.
   a. What is the probability that both John and Jim are selected?
   b. What is the probability that neither John nor Jim is selected?
   c. What is the probability that exactly 1 of the Morley brothers is selected?
   d. What is the probability that only John is selected?
6. Two urns colored red and blue contain different combinations of marbles. The red urn contains 3 red and 2 blue marbles. The blue urn contains 1 red, 1 blue, and 1 gold marble. A game is played in which a coin is flipped. If it comes up heads, you select a marble from the red urn. If it comes up tails, you select a marble from the blue urn.
   a. Find the probability that the marble is red.
   b. Find the probability that the marble is blue.
   c. Find the probability that the marble is gold.
7. Given the following tree diagram, find

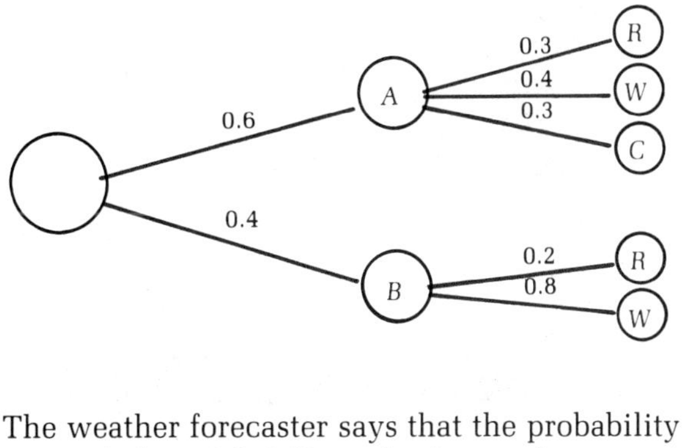

   a. $P(A)$.
   b. $P(C)$.
   c. $P(R)$.
   d. $P(R|A)$.
   e. $P(R|B)$.
   f. $P(A|R)$.
   g. $P(B|W)$.

8. The weather forecaster says that the probability of rain tomorrow is 0.7 and the probability of rain on the next day is 0.7 if it rains tomorrow. What is the probability that it rains both days?
9. There are 7 girls and 9 boys on a committee. A subcommittee of 4 people is to be formed by random drawing.
   a. What is the probability that no girls are selected?

b. What is the probability that at least 1 girl is selected?
c. What is the probability that exactly 2 boys and 2 girls are selected?

10. You play a game in which you first roll a pair of dice and then draw a card from a 52-card deck. If you roll a double (both dice show the same number of spots), you win if the card you draw is an *A*, *K*, *Q*, or *J*. If you do not roll a double, you win only if the card you draw is a jack (*J*). What is the probability that you win?

11. An urn contains 15 red, 8 blue, and 2 green marbles. You draw 2 marbles. What is the probability that both are blue if the drawing is
a. with replacement? **b.** without replacement?

12. The weather forecaster says that the odds are 3 to 2 that it will snow on Christmas day. What is the probability that it will not snow on Christmas day?

13. Given $P(E) = 0.7$, $P(F) = 0.5$, and $P(E \cup F) = 0.8$, find
a. $P(E \cap F)$.
b. $P(E|F)$.
c. $P(F|E)$.
d. $P(E \cap F')$.
e. $P(E' \cap F')$
f. $P(E|F')$.
g. $P(E' \cap F)$.
h. $P(E'|F)$.

14. Given $P(E) = 0.6$, $P(F) = 0.3$, $P(E \cup F) = 0.9$, find
a. $P(E \cap F)$.
b. $P(E \cap F')$.
c. $P(E' \cap F)$.
d. $P(E|F)$.
e. $P(E'|F)$.
f. $P(E|F')$.

15. What can be said about events E and F in problem 14?

16. Given $P(E) = 0.60$, $P(F) = 0.45$, and $P(E \cup F) = 0.78$, find
a. $P(E \cap F)$
b. $P[(E \cup F)']$
c. $P(E|F)$.
d. $P(F|E)$.
e. $P(E \cap F')$.
f. $P(E' \cup F')$.
g. $P(E|F')$.
h. $P(F'|E)$.

17. What can be said about the events E and F in problem 16?

18. If 60% of the people like dogs, 90% of the people who like cats also like dogs, and 60% of the people who like dogs also like cats, then what percent of the people like cats?

19. Given $P(A) = .4$,
a. Find the odds in favor of *A*. **b.** Find the odds against *A*.

20. If the odds in favor of a specific horse winning a race are 2 to 7, what is the probability that this specific horse will win?

21. You agree to play a game in which you select a card at random from a regular 52-card deck. If you select a face card (king, queen, or jack), you lose \$10. If you select any other card, you win the number of dollars indicated by the number of spots on your card. If it's an ace, you win \$1; if a deuce, \$2; if a 9, \$9; and so on. What is the expected value of this game?

22. Three girls—Tammy, Mandy, and Samantha—are good friends in the same first-grade class. The names of all 16 students in the class are placed in a hat. The names of 3 students are randomly selected to read a book to the principal.

a. What is the probability that Tammy, Mandy, and Samantha are all selected?

b. What is the probability that at least 1 of these 3 girls is selected?

23. Jamie and Kellie are playing a game in which Kellie draws a marble from a box containing 9 red marbles, 6 blue marbles, and 5 green marbles. If Kellie draws a red marble, she gives Jamie 3 pieces of candy. If she draws a blue marble, Jamie gives her 2 pieces of candy. How many pieces of candy should be "paid," and who should get them, so that the game is fair?

24. Six coins are tossed. What is the probability that exactly half show heads?

25. At a certain college, 30% of the students are freshmen, 25% are sophomores, 20% are juniors, 20% are seniors, and 5% are graduate students. In addition, 1% of the freshmen, 4% of the sophomores, 12% of the juniors, 8% of the seniors, and 2% of the graduate students are transfers.

a. What is the probability that a randomly selected student is a transfer?

b. If a randomly chosen student is known to be a transfer student, what is the probability that the student is a junior?

c. What is the probability that a randomly selected student is a sophomore, given the student is a transfer?

26. A company has 7 machines producing ball bearings. The following table gives the percent of production and the percent of bad ball bearings for each machine.

Machine	*Percent of production*	*Percent of bad ball bearings*
A	10%	1%
B	5%	4%
C	15%	2%
D	10%	3%
E	10%	1%
F	30%	1%
G	20%	2%

a. What is the probability that a randomly selected bearing is bad?

b. If a randomly selected bearing is bad, what is the probability that it was produced by machine *B*?

c. What is the probability that a randomly selected bearing was produced by machine *F*, given the ball bearing is bad?

d. What is the probability that a randomly selected bearing was produced by machine G, given the ball bearing is bad?

e. In a production run of 100,000 bearings, how many bad ones would you expect?

7 Descriptive Statistics

Introduction

Statistics is one of the most widely used branches of the mathematical sciences. Almost every aspect of daily life involves some use of statistics. This includes recreation and leisure, as well as occupations.

Statistics can be subdivided into two major classifications, descriptive statistics and statistical inference. Descriptive statistics involves, as the name suggests, describing data or information. Statistical inference involves many things, including estimation, prediction, and hypothesis testing.

In this chapter and the next, we will consider topics primarily associated with descriptive statistics. Statistical inference is beyond the scope of this text. We heartily recommend a course in statistics. Some of you may be in a discipline that requires statistics. However, everyone can benefit from a knowledge of how statistics is applied to make predictions and to test hypotheses.

7.1 Pictorial Representation of Data

In earlier chapters of this text, we have made use of pictures and diagrams to aid in problem solving. In this section, we will consider graphs and tables, as pictures, to represent statistical data. Graphs can be constructed from tables, and tables can be constructed from graphs.

We will begin with a discussion of tables and then consider several types of graphs. The most common type of table is a **frequency table**. The table consists of two columns. The first column lists, in some order, the possible data values. The second column gives the **frequency (f)** (number of times) of occurrence for each value. The order of the values can be from largest to smallest or smallest to largest. Some order is necessary for easy interpretation of the table.

■ Example 1. A die is tossed 20 times, and the following sequence is observed—2, 4, 3, 3, 1, 6, 2, 1, 3, 1, 2, 2, 4, 3, 4, 3, 6, 6, 4, and 1. Construct a frequency table.

Solution: The possible values are 1, 2, 3, 4, 5, and 6. We put these values in the left column, and in the right column we put the number of times each occurred (frequency).

Values	*Frequency*
1	4
2	4
3	5
4	4
5	0
6	3

Note here that 5 never occurred. It is, however, a possible value and should be included. The only time we would not include a possible value that didn't occur would be if it were smaller than the smallest value that occurred or larger than the largest value that occurred.

For example, if the lowest score on a 100-point exam were 57, we would not include 40, even though it is a possible value. Common sense will help in making decisions about whether or not to include some values outside the range of values that actually occur.

Some information can be easily obtained from a frequency table. From the table constructed for example 1, it is obvious that the most common value was 3 and the least common was 5.

■ Example 2. Given the following frequency table about grades on an exam, make some interpretations.

Grade	*Frequency*
A	1
B	6
C	20
D	15
F	8

Solution: Three facts are obvious.

1. C was the most common grade.
2. A was the least common grade.
3. The distribution of grades is "bottom heavy." That is, the number of Ds and Fs is quite a bit higher than the number of As and Bs. ■

It is often convenient to use a letter such as x,y, and so on to represent the values and the letter *f* to represent frequency as headings in the table. The tables in examples 1 and 2 could then be written as

x	*f*
1	4
2	4
3	5
4	4
5	0
6	3

and

x	*f*
A	1
B	6
C	20
D	15
F	8

respectively.

The only requirement is that you understand what the values represent. We will frequently use this type of abbreviation in the remainder of this chapter. The use of letters x, y, and so forth is convenient for graphing and indicates the fact that the values vary from observation to observation.

As the number of possible values increases, there is a corresponding increase in the size of the table. For example, if a large number of students take a 100-point exam and the grades range from a low of 17 to a high of 99, the left column should contain 83 values. The size of the table becomes unwieldy.

To handle this type of situation, a modification can be made in the table. Instead of listing each value, we list intervals of values. For example, we might use 15 through 19 to include the scores 15, 16, 17, 18, and 19. Normally, we would try to use 20 or fewer intervals and would want all of the intervals to be the same size. Tables of this type are called **grouped data frequency tables.**

When using grouped data frequency tables, we put the intervals in order, as before. One of the intervals should contain the smallest data value, and one should contain the largest data value. The intervals containing possible values outside the actual extremes are not normally included. In addition, the intervals are disjoint (no possible data value is in more than one interval).

The intervals are referred to as **classes.** When working with grouped data, we need to determine the number of classes we will use and the size or width of the classes. In the preceding illustration, when 15 through 19 was used as a class, the **class size** is determined by the number of values it includes, 5. If any value from 15 through 19 is possible (including fractions), we speak of the **class width** as $19 - 15 = 4$. We always write classes with the left end point as the smaller number, or **class boundary.**

When working with situations involving an infinite number of possible values, we sometimes use class boundaries such as 14.5 to 19.5, where one of the two end points is included and the other is not. That is, 14.5 is in this class, but 19.5 is in the next class. We will include the left (smaller) class boundary in these situations and exclude the right (larger) boundary.

If only integer values are possible, both class boundaries will typically be included in the class. The class from 15 to 19 includes both 15 and 19. The next class would be from 20 to 24, including both 20 and 24. Since no possible value exists between 19 and 20, no value is excluded. Remember, when class boundaries are determined, every possible data value (except those outside the actual extremes) must be included in exactly one class.

■ Example 3. Construct a grouped data frequency table for the following exam scores, using

a. a class size of 10.
b. a class size of 5.

Scores: 87, 55, 59, 48, 70, 62, 100, 70, 93, 67, 81, 68, 71, 78, 62, 56, 96, 70, 80, 66, 48, 80, 71, 87, 64, 84, 74, 79, 64, 87, 65, 100, 57, 94, 70, 92, 85, 45, 88, 75.

Solutions: There are actually several possibilities for each. We will give two choices in each case. Note that the smallest actual data value is 45 and the largest is 100. Then for part **a** we could make the smallest class 40 through 49 or 41 through 50. There are many other possibilities. If we use 40 through 49, then the last class will be 100 through 109, which contains only one possible data value, 100 (we assume the maximum possible is 100). Choosing 41 through 50 as the first class gives a final class of 91 through 100, and every class contains 10 possible data values. The latter is probably the preferred method. We will do parts **a** and **b** using both methods.

a.

Solution 1		**Solution 2** (preferred)	
Class	*f*	*Class*	*f*
40–49	3	41–50	3
50–59	4	51–60	4
60–69	8	61–70	12
70–79	10	71–80	8
80–89	9	81–90	7
90–99	4	91–100	6
100–109	2		

b.

Solution 1		**Solution 2** (preferred)	
Class	*f*	*Class*	*f*
45–49	3	41–45	1
50–54	0	46–50	2
55–59	4	51–55	1
60–64	4	56–60	3
65–69	4	61–65	5
70–74	7	66–70	7
75–79	3	71–75	4
80–84	4	76–80	4
85–89	5	81–85	3
90–94	3	86–90	4
95–99	1	91–95	3
100–104	2	96–100	3

The second solution is preferred in each case because each class contains the same number of possible data values and no class extends beyond the possible data values. ■

The use of **tally marks** can be a big help when working with large numbers of data values. As we go through the data, we put a tally mark beside the value or class corresponding to each data value. The frequency is the number of tally marks. Using the data for example 3 and solution 2 of part a, we put a tally mark beside the class 81 through 90 for the data value 87, we put a tally mark beside the class 51 through 60 for data value 55, and so on. The table would then appear as follows,

Class	*Tally*	*f*
41–50	///	3
51–60	////	4
61–70	~~////~~ ~~////~~ //	12
71–80	~~////~~ ///	8
81–90	~~////~~ //	7
91–100	~~////~~ /	6

When working with data where the possible values are infinite, say any rational number, we have available several alternatives for the choice of class boundaries. Suppose the data values are given to the nearest tenth. We could use classes such as 3.5 to 4.0 and 4.0 to 4.5, with the understanding that 4.0 is actually in only one of the intervals. That is, we might specify that the class 3.5 to 4.0 includes 3.5 but not 4.0 and the class 4.0 to 4.5 includes 4.0 but excludes 4.5.

Another possibility is to use end points that are not potential data values, such as 3.45 to 3.95, 3.95 to 4.45, 4.45 to 4.95, and so on. Since data values are given to the nearest tenth, no value of 3.95 would occur (it would be rounded off to 4.0), so having this value in two classes doesn't matter.

We could also use intervals such as 3.5 through 3.9, 4.0 through 4.4, 4.5 through 4.9, and so forth. These intervals do not overlap, and they are the same length, 0.4. Since values between 3.9 and 4.0 are rounded off to either 3.9 or 4.0, no potential values are actually omitted.

■ Example 4. The local radio station gives monthly precipitation amounts to the nearest hundredth of an inch. Suppose in the last 50 years, the smallest monthly precipitation has been 0.12 inches and the highest has been 6.76 inches. Construct classes to represent the 50-year data in a grouped data frequency table.

Solution: Several possibilities exist. Only one is given here. Consider the following classes: 0.00 through 0.99, 1.00 through 1.99, 2.00 through 2.99, 3.00

through 3.99, 4.00 through 4.99, 5.00 through 5.99, and 6.00 through 6.99. These intervals are all the same length, and every data value from 0.12 through 6.76 is included in exactly one interval. Since values are rounded off to the nearest hundredth, no values are excluded. ■

A second type of table, related directly to a frequency table, that is sometimes used to display data is a **cumulative frequency table.** Instead of giving the frequency for each value or class, it gives the total frequency up to, and including, the value or class under consideration. In other words, it accumulates the frequencies. Normally, for cumulative frequency tables, the table goes from the smallest value or class to the largest. We will use *cf* for cumulative frequency.

■ Example 5. Construct a cumulative frequency table for the die experiment of example 1.

Solution: We will include both the frequencies and the cumulative frequencies.

Value (x)	*f*	*cf*	
1	4	4	
2	4	8	(4 + 4 = 8)
3	5	13	(8 + 5 = 13)
4	4	17	(13 + 4 = 17)
5	0	17	(17 + 0 = 17)
6	3	20	(17 + 3 = 20)

■

■ Example 6. Construct a cumulative frequency table for the data of example 3 using a class size of 10 (solution 2 of part **a**).

Solution:

Class	*f*	*cf*
41–50	3	3
51–60	4	7
61–70	12	19
71–80	8	27
81–90	7	34
91–100	6	40

■

If a frequency table is given, the cumulative frequency table is easy to construct. Conversely, if the cumulative frequency table is given, it is just as easy to find the regular frequency table.

Consider the cumulative frequencies of example 6. Since *cf* for 71 through 80 is 27 and *cf* for 81 through 90 is 34, then *f* for 81 through 90 is $34 - 27 = 7$. Similarly, *f* for 61 through 70 is $19 - 7$ (*cf* for 61 through 70 minus *cf* for 51 through 60) $= 12$.

It is sometimes useful to work with **relative frequencies (rf)** instead of actual frequencies. Relative frequencies are just ratios of the frequency for a value or class to the total number of values. That is, relative frequency and **cumulative relative frequencies (crf),** are defined by

$$rf = \frac{f}{N} \text{ and } crf = \frac{cf}{N},$$

where N is the total number of values.

■ Example 7. Construct a table showing *rf* and *crf* for the data in example 6.

Solution:

Class	*rf*	*crf*
41–50	0.075	0.075
51–60	0.100	0.175
61–70	0.300	0.475
71–80	0.200	0.675
81–90	0.175	0.850
91–100	0.150	1.000

■

The relative frequencies are actually related to probabilities. Theoretically, the sum of the *rf*'s, and hence the last entry in the *crf* column, will be exactly 1. However, rounding off will sometimes give values slightly different from 1.

Another commonly used method for pictorial representation of data is a *graph*. Although many different kinds of graphs can be, and indeed are, used, we will restrict ourselves to a discussion of three types. Each of the three types uses frequency concepts like those used in the tables we have considered.

The first type to be considered is the **frequency polygon.** As is done in other graphing problems, two perpendicular lines are used for the *axes*. The horizontal axis will represent the values, and the vertical axis will represent the frequency. If the number of possible data values is fairly small, such as the face of a die showing, exact data values are used with corresponding frequencies. In situations where grouped data are more appropriate, a value is chosen to represent each of the various classes (intervals).

The value chosen to represent a class is called the **class mark.** The class mark is the middle value for the class. To find this middle value, add the end points of the class and divide by 2. Then for a class that goes from 61 to 70, the class mark is

$$\frac{61 + 70}{2} = \frac{131}{2} = 65.5.$$

Similarly, a class that goes from 4.0 to 4.4 would have a class mark of

$$\frac{4.0 + 4.4}{2} = \frac{8.4}{2} = 4.2.$$

To construct the frequency polygon, we put the values or class marks on the horizontal axis and possible frequencies on the vertical axis. Above each value or class mark we put a dot at a location to represent the appropriate frequency. The dots are then connected with line segments to complete the graph.

■ Example 8. Construct a frequency polygon for the die tossing experiment of example 1.

Solution:

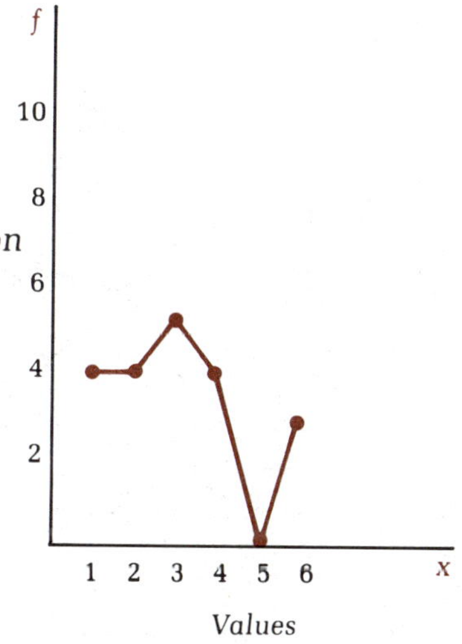

Frequency polygon

The possible values are 1, 2, 3, 4, 5, and 6, and the possible frequencies are 1 through 20 (the die was tossed 20 times). However, since the largest observed frequency was 5, there is no need to include all of the numbers up to 20 on the vertical scale. When we work with cumulative frequencies, however, it is necessary to go all the way to 20. ■

In statistical graphs, the units used on the two axes are frequently not the same. In addition, adjustments are sometimes made for a starting point. For example, if the smallest data value is 1,000, it is not necessary to include small values such as 1, 2, 3, and so on. When these adjustments are made, we will use the following type of representation to show that part of the axis is omitted.

The broken part indicates omitted values. The idea is to represent the data in a reasonable amount of space so adjustments are made in unit size and starting points to make the graph more compact.

■ Example 9. Construct a frequency polygon for the grouped data of example 3.

Solution: Since grouped data are being considered, we first find the class mark for each of the classes.

Class	*Class Mark* (x_i)	*f*
41–50	$\frac{41 + 50}{2} = 45.5$	3
51–60	$\frac{51 + 60}{2} = 55.5$	4
61–70	$\frac{61 + 70}{2} = 65.5$	12
71–80	$\frac{71 + 80}{2} = 75.5$	8
81–90	$\frac{81 + 90}{2} = 85.5$	7
91–100	$\frac{91 + 100}{2} = 95.5$	6

Frequency polygon

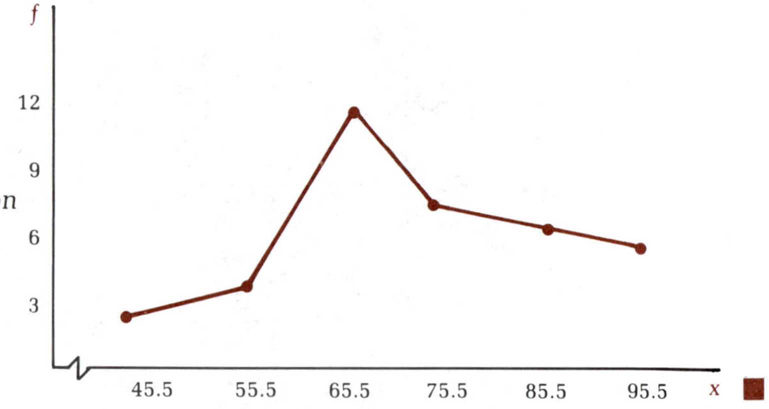

It is possible to construct **cumulative frequency polygons,** sometimes referred to as **ogives** (ō′jīves). For ungrouped data, no change is necessary in the set-up for a regular frequency polygon. The ogive will be nondecreasing.

When grouped data are worked with, however, a slight change is necessary. Since we are working with cumulative frequency, the value used for a class should be a value that actually accumulates all of the data for the class. Hence, we use the right end point of each class. In addition, the left end point of the first class (where the *cf* is zero) is used as a starting point. The following example illustrates.

■ Example 10. Construct cumulative frequency polygons for the data of examples 8 and 9.

Solutions:

Example 8

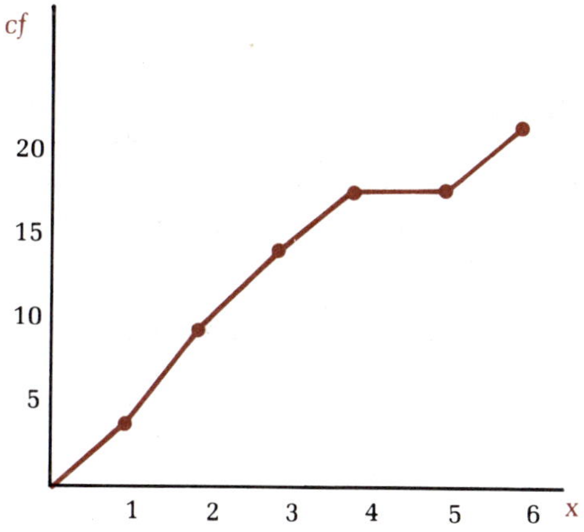

Example 9

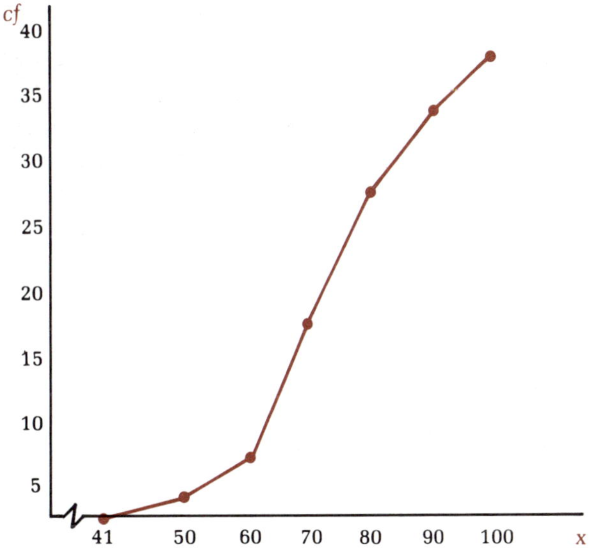

Relative frequency polygons are constructed in a similar manner, except the vertical axis corresponds to relative frequencies and, hence, only goes up to 1. Cumulative relative frequency polygons can also be constructed.

A second type of graph that is frequently used is the **histogram.** Histograms use rectangles above the values, or classes, where the height of the rectangle (vertical axis) is the frequency. The rectangles are of uniform width and are side by side (touching).

■ Example 11. Construct a histogram from the following data involving flipping 4 coins 30 times and observing the number of heads.

Number of heads (x)	f
0	2
1	7
2	10
3	8
4	3

Histogram

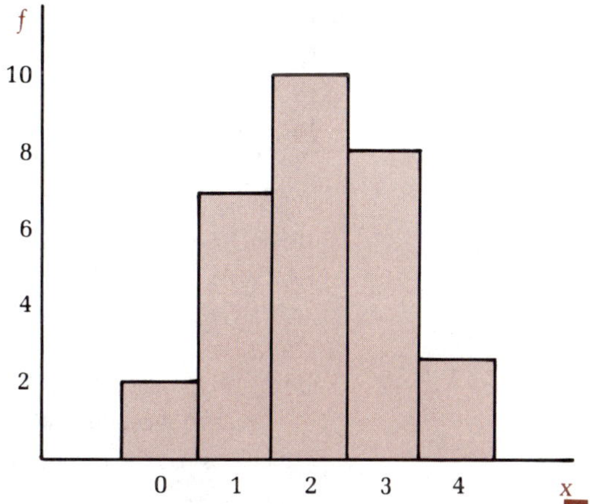

NOTE: The rectangles are centered on the possible values and are of equal width. ■

When grouped data are used, it is convenient to use slightly different intervals. For the distribution of grades given in example 9, since grades are to the nearest whole number, we construct classes in terms of tenths. Appropriate intervals would then be 40.5 to 50.5, 50.5 to 60.5, and so forth, up to 90.5 to 100.5. Using this idea, we can construct a histogram for the grade distribution data of example 9.

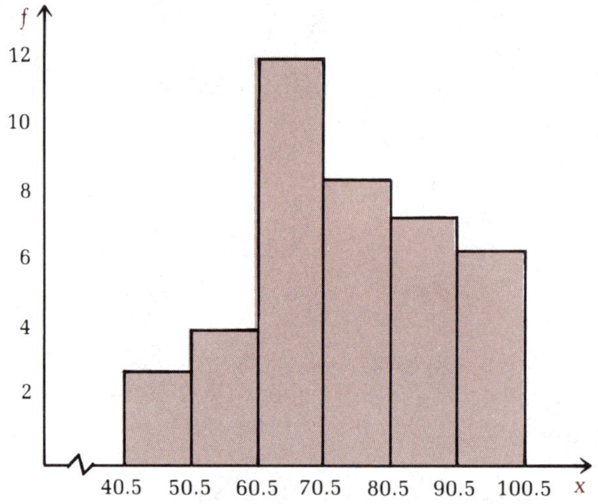

NOTE: The rectangles are the same width and are bounded by the end points of the classes. The rectangles are centered over the class marks.

Pictorial representations of data, such as tables and graphs, can be used to make observations about the data. It is fairly easy to see from either where concentrations of data occur. Graphs can be useful for deciding whether the distribution of values is about the same on both sides of some value (symmetrical) or has a concentration at one end and then tapers off toward the other (skewed).

It is good to remember, however, that tables and graphs are pictures. As is typical in mathematics, pictures help us see relationships, but we must be careful and not try to rely solely on the pictures. In succeeding sections of this chapter, we will consider ways of finding out additional information for describing data that have a firmer mathematical base than do graphs and tables. When we put all of these concepts together, we will be able to provide good descriptions of data.

When you see a graph of statistical data, note carefully the scale on the axes and the starting point. It is possible to construct graphs so that they are intentionally misleading. Facts about data can be glorified or camouflaged by making appropriate changes in units or starting points or both. The units on one axis may be drastically different in size from those on the other axis. That is, 1 unit on one axis may be the same length as 100 units on the other axis. To illustrate how graphs can be misleading, consider the following situation.

The sales manager of an auto dealership has a dilemma. The owner is upset about decreasing sales over the past 7 years, and now the Internal Revenue Service is also showing an interest in sales for the same period. The sales manager must prepare a graph showing sales for the years 1975 through 1981 to present to both the owner and the IRS auditor who is coming. He decides he should prepare separate graphs. The one for the owner should camouflage the decrease in sales, whereas the one for the IRS should exaggerate the decrease. Construct frequency polygons for the owner and the IRS based on the following sales figures.

Year	1975	1976	1977	1978	1979	1980	1981
Sales	410	390	350	300	310	275	225

For the owner

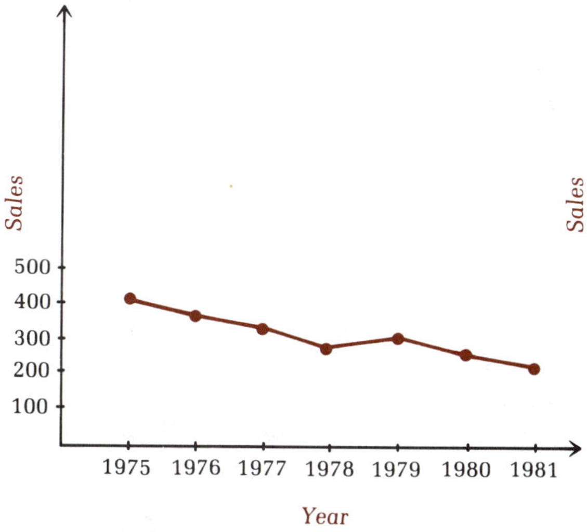

For the IRS

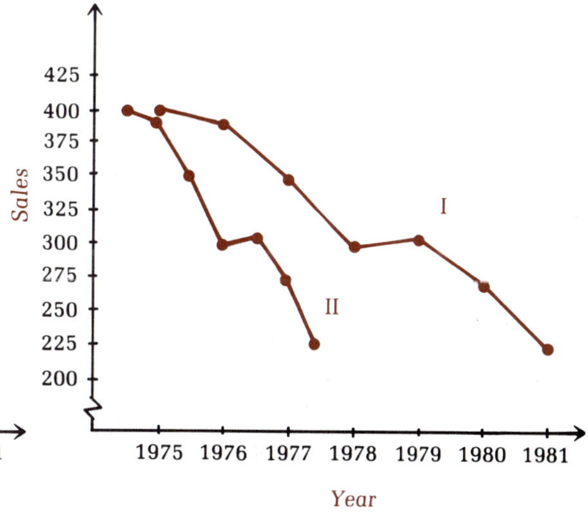

In the graph for the IRS, we have included two possibilities. Frequency polygon I uses the same horizontal scale as the graph for the owner. Frequency polygon II uses a smaller unit (actually one half of that used for I). The vertical scale for both IRS graphs uses a unit four times as large as that used for the owner's graph and doesn't start at zero.

Note that we can exaggerate the decrease by choosing smaller units on the horizontal scale or larger units on the vertical scale or both. The decrease can be camouflaged by choosing larger units on the horizontal scale or smaller units on the vertical scale or both.

Exercise Set 7.1

1. Five coins are flipped simultaneously, and the number of heads is observed. The experiment is performed 25 times, with the following numbers of heads recorded: 3, 4, 2, 2, 1, 4, 3, 1, 0, 2, 2, 4, 5, 3, 5, 4, 5, 2, 1, 3, 0, 2, 1, 4, and 3. Construct a table showing the frequency and cumulative frequency and draw both the frequency polygon and the cumulative frequency polygon.
2. An eight-sided (octahedron) die is tossed 30 times with the following sequence of faces showing: 6, 3, 1, 4, 4, 5, 7, 8, 2, 3, 2, 1, 5, 8, 7, 5, 6, 4, 3, 1, 1, 2, 2, 2, 4, 3, 1, 4, 7, and 8. Construct a table showing the frequencies and cumulative frequencies. Draw a frequency polygon and a cumulative frequency polygon.
3. Construct a histogram for the data of problem 1.
4. Construct a histogram for the data of problem 2.
5. For the data of problem 1, construct a relative frequency polygon.
6. Construct a relative frequency polygon for the data of problem 2.
7. Use grouped data to construct a table showing frequency and cumulative frequency for the following exam scores (100 possible on the exam): 83, 92, 64, 51, 49, 87, 75, 77, 62, 78, 71, 76, 27, 59, 81, 97, 83, 75, 77, 77, 76, 84, 95, 39, 69, 81, 53, 87, 73, 77, 69, 42, 75, 87, 91, 71, 70, 65, 85, and 97.
8. The I.Q.s of 30 children, measured by a particular exam, are 111, 123, 84, 79, 97, 135, 101, 107, 97, 103, 115, 106, 83, 97, 99, 113, 121, 87, 117, 83, 95, 78, 99, 141, 106, 102, 93, 89, 97, and 108. Construct a table showing frequencies and cumulative frequencies using grouped data.
9. Construct a frequency polygon, a cumulative frequency polygon, and a histogram for the data of problem 7.
10. Construct a frequency polygon, a cumulative frequency polygon, and a histogram for the data of problem 8.
11. Use the data in the table below to construct a frequency polygon and a histogram for the weights of 125 elementary school children, in pounds.

Weight (x_i)	44	45	46	47	48	49	50	51	52	53	54
f_i	1	0	2	5	9	12	31	33	20	7	5

12. A group of 150 adults is asked to complete a certain task. The times, in minutes, together with frequencies, are given in the following table. Construct a frequency polygon and a histogram for the data.

Time (t_i)	1	2	3	4	5	6	7
f_i	1	0	18	39	41	34	17

13. A soft drink machine is tested periodically to determine if it is functioning within certain limits. The table gives grouped data for 100 samples measured to the nearest hundredth of an ounce. Construct a relative frequency polygon and a relative frequency histogram for the data.

Class	f_i
5.00–5.49	1
5.50–5.99	1
6.00–6.49	2
6.50–6.99	5
7.00–7.49	13
7.50–7.99	36
8.00–8.49	31
8.50–8.99	10
9.00–9.49	1

14. A psychologist is studying the length of time it takes mice to go through a maze and reach the food at the other end. Time is measured to the nearest tenth of a second. A group of 200 mice is tested with the results given in the table. Use relative frequencies to construct a frequency polygon and a histogram for the data.

Time	f_i
2.0–2.9	3
3.0–3.9	7
4.0–4.9	15
5.0–5.9	29
6.0–6.9	81
7.0–7.9	50
8.0–8.9	10
9.0–9.9	5

7.2 Summation Notation

The study of descriptive statistics frequently involves adding up lists of numbers. This section will consider a notational convention that is commonly used as a short form of representing sums. The symbol used to represent sums is Σ, the capital Greek letter sigma. If it is desired to represent the sum $a_1 + a_2 + a_3 + \ldots + a_n$, we write $\sum_{i=1}^{n} a_i$. This is read "the sum from $i = 1$ to $i = n$ of a_i". Successive terms in the sum are obtained by replacing i by the numbers from 1 to n. We have then

$$\sum_{i=1}^{n} a_i = a_1 + a_2 + a_3 + \ldots + a_n.$$

In some situations, a_i is given in terms of i ($a_i = 2i$, $a_i = i + 1$, and so on), and sometimes a_i is just used to represent specific numbers. Suppose a student took 5 exams in a class and had scores of 67, 81, 77, 92, and 83 respectively. Then the total number of "points" for the student can be represented by $\sum_{i=1}^{5} a_i$ where a_i is the score on the ith exam.

■ **Example 12.** Evaluate $\sum_{i=1}^{5} a_i$ for the exam scores in the preceding paragraph.

Solution:

$$\sum_{i=1}^{5} a_i = 67 + 81 + 77 + 92 + 83 = 400.$$

■

■ **Example 13.** Find $\sum_{i=1}^{5} a_i$, if $a_i = 2i$.

Solution:

$$\begin{aligned}\sum_{i=1}^{5} a_i = \sum_{i=1}^{5} 2_i &= (2 \cdot 1) + (2 \cdot 2) + (2 \cdot 3) + (2 \cdot 4) + (2 \cdot 5)\\ &= 2 + 4 + 6 + 8 + 10\\ &= 30.\end{aligned}$$

■

■ **Example 14.** Find $\sum_{i=1}^{8} (i + 1)$.

Solution:

$$\begin{aligned}\sum_{i=1}^{8} (i + 1) &= (1 + 1) + (2 + 1) + (3 + 1) + (4 + 1)\\ &\quad + (5 + 1) + (6 + 1) + (7 + 1) + (8 + 1)\\ &= 2 + 3 + 4 + 5 + 6 + 7 + 8 + 9\\ &= 44.\end{aligned}$$

NOTE: We could have written

$$\sum_{i=1}^{8} (i + 1) = 2 + 3 + 4 + 5 + 6 + 7 + 8 + 9 = 44$$

and left out the first step. When you can do the simple arithmetic in your head, it shortens your work. ■

■ **Example 15.** Find $\sum_{i=1}^{6} a_i$ where $a_i = 2$ for every i.

Solution: Since a_i is always 2, we have

$$\sum_{i=1}^{6} a_1 = \sum_{i=1}^{6} 2 = 2 + 2 + 2 + 2 + 2 + 2 = 12.$$

NOTE: $\sum_{i=1}^{6} 2 = 12 = 6 \cdot 2 = 6 \cdot a_i$ where 6 is the upper limit of the sum. ■

When we work with summation notation, $\sum_{i=k}^{m}$, i is the index of summation (any other letter could be used instead of i). The number k is the lower limit of the sum (the starting point) and the number m is the upper limit of the sum (stopping point). Any letters not involving the index of summation are treated as constants.

■ **Example 16.** Find $\sum_{i=1}^{5} (ai + b)$.

Solution:

$$\sum_{i=1}^{5} (ai + b) = (a + b) + (2a + b) + (3a + b) + (4a + b) + (5a + b)$$
$$= 15a + 5b$$ ■

There are several properties of summation that help simplify computation. We state here, without proof, some of the more useful ones. The proofs are not difficult, but they are not necessary for our discussion.

> **Property 7.1** If c is a constant, then $\sum_{i=1}^{n} c = nc$.

■ **Example 17.** Find $\sum_{i=1}^{10} 5$.

Solution: $\sum_{i=1}^{10} 5 = 10 \cdot 5 = 50.$ ■

> **Property 7.2** If c is a constant, then $\sum_{i=1}^{n} ca_i = c \sum_{i=1}^{n} a_i$.

■ **Example 18.** Find **a.** $\sum_{i=1}^{10} 3a_i$. **b.** $\sum_{i=1}^{5} 7i$.

Solutions: **a.** $\sum_{i=1}^{10} 3a_i = 3\sum_{i=1}^{10} a_i$.

b. $$\begin{aligned}\sum_{i=1}^{5} 7i &= 7\sum_{i=1}^{5} i\\ &= 7(1 + 2 + 3 + 4 + 5)\\ &= 7 \cdot 15 = 105.\end{aligned}$$ ■

Property 7.3 $\sum_{i=1}^{n} (a_i + b_i) = \sum_{i=1}^{n} a_i + \sum_{i=1}^{n} b_i$.

■ **Example 19.** Find $\sum_{i=1}^{7} (2i + 3)$.

Solution:

$$\begin{aligned}\sum_{i=1}^{7} (2i + 3) &= \sum_{i=1}^{7} 2i + \sum_{i=1}^{7} 3 && \text{By property 7.3}\\ &= 2\sum_{i=1}^{7} i + \sum_{i=1}^{7} 3 && \text{By property 7.2}\\ &= 2\sum_{i=1}^{7} i + 7(3) && \text{By property 7.1}\\ &= 2(1 + 2 + 3 + 4 + 5 + 6 + 7) + 21\\ &= 2 \cdot 28 + 21\\ &= 56 + 21 = 77\end{aligned}$$ ■

Two special sums are worth including here. They help simplify some computations.

Property 7.4 **a.** $\sum_{i=1}^{n} i = \frac{n(n + 1)}{2}$. **b.** $\sum_{i=1}^{n} i^2 = \frac{n(n + 1)(2n + 1)}{6}$.

Combinations of these properties, as illustrated in the preceding example, can sometimes be used to simplify computations involving complicated looking sums. We close this section with one final example.

■ Example 20. Evaluate $\sum_{i=1}^{10} (2i - 3)^2$.

Solution:

$$\begin{aligned}\sum_{i=1}^{10} (2i - 3)^2 &= \sum_{i=1}^{10} (4i^2 - 12i + 9)\\ &= \sum_{i=1}^{10} 4i^2 - \sum_{i=1}^{10} 12i + \sum_{i=1}^{10} 9 && \text{By property 7.3}\\ &= 4\sum_{i=1}^{10} i^2 - 12\sum_{i=1}^{10} i + 90 && \text{By properties 7.2 and 7.1}\\ &= 4 \cdot \frac{10(11)(21)}{6} - 12\,\frac{10(11)}{2} + 90 && \text{By property 7.4}\\ &= 1540 - 660 + 90\\ &= 970\end{aligned}$$

■

Exercise Set 7.2

In exercises 1–20, evaluate the sum.

1. $\sum_{i=1}^{100} i$
2. $\sum_{i=1}^{18} (2i + 1)$
3. $\sum_{i=1}^{60} (3i - 2)$
4. $\sum_{i=1}^{20} (i^2 + 2i - 2)$
5. $\sum_{i=1}^{80} i^2$
6. $\sum_{i=1}^{6} (i^2 - 2)$
7. $\sum_{i=1}^{10} (i - 5)^2$
8. $\sum_{i=1}^{6} (i + 3)^2$
9. $\sum_{i=0}^{10} (2i + 1)$
10. $\sum_{i=0}^{5} (i - 3)^2$
11. $\sum_{i=1}^{10} \frac{(i - 1)^2}{10}$
12. $\sum_{i=0}^{7} \frac{(i + 2)^2}{3}$
13. $\sum_{i=1}^{50} 2i^2$
14. $\sum_{i=1}^{25} i(i + 1)$
15. $\sum_{i=1}^{5} (i - 1)^3$
16. $\sum_{i=1}^{10} (2i + 1)$
17. $\sum_{i=1}^{5} (i^3 - 2i^2 + 1)$
18. $\sum_{i=1}^{200} i$

19. $\sum_{i=1}^{20} 5$

20. $\sum_{i=1}^{5} \left(\frac{1}{2}\right)^i$

7.3 Numerical Descriptions of Data, I: Measures of Central Tendency

When analyzing data, the more information that is available, the more complete the analysis will be. Tables and graphs are useful as pictorial representations but do not, in general, provide precise information. In this section, and succeeding sections, numerical descriptors of data will be considered. These numerical descriptors help us with a more complete analysis of the data at hand, and provide a concise representation of data.

Three commonly used numerical descriptors are referred to as **measures of central tendency.** The idea is to give a numerical value to the middle, average, or common values of the data. Some books refer to all three of these measures as "averages" and, hence, the word average is not used specifically for any one of them. This is the approach we will use. As we discuss these measures we will point out some of the major strengths and weaknesses of each.

Perhaps the simplest of the measures to determine is the **mode.** The mode is the most common data value, i.e., the one that occurs most frequently. It is possible to have more than one mode. The following examples will help illustrate.

■ Example 21. A die is tossed 30 times. The table below gives the frequency with which each face appeared. Find the mode.

face (x)	1	2	3	4	5	6
f	2	5	6	8	5	4

Solution: Since 4 occurred the highest number of times, the mode is 4. ■

■ Example 22. A die is tossed 30 times with the results given in the following table. Find the mode.

face (x)	1	2	3	4	5	6
f	7	5	2	7	5	4

Solution: Since 1 and 4 each occurred 7 times and no other value occurred that often, 1 and 4 are both modes. ■

If two modes occur, the distribution is called **bimodal.** It is also possible to have more than two modes.

■ Example 23. In a class of 15 students, the scores on an exam were 83, 87, 72, 91, 95, 68, 84, 79, 77, 21, 73, 64, 93, 83, and 75. Find the mode.

Solution: Since the only score that occurred more than once was 83, the mode is 83. ■

Suppose in the previous example that the first 83 was an 82. What is the mode? Since each score occurs exactly the same number of times (1), no meaningful information is obtained by considering the frequency of scores. Hence, we will say there is no mode.

This is an illustration of a weakness of the mode. That is, the mode may not exist. We have also seen that the mode is not necessarily unique when it does exist (example 22).

Example 23 also points out another disadvantage of the mode. The score of 21 is an extreme value with no other value close to it. In general, extreme values have no effect on the mode. If you are told that the modal salary at a factory is $12,000 yearly, no information about extremes is included.

The mode does, however, provide some information about concentrations of data. If you own a shoe store, knowledge about the modal size or sizes in your area could certainly help you when you are placing an order.

The mode, then, is unaffected by extreme values, may not be unique, and may not reflect the middle of the data. In example 23, if one of the scores of 83 had been 21, then 21 would have been the mode. This value is in no way indicative of the middle of the scores.

The **median** is a measure of central tendency that truly reflects the middle of the scores. The median is the value such that the same number of values is above it as below it. To find the median, the values need to be arranged in either ascending or descending order.

If the number of data values is odd, the median is the actual middle value. When the number of values is even, the median is halfway between the two middle values.

■ Example 24. Find the median for the exam scores of example 23.

Solution: To start, we place the scores in order. We use ascending order, although descending order would also work: 21, 64, 68, 72, 73, 75, 77, 79, 83, 83, 84, 87, 91, 93, 95

Since the number of scores is 15 (odd), the median is the eighth score from the bottom (or top). Hence, the median is 79. Note that there are 7 scores above and 7 scores below the median. ■

■ **Example 25.** Find the median for the scores 61, 59, 81, 87, 93, 94, 47, 62, 68, and 73.

Solution: First arrange the scores in numerical order: 47, 59, 61, 62, 68, 73, 81, 87, 93, 94

The number of scores is 10 (even), so we first find the 2 middle scores by locating the fifth (10 ÷ 2 = 5) and sixth scores from the top or bottom. These scores are 68 and 73. The value 70.5 is halfway between 68 and 73, so 70.5 is the median. There are 5 scores above 70.5 and 5 scores below 70.5.

To find the score halfway between, we take the arithmetic average of the 2 middle scores. That is, we add the scores and divide by 2.

$$\frac{68 + 73}{2} = \frac{141}{2} = 70.5$$ ■

As with the mode, the median is not affected by extreme values. If the 21 in example 24 had been a 0 or a 78, the median would still have been 79.

The median will, however, be unique. If the number of data values is odd, the median will be one of the values. If the number of data values is even, the median may, or may not, be one of the data values. If the two middle values are equal, then their average is the same value, and the median is this value.

One major advantage of the median is its indication of the middle of the scores. If 45 students take an exam and the mode is 73, little knowledge is gained about the distribution of the scores. If, on the other hand, the median is 73, we know that about the same number of scores were above and below 73.

It should not always be construed as a disadvantage that the median is unaffected by extremes. If one value is way out of line with the remaining values, the median may still be a good representation of the central tendency of the data.

■ **Example 26.** A small company has nine employees. Their yearly salaries are \$9,000, \$10,000, \$11,000, \$12,000, \$13,000, \$14,000, \$14,000, \$14,000, and \$200,000. Find the mode and median.

Solution: The mode is \$14,000, and the median is \$13,000. The median is fairly indicative of the salaries, except for the one extreme. The median and mode together provide more information than either measure by itself. Remember, we want to obtain as much information as possible for our data analysis. ■

The final measure of central tendency to be considered is called the **mean.** The mean is the arithmetic average and is found by adding together all of the data values and dividing by the total number of data values.

Symbolically, we use either μ (small Greek letter mu) or $\bar{x}$ to represent the mean. The symbol μ is used for the mean of an entire population, and $\bar{x}$ is used when only a portion (*sample*) of the population is being considered.

If $x_1, x_2, x_3, \ldots, x_N$ are data values from a population of size N, then the mean is computed by

$$\mu = \frac{\sum_{i=1}^{N} x_i}{N}.$$

If a sample of size n is being considered, the mean is given by

$$\bar{x} = \frac{\sum_{i=1}^{n} x_i}{n}.$$

The concept of a *sample* plays a central role in statistics. Some populations are so large that it is not feasible to use the entire population, so a sample is used. In other situations, such as testing flash bulbs, the object is destroyed, and, hence, testing the entire population would destroy the population. Samples, then, must be used.

To compute the mean, it really doesn't matter whether we are working with a population or a sample, since the computational formulas are basically the same. However, in some situations involving other statistical calculations, as we will see later, there is a slight difference in the computational formulas, and the distinction needs to be clear.

To illustrate the mean, we consider some examples. It is sometimes interesting, and informative, to find the median and mode also and compare the three values.

■ Example 27. The 15 students in a class had exam grades of 21, 64, 68, 72, 73, 75, 77, 79, 83, 83, 84, 87, 91, 93, and 95. Find the mean.

Solution: It doesn't matter if we use μ or $\bar{x}$. We will use μ.

$$\mu = \frac{\sum_{i=1}^{15} x_i}{15} = \frac{21 + 64 + 68 + 72 + 73 + 75 + 77 + 79 + 83 + 83 + 84 + 87 + 91 + 93 + 95}{15}$$

$$= \frac{1145}{15} = 76.3\text{, or about } 76.$$

This value doesn't differ greatly from the median (79) or mode (83). The difference can be attributed, at least partially, to the extreme value of 21. If 21 is not included, the mean of the remaining 14 scores is about 80.3, which is even closer to the median of 81 $\left(\frac{79 + 83}{2} = \frac{162}{2} = 81\right)$. ■

The mean, unlike the mode and median, uses each data value. Hence, the mean is influenced by extreme values. The next example will help to illustrate.

■ **Example 28.** The yearly salaries of the nine employees of a company are \$9,000, \$10,000, \$11,000, \$12,000, \$13,000, \$14,000, \$14,000, \$14,000, and \$200,000. Find the mean and compare this with the median and mode from example 26.

Solution:

$$\mu = \frac{\sum_{i=1}^{9} x_i}{9}$$

$$= \frac{9{,}000 + 10{,}000 + 11{,}000 + 12{,}000 + 13{,}000 + 14{,}000 + 14{,}000 + 14{,}000 + 200{,}000}{9}$$

$$= \frac{297{,}000}{9} = \$33{,}000$$

The mean is more than twice as large as either the mode (\$14,000) or the median (\$13,000). This can be attributed to the one extreme value of \$200,000. ■

Which of the three measures provides the best information about the data? The median is probably the best choice in example 28. Sometimes the mode or mean is more informative. The measure used is somewhat dependent on the aspect being investigated. For most statistical work, however, the mean is the measure used.

When working with large quantities of data where specific values occur several times, we find that a slight modification of the formula for the mean is convenient. Suppose the values are $x_1, x_2, x_3, \ldots, x_k$ and they occur with frequencies of $f_1, f_2, f_3, \ldots, f_k$ respectively. Then the mean is computed by $\mu = \frac{\sum_{i=1}^{k} f_i x_i}{\sum_{i=1}^{k} f_i}$ or $\bar{x} = \frac{\sum_{i=1}^{k} f_i x_i}{\sum_{i=1}^{k} f_i}$. If, for example, a die is tossed 30 times and a 3 occurs 6 times, instead of using 3 + 3 + 3 + 3 + 3 + 3, we use $6 \cdot 3$. Both are equal to 18. Note that $\sum_{i=1}^{k} f_i$ is the total number of data values. That is, $\sum_{i=1}^{k} f_i = N$ for a population or $\sum_{i=1}^{k} f_i = n$ for a sample.

■ Example 29. A die is tossed 30 times with the results given below. Find the mean.

face	1	2	3	4	5	6
f	2	5	6	4	4	9

Solution:

$$\mu = \frac{\sum_{i=1}^{6} f_i x_i}{\sum_{i=1}^{6} f_i}$$

$$= \frac{\sum_{i=1}^{6} f_i x_i}{30}$$

$$= \frac{2 \cdot 1 + 5 \cdot 2 + 6 \cdot 3 + 4 \cdot 4 + 4 \cdot 5 + 9 \cdot 6}{30}$$

$$= \frac{120}{30} = 4$$

■

As seen in section 7.1, data are sometimes given in grouped form. That is, instead of being given the frequency for each specific value, we are given the frequency for an interval, or class. It is possible to calculate an approximate mean for this type of data.

The value we obtain is approximate, since we actually work with approximate data. To illustrate, suppose one class is 60 to 69, and the frequency for this class is 5. We don't know exactly what each of the 5 values is, so we pick a "representative" for the class and treat the data as if each of the 5 values was this "representative." The value we choose is the class mark, or midpoint, of the class.

If the actual data are spread out nicely through the classes, the class marks will serve as good representatives, and the value we obtain for the mean will be a good approximation. However, if the data pile up near the extremes of the classes, our approximation may not be as good. We always hope that the information tends to balance itself. That is, if the data in one class are piled up near one end, we hope this is balanced by the data in other classes.

When working with grouped data, or any data involving frequencies, we often find it helpful to construct a table. The table has columns to represent the class (or data values for ungrouped data), the frequencies (f_i), the class marks (x_i), and the product of the frequencies and class marks ($f_i x_i$). Tables of this type will be especially useful when we consider other statistical measures in the next section.

A hand-held calculator is also a useful tool for statistical computations. This is true for computing means as well as for computing other statistical measures to be discussed in the next sections. To be most effective, the calculator should include a square root key. Memory is also useful but is not as important in this work as the square root.

The following examples will help illustrate calculations for frequency data. The first is for ungrouped data, and the second is for grouped data.

Example 30. Thirty children are asked to complete a certain task, and the time, in minutes, each child takes is recorded. The data are given in the table. Find the mean time.

Time (x_i)	4	5	6	7	8	9	10
f_i	1	3	2	1	10	8	5

Solution: We begin by setting up a table.

Time (x_i)	f_i	$f_i x_i$
4	1	4
5	3	15
6	2	12
7	1	7
8	10	80
9	8	72
10	5	50
Totals (Σ)	30	240

The mean is $\bar{x} = \dfrac{\sum_{i=1}^{7} f_i x_i}{n} = \dfrac{240}{30} = 8.$ ■

If we include the last row of the table (totals), then the information needed for the calculation of $\bar{x}$ or μ is a part of the table. The mean time for completion of the task is 8 minutes.

Example 31. A local radio station has been keeping precipitation figures for 50 years. The monthly precipitation is recorded to the nearest tenth of an inch. The data for the month of September are given in the following table. Find the September mean precipitation.

Precipitation (inches)	f_i
0.0–0.9	1
1.0–1.9	5
2.0–2.9	4
3.0–3.9	9
4.0–4.9	20
5.0–5.9	4
6.0–6.9	7

Solution: We set up a table, including columns for the class marks (x_i) and the products ($f_i x_i$).

Class	f_i	x_i	$f_i x_i$
0.0–0.9	1	0.45	0.45
1.0–1.9	5	1.45	7.25
2.0–2.9	4	2.45	9.80
3.0–3.9	9	3.45	31.05
4.0–4.9	20	4.45	89.00
5.0–5.9	4	5.45	21.80
6.0–6.9	7	6.45	45.15
Totals (Σ)	50		204.50

The mean is then computed to be

$$\mu = \frac{\sum_{i=1}^{7} f_i x_i}{\sum_{i=1}^{7} f_i} = \frac{204.50}{50} = 4.09.$$

The mean is about 4.1 inches.

NOTE: The total (Σ) was not found for the x_i column. This total would have no real meaning. ■

How good is this approximation to the mean? We really don't know unless we have the actual data. In this case, the data values used by the authors actually had a mean of 4.16, which would round off to 4.2. The approximation from the grouped data is fairly close.

It is sometimes helpful, for computational purposes, to adjust data values or class marks prior to computation. That is, we add or subtract a constant to each value or multiply or divide each value by the same constant. This process is called **coding.**

How does coding affect the mean? We will consider only what happens when a constant is subtracted from each value, since this is perhaps the most common form of coding.

Suppose that we have data values, or class marks, $x_1, x_2, x_3, \ldots, x_n$. The mean is

$$\bar{x} = \frac{\sum_{i=1}^{n} x_i}{n}.$$

Suppose we subtract a constant, say M, from each value and call the new values $y_1, y_2, \ldots, y_n$. Hence, $y_1 = x_1 - M$, $y_2 = x_2 - M, \ldots,$ $y_n = x_n - M$ and

$$\bar{y} = \frac{\sum_{i=1}^{n} y_i}{n}.$$

How does $\bar{y}$ compare with $\bar{x}$. We use here some of the properties of summation.

$$\bar{y} = \frac{\sum_{i=1}^{n} y_i}{n} = \frac{\sum_{i=1}^{n} (x_i - M)}{n} \qquad y_i = x_i - M$$

$$= \frac{\sum_{i=1}^{n} x_i - \sum_{i=1}^{n} M}{n} \qquad \text{By property 7.3}$$

$$= \frac{\sum_{i=1}^{n} x_i - nM}{n} \qquad \text{By property 7.1}$$

$$= \frac{\sum_{i=1}^{n} x_i}{n} - M \qquad \frac{nM}{n} = M$$

$$= \bar{x} - M \qquad \text{By definition of } \bar{x}$$

Since $\bar{y} = \bar{x} - M$, it follows that $\bar{x} = \bar{y} + M$. After computing the mean for the coded values, we add the constant M to this mean to get the mean of the original data.

■ **Example 32.** Consider the data of example 31. Code this data by subtracting 0.45 from each class mark and compute the mean.

Solution:

Class	f_i	x_i	$y_i = x_i - 0.45$	$f_i y_i$
0.0–0.9	1	0.45	0	0
1.0–1.9	5	1.45	1	5
2.0–2.9	4	2.45	2	8
3.0–3.9	9	3.45	3	27
4.0–4.9	20	4.45	4	80
5.0–5.9	4	5.45	5	20
6.0–6.9	7	6.45	6	42
Totals (Σ)	50			182

$\bar{y} = \frac{182}{50} = 3.64$, so $\bar{x} = 3.64 + 0.45 = 4.09$. The coded values are easier to work with, particularly if no calculator is available.

The value 0.45 that was subtracted from each class mark was somewhat arbitrary, although it did give y_i values that are all whole numbers. Any other value could have been used. Had we chosen 4.45 instead of 0.45, the computation would appear as follows.

Class	f_i	x_i	$y_i = x_i - 4.45$	$f_i y_i$
0.0–0.9	1	0.45	−4	−4
1.0–1.9	5	1.45	−3	−15
2.0–2.9	4	2.45	−2	−8
3.0–3.9	9	3.45	−1	−9
4.0–4.9	20	4.45	0	0
5.0–5.9	4	5.45	1	4
6.0–6.9	7	6.45	2	14
Totals (Σ)	50			−18

$$\bar{y} = \frac{-18}{50} = -0.36, \text{ so } \bar{x} = -0.36 + 4.45 = 4.09.$$ ■

We conclude this section with some final comments about mode, median, and mean. The mode is the easiest of the measures to calculate. However, the mode is not affected by either the number of values or the extreme values. The mode can be used for nonnumerical data. Suppose, for example, 100 people are asked if they approve or disápprove or have no opinion regarding the president's foreign policy. If the results show that 55 approve, 35 disapprove, and 10 have no opinion, then the modal response is "approve." The median and mean would have no meaning in this case.

The median, unlike the mode, does consider the total number of values. Like the mode, the median is unaffected by extremes. If there are only one or two extremes, the median is a fairly good estimate of the center of the data. The median requires that the data be in ascending or descending order and is typically used for numerical data. Although the mode may not be unique and may not be close to the center of the data, the median is unique and is in the center of the data in terms of the number of data values.

The mean, which is the most commonly used of the three measures in statistical analysis, uses the actual data values. Hence, it is affected by both the number of values and any extremes. To compute the mean, we need numerical data values. In some situations, nonnumerical data can be assigned numerical values for the purpose of computing a mean. If, for instance, a questionnaire allows responses of "strongly agree," "agree," "no opinion," "disagree," or "strongly disagree," the responses could be assigned numerical values such as 1, 2, 3, 4, and 5 (or others), and a mean response could be obtained. In this case, a mean of 3.8 could be interpreted as "leaning toward disagree."

The mean, like the median, is a unique value. There are, however, situations involving numerical data in which the mean doesn't exist. The following example will serve as an illustration.

■ Example 33. Fifty people are given a task to perform. The researcher is interested in the mean time to complete the task. However, there is a 15-minute time limit. Any person not finished in 15 minutes is assigned a time "greater than 15 (>15)." Discuss the mean time for the data in the table.

Time (t_i)	f_i
10	9
11	7
12	19
13	0
14	8
15	1
>15	6

Solution: To compute the mean, we use

$$\bar{x} = \frac{\sum_{i=1}^{7} f_i t_i}{\sum_{i=1}^{7} t_i}.$$

But what value do we use for t_i for the six people who were not finished? Since the mean is affected by extremes, some of these six actual values could change the mean significantly if they are quite large. No true mean can be found for this data. ■

Exercise Set 7.3

For exercises 1–6, find the mean, median, and mode or modes for the given data.

1. A class had exam scores of 63, 87, 91, 72, 75, 69, 47, 71, 82, 67, 69, 84, 72, 75, 69, 75, 83, 93, 51, and 70.
2. A pair of dice are tossed 20 times, with the following sums of the faces recorded: 7, 8, 6, 2, 5, 3, 8, 11, 10, 11, 5, 7, 3, 12, 5, 5, 7, 6, 7, and 8.
3. The monthly salaries of the employees of a certain company are $850, $800, $1200, $950, $1100, $1500, $900, $900, $1350, $1450, $950, $900, $1750, $1650, and $1150.
4. The times, in seconds, that it takes a group of rats to go through a maze are 63, 71, 91, 43, 51, 47, 81, 69, 55, 44, 61, 82, 90, 59, and 65.

5. The net monthly earnings for a small store for a one-year period were \$3000, \$2500, \$2900, \$1900, \$2600, \$2300, \$2400, \$3200, \$2100, \$2200, \$3100, and \$3900.
6. The number of new cars sold per month for an auto dealer for a one-year period were 11, 9, 15, 20, 23, 21, 23, 17, 33, 30, 19, and 18.

In exercises 7–14, find the mean for the frequency data given in the table.

7. **Height in inches of first-graders**

x_i	f_i
47	7
48	3
49	17
50	13
51	10

8. **Weight in pounds of first-graders**

x_i	f_i
44	2
45	5
46	4
47	7
48	17
49	10
50	2
51	3

9. **Sum of faces of a pair of dice**

x_i	f_i
2	5
3	8
4	9
5	13
6	20
7	30
8	15
9	7
10	7
11	4
12	2

10. **Monthly car sales**

x_i	f_i
20	19
21	11
22	15
23	12
24	11
25	12
26	10
27	10
28	9
29	6
30	5

11. **IQ scores of third-graders**

Class	f_i
70–79	3
80–89	7
90–99	13
100–109	21
110–119	15
120–129	5
130–139	4
140–149	2

12. **Weekly mileage traveled by company sales people**

Class	f_i
500–599	7
600–699	11
700–799	12
800–899	12
900–999	15
1000–1099	12
1100–1199	11
1200–1299	9
1300–1399	8
1400–1499	3

13. **Amount of soft drink dispensed by a machine in hundredths of an ounce**

Class	f_i
5.00–5.49	1
5.50–5.99	1
6.00–6.49	2
6.50–6.99	5
7.00–7.49	13
7.50–7.99	36
8.00–8.49	31
8.50–8.99	10
9.00–9.49	1

14. **Grade point averages of members of a fraternity**

Class	f_i
0.01–0.50	2
0.51–1.00	5
1.01–1.50	5
1.51–2.00	11
2.01–2.50	18
2.51–3.00	30
3.01–3.50	17
3.51–4.00	8

7.4 Numerical Descriptions of Data, II: Measures of Dispersion

Consider the four sets of data $A = \{60,60,60,60,60\}$, $B = \{40,60,60,60,80\}$, $C = \{20,60,60,60,100\}$, and $D = \{0,20,60,60,160\}$. For each set of data, the mean, median, and mode are 60. Looking at the sets, however, we see that there are some distinct differences. The data in set D are more spread out than in the other three sets. Even though the mean is influenced by extremes, the two low values in D balance the one high extreme.

As indicated by the four sets of data in the preceding paragraph, knowing one or more of the measures of central tendency does not provide a complete description of the data. The central tendency measures do not distinguish between set A, where all data values are equal, and set D, where the data values are more spread out.

Other measures, then, that will identify how data is spread out would be useful for analyzing data. Three such measures of **dispersion,** or spread, are commonly used and will be discussed here.

The simplest measure of dispersion to calculate is the **range.** The range is defined as the difference between the largest data value and the smallest data value. That is, the range is the largest value minus the smallest value. The range is always a nonnegative number.

■ Example 34. Find the range for each of the sets of data A, B, C, and D given at the beginning of this section.

Solution: For set A, the largest and smallest values are both 60, so the range is $60 - 60 = 0$.
For set B, the range is $80 - 40 = 40$.
For set C, the range is $100 - 20 = 80$.
For set D, the range is $160 - 0 = 160$. ■

It is apparent that the larger the range, the more spread out the data values; or, at least, the extremes are more spread out. If the range is very small, then the extremes are close together, and, hence, so are all the data values.

If a company has 100 employees and the range for the annual salaries is \$200, we know that all of the employees have about the same annual salary. On the other hand, if a company has nine employees and the range of annual salaries is \$191,000, we know very little about the individual salaries. We know only that there is a large difference between the lowest and highest salaries.

The range considers only the highest and lowest data values. It does not consider any of the measures of central tendency and provides no information regarding nonextreme values.

Certainly, knowing the range and some of the measures of central tendency would be useful. However, changing one extreme value could make a drastic change in the range without affecting the mode or median at all. If the number of data values is large, the effect of changing one extreme may be drastic for the range but negligible for the mean.

As pointed out in the previous section, one of the advantages of the mean is that it uses all of the data values. It would seem reasonable, then, to consider measures of dispersion that also use all of the data values.

It also makes sense to have measures of dispersion that are somehow related to at least one of the measures of central tendency. Since the mean is the only measure of central tendency that uses all of the data values, it would appear to be the most appropriate choice.

The idea, then, is to measure dispersion, or spread, from the mean. Since we would prefer a single numerical value for the measure, we don't want to consider individually how far each data value is from the mean. The mean is an average, so we consider an **average deviation**, or difference, from the mean. Several types of average deviations could be considered. We will consider several possibilities, but we will finally settle on one of them as the most appropriate for statistical work.

Consider, first, the simplest average deviation from the mean. That is, consider $x_i - \mu$ for each data value x_i. The values $x_i - \mu$ will identify how far each value is from the mean and will also tell which direction (for values greater than the mean, $x_i - \mu$ will be positive, and for values less than the mean, $x_i - \mu$ will be negative) the values are from the mean. The average deviation from the mean will be the total deviation

$$\sum_{i=1}^{N} (x_i - \mu),$$

divided by the number of data values, N. The average (or mean) deviation from the mean is given by

$$\frac{\sum_{i=1}^{N} (x_i - \mu)}{N}.$$

■ Example 35. Find the average deviation from the mean for each of the data sets at the beginning of this section.

Solutions: We will use tables to find the appropriate information. For set A, $\mu = 60$.

x_i	$x_i - \mu$
60	0
60	0
60	0
60	0
60	0
Totals 300	0

$$\frac{\sum_{i=1}^{5} (x_i - \mu)}{5} = \frac{0}{5} = 0$$

For set B, $\mu = 60$,

x_i	$x_i - \mu$
40	−20
60	0
60	0
60	0
80	20
Totals 300	0

$$\frac{\sum_{i-1}^{5} (x_i - \mu)}{5} = \frac{0}{5} = 0$$

For set C, $\mu = 60$.

x_i	$x_i - \mu$
20	−40
60	0
60	0
60	0
100	40
Totals 300	0

$$\frac{\sum_{i-1}^{5} (x_i - \mu)}{5} = \frac{0}{5} = 0$$

For set D, $\mu = 60$,

x_i	$x_i - \mu$
0	−60
20	−40
60	0
60	0
160	100
Total 300	0

$$\frac{\sum_{i=1}^{5} (x_i - \mu)}{5} = \frac{0}{5} = 0$$

■

In each of these, the total deviation, and hence the average deviation, is zero. In fact, theoretically, the total deviation is always zero. This can be easily shown.

$$\sum_{i=1}^{N} (x_i - \mu) = \sum_{i=1}^{N} x_i - \sum_{i=1}^{N} \mu \qquad \text{By property 7.3}$$

$$= \sum_{i=1}^{N} x_i - N\mu \qquad \text{By property 7.1}$$

$$= N\mu - N\mu \qquad \mu = \frac{\sum_{i=1}^{N} x_i}{N} \text{ so } \sum_{i=1}^{N} x_i = N\mu$$

$$= 0$$

The only time the total deviation won't be zero is when the value for μ is obtained by rounding off. Even in these situations, though, the total deviation will be very close to zero. Obviously, this type of measure will not be useful since it doesn't distinguish between widely spread data and data with no spread at all.

The problem with average deviation from the mean is that some of the deviations are positive and some are negative. Hence, they tend to balance each other. To avoid this balancing, we wish to obtain deviations that are all positive or all negative. We actually choose to make them all positive. The deviations are then added and we get larger values for more widely spread data.

One way to make the deviations all positive is to use the absolute value of the deviations. That is, we could use $|x_i - \mu|$, which is never negative. For statistical purposes, however, this technique is not preferred.

The preferred technique in statistics is to work with the squares of the deviations, $(x_i - \mu)^2$, which are never negative. The *average square deviation* is called the **variance**. The *population variance* will be denoted σ^2 (sigma squared), and the *sample variance* will be denoted s^2.

The population variance is computed by the formula

$$\sigma^2 = \frac{\sum_{i=1}^{N} (x_i - \mu)^2}{N}.$$

For statistical reasons, a slight adjustment is made in the formula for sample variance. One explanation for this adjustment is that there is a tendency for the sample variance to underestimate the true variance because extremes may be missed in the sample. This is not, however, the statistical reason for the adjustment.

The sample variance is computed by the formula

$$s^2 = \frac{\sum_{i=1}^{n} (x_i - \bar{x})^2}{n - 1}.$$

Note that we divide by $n - 1$ instead of by n for sample variance.

We will assume that we are working with a population instead of a sample unless stated otherwise. Hence, we will normally divide by N. We will also continue to use tables for computational purposes.

■ Example 36. Compute the variance for the data values

a. 40, 60, 60, 60, and 80,

b. 0, 20, 60, 60, and 160.

Solutions: **a.**

x_i	$(x_i - \mu)$	$(x_i - \mu)^2$
40	−20	400
60	0	0
60	0	0
60	0	0
80	20	400
Totals 300	0	800

$$\mu = \frac{300}{5} = 60 \qquad \sigma^2 = \frac{800}{5} = 160$$

b.

x_i	$x_i - \mu$	$(x_i - \mu)^2$
0	−60	3600
20	−40	1600
60	0	0
60	0	0
160	100	10,000
Totals 300	0	15,200

$$\mu = \frac{300}{5} = 60 \qquad \sigma^2 = \frac{15{,}200}{5} = 3040$$

■

In part **b** of the preceding example, the data are more spread out than in part **a**, and the variance is correspondingly larger. This is a desirable property for our measure to have.

The major disadvantage in using the variance is that it is in square units. We can correct this disadvantage by considering a measure that converts back to the original unit. This measure of dispersion is called the **standard deviation** and is defined to be the square root of the variance.

Symbolically, we use **σ** for *population standard deviation* and **s** for *sample standard deviation*. The computational formulas for standard deviation are

$$\sigma = \sqrt{\sigma^2} = \sqrt{\frac{\sum_{i=1}^{N} (x_i - \mu)^2}{N}} \quad \text{and} \quad s = \sqrt{s^2} = \sqrt{\frac{\sum_{i=1}^{n} (x_i - \bar{x})^2}{n - 1}}.$$

For example 36(a), the variance was found to be 160. Hence the standard deviation is $\sigma = \sqrt{160} \cong 12.649$. The variance in example 36(b) was found to be 3040. Then the standard deviation is $\sigma = \sqrt{3040} \cong 55.136$. Recall that the symbol $\cong$ means "is approximately equal to." Typically, square roots will not be whole numbers, so we will have to round off. This also happens in other statistical computations. To simplify notation, we will agree to use an equal sign (=), even though we are rounding off.

We will also agree, in general, to use three more decimal places in the standard deviation than are used in the original data. That is, if data are to the nearest tenth, we will use four decimal places for the standard deviation.

In many instances, statistical data are given with frequencies. The formulas for variance can be easily adjusted to use the frequencies, just as the formula for the mean was. Suppose the values $x_1, x_2, x_3, \ldots, x_k$ occur with frequencies $f_1, f_2, f_3, \ldots, f_k$, respectively, and $f_1 + f_2 + \cdots + f_k = N$. Then

$$\sigma^2 = \frac{\sum_{i=1}^{k} f_1(x_i - \mu)^2}{N} \quad \text{and} \quad \sigma = \sqrt{\sigma^2}.$$

When grouped data are being used, the class mark is used for x_i in the computation of the variance and standard deviation. The following two examples will help illustrate computation of the variance and standard deviation for frequency data. The first is for ungrouped data, and the second is for grouped data.

■ Example 37. The heights of the 30 children in a particular class are measured to the nearest inch, with the results given in the table. Find the mean, variance, and standard deviation.

Height (x_i)	43	44	45	46	47	48	49	50
f_i	1	0	2	10	4	4	8	1

Solution: We put the information in a table and do the calculations. Note that the mean must be computed prior to completing the table.

x_i	f_i	$f_i x_i$	$x_i - \mu$	$(x_i - \mu)^2$	$f_i(x_i - \mu)^2$
43	1	43	−4	16	16
44	0	0	−3	9	0
45	2	90	−2	4	8
46	10	460	−1	1	10
47	4	188	0	0	0
48	4	192	1	1	4
49	8	392	2	4	32
50	1	50	3	9	9
Σ	30	1415			79

$$\mu = \frac{1415}{30} = 47.17, \text{ or about } 47.$$

$$\sigma^2 = \frac{79}{30} = 2.633333$$

$$\sigma = \sqrt{2.633333} = 1.623$$

■

■ Example 38. Find the mean, variance, and standard deviation for the data on September precipitation for a certain city given in the following table.

Class	*f*
0.0–0.9	1
1.0–1.9	5
2.0–2.9	4
3.0–3.9	9
4.0–4.9	20
5.0–5.9	4
6.0–6.9	7

Solution:

Class	f_i	x_i	$f_i x_i$	$x_i - \mu$	$(x_i - \mu)^2$	$f_i(x_i - \mu)^2$
0.0–0.9	1	0.45	0.45	−3.65	13.3225	13.3225
1.0–1.9	5	1.45	7.25	−2.65	7.0225	35.1125
2.0–2.9	4	2.45	9.80	−1.65	2.7225	10.8900
3.0–3.9	9	3.45	31.05	−0.65	0.4225	3.8025
4.0–4.9	20	4.45	89.00	0.35	0.1225	2.4500
5.0–5.9	4	5.45	21.80	1.35	1.8225	7.2900
6.0–6.9	7	6.45	45.15	2.35	5.5225	38.6575
Σ	50		204.50			111.5250

$$\mu = \frac{204.50}{50} = 4.09, \text{ or about } 4.1.$$

$$\sigma^2 = \frac{111.5250}{50} = 2.2305$$

$$\sigma = \sqrt{2.2305} = 1.4935$$

■

The use of tables is not critical in problems of this type. With a handheld calculator, the computation can be done directly. However, the table does help eliminate some errors caused by inadvertently skipping a value or not multiplying by a frequency.

When using a calculator, you must know how operations are performed on the specific model you are using. For example, some calculators will give $2 + 4 \cdot 5 = 30$, adding the 2 and 4 before multiplying by 5, whereas others will give $2 + 4 \cdot 5 = 22$, where the product is computed first. The latter answer, $2 + 4 \cdot 5 = 22$, involves the mathematical priority of operations. Know your calculator before trying to use it for complex problems.

An alternate formula is sometimes used to compute the variance. It is equivalent to the actual variance formula but does not require that we find $x_i - \mu$ for each of the x_i values. We will use properties of summation and the definition of the mean to derive this formula.

Recall that $\sigma^2 = \dfrac{\sum_{i=1}^{N} (x_i - \mu)^2}{N}$. We will work with the numerator of this formula.

$$\sum_{i-1}^{N} (x_i - \mu)^2 = \sum_{i=1}^{N} (x_i^2 - 2\, x_i\mu + \mu^2)$$

$$= \sum_{i=1}^{N} x_i^2 - \sum_{i=1}^{N} 2\, x_i\mu + \sum_{i=1}^{N} \mu^2 \qquad \text{By property 7.3}$$

$$= \sum_{i=1}^{N} x_i^2 - 2\mu \sum_{i=1}^{N} x_i + \sum_{i=1}^{N} \mu^2 \qquad \text{By property 7.2}$$

$$= \sum_{i=1}^{N} x_i^2 - 2\mu \sum_{i=1}^{N} x_i + N\mu^2 \qquad \text{By property 7.1}$$

$$= \sum_{i=1}^{N} x_i^2 - 2\mu \cdot N\mu + N\mu^2 \qquad \mu = \frac{\sum_{i=1}^{N} x_i}{N} \text{ so } \sum_{i=1}^{N} x_i = N\mu$$

$$= \sum_{i=1}^{N} x_i^2 - 2N\mu^2 + N\mu^2$$

$$= \sum_{i=1}^{N} x_i^2 - N\mu^2$$

Then the variance can be written as

$$\sigma^2 = \frac{\sum_{i=1}^{N} x_i^2 - N\mu^2}{N} \quad \text{or} \quad \frac{\sum_{i=1}^{N} x_i^2}{N} - \mu^2$$

Either form is equivalent to the original variance formula and can be used in its place. Which formula you use is not important. Choose one that you are comfortable with and use it instead of trying first one and then another.

As an illustration of the use of this formula, we will redo example 38 in table form. Since we have frequency data, the actual formula will be

$$\sigma^2 = \frac{\sum_{i=1}^{N} f_i x_i^2 - N\mu^2}{N}.$$

Class	f_i	x_i	$f_i x_i$	x_i^2	$f_i x_i^2$
0.0–0.9	1	0.45	0.45	.2025	.2025
1.0–1.9	5	1.45	7.25	2.1025	10.5125
2.0–2.9	4	2.45	9.80	6.0025	24.0100
3.0–3.9	9	3.45	31.05	11.9025	107.1225
4.0–4.9	20	4.45	89.00	19.8025	396.0500
5.0–5.9	4	5.45	21.80	29.7025	118.8100
6.0–6.9	7	6.45	45.15	41.6025	291.2175
Σ	50		204.50		947.9250

$$\mu = \frac{204.50}{50} = 4.1, \quad \sigma^2 = \frac{947.9250 - 50(4.1)^2}{50} = 2.1485, \quad \sigma = 1.4658$$

Note that the values for σ^2 and σ differ somewhat from the values obtained before. This happens because of the roundoff. Using the actual value of $\mu = 4.09$, we obtain

$\sigma^2 = \dfrac{947.9250 - 50(4.09)^2}{50} = 2.2304$ and $\sigma = 1.4935$. Either pair of answers is close enough for most statistical work.

Exercise Set 7.4

1. Find the range for each of problems 1–6 in exercise set 7.3.
2. Find the range for problems 7 and 8 in exercise set 7.3.
3. Find the variance and standard deviation for each of the following:
   a. problem 1, exercise set 7.3.
   b. problem 3, exercise set 7.3.
   c. problem 5, exercise set 7.3.
4. Find the variance and standard deviation for each of the following:
   a. problem 2, exercise set 7.3.
   b. problem 4, exercise set 7.3.
   c. problem 6, exercise set 7.3.
5. Find the variance and standard deviation for each of the following problems from exercise set 7.3.
   a. problem 7.
   b. problem 9.
   c. problem 11.
   d. problem 13 (treat the data as a *sample*).
6. Find the variance and standard deviation for each of the following problems from exercise set 7.3.
   a. problem 8.
   b. problem 10 (treat the data as a sample).
   c. problem 12.
   d. problem 14.
7. The ages of the residents of a nursing home are 81, 61, 92, 73, 67, 77, 73, 72, 85, 97, 69, 70, 72, 79, 83, 84, 79, 89, 92, 88, 72, 75, 73, 69, 84, 67, 63, 52, 73, 67, 75, 81, 83, 91, 80, 76, 77, 76, 82, 81, 73, 72, 91, 81, 67, 93, 82, 81, 80, and 77. Find μ (to the nearest tenth), σ^2, and σ (to the nearest hundredth).
8. The grades on a statistics exam were 86, 87, 81, 86, 75, 98, 97, 92, 80, 73, 84, 87, 44, 82, 39, 88, 88, 72, 62, 84, 100, 89, 95, 95, 95, 79, 54, and 88. Find μ (nearest whole number), σ^2, and σ (nearest hundredth).
9. The administration of a small college is interested in statistics relating to the age of entering freshmen. They gathered the following data regarding the current freshman class. Find the mean, variance, and standard deviation.

Age	f_i
16	2
17	0
18	119
19	63
20	17
21	0
22	5

10. Find the mean, variance, and standard deviation for the following data on course load (in semester hours) for the seniors at a certain college.

Semester hours	f_i
13	24
14	47
15	121
16	112
17	63
18	22
19	11
20	2

11. Find μ, σ^2, and σ for the following data on the length (in centimeters) of the babies born last year at a certain hospital.

Length	f_i
36–40	17
41–45	49
46–50	123
51–55	73
56–60	33
61–65	17

12. Mr. Smith is considering the purchase of a gift shop. The owner has provided him with data indicating the gross weekly sales for the last 5 years. Find the mean, variance, and standard deviation.

Sales (in dollars)	f_i
3,001–3,500	7
3,501–4,000	6
4,001–4,500	15
4,501–5,000	20
5,001–5,500	43
5,501–6,000	95
6,001–6,500	27
6,501–7,000	24
7,001–7,500	13
7,501–8,000	10

13. The college placement office is studying the beginning monthly salary of recent graduates. To obtain data, they select a *sample* of recent graduates and obtain the data in the table below. Find $\bar{x}$, s^2, and s for this data.

Monthly salary (in dollars)	f_i
801–875	2
876–950	0
951–1,025	3
1,026–1,100	7
1,101 1,175	15
1,176–1,250	37
1,251–1,325	52
1,326–1,400	31
1,401–1,475	17
1,476–1,550	12
1,551–1,625	4
1,626–1,700	1

14. To determine performance of their 60-watt light bulbs, a company selects a *sample* and tests them to see how long they light before burning out. Find $\overline{x}$, s^2, and s.

Lifetime (hours)	f_i
701–725	3
726–750	2
751–775	0
776–800	14
801–825	31
826–850	81
851–875	83
876–900	42
901–925	29
926–950	13
951–975	0
976–1,000	0
1,001–1,025	2

7.5 Numerical Descriptions of Data, III: Measures of Location

The measures of central tendency and dispersion discussed in the preceding sections are commonly used for data description. They are not, however, the only measures that can be used. In fact, some fields commonly use measures related to location or position. One such measure, the median, has already been considered. In this section, additional measures of location will be discussed.

One technique that is frequently used in the field of education is to arrange the data values from lowest to highest and then divide the data into 100 blocks, or groups, of the same size. That is, each block contains $\frac{1}{100}$ of the data values. The first block would contain the smallest data values and the last would contain the largest data values. Since the fraction $\frac{1}{100} = 0.01$ can be converted to 1%, the blocks are called **percentiles.**

The numerical value corresponding to the ath percentile is the number that has a% of the values below it and $(100 - a)$% above it. Thus, a score that exceeded 90% of the scores and was less than 10% of the scores is said to be in the 90th percentile.

Raw scores can be transformed into percentiles by using cumulative frequency. If a raw score of 50 corresponds to a percentile of 90, we interpret this to mean that 90% of the scores are less than or equal to 50. If N scores are arranged in order with frequencies f_i, then the percentile, P, of a score x_i is computed by the formula $P = \frac{cf_i}{N} \cdot 100$ where cf_i is the cumulative frequency for x_i. We multiply by 100 to convert the decimal to a percent.

To find the score, x_i, corresponding to the ath percentile, denoted P_a, we first find the necessary cumulative frequency, cf_i, and then find the score corresponding to this cf_i. That is, $cf_i = \frac{aN}{100}$. We then find the score which corresponds to this cf_i. If the cf_i value is not listed, we take the next higher value.

■ Example 39. The heights, in inches, of a group of 50 children are given in the following table. Find

a. the percentile rank of a height of 48 inches, and

b. the height corresponding to the 75th percentile.

Height (x_i)	45	46	47	48	49	50	51	52	53
f	2	0	5	7	13	12	6	3	2

Solutions: We first construct the cumulative frequency table.

x_i	f_i	cf_i
45	2	2
46	0	2
47	5	7
48	7	14
49	13	27
50	12	39
51	6	45
52	3	48
53	2	50

a. The percentile rank of 48 is computed by $P = \frac{14}{50} \cdot 100 = 28\%$. Hence 28% of the students are less than or equal to 48 inches tall.

CAUTION: When N is small or f_i is large the formula $P = \frac{cf_i - \frac{1}{2}f_i}{N} \cdot 100$ is more descriptive and usually preferred. The percentile rank of 48 using this new formula is 21%. We will use the formula $P = \frac{cf_i}{N} \cdot 100$ for consistency.

b. To find the value corresponding to the 75th percentile, P_{75}, we first find the appropriate cf_i. $cf_i = \frac{75(50)}{100} = 37.5$. Since this value is not in the cf_i column, we go to the next larger value, 39. The cf_i of 39 corresponds to a height of 50 inches, so 50 is the value corresponding to P_{75}. ■

Two other measures of location that are directly related to percentiles are sometimes used. If the data values are divided into 4 equal size blocks, the blocks are called **quartiles.** Each of 10 equal size blocks is called a **decile.** If we let Q_i be the ith quartile, then $Q_1 = P_{25}$, $Q_2 = P_{50}$, and $Q_3 = P_{75}$. That is, the third quartile is the 75th percentile, etc. Similarly, the first decile, D_1, is the 10th percentile, the second decile, D_2, is the 20th percentile, and so on.

Note that the 100th percentile is not used in reporting scores. In the field of education, where percentiles are commonly used for reporting scores, the highest percentile used is the 99th.

It is also important to remember that a score of 70% on an exam is not the same as being in the 70th percentile. A score of 70% is a raw score while the 70th percentile is a value denoting relative position.

Percentiles do not make any use of the measures of central tendency or dispersion discussed earlier. There is, however, one measure of location that is dependent on the mean and standard deviation. This measure is referred to as a ***z* score.**

Suppose John tells you that he got an 85 on a history exam and a 75 on a math exam. Did he really do better in history? On the surface, it may appear so, but if additional information is available, we can be more confident. The scores given do not relate any information about how the classes, as a whole, did. If, for example, the mean in history was 80 and the mean in math was 80, it is obvious John did better in history.

However, if the history mean was 80 and the math mean was 70, we know no more than we did before about John's relative standing. Suppose that we know the mean in history was 80 and the standard deviation was 10 and the mean in math was 70 with a standard deviation of 5. In which class is John's relative standing better?

Note that John's history score is $\frac{1}{2}$ standard deviation higher (better) than the mean while his math score is 1 standard deviation higher than the mean. His relative standing is better in math.

A *z score* is a measure of how far above or below the mean a given value is in terms of standard deviation units. A value of x in a distribution with mean μ and standard deviation σ has a z score of $z = \frac{x - \mu}{\sigma}$. A negative z score indicates below the mean and a positive z score indicates above the mean.

■ Example 40. Sally took five exams one week. Her scores, together with the mean and standard deviation for each class, are given in the table below. Use z scores to determine in which class Sally had the highest relative standing.

Class	*Sally*	μ	σ
English	88	80	6
Math	85	78	4
History	92	84	8
Accounting	90	75	12
Biology	78	70	4

Solutions: The z scores are:

English $z = \frac{88 - 80}{6} = \frac{8}{6} = 1.33$

Math $z = \frac{85 - 78}{4} = \frac{7}{4} = 1.75$

History $z = \frac{92 - 84}{8} = \frac{8}{8} = 1.00$

Accounting $z = \frac{90 - 75}{12} = \frac{15}{12} = 1.25$

Biology $z = \frac{78 - 70}{4} = \frac{8}{4} = 2.00$ ■

Sally's highest relative standing is in biology where she was 2 standard deviations above the mean. Her lowest relative standing is in history where she was only 1 standard deviation above the mean.

It is not always true that large positive z scores are better. If you are given a task to perform and you are timed, a time less than the mean (resulting in a negative z score) would normally be considered better. The "best" value is the one giving the z score farthest from the mean in the "best" direction; i.e., the side of the mean that would be interpreted as good.

We can also use z scores to compare values in totally different units. When values are converted to z scores, the z scores are in standard deviation units from the mean. The original units are no longer important in the discussion.

The following example will help illustrate both of the preceding ideas. We first decide which direction from the mean is best and then choose the value that is farthest from the mean in this direction.

■ Example 41. Mary Anne has been hired to work in an office for the summer. Her performance on a series of tasks is compared to means that the company has established over the years. Her assignment will be in the area where her performance is best. The tasks are typing, filing, duplicating, switchboard, and mail distribution. Typing is measured in words per minute (wpm), filing in terms of the number of minutes to file 50 folders, duplicating in number of copies of a seven-page report made and collated in one hour, switchboard in terms of length of time it takes to connect the caller with the right person, and mail distribution in terms of the length of time it takes to sort and distribute 50 pieces of mail. Which job should Mary Anne get based on the data in the table below?

Task	*Mary Anne*	μ	σ
Typing	95 wpm	89 wpm	4.3
Filing	8.5 min	9.5 min	0.8
Duplicating	80 copies	75 copies	9.3
Switchboard	45 sec	52 sec	3.4
Mail	64 min	68 min	2.8

Solution: All of Mary Anne's scores are "better" than the mean. In the three areas where speed is considered important (filing, switchboard, and mail distribution) her score is less than the mean and her other two scores are higher than the mean. The z scores are as follows.

Typing $$z = \frac{95 - 89}{4.3} = \frac{6}{4.3} = 1.40$$

Filing $$z = \frac{8.5 - 9.5}{0.8} = \frac{-1}{0.8} = -1.25$$

Duplicating $$z = \frac{80 - 75}{9.3} = \frac{5}{9.3} = 0.54$$

Switchboard $$z = \frac{45 - 52}{2.7} = \frac{-7}{2.7} = -2.59$$

Mail $$z = \frac{64 - 68}{2.8} = \frac{-4}{2.8} = -1.43$$

Mary Anne should get the job working on the switchboard since her score for this task is farthest from the mean in the "best" direction. ■

Exercise Set 7.5

1. For the data of example 39, find:
   a. the percentile rank of a height of 46 inches.
   b. the percentile rank of a height of 50 inches.
   c. the percentile rank of a height of 51 inches.
   d. the height corresponding to P_{60}.
   e. the height corresponding to P_{25}.
   f. the height corresponding to P_{85}.
2. The table below gives scores of students on a mathematics placement exam. Find:
   a. the percentile rank of a score of 8.
   b. the percentile rank of a score of 11.
   c. the percentile rank of a score of 16.
   d. the score corresponding to P_{75}.
   e. the score corresponding to P_{94}.
   f. the score corresponding to P_{60}.

Score	1	2	3	4	5	6	7	8	9	10	11	12	13
f_i	1	1	4	3	7	7	9	4	5	5	2	2	3

Score	14	15	16	17	18	19
f_i	0	0	1	1	2	1

3. For the data given in the table below, find
   a. the percentile rank of a score of 6.
   b. the percentile rank of a score of 12.
   c. the percentile rank of a score of 15.
   d. the score corresponding to P_{35}.
   e. the score corresponding to P_{50}.
   f. the score corresponding to P_{95}.

Score	0	1	2	3	4	5	6	7	8	9	10	11	12
f_i	12	5	15	15	35	49	53	59	62	61	47	42	34

Score	13	14	15	16	17	18	19	20
f_i	27	19	15	8	6	2	3	1

4. For the data in the table below, find
   a. the percentile rank of 6.
   b. the percentile rank of 9.
   c. the percentile rank of 15.
   d. the score corresponding to P_{35}.

e. the score corresponding to P_{80}.
f. the score corresponding to P_{95}.

Score	1	2	3	4	5	6	7	8	9	10	11	12	13
f_i	3	3	6	12	25	21	45	35	41	49	39	31	19

Score	14	15	16	17	18	19	20	21	22	23	24	25
f_i	20	8	5	4	3	3	0	3	0	1	0	1

5. On a 25-question exam, the mean was $\mu = 13.5$ and the standard deviation was $\sigma = 5.08$. Find the z score corresponding to each of the following scores.
a. 22 **b.** 17 **c.** 11 **d.** 6 **e.** 20 **f.** 9

6. On a 40-question exam, the mean was $\mu = 23$ and the standard deviation was $\sigma = 8.04$. Find the z score corresponding to each of the following scores.
a. 32 **b.** 26 **c.** 38 **d.** 20 **e.** 12 **f.** 6

7. Mark has decided to try out for the school track team but doesn't know what event he is best in. He decides to try five events. The table below gives Mark's performance and the school mean and standard deviation based on data kept by the coach. Which event is Mark's best? Which is his worst?

Event	*Mark*	*School mean*	σ
100-meter dash	11.3 sec	11.7 sec	0.5
200-meter dash	21.9 sec	22.6 sec	0.6
110-meter hurdles	17.1 sec	16.9 sec	0.8
long jump	18.3 ft	19.7 ft	1.2
high jump	5.6 ft	5.2 ft	0.3

8. Five girls are enrolled in different sections of the same course. After an exam they are discussing how they are doing. The table below gives each girl's score and the mean and standard deviation for their respective classes. Which girl has the highest relative standing and which has the lowest relative standing?

Student	*Score*	μ	σ
Ann	88	81	8.2
Betty	78	80	5.1
Carol	81	85	4.6
Denise	89	82	4.7
Ellen	92	88	8.5

7.6 Chapter Review

Important Terms and Symbols in Chapter 7.

frequency table (section 7.1)
frequency (f) (section 7.1)
grouped data (section 7.1)
classes (section 7.1)
class size (section 7.1)
class width (section 7.1)
class boundary (section 7.1)
tally marks (section 7.1)
cumulative frequency (*cf*) table (section 7.1)
relative frequency (*rf*) (section 7.1)
cumulative relative frequency (*crf*) (section 7.1)
frequency polygon (section 7.1)
class mark (section 7.1)
—√∧— (section 7.1)
cumulative frequency polygon (ogive) (section 7.1)
histogram (section 7.1)
Σ (section 7.2)
measures of central tendency (section 7.3)
mode (section 7.3)
bimodal (section 7.3)
median (section 7.3)
mean (section 7.3)
μ, $\bar{x}$ (section 7.3)
coding (section 7.3)
dispersion (section 7.4)
range (section 7.4)
average deviation (section 7.4)
variance (section 7.4)
σ^2, s^2 (section 7.4)
standard deviation (section 7.4)
σ, s (section 7.4)
percentile (section 7.5)
quartile (section 7.5)
decile (section 7.5)
z score (section 7.5)

Review Exercises

1. For each of the following, construct a frequency table and a frequency polygon.
   a. Five coins are flipped 40 times and the numbers of heads observed are 2, 3, 1, 2, 4, 0, 2, 1, 4, 3, 2, 5, 1, 5, 4, 3, 2, 2, 4, 5, 1, 1, 1, 0, 3, 2, 3, 4, 3, 2, 4, 1, 5, 3, 2, 3, 3, 0, 2, 4.
   b. A pair of dice are tossed 30 times and the sums of the faces are 6, 7, 5, 7, 8, 11, 3, 6, 7, 7, 2, 4, 5, 7, 6, 9, 12, 8, 11, 5, 3, 2, 8, 7, 8, 6, 7, 7, 8, 7.
2. Construct a cumulative frequency table and a cumulative frequency polygon (ogive) for each part of problem 1.
3. Construct a histogram for each part of problem 1.
4. For each of the following, construct a grouped data frequency table and corresponding frequency polygon.
   a. A psychologist observes 40 rats going through a maze and records the following completion times (nearest tenth of a second): 5.4,

7.3, 6.5, 4.9, 8.1, 6.2, 4.9, 4.9, 5.8, 5.3, 5.9, 8.0, 7.5, 7.3, 6.6, 5.9, 4.8, 5.9, 6.2, 4.9, 6.8, 5.8, 4.7, 5.8, 6.2, 7.4, 6.6, 5.3, 5.2, 7.1, 7.3, 4.5, 5.8, 5.8, 6.3, 5.8, 5.1, 5.6, 7.6, 5.8.

The grade point averages (nearest hundredth) for the 40 members of a fraternity are 2.74, 3.21, 1.57, 2.34, 2.45, 3.41, 3.87, 2.79, 2.65, 3.10, 3.51, 2.96, 1.65, 2.67, 3.51, 2.76, 4.00, 2.11, 3.82, 2.97, 2.96, 3.42, 1.88, 1.25, 3.61, 3.04, 2.76, 2.59, 2.96, 2.14, 3.25, 2.96, 3.21, 3.07, 2.87, 3.51, 3.15, 2.77, 3.81, 2.95.

5. Construct a cumulative frequency table and cumulative frequency polygon for the grouped data of problem 4(a).

6. Construct a cumulative frequency table and cumulative frequency polygon for the grouped data of problem 4(b).

7. Construct a histogram for the data in each part of problem 4.

8. For the data of problem 1(a), construct a table showing relative frequency and cumulative relative frequency.

9. The following table gives information on the scores of students on a 20 question multiple choice exam in a large lecture class.

Score (x_i)	6	7	8	9	10	11	12	13	14	15	16	17	18	19
f_i	25	37	40	32	32	20	20	17	10	4	4	3	4	2

a. Find the percentile rank of a score of 13.
b. Find the percentile rank of a score of 9.
c. Find the percentile rank of a score of 17.
d. Find the score corresponding to P_{75}.
e. Find the score corresponding to P_{65}.
f. Find the score corresponding to P_{40}.

10. Find the mode for each part of problem 1 and each part of problem 4.

11. Find the median for each part of problem 4.

12. For problem 4(a), find $\overline{x}$, s^2, and s (assume you are working with a random sample) for the given data
a. using the actual values,
b. using the grouped data of problem 4(a).

13. Find μ, σ^2, and σ for the data of problem 4(b) using
a. the actual values,
b. the grouped data you used in 4(b).

14. Find μ, σ^2, and σ for the exam data of problem 9.

15. The following scores were recorded on a 100 point math exam by one class: 74, 60, 94, 56, 70, 91, 79, 63, 53, 91, 80, 66, 58, 69, 72, 28, 65, 85, 62, 76, 88, 84, 60, 79, 81, 72, 84, 85, 71, 87, 35, 84, 62, 82, 76, 66, 90, 99, 77, 80.
a. Find the mode and median.
b. Find μ, σ^2, and σ.

16. The following scores were received on a 100 point exam by students in a statistics class: 93, 72, 89, 83, 46, 74, 90, 90, 67, 92, 85, 79, 58, 83, 80, 97, 90, 28, 98, 73, 55, 93, 83, 84, 73, 84, 82, 92, 86, 87, 67, 93, 60.

a. Find the mode and median.

b. Find μ, σ^2, and σ.

17. On a 100 point exam, the mean was $\mu = 77$ and the standard deviation was $\sigma = 9.21$. Find the z score corresponding to each of the following exam scores.

a. 79 **b.** 88 **c.** 63 **d.** 97 **e.** 57 **f.** 71

18. In a certain factory, the mean number of a certain type product that a person can inspect in one hour is $\mu = 80$ and the standard deviation is $\sigma = 7.24$. During one hour last week, 6 people were asked to indicate how many items they inspected. Find the z score corresponding to the following numbers turned in by the inspectors.

a. 87 **b.** 65 **c.** 93 **d.** 105 **e.** 73 **f.** 97

19. Sally had exams in all six of her classes last week. The table below gives Sally's scores together with the mean and standard deviation for each class.

Class	*Sally*	*Mean*	*Standard deviation*
Math	85	77	3.14
History	91	87	5.02
Biology	23	21	3.05
Accounting	79	80	4.95
English	68	65	3.02
Philosophy	91	84	7.56

a. In which class was Sally's relative standing highest?

b. In which class was Sally's relative standing lowest?

20. A nursery school teacher has a series of activities she has students do each year. She has kept records over the years so she knows the mean and standard deviation for each activity. The activities are stacking blocks, putting together a picture puzzle, carrying a full glass of water, hopping on one leg, and fitting shapes into proper holes. She measures the number of blocks stacked before the stack falls, the length of time it takes to put the puzzle together (in minutes), the amount of water spilled (in ounces), the length of time the child can hop on one leg without stopping, and the length of time the child takes to correctly place all of the shapes in the proper holes (in minutes).

The table below gives the mean and standard deviation for each activity together with Jay's results.

Activity	*Jay*	*Mean*	*Standard deviation*
blocks	18	15	3.5
puzzle	2.0	2.1	.5
water	1.7	1.1	.4
hopping	4.5	3.5	1.1
shapes	3.8	4.2	1.4

a. In which activity did Jay have his best performance?
b. In which activity was Jay's performance the poorest?

8 Probability Distributions

Introduction

As indicated in the introduction to chapter 6, probability theory plays an important role in statistics. In this chapter, this relationship between probability and statistics will be discussed.

We will begin with a general discussion of *probability distributions* and some of the terms that will be used. The remaining sections will consider specific probability distributions and some of their applications.

8.1 Probability Distributions

When an experiment is performed, the various outcomes of the experiment occur randomly. Prior to performing an experiment, such as tossing a die, we are not sure what outcome, or value, we are going to get. If we are tossing a die, we know in advance that we will get a face showing 1, 2, 3, 4, 5, or 6, but we don't know which of these will actually occur.

Similarly, if one student is chosen, by drawing, from a class of 15 girls and 10 boys, we don't know in advance whether the student will be a boy or a girl. We can, however, determine probabilities associated with the possible outcomes.

In statistics, we are normally interested in numerical results of experiments since we are interested in the mean and standard deviation. Some experiments, such as the drawing of a student's name or flipping a coin and observing a head or a tail, do not have numerical results. It's difficult to describe these experiments statistically without somehow using numerical values.

This can be accomplished by working with random variables. A **ran-**

dom variable is a rule that associates numbers with outcomes of an experiment. The association may be natural such as the face on a die, the sum of the faces on a pair of dice, or the number of heads when five coins are flipped. In other experiments, such as drawing the name of a student or observing a head or tail on one flip of a coin or the answers to opinion type questions, it may be necessary to assign numerical values. If, for example, a question could be answered "yes," "no," or "no opinion," we might use 1 for yes, 0 for no opinion, and -1 for no. Other values could also be used.

Random variables are normally divided into two types, discrete and continuous. If the possible values of the random variable form a finite set, the random variable is **discrete.** In addition, any random variable whose possible values can be matched one-to-one with the counting numbers $\{0,1,2, \ldots\}$ is discrete. For our purposes, any random variable whose possible values are a subset of the rational numbers is discrete. A **continuous random variable** can take on any real number value in an interval. For example, if a random variable can assume any value between 1 and 2, the random variable is continuous.

Notationally, we will use capital letters to denote random variables, normally $X,Y,Z, \ldots$, or $X_1,X_2, \ldots$. Small letters such as $x,y,z, \ldots$ or $x_1,x_2, \ldots$ will be used for the *values* of random variables.

When we consider the values of a random variable and the associated probabilities we have a **probability distribution.** In this section only discrete random variables will be considered. Hence, only **discrete probability distributions** will be discussed. In section 8.3, one very important continuous probability distribution will be discussed.

If X is a random variable assuming values $x_1,x_2,x_3, \ldots ,x_n$, then the associated probabilities will be denoted $P(X = x_1),P(X = x_2), \ldots ,P(X = x_n)$ or, sometimes, $P\{x_1\},P\{x_2\}, \ldots ,P\{x_n\}$. The probability distribution of X will include both the values and the probabilities.

■ Example 1. Find the probability distribution for the experiment of tossing a die.

Solution: We will give the distribution in table form.

x_i	$P(X = x_i)$
1	$\frac{1}{6}$
2	$\frac{1}{6}$
3	$\frac{1}{6}$
4	$\frac{1}{6}$
5	$\frac{1}{6}$
6	$\frac{1}{6}$

■

■ **Example 2.** Find the probability distribution for the experiment of picking a name at random from a class list consisting of 15 girls and 10 boys, if we are interested in the sex of the person whose name is selected.

Solution: We first must assign numerical values to the possible outcomes. We will use 0 for a boy and 1 for a girl. The random variable X, then, can assume either of these values.

x_i	$P\{x_i\}$
0	$\frac{10}{25} = 0.4$
1	$\frac{15}{25} = 0.6$

■

■ **Example 3.** Four coins are flipped. Let the random variable X denote the number of heads and find the probability distribution of X.

Solution: The values for X will be 0,1,2,3,4. The probability distribution of X is given in the table.

x_i	$P\{x_i\}$
0	$\frac{1}{16}$
1	$\frac{4}{16}$
2	$\frac{6}{16}$
3	$\frac{4}{16}$
4	$\frac{1}{16}$

■

For any probability distribution, the sum of the probabilities for all values of X will always be 1. That is, if the random variable X has values $x_1, x_2, \ldots, x_n$ and associated probabilities $P\{x_1\}, P\{x_2\}, \ldots, P\{x_n\}$, then

$$\sum_{i=1}^{n} P\{x_i\} = 1.$$

In chapter 7, the mean and variance were computed for given data values. It is also possible to show a relationship between these concepts and probability distributions. This relationship makes use of **expected value,** first considered in chapter 6.

If X is a random variable with values $x_1, x_2, \ldots, x_n$ and associated probabilities $P\{x_1\}, P\{x_2\}, \ldots, P\{x_n\}$, then the *expected value of* X, denoted $E[X]$, is given by $E[X] = \sum_{i=1}^{n} x_i \cdot P\{x_i\}$. The expected value is, then, the sum of all values multiplied by the corresponding probabilities. We have defined the **mean** to be $\mu = \dfrac{\sum_{i=1}^{k} f_i x_i}{N} = \sum_{i=1}^{k} x_i \cdot \dfrac{f_i}{N}$. If there are N data values and x_i has frequency f_i, then the probability of choosing the value x_i if one value is randomly chosen is $\dfrac{f_i}{N}$. That is, $\dfrac{f_i}{N} = P\{x_i\}$. Hence the mean is $\mu = \sum_{i=1}^{k} x_i P\{x_i\}$.

This is the same as the expected value of X. The mean of a random variable is, then, the expected value of the random variable and is computed by $\mu = E[X] = \sum_{i=1}^{n} x_i P\{x_i\}$.

The **variance** can also be treated as an expected value. If X is a random variable taking on values $x_1, x_2, \ldots, x_n$ with probabilities $P\{x_1\}, P\{x_2\}, \ldots, P\{x_n\}$, then X^2 is also a random variable and X^2 has values $x_1^2, x_2^2, \ldots, x_n^2$. The probabilities are the same as those for X. That is, if $P(x = x_i)$ is $P\{x_i\}$, then $P(x^2 = x_i^2)$ is also $P\{x_i\}$. If $P(x = 2) = \dfrac{1}{8}$, then $P(x^2 = 4) = \dfrac{1}{8}$. The following example will help illustrate.

■ Example 4. Four coins are flipped and the random variable X is the number of heads. Find the probability distribution of X^2.

Solution: The distribution of X is that given in example 3. It is also included here.

Distribution of X		Distribution of X^2	
x_i	$P\{x_i\}$	x_i^2	$P\{x_i^2\}$
0	$\frac{1}{16}$	0	$\frac{1}{16}$
1	$\frac{4}{16}$	1	$\frac{4}{16}$
2	$\frac{6}{16}$	4	$\frac{6}{16}$
3	$\frac{4}{16}$	9	$\frac{4}{16}$
4	$\frac{1}{16}$	16	$\frac{1}{16}$

Note that the probability of x_i^2 is the same as the probability of x_i for each i. That is,

$$P(X^2 = x_i^2) = P(X = x_i).$$ ■

If X is a random variable, then X^2, $X - A$ (A a constant), $(X - A)^2$ are also random variables. Many others could be constructed.

Let X be a random variable with mean $\mu(E[X] = \mu)$. Then the *variance of X*, denoted σ^2, is given by the formula $\sigma^2 = E[(X - \mu)^2] = \sum_{i=1}^{n} (x_i - \mu)^2 P\{x_i\}$.

As before, the **standard deviation** is defined to be the square root of the variance so $\sigma = \sqrt{\sigma^2} = \sqrt{\sum_{i=1}^{n} (x_i - \mu)^2 P\{x_i\}}$.

■ **Example 5.** Compute the mean, variance, and standard deviation for the random variable X denoting the face showing when a die is tossed.

Solution:

x_i	$P\{x_i\}$	$x_iP\{x_i\}$	$x_i - \mu$	$(x_i - \mu)^2$	$(x_i - \mu)^2P\{x_i\}$
1	$\frac{1}{6}$	$\frac{1}{6}$	$-\frac{5}{2}$	$\frac{25}{4}$	$\frac{25}{24}$
2	$\frac{1}{6}$	$\frac{2}{6}$	$-\frac{3}{2}$	$\frac{9}{4}$	$\frac{9}{24}$
3	$\frac{1}{6}$	$\frac{3}{6}$	$-\frac{1}{2}$	$\frac{1}{4}$	$\frac{1}{24}$
4	$\frac{1}{6}$	$\frac{4}{6}$	$\frac{1}{2}$	$\frac{1}{4}$	$\frac{1}{24}$
5	$\frac{1}{6}$	$\frac{5}{6}$	$\frac{3}{2}$	$\frac{9}{4}$	$\frac{9}{24}$
6	$\frac{1}{6}$	$\frac{6}{6}$	$\frac{5}{2}$	$\frac{25}{4}$	$\frac{25}{24}$
Totals	1	$\frac{21}{6}$			$\frac{70}{24}$

$$\mu = \sum_{i=1}^{6} x_i \cdot P\{x_i\} = \frac{21}{6} = \frac{7}{2} \text{ or } 3.5.$$

$$\sigma^2 = \sum_{i=1}^{6} (x_i - \mu)^2 P\{x_i\} = \frac{70}{24} = \frac{35}{12} = 2\frac{11}{12} \text{ or } 2.9167.$$

$$\sigma = \sqrt{\frac{35}{12}} = 1.708.$$

■

■ Example 6. Four coins are flipped and X denotes the number of heads showing. Find μ, σ^2, and σ for this random variable X.

Solution:

x_i	$P\{x_i\}$	$x_iP\{x_i\}$	$x_i - \mu$	$(x_i - \mu)^2$	$(x_i - \mu)^2P\{x_i\}$
0	$\frac{1}{16}$	0	−2	4	$\frac{4}{16}$
1	$\frac{4}{16}$	$\frac{4}{16}$	−1	1	$\frac{4}{16}$
2	$\frac{6}{16}$	$\frac{12}{16}$	0	0	0
3	$\frac{4}{16}$	$\frac{12}{16}$	1	1	$\frac{4}{16}$
4	$\frac{1}{16}$	$\frac{4}{16}$	2	4	$\frac{4}{16}$
Σ	1	$\frac{32}{16}$			$\frac{16}{16}$

$$\mu = \sum_{i=1}^{5} x_iP\{x_i\} = \frac{32}{16} = 2.$$

$$\sigma^2 = \sum_{i=1}^{5} (x_i - \mu)^2P\{x_i\} = \frac{16}{16} = 1.$$

$$\sigma = \sqrt{\sigma^2} = 1.$$

■

Exercise Set 8.1

1. A subcommittee of two people is to be chosen from a committee consisting of three women and two men. Let the random variable X denote the number of women on the subcommittee.

a. Find the probability distribution of X.

b. Find $\mu = E[X]$.

c. Find σ^2 and σ.

2. An urn contains 7 red balls and 3 blue balls. Two balls are to be drawn without replacement. Let X be a random variable denoting the number of red balls drawn.
   a. Find the probability distribution of X.
   b. Find $\mu = E[X]$.
   c. Find σ^2 and σ.
3. Three cards are drawn one at a time from a standard deck of 52 cards with cards replaced after each draw. The random variable X represents the number of hearts drawn.
   a. Find the probability distribution of X.
   b. Find $\mu = E[X]$.
   c. Find σ^2 and σ.
4. An urn contains 30 red balls and 70 green balls. Four balls are to be drawn one at a time with replacement. The random variable X represents the number of green balls drawn.
   a. Find the probability distribution for X.
   b. Find $\mu = E[X]$.
   c. Find σ^2 and σ.
5. Six fair coins are flipped. Let X denote the number of heads showing.
   a. Find the probability distribution of X.
   b. Find $\mu = E[X]$.
   c. Find σ^2 and σ.
6. A pair of fair dice is tossed. The random variable X denotes the sum of the faces.
   a. Find the probability distribution of X.
   b. Find $\mu = E[X]$.
   c. Find σ^2 and σ.
7. A store is having a sale with a special promotion where customers draw a card from a jar to show the amount of additional discount. In the jar are 30 cards showing a 10% additional discount, 15 cards showing a 25% additional discount, and 5 cards showing a 50% additional discount. Let the random variable X denote the percent of additional discount. (Note: Cards are replaced after each person draws.)
   a. Find the probability distribution of X.
   b. Find the mean discount percent.
   c. Find σ^2 and σ.
8. An investor has $1,000 to invest and she has a choice of two investment opportunities. Investment X has an estimated return of $5,000, $2,000, or $0 with probabilities 0.25, 0.50, and 0.25 respectively. The estimated return on investment Y is $4,000, $3,000, $2,000, or $1,000 with equal probabilities.
   a. Find $E[X]$ and $E[Y]$.
   b. Which investment has the better expected return?
9. Let X be the random variable of exercise 3.
   a. Find $E[Y]$ where $Y = X + 3$.

b. Find $E[X^2]$.
c. Evaluate $E[X^2] - \mu^2$ and compare your answer with the variance from exercise 3.

10. Let X be the random variable of exercise 4.
a. Find $E[Y]$ where $Y = 2X + 1$.
b. Find $E[X^2]$.
c. Evaluate $E[X^2] - \mu^2$ and compare your answer with the variance from exercise 4.

8.2 The Binomial Distribution

In chapter 6, probabilities associated with flipping coins were discussed. Whether the experiment consists of flipping one coin 3 times or 3 coins once really doesn't matter. The key is that the outcomes on successive flips do not depend on each other. That is, if one coin is flipped 3 times, the probability of a head on the third flip is not affected by what happened on the two preceding flips. If 3 coins are flipped, the probability of a head on any one of them is not affected by the face showing on the other two.

Flipping coins is one illustration of a special type of probability experiment called a **binomial experiment.** If one flip is called a trial (sometimes called a Bernouli trial), then 3 flips would be considered 3 successive trials. Flipping 3 coins simultaneously could also be considered 3 trials. In order to define a binomial experiment, one other aspect of this type of experiment must be noted. On each individual trial, there are only two possible outcomes. These outcomes are sometimes referred to as success or failure, and they must be mutually exclusive.

Definition 8.1 A ***binomial experiment*** is an experiment consisting of n independent trials, where each trial results in 1 of 2 mutually exclusive outcomes and the probability of each outcome is the same on every trial.

■ Example 7. Each of the following could be considered a binomial experiment.

a. Flip a coin 20 times and observe whether the coin shows heads or tails.
b. Toss a die ten times and observe whether the result is even or odd.

c. Shoot at a basket 50 times and observe whether the shot is made or missed.
d. Note 50 people entering a store and observe whether each is male or female. ■

The preceding example gives only a few situations in which the experiment could be considered binomial. You probably can think of others. The list seems endless.

In most binomial experiments, one of the two possible outcomes on a trial is thought of as a success. We are interested, then, in the number of successes in n trials and the probability of getting a certain number of successes.

For example, we know that if a coin is flipped 5 times, the number of heads is 0, 1, 2, 3, 4, or 5. What is the probability of exactly 3 heads? On the basis of previous work, there are $2^5 = 32$ different sequences of heads and tails if a coin is flipped 5 times. The number of sequences containing 3 heads is given by $\binom{5}{3} = 10$. Hence, the probability of getting 3 heads is $P(3 \text{ heads}) = \frac{10}{32} = \frac{5}{16}$.

■ Example 8. A die is tossed 10 times. What is the probability of obtaining exactly 8 even numbers?

Solution: Since even and odd are equally likely on each toss, there are $2^{10} = 1024$ equally likely outcomes for the experiment. The number of these containing exactly 8 evens is $\binom{10}{8} = 45$. The probability of exactly 8 evens is $\frac{45}{1024}$. ■

In both of the preceding examples, the two possible outcomes on each trial were equally likely. Suppose the coin that is flipped in the previous illustration is loaded in such a way that 9 out of 10 flips will result in a tail. Would the probability of getting 3 heads in 5 flips still be $\frac{5}{16}$? Certainly not!

In order to compute **binomial probabilities** when outcomes are not equally likely for each trial, some additional work is necessary. The end result will be related directly to the binomial theorem of section 5.6.

■ Example 9. Suppose a loaded coin which has a probability of 0.9 of showing a tail is flipped 5 times. Find the probability of getting exactly 3 heads.

Solution: There are still $2^5 = 32$ sequences of heads and tails. However, these sequences are not equally likely since (T,T,T,T,T) is more apt to occur than (H,H,H,H,H).

There are still $\binom{5}{3} = 10$ sequences containing 3 heads. Two such sequences are (H,H,H,T,T) and (T,H,H,T,H). The trials are independent so we compute the probabilities of these two sequences as follows:

$$\begin{aligned}
P[(H,H,H,T,T)] &= P(\{H\}) \cdot P(\{H\}) \cdot P(\{H\}) \cdot P(\{T\}) \cdot P(\{T\}) \\
&= (0.1)(0.1)(0.1)(0.9)(0.9) \\
&= (0.1)^3(0.9)^2 = (0.001)(0.81) = 0.00081 \\
P[(T,H,H,T,H)] &= P(\{T\}) \cdot P(\{H\}) \cdot P(\{H\}) \cdot P(\{T\}) \cdot P(\{H\}) \\
&= (0.9)(0.1)(0.1)(0.9)(0.1) \\
&= (0.1)^3(0.9)^2 = 0.00081
\end{aligned}$$

In fact, each of the 10 acceptable sequences has this same probability of occurring. The probability of 3 heads is, then, the product of the number of sequences times the common probability of each, or

$$\begin{aligned}
P(3 \text{ heads}) &= 10(0.00081) \\
&= 0.0081
\end{aligned}$$

This is much different than $\frac{5}{16} = 0.3125$ which we obtained in the previous illustration where the coin was fair. ■

We generalize, now, for any binomial experiment.

Property 8.1 If a binomial experiment consists of n trials where the probability of a success on each trial is $p (0 < p < 1)$, then the probability of k successes out of the n trials $(k \le n)$ is given by

$$P(k \text{ successes in } n \text{ trials}) = \binom{n}{k} p^k (1-p)^{n-k}.$$

NOTE: If the probability of a success is p, then $1 - p$ is the probability of a failure. Some authors use q as the probability of a failure, where, of course, $q = 1 - p$.

Proof:
We want k successes, so we consider sequences of n outcomes where k are successes and $n - k$ are failures. One such sequence is

$$(\underbrace{S,S,S,\ldots,S}_{k \text{ terms}},\underbrace{F,F,\ldots,F}_{(n-k) \text{ terms}})$$

and the independence of trials gives the probability of this sequence as

$$\underbrace{p \cdot p \cdot p \cdot \ldots \cdot p}_{k \text{ factors}} \cdot \underbrace{(1-p) \cdot (1-p) \ldots \cdot (1-p)}_{(n-k) \text{ factors}} = p^k(1-p)^{n-k}.$$

Any other sequence with k successes will have exactly the same probability. Since the number of acceptable sequences is $\binom{n}{k}$, the probability of k successes is

$$P(k \text{ successes in } n \text{ trials}) = \binom{n}{k} p^k(1-p)^{n-k}.$$

CAUTION: The independence of trials is crucial here. If the trials are not independent, conditional probability would be needed to obtain the probability of a sequence of outcomes. Different sequences may, then, have different probabilities, and the preceding result cannot be used.

If q is used for the probability of a failure in the property, then the probability of k successes is given by $\binom{n}{k} p^k q^{n-k}$. This is a term of the binomial expansion of $(p + q)^n$. The probability of k successes is the term in the binomial expansion of $(p + q)^n$ involving p^k. Note that

$$1 = [p + (1-p)]^n = \sum_{k=0}^{n} \binom{n}{k} p^k(1-p)^{n-k}.$$

■ **Example 10.** A professional basketball player makes, on the average, 80% of his free throws. If he shoots 10 free throws in tonight's game, what is the probability that he will make

a. exactly 8?
b. exactly 6?
c. either 6 or 8?

Solutions: Since he makes 80%, the probability of a success is 0.8.

a. $P(\text{exactly } 8) = \binom{10}{8}(0.8)^8(0.2)^2 = 0.302$

b. $P(\text{exactly } 6) = \binom{10}{6}(0.8)^6(0.2)^4 = 0.088$

c. The events "make exactly 8" and "make exactly 6" are mutually exclusive so

$$\begin{aligned} P(6 \text{ or } 8) &= P(\text{exactly } 6) + P(\text{exactly } 8) \\ &= 0.302 + 0.088 = 0.390. \end{aligned}$$

■

■ Example 11. An insurance salesman estimates, based on experience, that the probability of selling a policy to an individual who has contacted him is 0.7. If he calls on 12 such people one week, what is the probability that he sells

a. exactly 5 policies? b. at least 9 policies?

Solutions: (We assume independence.)

a. $$P(\text{exactly } 5) = \binom{12}{5}(0.7)^5(0.3)^7 = 0.0291$$

b. $$\begin{aligned} P(\text{at least } 9) &= P(\text{exactly } 9) + P(\text{exactly } 10) + P(\text{exactly } 11) + P(\text{exactly } 12) \\ &= \binom{12}{9}(0.7)^9(0.3)^3 + \binom{12}{10}(0.7)^{10}(0.3)^2 + \binom{12}{11}(0.7)^{11}(0.3)^1 + \binom{12}{12}(0.7)^{12}(0.3)^0 \\ &= 0.2397 + 0.1678 + 0.0712 + 0.0138 \\ &= 0.4925 \end{aligned}$$ ■

An experiment that consists of n independent trials where each trial results in one of two possible outcomes is called a binomial experiment. The two possible outcomes are sometimes referred to as success or failure. The random variable, X, denoting the number of successes in n trials is called a **binomial random variable.**

A binomial random variable has possible values $0,1,2, \ldots, n$. If p is the probability of a success (a success is the one outcome we are studying) on any one trial, then the probability of exactly k successes is given by the equation $P(X = k) = \binom{n}{k} p^k(1 - p)^{n-k}$. Remember, in a binomial experiment, the probability of a particular outcome is the same on each trial.

Frequently, the interest is not in the probability of an exact number of successes. Phrases like "at least k successes," "more than k successes," "at most k successes," "less than k successes," or "from j to k successes" are relatively common. Probabilities for statements like these are easily expressable using summation notation. Some examples follow.

1. Probability of at most k successes, $P(X \leq k)$.

$$P(X \leq k) = \sum_{i=0}^{k} \binom{n}{i} p^i(1 - p)^{n-i}.$$

2. Probability of less than k successes, $P(X < k)$.

$$P(X < k) = \sum_{i=0}^{k-1} \binom{n}{i} p^i (1-p)^{n-i}.$$

3. Probability of at least k successes, $P(X \geq k)$.

$$\begin{aligned} P(X \geq k) &= \sum_{i=k}^{n} \binom{n}{i} p^i (1-p)^{n-i} \\ &= 1 - P(X < k) \\ &= 1 - \sum_{i=0}^{k-1} \binom{n}{i} p^i (1-p)^{n-i}. \end{aligned}$$

4. Probability of more than k successes, $P(X > k)$.

$$\begin{aligned} P(X > k) &= \sum_{i=k+1}^{n} \binom{n}{i} p^i (1-p)^{n-i} \\ &= 1 - P(X \leq k) \\ &= 1 - \sum_{i=0}^{k} \binom{n}{i} p^i (1-p)^{n-i}. \end{aligned}$$

5. Probability of from j to k $(j < k)$ successes, $P(j \leq X \leq k)$ or $P(j < X < k)$ depending on whether j and k are included or excluded.

$$P(j \leq X \leq k) = \sum_{i=j}^{k} \binom{n}{i} p^i (1-p)^{n-i}$$

$$P(j < X < k) = \sum_{i=j+1}^{k-1} \binom{n}{i} p^i (1-p)^{n-i}.$$

We can also write

$$\begin{aligned} P(j \leq X \leq k) &= P(X \leq k) - P(X < j) \\ &= \sum_{i=0}^{k} \binom{n}{i} p^i (1-p)^{n-i} - \sum_{i=0}^{j-1} \binom{n}{i} p^i (1-p)^{n-i}. \end{aligned}$$

The preceding formulas are particularly useful when working with binomial tables. This is especially true when the cumulative binomial probabilities are given. That is, if the table values are for $P(X \leq k)$ instead of $P(X = k)$, the above formulas are helpful.

Binomial probability distributions for binomial random variables are completely determined by the number of trials (n) and the probability of a success on each trial (p). For relatively small values of n (say $n \leq 10$) binomial probabilities are not too difficult to compute using a calculator. For values of n from 11 to 20, we have provided tables in the appendix for some selected values of p.

The tables give $P(X = k)$ for each value of n and the appropriate value of p. Blanks in the table indicate that the probability is 0 when rounded off to 3 decimal places.

Example 12. Given the binomial random variable X with $n = 15$ and $p = 0.60$, find

a. $P(X = 11)$
b. $P(X = 5)$
c. $P(X \leq 7)$
d. $P(X > 10)$
e. $P(5 \leq X \leq 10)$
f. $P(3 < X \leq 7)$.

Solutions: We use the table for $n = 15$ and the column for $p = 0.60$.

a. $P(X = 11) = 0.127$
b. $P(X = 5) = 0.024$
c.
$$P(X \leq 7) = \sum_{k=0}^{7} P(X = k)$$
$$= 0 + 0 + 0 + 0.002 + 0.007 + 0.024 + 0.061 + 0.118$$
$$= 0.212$$
d.
$$P(X > 10) = \sum_{k=11}^{15} P(X = k)$$
$$= 0.127 + 0.063 + 0.022 + 0.005 + 0$$
$$= 0.217$$
e.
$$P(5 \leq x \leq 10) = \sum_{k=5}^{10} P(X = k)$$
$$= 0.024 + 0.061 + 0.118 + 0.177 + 0.207 + 0.186$$
$$= 0.773$$
f.
$$P(3 < X \leq 7) = \sum_{k=4}^{7} P(X = k)$$
$$= 0.007 + 0.024 + 0.061 + 0.118$$
$$= 0.210$$

Example 13. The State Highway Patrol has estimated that 70% of the drivers traveling a certain section of four-lane highway are within the speed limit.

a. What is the probability that of 18 cars traveling the section at a particular time, at least 15 of them are within the speed limit?
b. What is the probability that less than 8 are within the speed limit?

Solutions: We assume all of the conditions of a binomial distribution are met.

a. $$P(X \geq 15) = \sum_{k=15}^{18} P(X = k)$$
$$= 0.105 + 0.046 + 0.013 + 0.002$$
$$= 0.166$$

b. $$P(X < 8) = \sum_{k=0}^{7} P(X = k)$$
$$= 0.001 + 0.005$$
$$= 0.006$$

■ Example 14. City officials have claimed that 80% of the city's voters support a proposed city sales tax to help finance various projects. A group of citizens opposing the tax doubts this claim. They randomly select 20 voters and ask their opinion, finding only 11 people who favor the tax. Assuming the 80% figure is correct, and assuming that the only responses were "favor" or "do not favor" ("no opinion" counts as "do not favor"), what is the probability that in a randomly selected group of 20 people, no more than 11 favor the tax?

Solution: Using the table for $n = 20$ and $p = 0.8$, we want to find $P(X \leq 11)$.

$$P(X \leq 11) = \sum_{k=0}^{11} P(X = k)$$
$$= 0.002 + 0.007$$
$$= 0.009$$

With such a small probability, the citizens group might have grounds for debating the 80% figure if they truly did a random selection. However, 20 individuals is not a very large sample unless the city is awfully small. ■

We have restricted ourselves thus far to $n \leq 20$. In the last section of this chapter we will discuss one approach to binomial probabilities for $n > 20$. For the remainder of this section, the only requirement on n is that $n > 0$.

> If a binomial experiment consists of n trials with a probability of success on each trial of p, it can be shown that the mean or expected number of successes is given by
>
> $$\mu = E[X] = np.$$
>
> The variance and standard deviation for a binomial random variable are given by
>
> $$\sigma^2 = np(1 - p) \quad \text{and} \quad \sigma = \sqrt{np(1 - p)}, \text{ respectively.}$$

■ Example 15. Referring to example 14, how many of the 20 people would be expected to favor the tax assuming the 80% figure given by city officials? Also, compute the variance and standard deviation.

Solutions:

$\mu = np = 20 \cdot (0.8) = 16$

$\sigma^2 = np(1 - p) = 20\,(0.8)(0.2) = 3.2$

$\sigma = \sqrt{3.2} = 1.789$ ■

■ Example 16. Find the mean, variance, and standard deviation for a binomial distribution with $n = 100$, $p = 0.45$.

Solutions:

$\mu = np = 100(0.45) = 45$

$\sigma^2 = np(1 - p) = 100(0.45)(0.55) = 24.75$

$\sigma = \sqrt{24.75} = 4.975$ ■

Exercise Set 8.2

1. If a fair coin is flipped 8 times, what is the probability of getting
- **a.** exactly 4 heads?
- **b.** exactly 6 heads?
- **c.** at least 6 heads?
- **d.** less than 6 heads?

2. A die is tossed 7 times. What is the probability of getting an even number
- **a.** exactly 3 times?
- **b.** exactly 5 times?
- **c.** at least 5 times?
- **d.** less than 5 times?

3. An expert marksman has probability 0.9 of hitting a target on any one shot. What is the probability he will hit the target at least 8 times out of 10 shots?

4. What is the probability that a professional basketball player with a free throw percentage of 80% will make at least 6 out of 8 shots in a playoff game?

5. In producing flashbulbs, a company estimates that 1% of the bulbs are defective. If the company picks a random sample of 4 bulbs, what is the probability that 2 of the bulbs are defective?

6. If it is assumed that boy and girl babies are equally likely, what is the probability that a couple with 6 grandchildren will have 5 granddaughters? Assume births are independent events.

7. Under the same assumption as in problem 6, what is the probability that a couple with 5 grandchildren will have 4 grandsons?

8. The local police department estimates that 40% of the calls they respond to result in an arrest. If they respond to 10 calls during a 2-hour period, what is the probability that they make at least 8 arrests?

9. According to a store manager, 60% of the people who enter her store actually make a purchase. How many of the next 10 people entering would be expected to make a purchase? What is the probability that exactly this number make a purchase?
10. An encyclopedia salesman estimates that 25% of the people who allow him into the house will make a purchase. How many sales should he expect to make in the next 12 homes he enters? What is the probability that he will make exactly this number of sales?
11. If the probability of success is $p = 0.6$, then the probability of failure is $q = 0.4$. Given there are 5 trials, expand $(p + q)^5$ and analyze your results.
12. On the basis of past performance, your team is expected to win 60% of its games.
    a. If you play 4 games this weekend, what are your possible results?
    b. What is the probability of each result?
    c. What is the probability that you win more games of the 4 than you lose?

For each of exercises 13–20 use Table V in the appendix. Assume a binomial distribution in each case.

13. A fair coin is flipped 20 times. Find the probability of getting
    a. at least 15 heads.
    b. no more than 8 heads.
    c. at least 5, but not more than 12, heads.
    d. more than 10, but less than 16, heads.
14. A fair coin is flipped 18 times. Find the probability of getting
    a. more than 12 heads.
    b. less than 9 heads.
    c. more than 7, but less than 13, heads.
    d. at most 11, but at least 6, heads.
15. City officials claim that 90% of the taxpayers support the proposed increase in sales taxes to be used for street improvements. A group of 20 taxpayers is randomly chosen and questioned about the tax.
    a. What is the expected number favoring the tax?
    b. What is the probability that at least 16 of the 20 favor the tax?
    c. What is the probability that at least 7 oppose the tax?
16. A manufacturer claims that 95% of its electric toasters work perfectly. The Brown Hardware Store has just purchased 12 of the toasters:
    a. What is the expected number of defective toasters?
    b. What is the probability that all 12 toasters work properly?
    c. What is the probability of at least 3 defective toasters?
17. The local commuter train is on time 70% of the time on its morning run. For the next three work weeks (15 mornings), what is the probability that it is on time
    a. at least 11 times?
    b. at most 7 times?
    c. at least 10, but at most 12, times?

18. Harvey makes, on the average, 60% of his free throw attempts in basketball. If he has 17 attempts during two weekend games, what is the probability that he makes
a. at least 10?
b. at most 8?
c. at least 11, but at most 15?

19. The Useless Toy Company claims that 99% of their talking dolls work properly. The Smith Department Store has sold 19 of the dolls. What is the probability that
a. at least 18 work properly?
b. at least 15, but not more than 18, work properly?
c. at least 3 are defective?

20. An auto manufacturer has determined that 20% of the cars of a certain model have a minor steering defect. The local phone company has 11 cars of this model in their fleet. What is the probability that
a. at least 5 of the cars have the defect?
b. at most 8 of the cars have the defect?
c. more than 3, but less than 7, have the defect?

For exercises 21–26, X is a binomial random variable with the values of n and p given. Use Table V in the appendix to find the probabilities.

21. $n = 15$, $p = 0.30$
a. Find μ, σ^2, and σ.
b. Find $P(X < 6)$.
c. Find $P(X \geq 4)$.
d. Find $P(5 < X \leq 8)$.

22. $n = 13$, $p = 0.10$
a. Find μ, σ^2, and σ.
b. Find $P(X > 3)$.
c. Find $P(X \leq 4)$.
d. Find $P(2 \leq X < 6)$.

23. $n = 16$, $p = 0.40$
a. Find μ, σ^2, and σ.
b. Find $P(X < 7)$.
c. Find $P(X > 4)$.
d. Find $P(3 < X < 9)$.

24. $n = 18$, $p = 0.60$
a. Find μ, σ^2, and σ.
b. Find $P(X \leq 7)$.
c. Find $P(X \geq 11)$.
d. Find $P(3 \leq X \leq 7)$.

25. $n = 19$, $p = 0.30$
a. Find μ, σ^2, and σ.
b. Find $P(X > 8)$.
c. Find $P(X \leq 5)$.
d. Find $P(7 < X \leq 14)$.

26. $n = 17$, $p = 0.70$
a. Find μ, σ^2, and σ.
b. Find $P(X < 9)$.
c. Find $P(X \geq 14)$.
d. Find $P(11 \leq X < 17)$.

27. Given a binomial distribution with $n = 200$ and $p = 0.75$, find μ, σ^2, and σ.

28. Given a binomial distribution with $n = 400$ and $p = 0.55$, find μ, σ^2, and σ.

8.3 Normal Distributions

Perhaps the most commonly used of all probability distributions is the so-called **normal distribution.** The normal distribution is a **continuous probability distribution** and has applications in a wide range of fields.

A complete formal discussion of the normal distribution requires techniques developed in calculus. We will adopt a more informal approach based on description rather than formal development.

A random variable X having a normal distribution can assume any real number in some interval as a value. Since, for any random variable X, the sum of the probabilities for all values that X can assume must equal 1, special considerations must be made for the normal random variable (actually for any continuous random variable). In particular, if X is a normal random variable and r is a potential value of X, we have $P(X = r) = 0$. That is, the probability that X assumes any one of its values is always 0. This does not mean the value cannot occur. If each value had a probability greater than 0, the sum of all of the probabilities would exceed 1.

Probabilities associated with the normal distribution are probabilities that X takes on some value in an interval. For example, we may want to find the probability that X is between a and b or the probability that X is greater than b. We write probabilities of this type as $P(a < X < b)$ or $P(X > b)$.

Before considering normal probabilities, some general ideas associated with normal distributions should be discussed. Graphs of normal distribution are the familiar "bell-shaped" graphs such as that in figure 8.1.

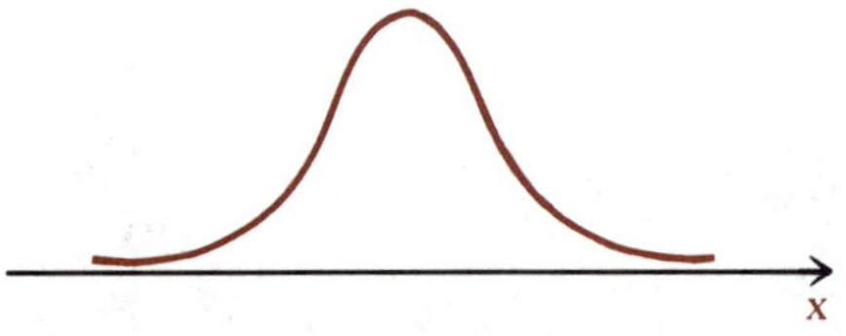

Figure 8.1

Any normal distribution can be related to a graph with this general shape. Normal distributions, and the specific shape of the graph, are completely determined by the mean μ and standard deviation σ of the distribution.

The mean is the value of X for which the curve has its highest point. The value of σ determines how sharply the curve falls on the two sides of the mean. The smaller the value of σ, the more the values concentrate close to μ.

Figure 8.2 shows normal distributions with the same mean and different values of σ. Notice how much flatter the center portion of the graph is for larger values of σ.

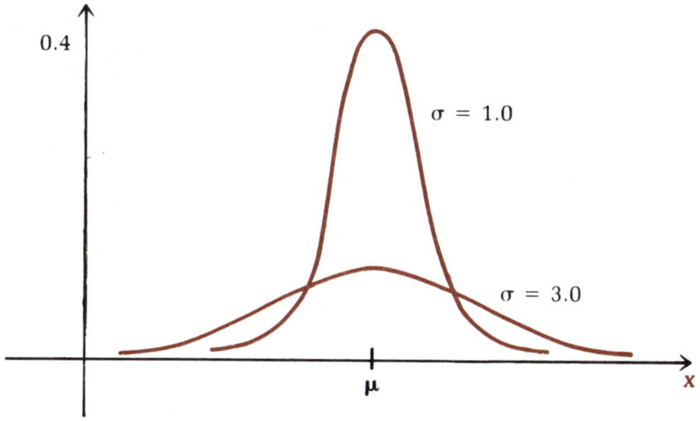

The values on the vertical scale give the height of the curve at $X = \mu$. The smaller σ is, the higher the curve at $X = \mu$ and the narrower the center part of the curve.

Figure 8.2

Normal probabilities are associated with areas under a normal curve like those in figure 8.2. The total area under the curve is 1. (This may not seem reasonable to you but is, in fact, true.) Hence, the area under any section of the curve is a value between 0 and 1.

Suppose X is a normal random variable with mean μ and standard deviation σ. Referring to figure 8.3, $P(a < X < b)$ is the area of the shaded region in (a). In addition, $P(X > c)$ is the area of the shaded region in (b). Note here that $P(X = c)$ would be the area of the line segment from the curve to c. This area is 0. Hence, $P(X = c) = 0$ as we indicated before.

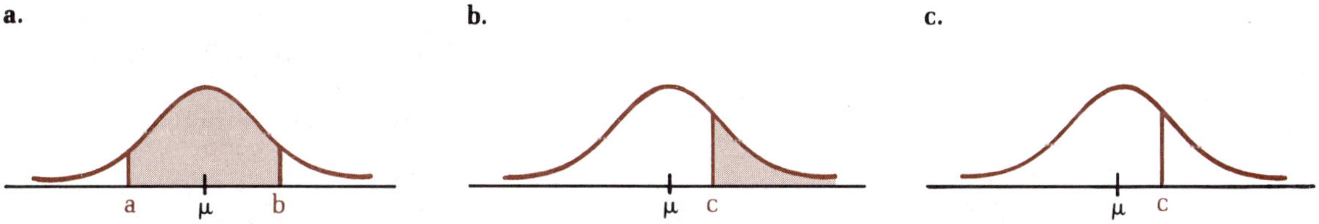

Figure 8.3

Some normal probabilities are of special interest. These involve the probability that X is within some number of standard deviations of the mean. For example, what is the probability X is within 1 standard deviation of the mean? This is equivalent to the proportion of the area within 1 standard deviation of the mean and can be expressed as $P(\mu - \sigma < X < \mu + \sigma)$.

It can be shown that for any normal random variable X, the following probabilities are approximately true.

$\pm 1\sigma$; $P(\mu - \sigma < X < \mu + \sigma) = 0.6826$

$\pm 2\sigma$; $P(\mu - 2\sigma < X < \mu + 2\sigma) = 0.9544$

$\pm 3\sigma$; $P(\mu - 3\sigma < X < \mu + 3\sigma) = 0.9974$

Hence, in a normal distribution, about 68% of the values are within 1 standard deviation of the mean, 95% within 2 standard deviations of the mean, and over 99% within 3 standard deviations of the mean.

■ Example 17. Suppose scores on a certain I.Q. test are normally distributed with a mean of 100 and a standard deviation of 15. About what percent of the scores are

a. between 85 and 115?
b. between 70 and 130?

Solutions:

a. Since 85 is 1 standard deviation below the mean and 115 is 1 standard deviation above the mean, about 68% of the scores are between 85 and 115.
b. Since 70 and 130 are 2 standard deviations below and above the mean respectively, about 95% of the scores are between 70 and 130. ■

Other properties of the normal distribution are useful for computing probabilities. One very important property is the **symmetry** of the distribution. If the graph is considered, the portion to the right of μ is the mirror image of the portion to the left of μ. Then $P(X < \mu) = P(X > \mu) = 0.500$. If a is a positive number, $P(X > \mu + a) = P(X < \mu - a)$.

Since $P(X = r) = 0$, we have $P(X \leq r) = P(X < r)$. The single value r doesn't contribute to the probability. Remember, this is not true for discrete random variables. It is also true that, for any value a, $P(X > a) = 1 - P(X < a)$.

Armed with this information, we are now prepared to consider normal probabilities. We first consider the case where $\mu = 0$ and $\sigma = 1$. This is sometimes referred to as the **standard normal distribution** and the random variable is called Z.

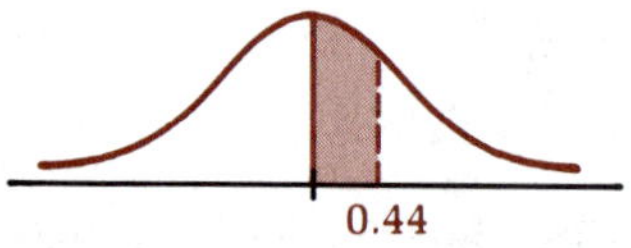

$\mu = 0$
$P(0.00 < Z < 0.44)$

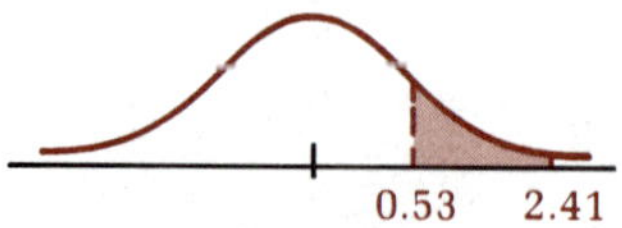

$\mu = 0$
$P(0.53 < Z < 2.41)$

Table VI in the appendix gives probabilities for the standard normal distribution for values of z from 0.00 to 4.00, in hundredths. The table value is $P(0.0 < Z < z)$. Hence, if we take the fifth entry in the z column (0.4) and move over to the column headed by .04, we find a value of .1700. This means that $P(0.00 < Z < 0.44) = 0.1700$. If we want to find $P(Z < 0.44)$, we find $z = 0.44$ in the table, as before, and add 0.5000 to the table value since $P(Z < 0.44) = P(Z \leq 0) + P(0 < Z < 0.44)$, and

$P(Z \leq 0) = 0.5000$ because $\mu = 0$ and $P(X \leq \mu) = 0.5000$ for any normal variable. Then $P(Z < 0.44) = 0.5000 + 0.1700 = 0.6700$.

To find $P(0.53 < Z < 2.41)$, we reason as follows: The area between 0.53 and 2.41 is the area between 0 and 2.41 minus the area between 0 and 0.53. Hence, using Table VI in the appendix, we have

$$\begin{aligned} P(0.53 < Z < 2.41) &= P(0 < Z < 2.41) \\ &\quad - P(0 < Z < 0.53) \\ &= 0.4920 - 0.2019 \\ &= 0.2901. \end{aligned}$$

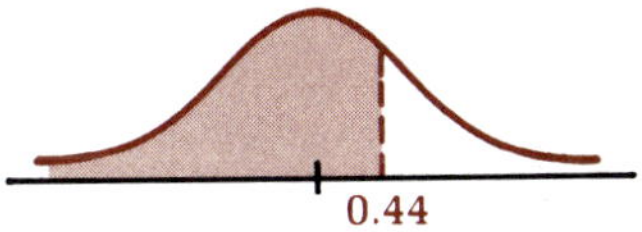

$\mu = 0$
$P(Z < 0.44)$

To find $P(Z < -1.25)$, we proceed in the following way.

$$\begin{aligned} P(Z < -1.25) &= P(Z > 1.25) \quad \text{Symmetry} \\ &= 1 - P(Z < 1.25) \\ &= 1 - [P(Z < 0) \\ &\quad + P(0 < Z < 1.25)] \\ &= 1 - (0.5000 + .3944) \\ &= 1 - 0.8944 \\ &= 0.1056 \end{aligned}$$

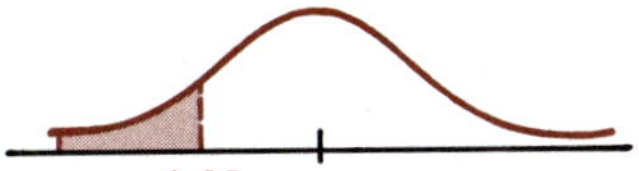

$\mu = 0$
$P(Z < -1.25)$

Statements like $P(Z < 1.25)$ do not have to be broken up as we did above if you will just remember to add 0.5000. That is, remember that for any $a > 0$, $P(Z < a) = 0.5000 + P(0 < Z < a)$ which is 0.5000 plus the table value for a.

Using this idea, $P(0.53 < Z < 2.41)$ can be computed as

$$\begin{aligned} P(0.53 < Z < 2.41) &= P(Z < 2.41) - P(Z < 0.53) \\ &= 0.9920 - 0.7019 \\ &= 0.2901 \end{aligned}$$

Example 18. Use appendix Table VI and properties of normal distributions to find

a. $P(Z > -1.11)$
b. $P(-0.93 < Z < 1.21)$.

Solutions: **a.**

$$\begin{aligned} P(Z > -1.11) &= P(Z < 1.11) \qquad \text{Symmetry} \\ &= 0.5000 + 0.3665 \\ &= 0.8665 \end{aligned}$$

b.

$$\begin{aligned} P(-0.93 < Z < 1.21) &= P(Z < 1.21) - P(Z < -0.93) \\ &= P(Z < 1.21) - P(Z > 0.93) \qquad \text{Symmetry} \\ &= P(Z < 1.21) - [1 - P(Z < 0.93)] \\ &= 0.8869 - (1 - 0.8238) \\ &= 0.8869 - 0.1762 \\ &= 0.7107 \end{aligned}$$

Alternative method for part **b:**

$$\begin{aligned} P(-0.93 < Z < 1.21) &= P(-0.93 < Z < 0) + P(0 < Z < 1.21) \\ &= P(0 < Z < 0.93) + P(0 < Z < 1.21) \\ &= 0.3238 + 0.3869 \\ &= 0.7107 \end{aligned}$$

The above examples illustrate the use of the probability table for the normal random variable with $\mu = 0$ and $\sigma = 1$. However, the mean of a normal random variable can be any real number and the standard deviation can be any positive real number. Obviously, we cannot include normal tables for all of these since there are an infinite number of possibilities. Fortunately it is not necessary.

Suppose X is a normal random variable with mean μ and standard deviation σ. It can be shown that the random variable defined by $Z = \dfrac{X - \mu}{\sigma}$ is the standard normal random variable. That is, Z is normal with $\mu = 0$ and $\sigma = 1$. To find probabilities for the random variable X, we convert to Z and use Table VI.

In particular, if X is normal with mean μ and standard deviation σ and x_1 and x_2 are real numbers, the following are true.

1. $P(X < x_1) = P\left(Z < \dfrac{x_1 - \mu}{\sigma}\right)$.

2. $P(X > x_2) = P\left(Z > \dfrac{x_2 - \mu}{\sigma}\right)$.

3. $P(x_1 < X < x_2) = P\left(\dfrac{x_1 - \mu}{\sigma} < Z < \dfrac{x_2 - \mu}{\sigma}\right)$.

The following examples will help illustrate this method of computing normal probabilities.

■ Example 19. Suppose for an I.Q. test, scores are normally distributed with a mean of 100 and a standard deviation of 15. Let X be a random variable representing a randomly selected score. Find

a. $P(X < 76)$
b. $P(X > 109)$
c. $P(91 < X < 124)$

Solutions: Since $\mu = 100$ and $\sigma = 15$,

a.
$$P(X < 76) = P\left(Z < \frac{76 - 100}{15}\right) = P\left(Z < \frac{-24}{15}\right)$$
$$= P(Z < -1.60) = P(Z > 1.60)$$
$$= 1 - P(Z < 1.60)$$
$$= 1 - 0.9452 = 0.0548$$

b.
$$P(X > 109) = P\left(Z > \frac{109 - 100}{15}\right)$$
$$= P\left(Z > \frac{9}{15}\right) = P(Z > 0.6)$$
$$= 1 - P(Z < 0.6)$$
$$= 1 - 0.7257 = 0.2743$$

c.
$$P(91 < X < 124) = P\left(\frac{91 - 100}{15} < Z < \frac{124 - 100}{15}\right)$$
$$= P(-0.6 < Z < 1.6)$$
$$= P(0 < Z < 0.6) + P(0 < Z < 1.6)$$
$$= 0.2257 + 0.4452 = 0.6709$$ ■

■ Example 20. Suppose the useful life of a certain brand of light bulb is normally distributed with a mean of 800 hours and a variance of 650. If a bulb is selected at random, find the following probabilities associated with its useful life.

a. $P(X < 770)$
b. $P(X > 860)$
c. $P(780 < X < 840)$
d. $P(820 < X < 880)$

Solutions: We first must find σ since σ^2 is given. $\sigma = \sqrt{650} = 25.495$ to three decimal places. To use the table we will round $\dfrac{x_1 - \mu}{\sigma}$ off to two decimal places.

a. $$P(X < 770) = P\left(Z < \frac{770 - 800}{25.495}\right)$$
$$= P(Z < -1.18)$$
$$= 1 - P(Z < 1.18)$$
$$= 0.1190$$

b. $$P(X > 860) = P\left(Z > \frac{860 - 800}{25.495}\right)$$
$$= 1 - P(Z < 2.35) = 0.0094$$

c. $$P(780 < X < 840) = P\left(\frac{780 - 800}{25.495} < Z < \frac{840 - 800}{25.495}\right)$$
$$= P(-0.78 < Z < 1.57)$$
$$= P(0 < Z < 0.78) + P(0 < Z < 1.57)$$
$$= 0.2823 + 0.4418$$
$$= 0.7241$$

d. $$P(820 < X < 880) = P\left(\frac{820 - 800}{25.495} < Z < \frac{880 - 800}{25.495}\right)$$
$$= P(0.78 < Z < 3.14)$$
$$= P(0 < Z < 3.14) - P(0 < Z < 0.78)$$
$$= 0.4992 - 0.2823$$
$$= 0.2169$$ ■

Exercise Set 8.3

For exercises 1–16, Z is a standard normal random variable ($\mu = 0, \sigma = 1$). Find the indicated probabilities using Table VI in the appendix.

1. $P(Z < 2.34)$
2. $P(Z < 1.78)$
3. $P(Z > 1.05)$
4. $P(Z > 1.76)$
5. $P(Z < -1.01)$
6. $P(Z < -0.85)$
7. $P(Z > -0.79)$
8. $P(Z > -1.47)$
9. $P(0.47 < Z < 2.07)$
10. $P(1.21 < Z < 3.09)$
11. $P(-2.12 < Z < 0.79)$
12. $P(-0.89 < Z < 1.75)$
13. $P(-2.61 < Z < -0.59)$
14. $P(-0.97 < Z < -0.07)$
15. $P(-1.9 < Z < 1.9)$
16. $P(-0.7 < Z < 0.7)$
17. Use Table VI in the appendix to find what percent of the values in a normal distribution are
a. between $\mu - 0.5\sigma$ and $\mu + 0.5\sigma$.
b. between $\mu + 0.5\sigma$ and $\mu + 1.5\sigma$.
c. between $\mu - 1.5\sigma$ and $\mu - 0.5\sigma$.
d. greater than $\mu + 1.5\sigma$.

e. less than $\mu - 1.5\sigma$.

These are the percents sometimes used for the "normal grading curve."

18. Use appendix Table VI to find what percent of the values in a normal distribution are
    a. within 1.5 standard deviations of the mean.
    b. between $\mu + \sigma$ and $\mu + 2\sigma$.
    c. between $\mu + 2\sigma$ and $\mu + 3\sigma$.
    d. less than $\mu - 2\sigma$.

In each of exercises 19–22, assume the distribution is normal.

19. A light bulb manufacturer claims that the mean life of its new 60-watt bulbs is 2000 hours with a standard deviation of 42.5. If a bulb is selected at random, what is the probability it lasts
    a. at least 2050 hours?
    b. at least 2100 hours?
    c. between 1980 and 2020 hours?
    d. less than 1925 hours?
    e. at least 1900 hours?
20. Scores on a certain I.Q. test follow a normal distribution with $\mu = 100$ and $\sigma = 12$. If a person who has taken the test is randomly selected and the score is X, find
    a. $P(X > 120)$.
    b. $P(X < 85)$.
    c. $P(90 < X < 110)$.
    d. $P(110 < X < 120)$.
    e. $P(X > 130)$.
21. According to a manufacturer, its new radial tire has a normally distributed lifetime with $\mu = 40{,}000$ miles and $\sigma = 900$ miles. A randomly selected tire is tested on a special machine that simulates actual driving conditions. If the lifetime of the sample tire is X, find
    a. $P(X < 39{,}000)$.
    b. $P(X > 42{,}000)$.
    c. $P(39{,}000 < X < 42{,}000)$.
    d. $P(41{,}000 < X < 43{,}000)$.
    e. $P(X < 38{,}000)$.
22. The Ace Vending Company claims the mean amount of soft drink dispensed by their machines is 8 ounces with a standard deviation of 0.4. What is the probability that a randomly selected soft drink will
    a. contain less than 7.8 ounces?
    b. contain more than 8.1 ounces?
    c. contain between 7.7 and 8.3 ounces?
    d. contain between 7.4 and 7.9 ounces?
    e. overflow the 9 ounce cup used in the machines?

8.4 Normal Approximation to Binomial Distributions

In section 8.2, the binomial distribution was discussed as a special discrete probability distribution. We restricted ourselves to $n \leq 20$. For smaller values of n, say $n \leq 10$, hand calculators with $x!$ and y^x keys are fairly efficient for computing binomial probabilities. A table of binomial probabilities has been included for values of n from 11 to 20 and selected values of p.

For larger values of n, the computations become rather tedious. That is, we must compute $P(X = k) = \dfrac{n!}{k!(n - k)!} p^k(1 - p)^{n-k}$. Tables can be included, but we can't consider every possible value of n (there are an infinite number of possibilities) with all possible values of p. How can binomial probabilities be efficiently computed for large values of n?

Figure 8.4 pictures the relative frequency histogram for selected values of n and p. Observe the general shape of the outline of the histogram.

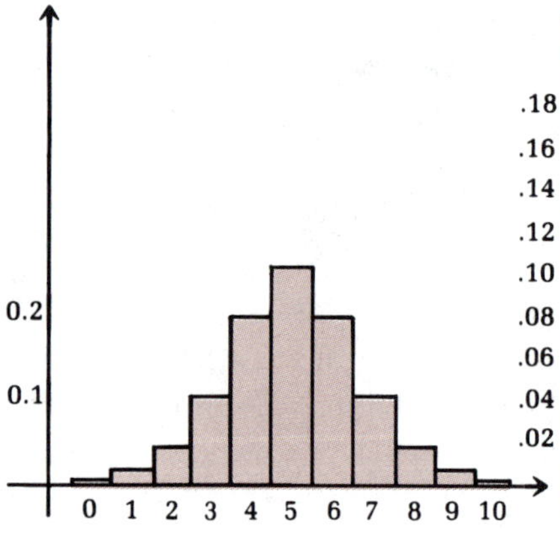

Figure 8.4(a) n = 10, p = 0.5

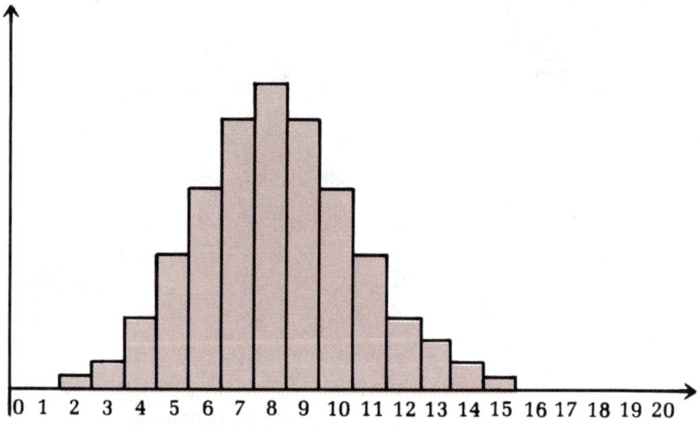

Figure 8.4(b) n = 20, p = 0.4

In figure 8.4(a), the histogram is symmetric with respect to the mean $\mu = np = 10(0.5) = 5$. The histogram for any binomial distribution with $p = 0.5$ will be symmetric with respect to its mean. The size of n doesn't matter.

For the binomial distribution with $n = 20$ and $p = 0.4$ in figure 8.4(b), the mean is $\mu = 20(0.4) = 8$. The histogram is not exactly symmetric since the relative frequency for $k = 7(0.166)$ is slightly larger than the relative frequency for $k = 9(0.160)$. There is still an appearance of balance around the mean.

In figure 8.5, the same histograms have been drawn. This time, though, a smooth curve is drawn connecting endpoints of the intervals. Notice the bell-shaped appearance of the smooth curve.

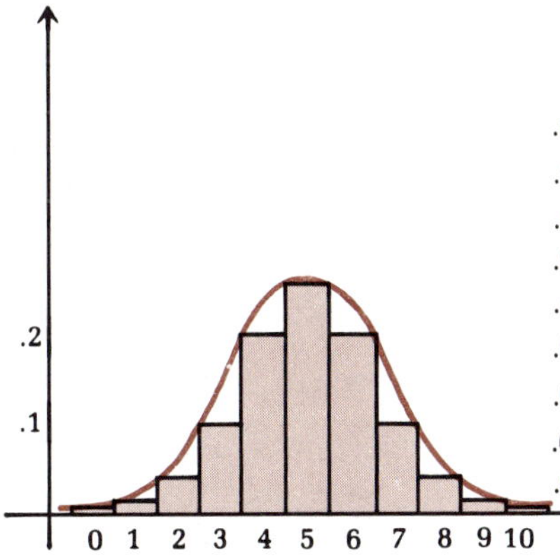

Figure 8.5(a) n = 10, p = 0.5

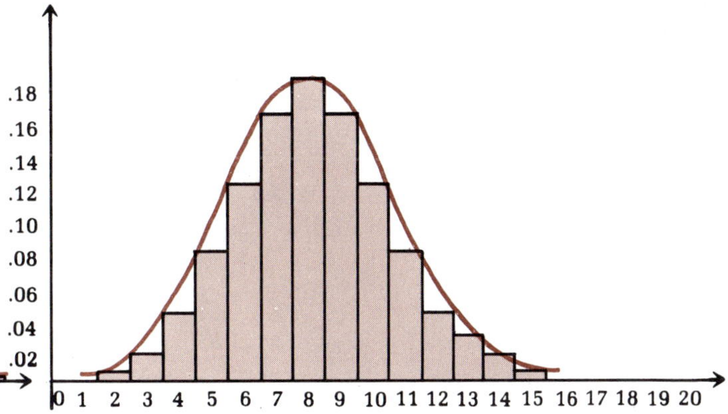

Figure 8.5(b) n = 20, p = 0.4

This bell-shaped appearance suggests the possibility of using a *normal distribution* to *approximate a binomial distribution* **(normal approximation to binomial).** The approximation works amazingly well. The approximation is better for large values of n and for values of p close to 0.5, but it is used for relatively small values of n and values of p quite different from 0.5.

Since the normal distribution is continuous and the binomial is discrete, a **continuity correction** must be made. That is, for the normal distribution, $P(X = k) = 0$ for each k but for the binomial, $P(X = k) > 0$ for at least some of the values of k. When the binomial distribution is approximated using the normal distribution, instead of $P(X = k)$ we use $P(k - \frac{1}{2} < X < k + \frac{1}{2})$ which considers a continuous scale.

To approximate a binomial distribution with a normal distribution we use $\mu = np$ and $\sigma = \sqrt{np(1 - p)}$, the mean and standard deviation of the actual binomial. Suppose we have a binomial distribution with $n = 20$ and $p = 0.4$ and we want to find $P(X = 12)$. The approximating normal has mean $\mu = 20(0.4) = 8$ and $\sigma = \sqrt{20(0.4)(0.6)} = 2.191$. Then $P(X = 12) = P(11.5 < Y < 12.5)$ where Y is normal with $\mu = 8$ and $\sigma = 2.191$. Y values are converted to the standard normal and probabilities are determined from the table.

$$
\begin{aligned}
P(X = 12) &\doteq P(11.5 < Y < 12.5) \\
&= P\left(\frac{11.5 - 8}{2.191} < Z < \frac{12.5 - 8}{2.191}\right) \\
&= P(1.60 < Z < 2.05) \\
&= 0.0346
\end{aligned}
$$

The binomial table value for $P(X = 12)$ where $n = 20$ and $p = 0.4$ is 0.036 so the approximation is very good.

■ **Example 21.** Given a binomial random variable X with $n = 40$, $p = 0.4$, find $P(17 \le X \le 20)$

a. Using the binomial probabilities, and
b. Using a normal approximation to the binomial.

Solutions: **a.**

$$\begin{aligned} P(17 \le X \le 20) &= P(X = 17) + P(X = 18) \\ &\quad + P(X = 19) + P(X = 20) \\ &= \binom{20}{17}(.4)^{17}(.6)^{23} + \binom{20}{18}(.4)^{18}(.6)^{22} \\ &\quad + \binom{20}{19}(.4)^{19}(.6)^{21} + \binom{20}{20}(.4)^{20}(.6)^{20} \\ &= 0.1203 + 0.1025 + 0.0789 + 0.0554 \\ &= 0.3571 \text{ using a calculator.} \end{aligned}$$

b. $P(17 \le X \le 20) = P(16.5 < Y < 20.5)$ where Y is a normal random variable with $\mu = 40(0.4) = 10.0$ and $\sigma = \sqrt{np(1-p)} = \sqrt{9.6} = 3.098$. Then

$$\begin{aligned} P(16.5 < Y < 20.5) &= P\left(\frac{16.5 - 16}{3.098} < Z < \frac{20.5 - 16}{3.098}\right) \\ &= P(.16 < Z < 1.45) \\ &= 0.3629. \end{aligned}$$

The two answers agree to 2 decimal places. ■

■ **Example 22.** The estimated probability of a newborn baby being a boy is 0.53. What is the probability that, out of 500 newborn babies, at least 290 of them will be boys?

Solution: Use a normal approximation with $\mu = 500(0.53) = 265$ and $\sigma = \sqrt{500(.53)(.47)} = 11.160$. We want $P(X > 289.5)$, using the correction for continuity.

$$\begin{aligned} P(X > 289.5) &= P\left(Z > \frac{289.5 - 265}{11.160}\right) \\ &= P(Z > 2.20) \\ &= 1 - P(Z < 2.20) = 0.0139 \end{aligned}$$ ■

■ **Example 23.** It is estimated that 3% of brand X flashbulbs fail to work. In a randomly selected sample of 1000 flashbulbs, what is the probability that less than 25 fail.

Solution: Use a normal approximation with $\mu = 1000(0.03) = 30$ and $\sigma = \sqrt{1000(0.03)(0.97)} = 5.394$. Then $P(X < 25) = P(Y < 24.5)$ where Y is the approximating normal distribution and we use the continuity correction. (Note that $X < 25$ means $X \le 24$, so we use $Y < 24.5$).

$$\begin{aligned} \text{So } P(X < 25) &= P(Y < 24.5) \\ &= P\left(Z < \frac{24.5 - 30}{5.394}\right) \\ &= P(Z < -1.02) \\ &= 0.1539 \end{aligned}$$

■

Exercise Set 8.4

For exercises 1–8, X is a binomial random variable with $n = 150$ and $p = 0.4$. Use a normal approximation to the binomial to find the indicated probability.

1. $P(X \leq 69)$
2. $P(X \geq 63)$
3. $P(X < 75)$
4. $P(X > 67)$
5. $P(X < 55)$
6. $P(X \geq 49)$
7. $P(45 \leq X \leq 70)$
8. $P(57 \leq X \leq 67)$

For exercises 9–12, assume the given distribution is binomial and use a normal approximation to find the probabilities.

9. The Betterbilt Battery Company claims that only 1% of their batteries are defective. A random sample of 6000 batteries is tested and X represents the number of defective batteries. Find
  a. $P(X \geq 70)$.
  b. $P(X \leq 55)$.
  c. $P(50 \leq X \leq 65)$.
  d. $P(60 \leq X \leq 65)$.
  e. $P(40 \leq X \leq 55)$.
  f. $P(65 \leq X \leq 80)$.

10. The probability of a newborn baby being a boy is 0.53. Let X be the number of boys born out of the next 1000 births at County Hospital. Find
  a. $P(X \geq 550)$.
  b. $P(X \leq 500)$.
  c. $P(525 \leq X \leq 535)$.
  d. $P(490 \leq X \leq 520)$.
  e. $P(535 \leq X \leq 575)$.

11. According to American Red Cross estimates, about 85% of the American people have Rh+ blood. In a randomly selected group of 5000 Americans, let X represent the number having Rh+ blood. Find
  a. $P(X \geq 4300)$.
  b. $P(X \leq 4175)$.
  c. $P(4230 \leq X \leq 4260)$.
  d. $P(4190 \leq X \leq 4240)$.
  e. $P(4270 \leq X \leq 4315)$.

12. The quarterback of a pro football team completes 55% of his passes. If he throws 480 passes this season, find the probability that he completes
  a. at least 280.
  b. at most 240.
  c. at least 250 but not more than 270.
  d. between 260 and 290, inclusive.
  e. between 245 and 260, inclusive.

8.5 Chapter Review

Important Terms and Symbols in Chapter 8.

random variable (section 8.1)
discrete random variable (section 8.1)
continuous random variable (section 8.1)
probability distribution (section 8.1)
discrete probability distributions (section 8.1)
expected value (section 8.1)
mean (section 8.1)
variance (section 8.1)
standard deviation (section 8.1)
binomial experiment (section 8.2)
binomial probabilities (section 8.2)
binomial random variable (section 8.2)
$\binom{n}{k} p^k(1 - p)^{n-k}$ (section 8.2)
normal distribution, continuous probability distribution (section 8.3)
symmetry (section 8.3)
standard normal distribution (section 8.3)
normal approximation to binomial (section 8.4)
continuity correction (section 8.4)

Review Exercises

1. You walk into a local store and the clerk tells you they are having a "determine your discount" sale. In a box they have 25 tags marked 10%, 15 tags marked 20%, 6 tags marked 30%, 3 tags marked 40%, and 1 tag marked 50%. You draw a tag and you get that discount on anything you buy. Find the probability distribution for this experiment.
2. An urn contains 11 red marbles and 9 green marbles. You reach in and take 3 marbles simultaneously, noting the number of red marbles you get. Find the probability distribution for this experiment.
3. Suppose instead of drawing 3 marbles simultaneously as in problem 2, you draw three marbles, one at a time, replacing each marble after it is drawn. Find the probability distribution for the random variable denoting the number of red marbles obtained.
4. Five fair coins are flipped simultaneously with the random variable X denoting the number of heads.
   a. Find the probability distribution of X.
   b. Find the probability distribution of the random variable $Y = X + 5$.
   c. Find the probability distribution of $Z = X^2$.
5. If X is a binomial random variable with $n = 20$ and $p = 0.7$, find
   a. μ
   b. σ^2
   c. σ
   d. $P(X \geq 15)$
   e. $P(X \leq 10)$
   f. $P(8 \leq X < 12)$
   g. $P(10 < X < 15)$
   h. $P(X \leq 17)$
   i. $P(12 \leq X \leq 16)$
   j. $P(15 \leq X \leq 17)$

6. If X is a binomial random variable with $n = 19$ and $p = 0.4$, find

a. μ

b. σ^2

c. σ

d. $P(X = 8)$

e. $P(X = 7)$

f. $P(7 \leq X < 12)$

g. $P(X \leq 10)$

h. $P(X \geq 12)$

i. $P(6 \leq X \leq 11)$

j. $P(3 \leq X \leq 9)$

7. If 80% of all boys in your area have had shots for mumps, what is the probability that at least 1 of 5 randomly selected boys has not had his shot?

8. A baseball player gets on base by some method 40% of the time. If he comes to bat 5 times in today's game, what is the probability that he gets on base

a. exactly 2 times?

b. at least 4 times?

c. at least 2 times?

d. no times?

9. A basketball player makes 70% of his free throws. In a game he gets 8 free throws. What is the probability that he makes

a. at least 6?

b. at most 5?

c. at least 4 but not more than 7?

d. fewer than 3?

10. If 60% of the voters in a community oppose a school tax proposal and the rest support the proposal, what is the probability that in a random sample of 10 voters

a. more than half favor the proposal?

b. at least 3 favor the proposal?

c. at least 6 oppose the proposal?

For exercises 11–14, assume all conditions for a binomial distribution are satisfied.

11. The school health nurse has estimated that 90% of the students have had their required immunizations. In a randomly selected group of 15 students, what is the probability that

a. at least 10 have had their immunizations?

b. less than 12 have had their immunizations?

c. at least 3 have not had their immunizations?

12. The county health department has estimated that 20% of the people who have received flu shots will still get flu because of the seriousness of the outbreak. If a group of 14 people who have had the shot is randomly chosen, what is the probability that

a. at least 6 will get flu?

b. no more than 3 will get flu?

c. at least 2 but no more than 5 will get flu?

d. none of the 14 will get flu?

13. Based on statistics from the last 5 years, the local police have estimated that the probability of a conviction on a charge of child abuse is 0.4. In a randomly selected group of 12 people charged with child abuse, what is the probability that

a. at least 8 are convicted?

b. less than 5 are convicted?

c. at least 4 but less than 9 are convicted?
d. at least 2 are convicted?

14. The conservation department has estimated that 30% of the deer in a certain county will not survive the extremely severe winter. What is the probability that, in a herd of 11 deer

a. at least 8 survive?
b. less than 5 survive?
c. less than 4 die?
d. none survive?

15. Let Z be the standard normal random variable ($\mu = 0$, $\sigma = 1$). Find the indicated probabilities.

a. $P(Z > 1.35)$
b. $P(Z < 1.01)$
c. $P(Z > -1.31)$
d. $P(Z < -0.75)$
e. $P(1.01 < Z < 1.79)$
f. $P(-2.02 < Z < -1.00)$
g. $P(-0.43 < Z < 1.70)$
h. $P(-1.57 < Z < 1.57)$
i. $P(-2.01 < Z < 1.21)$
j. $P(0.00 < Z < 1.20)$
k. $P(-1.71 < Z < 0.00)$
l. $P(-0.05 < Z < 2.05)$

16. If X is a normal random variable with $\mu = 150$ and $\sigma = 15$, find

a. $P(X < 175)$
b. $P(X < 140)$
c. $P(X > 130)$
d. $P(X > 168)$
e. $P(139 < X < 169)$
f. $P(110 < X < 145)$
g. $P(145 < X < 195)$
h. $P(160 < X < 190)$

In problems 17–22, assume the distribution is normal.

17. The state police have estimated that the average speed on a section of interstate highway is 58 miles per hour (mph) with a variance of $\sigma^2 = 20$. If a randomly selected vehicle is observed, what is the probability that

a. it is exceeding 60 mph?
b. its speed is between 55 and 65 mph?
c. it is within the legal speed limits of 40 and 55 mph?

18. The mean weight of a company's bag of sugar is 5 pounds with a standard deviation of $\sigma = 0.08$. What is the probability that a randomly selected bag weighs

a. less than 4.8 pounds?
b. more than 5.2 pounds?
c. between 4.9 and 5.1 pounds?
d. between 4.8 and 5.2 pounds?

19. A vending company claims that their coffee machine dispenses, on the average, 8 ounces of coffee with a standard deviation of $\sigma = 0.2$. If the company's figures are correct, what is the probability that you will get a cup with

a. more than 8.3 ounces?
b. less than 8.1 ounces?
c. less than 7.6 ounces?
d. between 7.9 and 8.1 ounces?

20. A nationally standardized exam has a mean of 22 with a standard deviation of 2.05. If a student who has taken the exam is randomly chosen, what is the probability this student scored

a. below 16?
b. above 26?
c. between 17 and 27?
d. between 20 and 24?

21. A company claims that the mean life of their deluxe car battery is 60 months with a standard deviation of $\sigma = 4.8$. What is the probability that a randomly selected test battery will last
    a. less than 48 months?
    b. more than 66 months?
    c. more than 75 months?
    d. between 48 and 74 months?
22. According to government tests, a certain auto model has a mean gas mileage of 36 miles per gallon (mpg) on the highway with a variance of $\sigma^2 = 10.24$. If a randomly selected car is selected and driven, what is the probability it will get
    a. at least 32 mpg?
    b. less than 30 mpg?
    c. more than 39 mpg?
    d. between 33 and 40 mpg?

In exercises 23 and 24, use a normal approximation to the binomial to find the probabilities.

23. Mayor Horace Horrible announced at a press conference that 80% of the city's voters will support his proposed city income tax so government officials can get a salary increase. The local radio station decides to do a random survey to check the mayor's claim. They question 400 voters. Assuming the mayor's claim is correct,
    a. how many of the 400 voters would be expected to support the tax?
    b. what is the probability that less than 300 of the voters in the sample support the tax?
    c. what is the probability that at least 340 people in the sample support the tax?
    d. what is the probability that at least 310 but at most 340 of the voters in the sample support the tax?
24. A medical group has estimated that 30% of all school children have some type of allergy. Assuming this claim is true, in a random sample of 1,000 children,
    a. how many would be expected to have some type of allergy?
    b. what is the probability that at least 275 have an allergy?
    c. what is the probability that less than 260 have an allergy?
    d. what is the probability that at least 290 but less than 330 have an allergy?

9 Markov Chains

Introduction

Various problems involving shifting and sorting of fixed populations began to appear with increasing frequency in the various disciplines around the turn of the century. These included practical problems involving the business and commercial communities of the world as well as the scientific community. These problems were approached for a very long time using only probability urn models and tree diagrams. A. A. Markov (1856–1922) published his pioneer work in this area in 1906. However it was not until the Russian, A. Kolmogorov, published a paper in 1936 that this process drew wide public attention. Kolmogorov's paper contained no proofs, a rare thing in mathematics, but it was so powerful that a complete exposition appeared in a French journal one year later, in 1937.

While we will not address ourselves to a full investigation of **Markov chains** it is appropriate to look at some problems from this important branch of mathematics. Recognizing that problems of the type studied here have solutions may be the practical result.

9.1 Transition Matrices

Suppose a farmer uses a regular crop rotation system for his tillable land. This year, he planted 50% in corn, 30% in soybeans, and the remaining 20% was used for other purposes. Each year, $\frac{1}{2}$ of the previous year's corn land is planted in corn again, $\frac{1}{6}$ is converted to soybeans, and $\frac{1}{3}$ is converted to other use. The previous year's soybean land is divided as $\frac{1}{4}$ in corn, $\frac{1}{2}$ in soybeans, and $\frac{1}{4}$ for other uses. In addition, of the land used last year

for other purposes, $\frac{1}{6}$ is converted to corn, $\frac{1}{3}$ is converted to soybeans, and $\frac{1}{2}$ is used for other purposes again. What will the farmer's land distribution be next year if he follows this plan?

We are dealing here with an initial population, the farm land, and the way it is partitioned according to its use. These different uses (corn, soybeans, other) are referred to as **states.** The row matrix $D = [.50 \quad .30 \quad .20]$ indicating the proportion of land allocated for each use is called the **distribution matrix.** The distribution matrix can be given in percentages or decimals as we have done here, or it can be given in terms of the actual number of acres set aside for each type of use.

The change, or transition, in land use can be represented in table form. Each row of the table indicates the portion of land going from the initial state to a given state.

From	**To:** *corn*	*soybeans*	*other*
corn	$\frac{1}{2}$	$\frac{1}{6}$	$\frac{1}{3}$
soybeans	$\frac{1}{4}$	$\frac{1}{2}$	$\frac{1}{4}$
other	$\frac{1}{6}$	$\frac{1}{3}$	$\frac{1}{2}$

If the column and row headings are deleted, this table can be written as the matrix

$$T = \begin{bmatrix} \frac{1}{2} & \frac{1}{6} & \frac{1}{3} \\ \frac{1}{4} & \frac{1}{2} & \frac{1}{4} \\ \frac{1}{6} & \frac{1}{3} & \frac{1}{2} \end{bmatrix}.$$

This matrix is called the **transition matrix.** This matrix describes the transition, or change, in land use.

We can now answer the question posed in a previous paragraph. That is, what will the land use distribution be next year under this plan? To find the distribution, we compute the matrix product DT where

$$D = [.50 \quad .30 \quad .20] = [\tfrac{1}{2} \quad \tfrac{3}{10} \quad \tfrac{1}{5}].$$

$$DT = [\tfrac{1}{2} \quad \tfrac{3}{10} \quad \tfrac{1}{5}]\begin{bmatrix} \frac{1}{2} & \frac{1}{6} & \frac{1}{3} \\ \frac{1}{4} & \frac{1}{2} & \frac{1}{4} \\ \frac{1}{6} & \frac{1}{3} & \frac{1}{2} \end{bmatrix}$$

$$= [\tfrac{43}{120} \quad \tfrac{36}{120} \quad \tfrac{41}{120}].$$

The farmer will have about 35.8% of his land in corn, 30% in soybeans, and 34.2% in other uses for next year.

We are now ready to describe the model in more general terms. Important terms are repeated for emphasis.

In this model we start and end with the same population which is partitioned into several well-defined subsets. These subsets, called *states*, will contain all the elements of the population. The original partition is presented as a **distribution matrix.** A **stochastic process** is the periodic rearranging of the elements into different states. Each rearrangement may result in any possible permutation. The actual device which defines the change is called a **transition matrix.**

The entries in any transition matrix refer to the percent of the population shifted from state to state. It is common practice to present these entries as fractions or decimals. In certain problems the matrix entries, being fractions, are used as the probability of a change. If, for example, a matrix entry is $\frac{1}{3}$, indicating $\frac{1}{3}$ of the elements in state A change to state B then it would be equally correct to interpret this as "the probability is $\frac{1}{3}$ that one element, selected at random from state A, changes to state B." Every transition matrix must have the following properties.

1. It is a square matrix which contains only non-negative entries.
2. Every entry is a real number p such that $0 \leq p \leq 1$.
3. The sum of the entries in every row equals 1.

These properties assure the following two results.

1. The product of any distribution matrix with a transition matrix produces a new distribution matrix.
2. The product of two transition matrices is a transition matrix.

■ **Example 1.** Given a distribution matrix $D = [500 \quad 300 \quad 200]$ and transition matrices

$$T_1 = \begin{bmatrix} 1 & 0 & 0 \\ 0 & \frac{1}{2} & \frac{1}{2} \\ \frac{1}{5} & 0 & \frac{4}{5} \end{bmatrix} \text{ and } T_2 = \begin{bmatrix} 0 & 1 & 0 \\ \frac{1}{3} & 0 & \frac{2}{3} \\ 0 & 0 & 1 \end{bmatrix}$$

find and analyze

a. DT_1 **b.** DT_2 **c.** T_1^2 **d.** DT_1T_2 **e.** DT_2T_1.

Solutions:

a. $$DT_1 = [500 \quad 300 \quad 200]\begin{bmatrix} 1 & 0 & 0 \\ 0 & \frac{1}{2} & \frac{1}{2} \\ \frac{1}{5} & 0 & \frac{4}{5} \end{bmatrix} = [540 \quad 150 \quad 310]$$

$D = [500 \quad 300 \quad 200]$ indicated that the original distribution had 500 elements in state 1, 300 elements in state 2, and 200 elements in state 3. $DT_1 = [540 \quad 150 \quad 310]$ is the resulting rearrangement induced by the transition matrix. State 1 now contains 540 elements, state 2 now contains 150 elements, and state 3 now contains 310 elements. Observe that this is only a rearrangement of the 1000 elements, i.e. the sum of the elements in both D and DT_1 is 1000.

b. $$DT_2 = [500 \quad 300 \quad 200]\begin{bmatrix} 0 & 1 & 0 \\ \frac{1}{3} & 0 & \frac{2}{3} \\ 0 & 0 & 1 \end{bmatrix} = [100 \quad 500 \quad 400]$$

DT_2 is yet another rearrangement of the initial distribution.

c. $$T_1^2 = T_1 \cdot T_1 = \begin{bmatrix} 1 & 0 & 0 \\ 0 & \frac{1}{2} & \frac{1}{2} \\ \frac{1}{5} & 0 & \frac{4}{5} \end{bmatrix}\begin{bmatrix} 1 & 0 & 0 \\ 0 & \frac{1}{2} & \frac{1}{2} \\ \frac{1}{5} & 0 & \frac{4}{5} \end{bmatrix} = \begin{bmatrix} 1 & 0 & 0 \\ \frac{1}{10} & \frac{1}{4} & \frac{13}{20} \\ \frac{9}{25} & 0 & \frac{16}{25} \end{bmatrix}$$

$T_1^2 = T_1T_1$ is itself a transition matrix, since

1. it is a square matrix whose entries are non-negative,
2. each entry, p, is such that $0 \leq p \leq 1$, and
3. the sum of the entries in each row is 1.

i.e. row 1: $1 + 0 + 0 = 1$

row 2: $\frac{1}{10} + \frac{1}{4} + \frac{13}{20} = 1$

and row 3: $\frac{9}{25} + 0 + \frac{16}{25} = 1$

d. $D(T_1T_2) = [500 \quad 300 \quad 200]\begin{bmatrix} 1 & 0 & 0 \\ 0 & \frac{1}{2} & \frac{1}{2} \\ \frac{1}{5} & 0 & \frac{4}{5} \end{bmatrix}\begin{bmatrix} 0 & 1 & 0 \\ \frac{1}{3} & 0 & \frac{2}{3} \\ 0 & 0 & 1 \end{bmatrix}$

$$= [500 \quad 300 \quad 200]\begin{bmatrix} 0 & 1 & 0 \\ \frac{1}{6} & 0 & \frac{5}{6} \\ 0 & \frac{1}{5} & \frac{4}{5} \end{bmatrix} = [50 \quad 540 \quad 410]$$

$$(DT_1)T_2 = [540 \quad 150 \quad 310]\begin{bmatrix} 0 & 1 & 0 \\ \frac{1}{3} & 0 & \frac{2}{3} \\ 0 & 0 & 1 \end{bmatrix} = [50 \quad 540 \quad 410]$$

Observe that the single transition matrix $T_1T_2 = \begin{bmatrix} 0 & 1 & 0 \\ \frac{1}{6} & 0 & \frac{5}{6} \\ 0 & \frac{1}{5} & \frac{4}{5} \end{bmatrix}$ has the same result as the transition matrices T_1 and T_2 applied one at a time.

e. $(DT_2)T_1 = [180 \quad 250 \quad 570]$

Observe in parts **d** and **e** that $DT_2T_1 \neq DT_1T_2$. This means that the order in which transition matrices are applied may make a big difference in the result. ■

■ Example 2. The 1000 employees of a certain factory took their vacations in the months of June, July, and August last year according to table D.

Table D

	June	*July*	*August*
No. of people on vacation	500	300	200

This year the following article appears in the company paper.

Due to production requirements it is necessary to rearrange the vacation schedule for this year. Management will try to give everyone their first choice so turn in your request as soon as possible. The following guidelines will be used. Everyone who took their vacation in June last year will take it in June this year. Half of those who were on vacation in July last year will still be off in July but the other half will have their vacation moved to August. Also, 80% of those who were off in August last year will still be off in August while the remaining 20% must take their vacations in June.

a. How many people will be on vacation in each of the months June, July, and August this year?
b. If the company does the same thing next year what will be the result?

Solution: The transition matrix is determined from the following table.

From \ To	*June*	*July*	*August*
June	100%	0	0
July	0	50%	50%
August	20%	0	80%

$$T_1 = \begin{bmatrix} 1 & 0 & 0 \\ 0 & \frac{1}{2} & \frac{1}{2} \\ \frac{1}{5} & 0 & \frac{4}{5} \end{bmatrix}$$

$DT_1 = [540 \quad 150 \quad 310]$ (see part **a** of example 1). Hence, the company will have 540 employees vacationing in June, 150 in July, and 310 in August.

b. $$DT_1^2 = [500 \quad 300 \quad 200]\begin{bmatrix} 1 & 0 & 0 \\ \frac{1}{10} & \frac{1}{4} & \frac{13}{20} \\ \frac{9}{25} & 0 & \frac{16}{25} \end{bmatrix} = [602 \quad 75 \quad 323]$$ ■

A **Markov process** is a stochastic process where the transition matrix depends on the preceding outcome. Specifically, this means that the entries in the transition matrix are conditional probabilities. A **Markov chain** is a Markov process with a fixed transition matrix. That is, the transition matrix is the same at every stage. This fixed matrix is often called a **Markov matrix.**

■ Example 3. In a certain county in the year 1972, 50% of the voters who voted Democratic in the last election again voted Democratic; 20% switched to vote Republican and 30% did not vote at all. 60% of the voters who voted Republican in the last election voted Republican again while 20% switched to vote Democratic and 20% did not vote. It is also known that 40% of the people who did not vote in the last election voted Democratic in this one, 30% voted Republican, and the other 30% did not vote.

a. Construct a transition matrix reflecting this data.

Solution: The data are first expressed as a table with well defined row and column headings and then simplified into matrix form.

From \ To	D	R	No vote
D	.50	.20	.30
R	.20	.60	.20
No vote	.40	.30	.30

$$T = \begin{bmatrix} .5 & .2 & .3 \\ .2 & .6 & .2 \\ .4 & .3 & .3 \end{bmatrix}$$

b. In the election before 1972, 500 voted Democratic, 400 voted Republican, and 600 did not vote. What was the voting pattern in the 1972 election?

Solution: First construct a distribution matrix, then multiply by the transition matrix to obtain the new distribution matrix.

$$[500 \quad 400 \quad 600]\begin{bmatrix} .5 & .2 & .3 \\ .2 & .6 & .2 \\ .4 & .3 & .3 \end{bmatrix} = [570 \quad 520 \quad 410]$$

c. Assume 1976 and 1980 produced voting patterns described by the transition matrices R and S respectively. What was the result of these elections?

$$R = \begin{bmatrix} .4 & .4 & .2 \\ .3 & .5 & .2 \\ .4 & .3 & .3 \end{bmatrix}, \; S = \begin{bmatrix} .4 & .1 & .5 \\ .2 & .6 & .2 \\ .2 & .2 & .6 \end{bmatrix}.$$

Solutions: In 1976 the Republicans win.

$$[570 \quad 520 \quad 410]R = [548 \quad 611 \quad 341]$$

In 1980 the Republicans win.

$$[548 \quad 611 \quad 341]S = [409 \quad 490 \quad 601]$$

d. If the transition matrix T had remained the same in all three elections this would have been a Markov chain. What would have been the result of the 1980 election?

Solution: The voting pattern in 1980 would be the product of the initial distribution and the cube of the Markov matrix T.

$$DT^3 = [500 \quad 400 \quad 600]\begin{bmatrix} .5 & .2 & .3 \\ .2 & .6 & .2 \\ .4 & .3 & .3 \end{bmatrix}^3$$

$$= [500 \quad 400 \quad 600]\begin{bmatrix} .379 & .352 & .269 \\ .338 & .408 & .254 \\ .368 & .367 & .265 \end{bmatrix}$$

$$= [546 \quad 559 \quad 395]$$

It looks as if the Republicans have won a very close election. However, remember that in the real world the entries in the transition matrix were statistically approximated and may vary from sample to sample. The results are, for the most part, used only to indicate trends. This election would have been too close to call! ■

Brand switching patterns provide another illustration of the use of Markov processes. A company may invest much time, money and effort on "consumer juries" to try to predict market patterns resulting from brand switching.

When these pattern changes are used together with the Markov model, predictions can be made regarding future market behavior. The information contained in the transition matrix may indicate where to attack a problem, if there is one, and give some indication as to how much time is needed to eliminate the problem.

When the Markov model is used directly on a probability problem the entries in the distribution matrix as well as the transition matrix may be themselves probabilities.

■ **Example 4.** The initial probability that an element of the population is in state A is $\frac{1}{2}$, in state B is $\frac{1}{4}$, and in state C is $\frac{1}{4}$. The transition matrix is known to be

$$\begin{bmatrix} \frac{1}{2} & 0 & \frac{1}{2} \\ 1 & 0 & 0 \\ \frac{1}{3} & \frac{1}{3} & \frac{1}{3} \end{bmatrix}$$

What is the probability that an element is in each of the states after the first transition?

Solution:

$$\begin{bmatrix} \frac{1}{2} & \frac{1}{4} & \frac{1}{4} \end{bmatrix} \begin{bmatrix} \frac{1}{2} & 0 & \frac{1}{2} \\ 1 & 0 & 0 \\ \frac{1}{3} & \frac{1}{3} & \frac{1}{3} \end{bmatrix} = \begin{bmatrix} \frac{7}{12} & \frac{1}{12} & \frac{4}{12} \end{bmatrix}.$$

The probability of being in state $A = \frac{7}{12}$,

in state $B = \frac{1}{12}$,

and in state $C = \frac{4}{12}$. ■

Exercise Set 9.1

1. Which of the following is not a transition matrix? Why?

a. $\begin{bmatrix} \frac{1}{2} & \frac{1}{2} \\ 1 & 0 \end{bmatrix}$ **b.** $\begin{bmatrix} \frac{1}{3} & \frac{2}{3} \\ 0 & 0 \end{bmatrix}$ **c.** $\begin{bmatrix} \frac{1}{2} & \frac{1}{4} & \frac{1}{4} \\ \frac{1}{3} & \frac{1}{3} & \frac{1}{3} \end{bmatrix}$ **d.** $\begin{bmatrix} -\frac{1}{2} & 1 & \frac{1}{2} \\ \frac{1}{3} & 0 & \frac{2}{3} \\ 1 & 0 & 0 \end{bmatrix}$ **e.** $\begin{bmatrix} .2 & .3 & .4 \\ 1 & 0 & 0 \\ 0 & .6 & .4 \end{bmatrix}$

2. Given the distribution matrix $D = [72 \quad 64 \quad 36]$ and the Markov matrix $T = \begin{bmatrix} \frac{1}{2} & \frac{1}{2} & 0 \\ 0 & \frac{3}{4} & \frac{1}{4} \\ \frac{1}{2} & 0 & \frac{1}{2} \end{bmatrix}$, compute

   a. DT **b.** DT^2 **c.** DT^4

3. A survey reveals the following patterns of change in the buying habits of customers in a fast food store. 70% of the customers who buy a big-burger will order one the next time they come in while 20% will try the chicken-wich and 10% try the fish-wich. 50% of those who eat chicken-wich will order one again but 30% will order a big-burger and the remaining 20% will order a fish-wich. 40% who had fish before will order a burger, 40% will have chicken, and the remaining 20% will have fish again.
   a. Express the data as a transition matrix.
   b. The fast food store sales last week were 40,000 big-burgers, 30,000 chicken-wiches and 20,000 fish-wiches. Express this as a distribution matrix and predict next week's sales subject to the transition matrix in part **a.**
   c. Express the distribution matrix as a probability matrix, i.e., the sum of the entries equals 1. Interpret this matrix in words.
   d. Use **c.** to find the probability that if a customer walks in the door he will order **1.** big-burger, **2.** chicken-wich, **3.** fish-wich.

4. Market research indicates the brand-switching trends indicated in the charts below. The symbol in figure 9.1 is read 20% change from A to B, 30% change from A to C and 50%, not shown, remained with A. Express each as a transition matrix.

a.

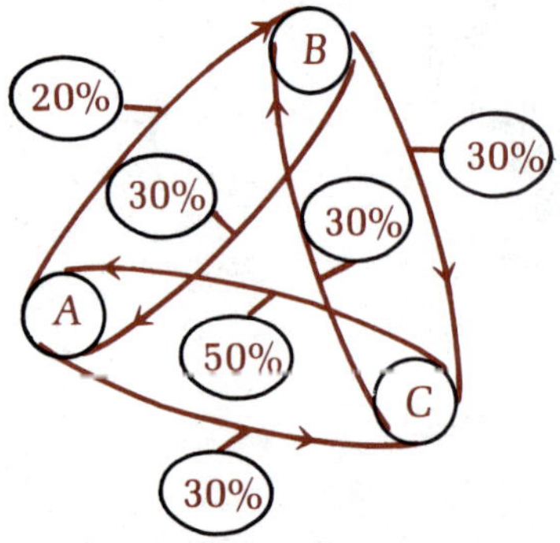

Figure 9.1

b.

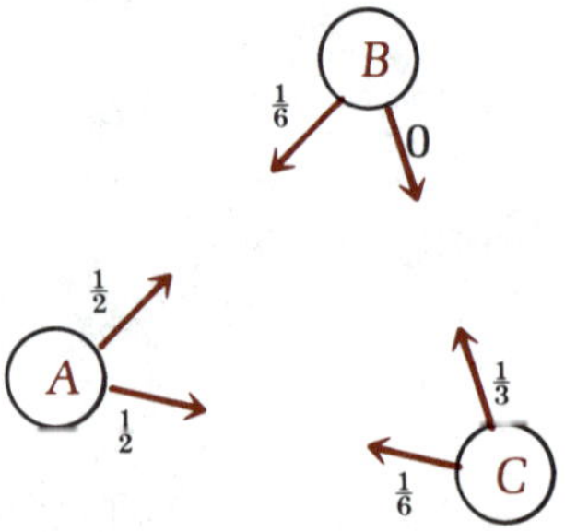

Figure 9.2

c.

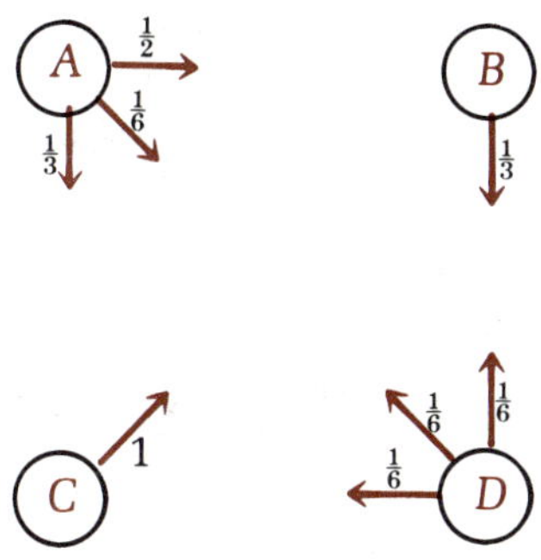

Figure 9.3

d.

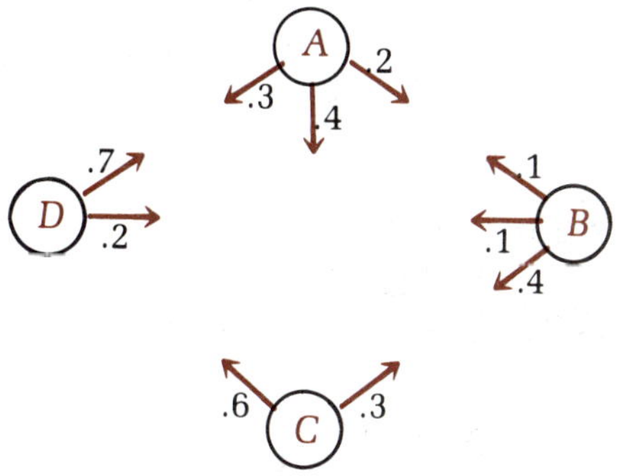

Figure 9.4

5. If $A = \begin{bmatrix} \frac{1}{2} & \frac{1}{2} & 0 \\ 0 & 1 & 0 \\ \frac{1}{6} & \frac{1}{3} & \frac{1}{2} \end{bmatrix}$, $B = \begin{bmatrix} \frac{1}{6} & 0 & \frac{5}{6} \\ 1 & 0 & 0 \\ 0 & \frac{1}{3} & \frac{2}{3} \end{bmatrix}$ and $D = [\frac{1}{3} \quad \frac{1}{2} \quad \frac{1}{6}]$

compute

a. DA **b.** DA^2 **c.** DB **d.** DAB **e.** DBA

6. Initially $\frac{1}{2}$ of the members of a local chapter of Alpha Beta Zip agree to give blood. Three months later $\frac{3}{4}$ of those who gave before agree to give again and $\frac{1}{2}$ of those who did not give last time agree to give this time. What is the probability that a member chosen at random will give blood the second time?

7. Assuming the transition matrix in problem 6 remains unchanged and the blood drive repeats every three months, what will be the distribution matrix one year from now?

8. In a three channel TV market, a survey reveals the following viewing habits from 8 to 11 p.m. on Fridays. At 8 o'clock $\frac{1}{4}$ of the sets are tuned to channel 2, $\frac{1}{4}$ are tuned to channel 3, $\frac{1}{6}$ to channel 7 and $\frac{1}{3}$ are not on. Every hour $\frac{1}{6}$ of the channel 2 viewers change to channel 3, $\frac{1}{6}$ change to channel 7, and $\frac{1}{4}$ turn the set off. During the same hour $\frac{1}{4}$ of the channel 3 viewers switch to channel 2, $\frac{1}{4}$ switch to channel 7 and $\frac{1}{4}$ turn off the set. $\frac{1}{3}$ of the channel 7 viewers switch to channel 2, $\frac{1}{6}$ switch to channel 3, and $\frac{1}{3}$ turn off the set. Three-fourths of the sets that are off remain off while those which are turned on are equally divided among the three stations.

a. Construct an initial distribution matrix and a transition matrix.
b. What is the distribution at 9 o'clock?
c. What is the distribution at 10 o'clock?
d. What is the distribution at 11 o'clock?

9.2 Steady States

A transition matrix T is called a **regular transition matrix** if there exists some power of T which has no zero entries. Intuitively, then, in a regular transition matrix the elements may move freely from state to state.

■ Example 5. Which of the following matrices are regular?

$$A = \begin{bmatrix} \frac{1}{2} & \frac{1}{2} \\ 1 & 0 \end{bmatrix} \quad B = \begin{bmatrix} \frac{1}{2} & \frac{1}{2} \\ 0 & 1 \end{bmatrix} \quad C = \begin{bmatrix} 0 & 1 & 0 \\ 0 & \frac{1}{2} & \frac{1}{2} \\ 0 & \frac{1}{2} & \frac{1}{2} \end{bmatrix}$$

Solution: A is regular since $A^2 = \begin{bmatrix} \frac{3}{4} & \frac{1}{4} \\ \frac{1}{2} & \frac{1}{2} \end{bmatrix}$ has no zero entries. B is not regular since the (2,1) entry of B^N equals zero for all N. C is not regular since $C^N = C^2$ for all $N > 1$ and C^2 has zero entries. ■

Theorem. If a Markov chain has a regular transition matrix then it will produce, finally, a distribution which is in equilibrium. This state of equilibrium is referred to as a **steady state.**

For an example of a system which is in a steady state consider the distribution matrix $[10 \quad 30]$ together with the transition matrix $\begin{bmatrix} .1 & .9 \\ .3 & .7 \end{bmatrix}$.

$[10 \quad 30] \begin{bmatrix} .1 & .9 \\ .3 & .7 \end{bmatrix} = [10 \quad 30]$. In simplistic terms, if I have 10 cows and you have 30 and every day I give you 90% of my cows and receive in return 30% of your cows we will each have 10 and 30 cows respectively every day forever. The theorem states that no matter how we start out, we will eventually end up with my having 10 cows and your having 30. I could have had all 40 cows to start with and you none or we could have had 20 cows each or you could have had 38 and I 2 cows. In each case we would have found this state of equilibrium (as illustrated in the following examples). The number of cows has been rounded off to eliminate fractions.

$$[40 \quad 0]\begin{bmatrix} .1 & .9 \\ .3 & .7 \end{bmatrix} = [4 \quad 36]$$

$$[4 \quad 36]\begin{bmatrix} .1 & .9 \\ .3 & .7 \end{bmatrix} = [11.2 \quad 28.8] \simeq [11 \quad 29]$$

$$[11 \quad 29]\begin{bmatrix} .1 & .9 \\ .3 & .7 \end{bmatrix} = [9.8 \quad 30.2] \simeq [10 \quad 30]$$

The steady state is reached after 3 transitions (days).

$$[20 \quad 20]\begin{bmatrix} .1 & .9 \\ .3 & .7 \end{bmatrix} = [8 \quad 32]$$

$$[8 \quad 32]\begin{bmatrix} .1 & .9 \\ .3 & .7 \end{bmatrix} = [10.4 \quad 29.6] \simeq [10 \quad 30]$$

The steady state is reached after 2 transitions (days).

$$[0 \quad 40]\begin{bmatrix} .1 & .9 \\ .3 & .7 \end{bmatrix} = [12 \quad 28]$$

$$[12 \quad 28]\begin{bmatrix} .1 & .9 \\ .3 & .7 \end{bmatrix} = [9.6 \quad 30.4] \simeq [10 \quad 30]$$

The steady state is reached after 2 transitions (days).

In most mathematical computations, rounding off decimal fractions would produce arithmetic errors. However, since our theorem indicates that a unique steady state will be reached, changing the distribution matrix does not introduce any error. If these fractions were not rounded off successive permutations could go on endlessly, only approaching the steady state. In any case there is a better way to determine exactly the equilibrium distribution. When the system is in equilibrium the following condition will exist.

$$[x_1 \quad x_2]\begin{bmatrix} .1 & .9 \\ .3 & .7 \end{bmatrix} = [x_1 \quad x_2]$$

To find our solution $[x_1 \quad x_2] = [10 \quad 30]$ we solve this matrix equation.

$$[x_1 \quad x_2]\begin{bmatrix} .1 & .9 \\ .3 & .7 \end{bmatrix} - [x_1 \quad x_2] = [0 \quad 0]$$

$$[x_1 \quad x_2]\left(\begin{bmatrix} .1 & .9 \\ .3 & .7 \end{bmatrix} - \begin{bmatrix} 1 & 0 \\ 0 & 1 \end{bmatrix}\right) = [0 \quad 0]$$

$$[x_1 \quad x_2]\begin{bmatrix} -.9 & .9 \\ .3 & -.3 \end{bmatrix} = [0 \quad 0]$$

The last equation is a 2 × 2 system of equations $\begin{cases} -.9x_1 + .3x_2 = 0 \\ .9x_1 - .3x_2 = 0 \end{cases}$. This system of equations is *not* independent. This is not an accident; these systems will always be dependent. There is however one additional constraint which must be met, namely the sum $(x_1 + x_2)$ remains constant. In this example $x_1 + x_2 = 40$. To find x_1 and x_2 we solve the system $\begin{cases} x_1 + x_2 = 40 \\ -.9x_1 + .3x_2 = 0 \\ .9x_1 - .3x_2 = 0 \end{cases}$. Using techniques already mastered in chapter 2, we obtain

$$\left[\begin{array}{cc|c} 1 & 1 & 40 \\ -.9 & .3 & 0 \\ .9 & -.3 & 0 \end{array}\right] \sim \left[\begin{array}{cc|c} 1 & 1 & 40 \\ -9 & 3 & 0 \\ 0 & 0 & 0 \end{array}\right]$$

$$\sim \left[\begin{array}{cc|c} 1 & 1 & 40 \\ 0 & 12 & 360 \end{array}\right] \sim \left[\begin{array}{cc|c} 1 & 1 & 40 \\ 0 & 1 & 30 \end{array}\right] \sim \left[\begin{array}{cc|c} 1 & 0 & 10 \\ 0 & 1 & 30 \end{array}\right]$$

Therefore $x_1 = 10$ and $x_2 = 30$ and the steady state becomes $[10 \quad 30]$.

It has become standard practice to let x_1 and x_2 be the fractional parts of the population, i.e., in the preceding example, $[x_1 \quad x_2] = [\frac{1}{4} \quad \frac{3}{4}]$. In this way x_1 is $\frac{1}{4}$ of the total population and x_2 is the other $\frac{3}{4}$. Looking at our problem in this way our system of equations would have become $\left[\begin{array}{cc|c} 1 & 1 & 1 \\ -.9 & .3 & 0 \\ .9 & -.3 & 0 \end{array}\right]$. Here the equation $x_1 + x_2 = 1$ is used. The solution to this system $\left[\begin{array}{cc|c} 1 & 0 & \frac{1}{4} \\ 0 & 1 & \frac{3}{4} \end{array}\right]$, would define the steady state distribution for the given regular transition matrix. Regardless of the size of the population, x_1 would have $\frac{1}{4}$ of the total while x_2 would have $\frac{3}{4}$ of the total.

■ Example 6. Find the steady state obtained by any conformable distribution, subject to the regular transition matrix.

$$T = \begin{bmatrix} \frac{1}{2} & \frac{1}{4} & \frac{1}{4} \\ 0 & \frac{1}{2} & \frac{1}{2} \\ \frac{1}{6} & \frac{1}{2} & \frac{1}{3} \end{bmatrix}.$$

Solution: The system is defined by the matrix equation $[x_1 \quad x_2 \quad x_3] \cdot T = [x_1 \quad x_2 \quad x_3]$ and $x_1 + x_2 + x_3 = 1$, or $[x_1 \quad x_2 \quad x_3]\{T - I\} = [0 \quad 0 \quad 0]$ and $x_1 + x_2 + x_3 = 1$.

Then $[x_1 \quad x_2 \quad x_3]\begin{bmatrix} -\frac{1}{2} & \frac{1}{4} & \frac{1}{4} \\ 0 & -\frac{1}{2} & \frac{1}{2} \\ \frac{1}{6} & \frac{1}{2} & -\frac{2}{3} \end{bmatrix} = [0 \quad 0 \quad 0]$ and

$x_1 + x_2 + x_3 = 1$ becomes the system $\begin{cases} x_1 + x_2 + x_3 = 1 \\ -\frac{1}{2}x_1 + \frac{1}{6}x_3 = 0 \\ \frac{1}{4}x_1 - \frac{1}{2}x_2 + \frac{1}{2}x_3 = 0 \\ \frac{1}{4}x_1 + \frac{1}{2}x_2 - \frac{2}{3}x_3 = 0 \end{cases}$.

In matrix form we have $\left[\begin{array}{ccc|c} 1 & 1 & 1 & 1 \\ -\frac{1}{2} & 0 & \frac{1}{6} & 0 \\ \frac{1}{4} & -\frac{1}{2} & \frac{1}{2} & 0 \\ \frac{1}{4} & \frac{1}{2} & -\frac{2}{3} & 0 \end{array}\right]$. Solving, complete with check column, we have

$$\left[\begin{array}{ccc|c} \textcircled{1} & 1 & 1 & 1 \\ -3 & 0 & 1 & 0 \\ 1 & -2 & 2 & 0 \\ 3 & 6 & -8 & 0 \end{array}\right] \begin{array}{c} 4 \\ -2 \\ 1 \\ 1 \end{array} \left[\begin{array}{ccc|c} 1 & 1 & 1 & 1 \\ 0 & \textcircled{3} & 4 & 3 \\ 0 & -3 & 1 & -1 \\ 0 & 3 & -11 & -3 \end{array}\right] \begin{array}{c} 4 \\ 10 \\ -3 \\ -11 \end{array}$$

$$\left[\begin{array}{ccc|c} 3 & 0 & -1 & 0 \\ 0 & 3 & 4 & 3 \\ 0 & 0 & \textcircled{5} & 2 \\ 0 & 0 & -15 & -6 \end{array}\right] \begin{array}{c} 2 \\ 10 \\ 7 \\ -21 \end{array} \left[\begin{array}{ccc|c} 3 & 0 & -1 & 0 \\ 0 & 3 & 4 & 3 \\ 0 & 0 & \textcircled{5} & 2 \end{array}\right] \begin{array}{c} 2 \\ 10 \\ 7 \end{array}$$

$$\left[\begin{array}{ccc|c} 15 & 0 & 0 & 2 \\ 0 & 15 & 0 & 7 \\ 0 & 0 & 5 & 2 \end{array}\right] \text{ or } [x_1 \quad x_2 \quad x_3] = [\tfrac{2}{15} \quad \tfrac{7}{15} \quad \tfrac{6}{15}]$$

Checking:

Suppose the transition matrix is applied to the distribution whose sum is 150. Then 150 times $[\frac{2}{15} \quad \frac{7}{15} \quad \frac{6}{15}] = [20 \quad 70 \quad 60]$ which will be the steady state.

Note: $[20 \quad 70 \quad 60]\begin{bmatrix} \frac{1}{2} & \frac{1}{4} & \frac{1}{4} \\ 0 & \frac{1}{2} & \frac{1}{2} \\ \frac{1}{6} & \frac{1}{2} & \frac{1}{3} \end{bmatrix} = [20 \quad 70 \quad 60]$ ■

This example suggests a method for finding the steady state produced by any regular transition matrix. Construct the following echelon matrix where $M = (T - I)^T$, the transpose of the matrix $(T - I)$ (see chapter 1). In effect one simply subtracts the number '1' from the entries on the main diagonal of the transition matrix T and interchanges the rows and columns. Then solve the resulting echelon matrix.

$$\left[\begin{array}{cccccccc|c} 1 & 1 & 1 & . & . & . & . & 1 & 1 \\ \hline & & & & & & & & 0 \\ & & & & & & & & 0 \\ & & & M & & & & & \vdots \\ & & & & & & & & 0 \end{array}\right]$$

■ Example 7. Find the steady state determined by $T = \begin{bmatrix} \frac{1}{3} & \frac{2}{3} \\ \frac{1}{2} & \frac{1}{2} \end{bmatrix}$.

Step 1. Since T is 2×2, the resulting matrix is 3×3.

$$\left[\begin{array}{cc|c} 1 & 1 & 1 \\ \hline & M & 0 \\ & & 0 \end{array}\right]$$

Step 2. Compute the matrix $(T - I)^T$.

$$M = \left(\begin{bmatrix} \frac{1}{3} & \frac{2}{3} \\ \frac{1}{2} & \frac{1}{2} \end{bmatrix} - \begin{bmatrix} 1 & 0 \\ 0 & 1 \end{bmatrix}\right)^T$$

$$= \begin{bmatrix} -\frac{2}{3} & \frac{2}{3} \\ \frac{1}{2} & -\frac{1}{2} \end{bmatrix}^T = \begin{bmatrix} -\frac{2}{3} & \frac{1}{2} \\ \frac{2}{3} & -\frac{1}{2} \end{bmatrix}$$

Step 3. Complete the matrix and solve for x_1 and x_2 in the implied system of equations.

$$\left[\begin{array}{cc|c} 1 & 1 & 1 \\ -\frac{2}{3} & \frac{1}{2} & 0 \\ \frac{2}{3} & -\frac{1}{2} & 0 \end{array}\right] \sim \left[\begin{array}{cc|c} 1 & 1 & 1 \\ -4 & 3 & 0 \end{array}\right] \sim \left[\begin{array}{cc|c} 1 & 1 & 1 \\ 0 & 7 & 4 \end{array}\right] \sim \left[\begin{array}{cc|c} 1 & 0 & \frac{3}{7} \\ 0 & 1 & \frac{4}{7} \end{array}\right]$$

Check

$$\begin{bmatrix} \frac{3}{7} & \frac{4}{7} \end{bmatrix} \begin{bmatrix} \frac{1}{3} & \frac{2}{3} \\ \frac{1}{2} & \frac{1}{2} \end{bmatrix} = \begin{bmatrix} \frac{3}{7} & \frac{4}{7} \end{bmatrix}$$

hence $[x_1 \quad x_2] = [\frac{3}{7} \quad \frac{4}{7}]$

Exercise Set 9.2

1. Compute the steady state distribution associated with each of the following regular transition matrices.

a. $\begin{bmatrix} \frac{1}{2} & \frac{1}{4} & \frac{1}{4} \\ 1 & 0 & 0 \\ 0 & \frac{1}{2} & \frac{1}{2} \end{bmatrix}$ **b.** $\begin{bmatrix} .2 & .8 \\ .6 & .4 \end{bmatrix}$ **c.** $\begin{bmatrix} \frac{5}{16} & \frac{7}{16} & \frac{1}{4} \\ 1 & 0 & 0 \\ 0 & 1 & 0 \end{bmatrix}$

d. $\begin{bmatrix} \frac{1}{6} & \frac{5}{6} & 0 \\ \frac{1}{2} & \frac{1}{4} & \frac{1}{4} \\ \frac{7}{8} & 0 & \frac{1}{8} \end{bmatrix}$ **e.** $\begin{bmatrix} .1 & .2 & .3 & .4 \\ .5 & 0 & .5 & 0 \\ .1 & 0 & .9 & 0 \\ .2 & .2 & .2 & .4 \end{bmatrix}$

2. Assuming there were 60 members of the fraternity in problem 7 exercise set 9.1, what is the steady state?
3. Soap companies *A*, *B*, *C*, and *D* have a monopoly on a certain type of face soap. The current market has *A*, *B*, *C*, and *D* with 60%, 10%, 25%, and 5% of the market respectively. Current brand switching trends are reflected in the table below obtained by market research. What will be the eventual result if these trends go unchanged?

	To:			
Change from	*A*	*B*	*C*	*D*
A	.9	0	.1	0
B	.2	.6	.1	.1
C	.1	.1	.7	.1
D	.2	.3	.1	.4

4. What would be the long-range effect of the brand-switching trends associated with the transition matrix

$$\begin{bmatrix} .2 & .5 & .3 \\ .1 & 0 & .9 \\ 1 & 0 & 0 \end{bmatrix}?$$

5. What would be the long-range effect of the following transition matrix?

$$\begin{bmatrix} 1 & 0 & 0 & 0 \\ 0 & \frac{1}{2} & \frac{1}{2} & 0 \\ 0 & \frac{1}{4} & 0 & \frac{3}{4} \\ \frac{1}{5} & 0 & 0 & \frac{4}{5} \end{bmatrix}$$

6. Three food stores in a small midwestern town run weekly specials in the Thursday evening paper. It is estimated that 1,100 customers are affected by the ads. Most of these "smart buyers" stock up on bargains and switch stores the following week according to the table below. What is the eventual distribution of these 1,100 smart buyers?

	Next week		
Last week	A	B	C
A	0	$\frac{1}{3}$	$\frac{2}{3}$
B	$\frac{1}{2}$	0	$\frac{1}{2}$
C	$\frac{1}{4}$	$\frac{3}{4}$	0

7. The television sets in a large hospital fall into one of three states: working, out of service, or being repaired. One out of eight television sets goes out every day, one out of sixteen of the sets reported not working is sent to repair. The repair shop is able to return to service seven out of eight sets it has every day. If there are 220 sets at the hospital what will be the distribution when steady state is reached?

8. The Army has 5 missile sites. In an attempt to impress the world, win friends in Congress, and fool the Russians, a large number, 600 to be exact, of dummy missiles are shipped each month from site to site according to the transition matrix below. Eventually how many dummy missiles will be at each base?

$$\begin{bmatrix} .8 & 0 & .1 & .1 & 0 \\ .1 & .7 & .1 & .1 & 0 \\ 0 & 0 & .7 & .1 & .2 \\ .1 & 0 & 0 & .9 & 0 \\ 0 & .1 & 0 & .1 & .8 \end{bmatrix}$$

9.3 Absorbing Markov Chains

If a 1 appears in the main diagonal of a Markov matrix then the chain is known as an **absorbing Markov chain.** Consider the Markov matrix

$$T = \begin{array}{c} \\ A \\ B \\ C \\ D \\ E \end{array} \begin{array}{c} \begin{array}{ccccc} A & B & C & D & E \end{array} \\ \begin{bmatrix} \frac{1}{2} & \frac{1}{3} & 0 & 0 & \frac{1}{6} \\ 0 & 1 & 0 & 0 & 0 \\ 0 & 0 & 1 & 0 & 0 \\ 0 & 0 & 0 & 1 & 0 \\ \frac{1}{2} & 0 & \frac{1}{3} & 0 & \frac{1}{6} \end{bmatrix} \end{array}$$

At every level, when the transitions occur, State A will keep $\frac{1}{2}$ of its holding, send $\frac{1}{3}$ to B, nothing to C and D, and $\frac{1}{6}$ to E. States B, C and D are

absorbing states. Once an element gets into any one of these states it stays there. State E sends $\frac{1}{2}$ of its holding to A, $\frac{1}{3}$ to C and keeps $\frac{1}{6}$. Under these conditions states A and E will finally become empty, their contents absorbed by states B, C and D. Unlike regular chains the final distribution in an *absorbing Markov chain* depends both on the initial distribution and the transition matrix. The nature of the absorbing matrix does not allow for the free flow from state to state that is characteristic of a regular matrix.

From an initial distribution

$$D = [20{,}736 \quad 41{,}472 \quad 41{,}472 \quad 62{,}208 \quad 0]$$

and the transition matrix above the first 5 distributions in the chain and the steady state are given below.

$$\begin{array}{rcrrrrrl}
\text{1st level is } D \text{ itself} & = & [20{,}736 & 41{,}472 & 41{,}472 & 62{,}208 & 0] \\
\text{2nd level} - DT & = & [10{,}368 & 48{,}384 & 41{,}472 & 62{,}208 & 3{,}456] \\
\text{3rd level} - DT^2 & = & [\ 6{,}912 & 51{,}840 & 42{,}624 & 62{,}208 & 2{,}304] \\
\text{4th level} - DT^3 & = & [\ 4{,}608 & 54{,}144 & 43{,}392 & 62{,}208 & 1{,}536] \\
\text{5th level} - DT^4 & = & [\ 3{,}072 & 55{,}680 & 43{,}904 & 62{,}208 & 1{,}024] \\
\vdots & & \vdots & \vdots & \vdots & \vdots & \vdots \\
\text{nth level is } DT^n & = & [\qquad 0 & 58{,}752 & 44{,}928 & 62{,}208 & 0]
\end{array}$$

where $n \to \infty$.

NOTE: The size of the population (165,888) remains constant. Also, once the absorbing states have taken in all possible elements, the nonabsorbing states A and E are empty and the distribution is in equilibrium. To evaluate the final distribution of a Markov chain where absorbing states are involved, it is necessary to reorder the transition matrix in a **canonical form*** by shifting the rows and columns so that the absorbing states form the first rows and first columns. The transition matrix T would be rewritten

$$T = \begin{array}{c} B \\ C \\ D \\ A \\ E \end{array} \overset{\begin{array}{ccccc} B & C & D & A & E \end{array}}{\left[\begin{array}{ccc|cc} 1 & 0 & 0 & 0 & 0 \\ 0 & 1 & 0 & 0 & 0 \\ 0 & 0 & 1 & 0 & 0 \\ \hline \frac{1}{3} & 0 & 0 & \frac{1}{2} & \frac{1}{6} \\ 0 & \frac{1}{3} & 0 & \frac{1}{2} & \frac{1}{6} \end{array}\right]} \quad \text{instead of}$$

$$T = \begin{array}{c} A \\ B \\ C \\ D \\ E \end{array} \overset{\begin{array}{ccccc} A & B & C & D & E \end{array}}{\left[\begin{array}{ccccc} \frac{1}{2} & \frac{1}{3} & 0 & 0 & \frac{1}{6} \\ 0 & 1 & 0 & 0 & 0 \\ 0 & 0 & 1 & 0 & 0 \\ 0 & 0 & 0 & 1 & 0 \\ \frac{1}{2} & 0 & \frac{1}{3} & 0 & \frac{1}{6} \end{array}\right]}.$$

*Reorder the rows and columns of a matrix to group similar rows and columns together.

These two transition matrices define exactly the same rearrangement. It is instructive, for example, to observe that in both matrices A distributes $\frac{1}{2}$ of its holdings to A, $\frac{1}{3}$ to B, and $\frac{1}{6}$ to E.

This canonical form identifies two matrices, R and Q. The general form is $\left[\begin{array}{c|c} I & O \\ \hline R & Q \end{array}\right]$.

The matrix R defines the fractional amounts that are taken in by the absorbing states at every level where the transition matrix is applied. Q defines the rearrangement of the amounts retained by the nonabsorbing states.

In our example, when the first level is achieved we have

$$\textbf{1.}\quad \overset{\quad A \qquad\quad E}{[20{,}736 \quad 0]} \begin{bmatrix} \frac{1}{3} & 0 & 0 \\ 0 & \frac{1}{3} & 0 \end{bmatrix} = \overset{\quad B \qquad C \quad D}{[6{,}912 \quad 0 \quad 0]}$$

$$\textbf{2.}\quad \overset{\quad A \qquad\quad E}{[20{,}736 \quad 0]} \begin{bmatrix} \frac{1}{2} & \frac{1}{6} \\ \frac{1}{2} & \frac{1}{6} \end{bmatrix} = \overset{\quad A \qquad\quad E}{[10{,}368 \quad 3{,}456]}$$

Equation **1** determines the amounts added to the absorbing states B, C and D, respectively, at this first level. Equation **2** gives the respective amounts left in the nonabsorbing states. At the next level part of this will be absorbed.

The second level's arithmetic is reflected in equations **3** and **4** below.

$$\textbf{3.}\quad \overset{\quad A \qquad\quad E}{[10{,}368 \quad 3{,}456]}R = [10{,}368 \quad 3{,}456] \begin{bmatrix} \frac{1}{3} & 0 & 0 \\ 0 & \frac{1}{3} & 0 \end{bmatrix} = \overset{\quad B \qquad\quad C \qquad D}{[3{,}456 \quad 1{,}152 \quad 0]}$$

and

$$\textbf{4.}\quad \overset{\quad A \qquad\quad E}{[10{,}368 \quad 3{,}456]}Q = [10{,}368 \quad 3{,}456] \begin{bmatrix} \frac{1}{2} & \frac{1}{6} \\ \frac{1}{2} & \frac{1}{6} \end{bmatrix}$$

$$= \left([20{,}736 \quad 0] \begin{bmatrix} \frac{1}{2} & \frac{1}{6} \\ \frac{1}{2} & \frac{1}{6} \end{bmatrix} \right) \begin{bmatrix} \frac{1}{2} & \frac{1}{6} \\ \frac{1}{2} & \frac{1}{6} \end{bmatrix}$$

$$= [20{,}736 \quad 0]\, Q^2 = \overset{\quad A \qquad\quad E}{[6{,}912 \quad 2{,}304]}$$

At the next level

$$\textbf{5.}\quad \overset{\quad A \qquad\quad E}{[6{,}912 \quad 2{,}304]}R = \overset{\quad B \qquad C \quad D}{[2{,}304 \quad 768 \quad 0]}$$

$$\textbf{6.}\quad \overset{A}{[6{,}912}\ \ \overset{E}{2{,}304]}Q = [10{,}368\ \ 3{,}456]\,Q^2 = [20{,}736\ \ 0]\,Q^3$$
$$= \overset{A}{[4{,}608}\ \ \overset{E}{1{,}536]}$$

and so on until the absorbing states have everything. The exact distribution then is $[20{,}736\ \ 0]\,(I + Q + Q^2 + Q^3 + \cdots)(R)$. This infinite series $I + Q + Q^2 + Q^3 + \cdots$ is equal to $(I - Q)^{-1}$ and is denoted by N. N is called the **fundamental matrix.** Using notation from chapter 7, $I + Q + Q^2 + Q^3 + \cdots = \sum_{i=0}^{\infty} Q^i = N$.

Under specific conditions this series can be evaluated by simple algebraic methods. Since Q is a Markov matrix it satisfies those conditions and the algebraic trick of computing $N - QN$ produces the desired result.

Since $N - QN = \sum_0^{\infty} Q^i - Q\sum_0^{\infty} Q^i$ then

$$\sum_0^{\infty} Q^i - Q\sum_0^{\infty} Q^i = (I + Q + Q^2 + \cdots) - (Q + Q^2 + Q^3 + \cdots)$$

$$(I - Q)\left(\sum_0^{\infty} Q^i\right) = I + (Q - Q) + (Q^2 - Q^2) + (Q^3 - Q^3) + \cdots$$

$$(I - Q)\left(\sum_0^{\infty} Q^i\right) = I$$

therefore $\sum_0^{\infty} Q^i = (I - Q)^{-1} I = (I - Q)^{-1}$.

The amount added to each of the absorbing states is the matrix B where $B = DNR$, $N = (I - Q)^{-1}$, and D is the original distribution in the nonabsorbing states. In our example, since

$$Q = \begin{bmatrix} \frac{1}{2} & \frac{1}{6} \\ \frac{1}{2} & \frac{1}{6} \end{bmatrix},\ I - Q = \begin{bmatrix} \frac{1}{2} & -\frac{1}{6} \\ -\frac{1}{2} & \frac{5}{6} \end{bmatrix},$$

and $N = (I - Q)^{-1} = \dfrac{1}{\frac{5}{12} - \frac{1}{12}}\begin{bmatrix} \frac{5}{6} & \frac{1}{6} \\ \frac{1}{2} & \frac{1}{2} \end{bmatrix}$ (See property 1.4)

$$= 3\begin{bmatrix} \frac{5}{6} & \frac{1}{6} \\ \frac{1}{2} & \frac{1}{2} \end{bmatrix} = \begin{bmatrix} \frac{5}{2} & \frac{1}{2} \\ \frac{3}{2} & \frac{3}{2} \end{bmatrix}$$

Then $NR = \begin{bmatrix} \frac{5}{2} & \frac{1}{2} \\ \frac{3}{2} & \frac{3}{2} \end{bmatrix}\begin{bmatrix} \frac{1}{3} & 0 & 0 \\ 0 & \frac{1}{3} & 0 \end{bmatrix} = \begin{bmatrix} \frac{5}{6} & \frac{1}{6} & 0 \\ \frac{1}{2} & \frac{1}{2} & 0 \end{bmatrix}$

This matrix is understood to have the following headings.

$$\begin{array}{c} \\ A \\ E \end{array}\begin{array}{c} \begin{array}{ccc} B & C & D \end{array} \\ \begin{bmatrix} \frac{5}{6} & \frac{1}{6} & 0 \\ \frac{1}{2} & \frac{1}{2} & 0 \end{bmatrix} \end{array}$$

The first row indicates that A gives $\frac{5}{6}$ of its original holdings to state B and $\frac{1}{6}$ to state C. The second row indicates that E divides its holding equally between states B and C. The specific amount gained by each absorbing state is found by multiplying the initial distribution, D, of the nonabsorbing states by the matrix NR. That is,

$$\begin{matrix} A & E \\ [20{,}736 & 0] \end{matrix} \begin{matrix} & B & C & D \\ \begin{matrix} A \\ E \end{matrix} & \left[\begin{matrix} \frac{5}{6} \\ \frac{1}{2} \end{matrix}\right. & \begin{matrix} \frac{1}{6} \\ \frac{1}{2} \end{matrix} & \left.\begin{matrix} 0 \\ 0 \end{matrix}\right] \end{matrix} = \begin{matrix} B & C & D \\ [17{,}280 & 3{,}456 & 0]. \end{matrix}$$

This is the amount absorbed by states B, C, and D, respectively. The final distribution in states B, C, and D is given by

$$\begin{aligned} [B \quad C \quad D] &= [41{,}472 \quad 41{,}472 \quad 62{,}208] \\ &+ [17{,}280 \quad 3{,}456 \quad 0] \\ &= [58{,}752 \quad 44{,}928 \quad 62{,}208]. \end{aligned}$$

Exercise Set 9.3

In problems 1–10 use the transition matrices below.

$$A = \begin{bmatrix} 1 & 0 & 0 \\ 0 & \frac{1}{2} & \frac{1}{2} \\ 0 & 0 & 1 \end{bmatrix} \qquad B = \begin{bmatrix} 1 & 0 & 0 \\ \frac{1}{3} & 0 & \frac{2}{3} \\ 0 & 0 & 1 \end{bmatrix}$$

$$C = \begin{bmatrix} 1 & 0 & 0 \\ 0 & 1 & 0 \\ \frac{1}{4} & \frac{1}{3} & \frac{5}{12} \end{bmatrix} \qquad D = \begin{bmatrix} \frac{1}{2} & 0 & \frac{1}{2} \\ 0 & 1 & 0 \\ 0 & 0 & 1 \end{bmatrix}$$

$$E = \begin{bmatrix} 1 & 0 & 0 & 0 \\ 0 & \frac{1}{2} & \frac{1}{4} & \frac{1}{4} \\ 0 & 0 & 1 & 0 \\ 0 & 0 & 0 & 1 \end{bmatrix} \qquad F = \begin{bmatrix} 1 & 0 & 0 & 0 \\ 0 & 1 & 0 & 0 \\ 0 & 0 & 1 & 0 \\ .3 & .4 & .1 & .2 \end{bmatrix}$$

$$G = \begin{bmatrix} 1 & 0 & 0 & 0 \\ .2 & .5 & .2 & .1 \\ .3 & .1 & .5 & .1 \\ 0 & 0 & 0 & 1 \end{bmatrix} \qquad H = \begin{bmatrix} \frac{1}{2} & \frac{1}{4} & 0 & \frac{1}{4} \\ 0 & 0 & \frac{2}{3} & \frac{1}{3} \\ 0 & 0 & 1 & 0 \\ 0 & 0 & 0 & 1 \end{bmatrix}$$

$$J = \begin{bmatrix} \frac{1}{2} & 0 & 0 & 0 & \frac{1}{2} \\ 0 & 1 & 0 & 0 & 0 \\ \frac{1}{3} & 0 & \frac{1}{3} & 0 & \frac{1}{3} \\ 0 & 0 & 0 & 1 & 0 \\ \frac{1}{2} & 0 & \frac{1}{4} & 0 & \frac{1}{4} \end{bmatrix}$$

1. Put matrices A, C, E, and F in canonical form and find the resulting matrices R and Q.
2. Put matrices B, D, G, and H in canonical form and find the resulting matrices R and Q.
3. Compute $N = (I - Q)^{-1}$ for the matrices in problem 1.
4. Compute $N = (I - Q)^{-1}$ for the matrices in problem 2.
5. Compute the final distribution of $[\frac{1}{2} \quad \frac{1}{3} \quad \frac{1}{6}]$
   a. subject to matrix A.
   b. subject to matrix B.
   c. subject to matrix C.
   d. subject to matrix D.
6. Compute the final distribution of $[\frac{1}{2} \quad 0 \quad \frac{1}{2}]$
   a. subject to matrix A.
   b. subject to matrix B.
   c. subject to matrix C.
   d. subject to matrix D.
7. Compute the final distribution of the 1000 members initially arranged $[100 \quad 200 \quad 700]$
   a. subject to matrix C.
   b. subject to matrix D.
8. Compute the final distribution of $[.2 \quad .2 \quad .4 \quad .2]$
   a. subject to matrix E.
   b. subject to matrix F.
   c. subject to matrix G.
   d. subject to matrix H.
9. Compute the final distribution of $[.8 \quad .1 \quad 0 \quad .1]$
   a. subject to matrix E.
   b. subject to matrix F.
   c. subject to matrix G.
   d. subject to matrix H.
10. a. Put matrix J in canonical form.
    b. Find matrices R and Q defined by J.
    c. Compute the fundamental matrix N defined by J.
    d. Compute the final distribution of J subject to $[.18 \quad .17 \quad .22 \quad .20 \quad .23]$
    e. Is J an absorbing Markov chain?
11. What would be the long-range effect of the matrix
$$\begin{bmatrix} \frac{1}{2} & 0 & \frac{1}{4} & \frac{1}{4} \\ \frac{1}{3} & 0 & \frac{1}{2} & \frac{1}{6} \\ \frac{1}{2} & 0 & \frac{1}{2} & 0 \\ 0 & 0 & \frac{1}{3} & \frac{2}{3} \end{bmatrix}$$
? Note that this transition matrix is not an absorbing matrix even though "state 2" is being absorbed entirely at the first level.
12. Construct a 3×3 regular transition matrix which would result in the same result as the 4×4 matrix in problem 11.

9.4 Chapter Review

Important Terms and Symbols in Chapter 9.

states (section 9.1)
distribution matrix (section 9.1)
stochastic process (section 9.1)
transition matrix (section 9.1)

Markov process (section 9.1)
Markov chain (section 9.1)
Markov matrix (section 9.1)
regular transition matrix (section 9.2)
steady state (section 9.2)
absorbing Markov chain (section 9.3)
absorbing states (section 9.3)
canonical form (section 9.3)
$\left[\begin{array}{c|c} I & O \\ \hline R & Q \end{array}\right]$ (section 9.3)
fundamental matrix, $N = (I - Q)^{-1}$ (section 9.3)

Review Exercises

1. Find the value of x, y, and z which will make each of the following a transition matrix.

   a. $\begin{bmatrix} \frac{1}{2} & x \\ \frac{1}{3} & y \end{bmatrix}$ b. $\begin{bmatrix} .2 & x & .7 \\ y & .7 & 0 \\ .2 & .4 & z \end{bmatrix}$ c. $\begin{bmatrix} \frac{1}{2} & \frac{1}{3} & \frac{1}{6} \\ x & \frac{1}{5} & \frac{2}{5} \\ 0 & 1 & z \end{bmatrix}$

2. Given the distribution $D = [12 \quad 8 \quad 16]$ and the transition matrix $T = \begin{bmatrix} \frac{1}{2} & \frac{1}{4} & \frac{1}{4} \\ 0 & 1 & 0 \\ \frac{1}{2} & 0 & \frac{1}{2} \end{bmatrix}$ compute

   a. DT b. DT^2

3. In a small town the teenagers are divided into two states, employed and unemployed. It is determined that the probability that a teenager stays employed a full month is $\frac{1}{2}$. It is also known that $\frac{1}{3}$ of the unemployed teenagers find work each month.
   a. Express this information as a transition matrix.
   b. In March our town had 360 working teens and 420 nonworking teenagers. How many will be working in April? May?
   c. Find the steady state.
   d. Express in terms of percentage working.

4. Compute the steady state distribution associated with each of the following regular transition matrices.

   a. $\begin{bmatrix} .3 & .7 \\ .1 & .9 \end{bmatrix}$ b. $\begin{bmatrix} .1 & .1 & .8 \\ .2 & .8 & 0 \\ .9 & 0 & .1 \end{bmatrix}$

5. What would the long-range effect be of the transition matrix

   $$T = \begin{bmatrix} 0.2 & .01 & .97 \\ 0 & 1 & 0 \\ 0 & .1 & .99 \end{bmatrix}?$$

6. Is the transition matrix in problem 5 a *regular* transition matrix?

7. Find the final distribution of $[\frac{1}{2} \quad \frac{1}{2} \quad 0]$ subject to the transition matrix

$$T = \begin{bmatrix} 1 & 0 & 0 \\ 0 & \frac{2}{3} & \frac{1}{3} \\ 0 & \frac{1}{2} & \frac{1}{2} \end{bmatrix}.$$

8. Put the matrix

$$\begin{array}{c} \\ A \\ B \\ C \\ D \end{array} \begin{array}{c} \begin{array}{cccc} A & B & C & D \end{array} \\ \begin{bmatrix} .4 & .3 & .2 & .1 \\ 0 & 1 & 0 & 0 \\ .2 & .1 & .7 & 0 \\ 0 & 0 & 0 & 1 \end{bmatrix} \end{array}$$

a. in canonical form.
b. find the resulting matrices R and Q.
c. compute the fundamental matrix $N = (I - Q)^{-1}$.
d. find the final distribution of the 1,000 members initially arranged [140 300 420 140].

9. A systems design engineer wishes to know the exact minimum size of four holding tanks which will transfer 24 gallons of fluid according to the transfer table below.

From	To: 1	2	3	4
1	$\frac{1}{2}$	$\frac{1}{6}$	$\frac{1}{6}$	$\frac{1}{6}$
2	0	$\frac{1}{4}$	$\frac{1}{4}$	$\frac{1}{2}$
3	$\frac{1}{6}$	$\frac{1}{3}$	$\frac{1}{3}$	$\frac{1}{6}$
4	$\frac{1}{4}$	0	$\frac{1}{4}$	$\frac{1}{2}$

It is understood that row 1 is the transfer of holding tank 1, i.e., it keeps $\frac{1}{2}$ its volume and pumps $\frac{1}{6}$ of its volume into tanks 2, 3, and 4 respectively while receiving (column 1) nothing from tank 2, $\frac{1}{6}$ of tank 3 quantity, and $\frac{1}{4}$ of tank 4 quantity. This transfer is made in a very short period of time (about one minute) and it runs continuously. What size tanks should the engineer have?

10. What size tanks should the design engineer have in problem 9 if the system were to contain 36 liters?

11. What size tanks should the design engineer in problem 9 have if the transfer matrix were changed to

From	To: 1	2	3	4
1	0	$\frac{1}{2}$	$\frac{1}{3}$	$\frac{1}{6}$
2	$\frac{1}{2}$	0	$\frac{1}{4}$	$\frac{1}{4}$
3	$\frac{1}{6}$	$\frac{1}{3}$	0	$\frac{1}{2}$
4	0	$\frac{1}{4}$	$\frac{3}{4}$	0

10 Game Theory

Introduction

Game theory is a branch of mathematics developed in part by the eminent mathematician John von Neumann in 1928. The first major work designed for consumption by the general public was published by John von Neumann and Oskar Morgenstern in 1944 (*Theory of Games and Economic Behavior,* Princeton, N.J.: Princeton University Press, 1944).

This new branch of mathematics is introduced here because **1.** it is an elementary blend of matrices, the simplex techniques, and probabilities, **2.** it is an example of a developing area of applied mathematics in our time, and **3.** the present and future impact on our lives resulting from its many applications is impressive.

The theory of games, like all young sciences, is incomplete. There are many unanswered questions. Many of these questions are being asked and some answered outside of the ivy towers of academia. Indeed, game theory is being applied in computer design, business decisions, parlor games, psychology, sociology, political science, marketing and advertising, and even military strategy.

The most important single thing a student might learn in this chapter is to recognize problems which have their solutions in game theory. For that reason, this chapter is presented in an informative manner designed to expose the reader to the possibilities of how game theory might be applied. It is not designed to equip the reader with tools for solving real world problems since these, in general, are too complicated for existing theory. As we make these remarks, we are aware that computer software has been designed to play very strong games of checkers, chess, and bridge and with some minor refinement can solve many game theory problems.

The sobering central idea is that game theory is, in effect, decision making. And decision making and logical thinking are very similar. Machines may never be made that can develop new ideas, but they will be programmed to make logical decisions, design strategies, and formulate winning action plans.

10.1 Two-Person Zero-Sum Games

The word *game* is not well defined. The number of players in some "games" may be as few as one or more than a dozen. The players in a game may be adversaries and strive to achieve a position at the expense of the other players in the game while in other games the players may be mutually cooperative. Games may be simple parlor games between two players with the result of win, lose, or tie. Or games may be between two or more companies competing for their share of a fixed market. In some games, players play without any preplay communication while other games have their strategies known.

Some examples of games where the players cooperate with each other to their mutual advantage are new market acquisitions, trade regulation between countries, collective bargaining, so-called sweetheart contracts between labor and management, and coalitions and cartels. OPEC is an outstanding example where players (suppliers of oil) cooperate to their mutual advantage even while they are competing for the same market.

In this book we will restrict ourselves to games where there are exactly two players and these players are adversaries. To be more specific, the gain by one player is exactly balanced by the loss to his opponent. We will assume that losing is the same as winning a negative amount, thus the sum of the "winnings" of the two players will always equal zero. For this reason, these games are called **two-person zero-sum games.**

The plan of action taken by a player against his opponent is called a **strategy.** When a player intends to respond to his opponent's action in exactly one way, this action is called a **pure strategy.** In more complicated games a player may respond in various ways. This action is called a **mixed strategy.**

The basic problem in game theory is how each player should plan his strategy in order to optimize the result of the game from his own point of view. It is understood that each player in a game will know all the strategies available to himself and to his opponent while the game itself will consist of selecting a strategy without knowing which strategy is being selected by the opponent.

A **payoff matrix** can be written for one of the contestants with each row of the matrix as a pure strategy. It is only necessary to have one payoff matrix since the payoff of any outcome for the opponent is exactly the same as that of the player but opposite in sign.

Player A and his opponent, B, play the game of matching pennies. If player A wins 1¢, his opponent, B, wins minus 1¢. Player A has two strategies. He can show a head (H) or a tail (T). In this game the opponent, player B, has exactly the same two strategies. Below is a payoff table in which a '1' indicates a win for player A and a '-1' indicates a loss for player A.

	H	T
H	1	-1
T	-1	1

The payoff table is given for one player (player A) only and the entries in the table are his respective gains or losses. The matrix associated with this game is $\begin{bmatrix} 1 & -1 \\ -1 & 1 \end{bmatrix}$ and it is understood that this matrix is the number of pure strategies (rows of the matrix) for player A against player B. The columns of the payoff matrix are the possible strategies for the opponent, player B.

All two-person zero-sum games can be represented in matrix form. Two-person zero-sum games are sometimes referred to as **matrix games.** From a mathematical viewpoint, it is only necessary to study these payoff matrices and not the games from which they originate.

■ Example 1. Construct the payoff matrix for the child's game, "Rock, Paper, Scissors."

Two contestants simultaneously show a fist (rock), an open hand (paper), or two extended fingers (scissors). It is understood that paper covers rock, rock breaks scissors, and scissors cut paper.

Solution: The payoff matrix entries are 1 for a win, −1 for a loss, and 0 for a tie. Thus the payoff matrix is

$$\begin{array}{c} \\ S \\ P \\ R \end{array}\begin{array}{c} \begin{array}{ccc} S & P & R \end{array} \\ \begin{bmatrix} 0 & 1 & -1 \\ -1 & 0 & 1 \\ 1 & -1 & 0 \end{bmatrix} \end{array}$$

Note that the "player" has three pure strategies as follows:

Strategy 1 (row 1): Extend two fingers (scissors) and he will tie if the opponent also shows scissors, win if his opponent shows paper (open hand), or lose if his opponent shows rock (fist).

Strategy 2 (row 2): Show an open hand (paper) and he will lose if the opponent shows scissors, tie if the opponent shows paper, and win if the opponent shows rock.

Strategy 3 (row 3): Show rock and he will win if the opponent shows scissors, lose if the opponent shows paper, and tie if the opponent shows rock.

Observe that his opponent cannot be allowed to know when, or if, the player will use a particular strategy, say strategy 1. If he does have this information, his counter strategy will be to show rock and inflict the player with a loss. The criteria assumed in two-person zero-sum games is that both contestants are thinking, rational, and reacting opponents. This assumes that no player will take any action that is not in his own best interest. ■

In some games the best strategy may be obtained directly from this rational criteria and the concept of a **dominant strategy.** Strategy X is said to dominate strategy Y if strategy X will always result in a greater return, regardless of the counterstrategy selected by the opponent. When this occurs, the payoff matrix is reduced in size by omitting that strategy which is dominated. The eliminated strategy may be a row or a column. The following example will help illustrate.

■ **Example 2.** Apply the concept of dominant strategies to reduce the following matrix game.

Player B Strategies

Player A Strategies	1	2	3	4
1	1	3	4	2
2	1	0	5	1
3	0	2	1	1

Solution: It is apparent that player *A*'s strategy 3 is dominated by his strategy 1 since, regardless of the choice of strategy player *B* selects, *A* will be better placed with strategy 1. We eliminate row 3 and our payoff matrix is reduced from

$$\begin{array}{c} \\ 1 \\ 2 \\ 3 \end{array}\begin{array}{c} \begin{array}{cccc} 1 & 2 & 3 & 4 \end{array} \\ \left[\begin{array}{cccc} 1 & 3 & 4 & 2 \\ 1 & 0 & 5 & 1 \\ 0 & 2 & 1 & 1 \end{array}\right] \end{array} \quad \text{to the payoff matrix} \quad \begin{array}{c} \\ 1 \\ 2 \end{array}\begin{array}{c} \begin{array}{cccc} 1 & 2 & 3 & 4 \end{array} \\ \left[\begin{array}{cccc} 1 & 3 & 4 & 2 \\ 1 & 0 & 5 & 1 \end{array}\right] \end{array}$$

(Row 3 of the first matrix and columns 3 and 4 of the second matrix are struck through.)

However, from player *B*'s point of view, he can cut his losses by never selecting his strategies 3 or 4 since both of these columns are dominated by his strategy 1 (column 1). These strategies should be eliminated. The resulting payoff matrix becomes $\begin{array}{c} \\ 1 \\ 2 \end{array}\begin{array}{c} \begin{array}{cc} 1 & 2 \end{array} \\ \left[\begin{array}{cc} 1 & 3 \\ 1 & 0 \end{array}\right] \end{array}$.

We are now back to a position where *A*'s strategy 2 is dominated by his strategy 1. We again reduce and the result is $\begin{array}{c} \\ 1 \end{array}\begin{array}{c} \begin{array}{cc} 1 & 2 \end{array} \\ [\begin{array}{cc} 1 & 3 \end{array}] \end{array}$. However, *B*'s counterstrategy now would be to never select his strategy 2 since it is dominated by strategy 1. The final payoff matrix is $\begin{array}{c} \\ 1 \end{array}\begin{array}{c} 1 \\ [1] \end{array}$. Each player is best placed to select his own strategy 1. ■

It is apparent by looking at the original payoff matrix in example 2 that, even if the opponent, player *B*, were known to select strategy 1, player *A* cannot do better than his final choice. The same statement holds true if the opponent knows that player *A* has selected strategy 1. His best counter is still to select his strategy 1.

With this solution to this game, player *A* will receive value '1' from his opponent, player *B*. This amount is referred to as the **value of the game.** It is understood that the value of the game depends on both players using their optimum strategy.

CAUTION: If one of the players selects his strategy incorrectly, his opponent will, over the long run, win more than is necessary. The value of a game is the result of the game when all players play perfectly and is the long-range or average result.

When the value of a game is zero, we say the game is a **fair game.** In example 1 the outcome of any one trial can produce different results. In the long run, however, the contestants would come out even. Hence, "Rock, Paper, and Scissors" is a fair game.

Most parlor games, as well as board games like checkers and chess, are fair games. Matching coins, tossing dice, and playing cards are usually fair games. Players who spend time at these games and expect to win do so because they believe in luck, develop great skill in strategy selection, or rely on guile.

A second type of game which has an easy, direct solution is a game where one single entry in the payoff matrix is simultaneously the minimum possible loss for one player and the maximum possible gain for his opponent. By minimum possible loss and maximum possible gain, we mean to minimize the maximum losses and maximize the minimum gains in all pure strategies. The rational approach is called the **minimax** strategy and identifies a single entry in the payoff matrix called a **saddle point.**

It can be shown that in any game where a saddle point exists,

1. the best strategy for both players is to select the pure strategy indicated by the saddle point, and
2. the saddle point is the value of the game.

The minimax strategy is self-evident when a saddle point exists, and it may seem we have overstated its importance here. However, in the next section where we look at mixed strategies to solve more complex games, we will use the minimax concept, and its logical applications there are not so straightforward.

When repeated use of a dominant strategy, or location of a saddle point, indicates a pure strategy for each player, it is said to be a **strictly determined game.**

To locate a saddle point, we first construct a new column whose entries are the respective minimum values in each row. The largest of these is called the **maximin** for player A. Next, we construct a new row whose entries are the maximums of each respective column. The smallest of these is called the *minimax*.

When the maximin of player A is the same as the minimax of his opponent, player B, a saddle point has been found. When a game has a saddle point, it is in both players' interest to repeatedly select their strategy to include this point. Remember, the saddle point becomes the value of the game.

■ Example 3. Examine the following payoff matrix for a saddle point.

$$\begin{bmatrix} 4 & 2 & 3 \\ -2 & 2 & 5 \\ 3 & 1 & -6 \end{bmatrix}$$

Solution:

1. Add a new row along the bottom of the payoff matrix whose entries are the maximums for each respective column. Circle the smallest of these, the minimax.

$$\begin{bmatrix} 4 & 2 & 3 \\ -2 & 2 & 5 \\ 3 & 1 & -6 \end{bmatrix}$$
$$\begin{matrix} 4 & \textcircled{2} & 5 \end{matrix}$$

2. Add a new column to the right of the matrix whose entries are the smallest numbers in each respective row.

$$\begin{bmatrix} 4 & 2 & 3 \\ -2 & 2 & 5 \\ 3 & 1 & -6 \end{bmatrix} \begin{matrix} \textcircled{2} \\ -2 \\ -6 \end{matrix}$$

Circle the largest of these entries, the maximin. Player A should always select strategy 1 (row 1) and player B should always select strategy 2 (column 2). Note that if a player's strategy is known, his opponent cannot do better than to pick the pure strategy which contains the saddle point. ■

■ Example 4. Bill and Joe have been in business together for five years. The business is in trouble and Bill believes Joe is the problem. After a heated discussion, Bill accepts Joe's suggestion that Bill buy Joe's end of the business. Bill offers $42,000 and Joe counters with an asking price of $48,000. After a long, fruitless negotiation, it is decided that each will make a final sealed bid somewhere between $42,000 and $48,000 in $2,000 intervals. The business will be sold at a selling price bid by the partner who compromised the most from his original offer. For example, if Joe comes down $2,000 to $46,000 and Bill again bids $42,000, the selling price is set at $46,000. If both move the same, an average is taken.

Construct a payoff matrix for Joe where the pure strategies are rows = Joe's bids, columns = Bill's bids, and the matrix entries are the respective agreed selling prices according to the conditions of the game. Use the minimax to locate the saddle point.

Solution:

Joe's Bid \ Bill's Bid	42	44	46	48	
48	45	44	46	48	44
46	46	45	46	48	(45)
44	44	44	45	48	44
42	42	42	42	45	42
		(45)			

Bill's best strategy is to offer $44,000 and Joe's best strategy is to ask $46,000. The saddle point is $45,000. ■

Exercise Set 10.1

1. Use the concept of dominant strategies to reduce, as far as possible, each of the following matrix games.

a. $\begin{bmatrix} 2 & 3 \\ 1 & 4 \end{bmatrix}$ **b.** $\begin{bmatrix} 0 & 1 & 2 \\ 3 & 1 & 2 \\ 1 & 2 & 3 \end{bmatrix}$ **c.** $\begin{bmatrix} 1 & 0 & 2 & -2 \\ -1 & 0 & 3 & -1 \end{bmatrix}$

d. $\begin{bmatrix} 2 & 0 & -1 & -2 \\ 1 & 3 & 2 & -4 \\ 3 & -1 & 2 & 4 \\ 0 & 1 & 0 & 1 \end{bmatrix}$ **e.** $\begin{bmatrix} 2 & 1 & 4 \\ -2 & 3 & 4 \\ 3 & 2 & 5 \end{bmatrix}$

2. In problem 1, matrix games (a) and (c) have optimum pure strategies. What are these strategies and what is the value of the game?

3. Each of the following games has a saddle point. Find it.

a. $\begin{bmatrix} 2 & 1 \\ 1 & 0 \end{bmatrix}$ **c.** $\begin{bmatrix} 0 & -1 & -1 & 2 \\ 1 & 3 & 0 & 1 \\ 1 & 2 & 1 & 2 \\ 2 & 1 & -2 & 3 \end{bmatrix}$

b. $\begin{bmatrix} 5 & 6 & 4 \\ -2 & 3 & 4 \\ 5 & 2 & 1 \end{bmatrix}$ **d.** $\begin{bmatrix} 8 & 4 & 5 & 6 \\ 6 & 4 & 4 & 5 \\ 5 & 6 & 5 & 8 \\ 5 & 3 & 3 & 9 \end{bmatrix}$

4. One version of morra is a game played by two players who simultaneously display a number of fingers and shout odd or even. A player wins if the total number of fingers of both players agrees with his shouted expectation, odd or even, and at the same time does not agree with his opponent's shouted expectation. He loses if his shout is wrong and his opponent's is right. If both players shout the same word, the game is a draw. Write a payoff matrix for this game.

5. A second version of morra is played where each player extends 0, 1, or 2 fingers and the shout is, hopefully, the sum of the fingers extended by both players. Write a payoff matrix for this game.

6. Two players agree to gamble. Player *A* has the three cards shown below.

Player *B* has the following two cards.

PAY	REVERSE PAYMENT

The game is played with each player selecting one of his cards and then both cards are shown simultaneously. For example, if player *A* selects his WIN $2 card and player *B* selects his REVERSE PAYMENT card, player *A* must *pay* player *B* $2.

a. Write a payoff matrix for this game.

b. Is this a strictly determined game?

7. Is the game Tic, Tac, Toe a fair game?

8. Two companies are after the same fixed market. In the coming year, each company will try to increase its (percentage) share of this fixed market. Each company can select from any of three strategies: (1) reduce price; (2) increase quality; or (3) increase advertising. Market research indicates that the following payoff matrix will be the result, in percent, of the market lost or gained, by company *A*.

Company *A* \ Company *B*	1	2	3
1	8	−2	−4
2	5	−1	0
3	6	−2	4

a. Use the theory of dominant strategies to reduce this payoff matrix.

b. Use the minimax approach to locate a saddle point, if one exists.

c. What is the result of this game?

9. Labor and management have each made their final offer. The two offers are close, but since they cannot agree, the dispute will be turned over to arbitration. Each side has agreed to accept the arbitrator's assessment. The arbitrator requests that each party make a final concession in $.04 units off of their last publicly stated figures of $5.80 per hour demand from labor and a $5.56 offer from management. The arbitrator is known to take the offer of that side which moves the most from its last position. If they both move the same amount, he will average the two.

a. Find a payoff matrix for this game.

b. Use dominant strategies to reduce this payoff matrix.

c. Does this matrix have a saddle point?

10.2 Games with Mixed Strategies

In the discussion so far, we have considered games where the player's best choice was a pure strategy. We will now look at games where a player's optimal plan is to mix his strategies. We will assume that a rational opponent will play the game using his best counterstrategy.

Optimal strategy is defined as that combination of pure strategies which will result in the greatest expected value (see section 6.6). Before we investigate the mathematics involved in finding the optimal strategy for any game, let us look at two unskilled players A and B as they become more and more proficient playing the matrix game defined by

		Player B 1	Player B 2
Player A	1	8	−2
	2	−4	4

If either player A or B assumes a pure strategy, it will become apparent to his thinking opponent who will then elect the proper counterstrategy. To prevent this from happening, both players must mix their strategies.

Player A decides he cannot ever let his opponent B know which of his strategies he is about to select. He likes strategy 1 best because he can win as much as \$8 so he decides to select strategy 1 two-thirds of the time and strategy 2 the other third. Further, to make sure there is not a pattern to his choices, player A rolls a die and if the face shows a 5 or a 6 he uses strategy 2. Otherwise, he uses his strategy 1. Player A now has a **strategy matrix** $[\frac{2}{3} \quad \frac{1}{3}]$.

Meanwhile, player B has similar thoughts and decides to select his strategy 1 or 2 by the flip of a coin. The corresponding strategy matrix for player B is $\begin{bmatrix} \frac{1}{2} \\ \frac{1}{2} \end{bmatrix}$. The reason A's matrix was a row and B's was a column matrix will be apparent soon.

The probability that player A will win \$8 is $\left(\frac{2}{3}\right)\left(\frac{1}{2}\right) = \frac{1}{3}$ since the choice of rows by player A and columns by player B are independent events produced by the roll of one die and the flip of one coin. Similarly, the probability that player A will lose \$4 is $\left(\frac{1}{3}\right)\left(\frac{1}{2}\right) = \frac{1}{6}$.

Since the expected value V is, by definition, the sum of the four numerical values times their respective probabilities, we have

$$V = 8\left(\frac{2}{3}\right)\left(\frac{1}{2}\right) + (-4)\left(\frac{1}{3}\right)\left(\frac{1}{2}\right) + (-2)\left(\frac{2}{3}\right)\left(\frac{1}{2}\right) + 4\left(\frac{1}{3}\right)\left(\frac{1}{2}\right)$$

$$= \frac{16 - 4 - 4 + 4}{6} = \frac{12}{6} = 2$$

Using matrix notation

$$V = \begin{bmatrix} \frac{2}{3} & \frac{1}{3} \end{bmatrix} \begin{bmatrix} 8 & -2 \\ -4 & 4 \end{bmatrix} \begin{bmatrix} \frac{1}{2} \\ \frac{1}{2} \end{bmatrix} = \begin{bmatrix} \frac{12}{3} & 0 \end{bmatrix} \begin{bmatrix} \frac{1}{2} \\ \frac{1}{2} \end{bmatrix} = [2] = 2$$

This can be generalized. If A is the strategy matrix for a player A in a game defined by payoff matrix G in which his opponent counters with a mixed strategy defined by the column matrix B, then the average result per game, or the **expected value of the game,** is given by the matrix product AGB.

From the previous illustration, $V = 2$ means that player A is winning, on the average, $2 per game.

Player B is not happy and decides to change his strategy matrix to $\begin{bmatrix} \frac{1}{4} \\ \frac{3}{4} \end{bmatrix}$. The expected value of our game now changes.

$$V_1 = \begin{bmatrix} \frac{2}{3} & \frac{1}{3} \end{bmatrix} \begin{bmatrix} 8 & -2 \\ -4 & 4 \end{bmatrix} \begin{bmatrix} \frac{1}{4} \\ \frac{3}{4} \end{bmatrix} = \begin{bmatrix} 4 & 0 \end{bmatrix} \begin{bmatrix} \frac{1}{4} \\ \frac{3}{4} \end{bmatrix} = 1.$$

This means that B has cut his losses to $1 per game. Player A reacts by changing his strategy matrix to $[\frac{1}{2} \quad \frac{1}{2}]$ since he has sensed that player B is no longer selecting column 1 half the time. The expected value now is

$$V_2 = \begin{bmatrix} \frac{1}{2} & \frac{1}{2} \end{bmatrix} \begin{bmatrix} 8 & -2 \\ -4 & 4 \end{bmatrix} \begin{bmatrix} \frac{1}{4} \\ \frac{3}{4} \end{bmatrix} = \begin{bmatrix} 2 & 1 \end{bmatrix} \begin{bmatrix} \frac{1}{4} \\ \frac{3}{4} \end{bmatrix} = \frac{5}{4}.$$

Player A now is winning on the average $1.25 per game. He tries some other strategies and finally comes up with $[\frac{4}{9} \quad \frac{5}{9}]$. The expected value is now

$$V_3 = \begin{bmatrix} \frac{4}{9} & \frac{5}{9} \end{bmatrix} \begin{bmatrix} 8 & -2 \\ -4 & 4 \end{bmatrix} \begin{bmatrix} \frac{1}{4} \\ \frac{3}{4} \end{bmatrix} = \begin{bmatrix} \frac{4}{3} & \frac{4}{3} \end{bmatrix} \begin{bmatrix} \frac{1}{4} \\ \frac{3}{4} \end{bmatrix} = \frac{4}{3}.$$

Player A is now winning about $1.33 per game. Player B tries some counterstrategies:

$$V_4 = \begin{bmatrix} \frac{4}{9} & \frac{5}{9} \end{bmatrix} \begin{bmatrix} 8 & -2 \\ -4 & 4 \end{bmatrix} \begin{bmatrix} \frac{7}{8} \\ \frac{1}{8} \end{bmatrix} = \begin{bmatrix} \frac{4}{3} & \frac{4}{3} \end{bmatrix} \begin{bmatrix} \frac{7}{8} \\ \frac{1}{8} \end{bmatrix} = \frac{4}{3}.$$

$$V_5 = \begin{bmatrix} \frac{4}{9} & \frac{5}{9} \end{bmatrix} \begin{bmatrix} 8 & -2 \\ -4 & 4 \end{bmatrix} \begin{bmatrix} \frac{1}{2} \\ \frac{1}{2} \end{bmatrix} = \begin{bmatrix} \frac{4}{3} & \frac{4}{3} \end{bmatrix} \begin{bmatrix} \frac{1}{2} \\ \frac{1}{2} \end{bmatrix} = \frac{4}{3}.$$

In fact, it becomes apparent that player A has a fix on the game since the expected value

$$V = \begin{bmatrix} \frac{4}{9} & \frac{5}{9} \end{bmatrix} \begin{bmatrix} 8 & -2 \\ -4 & 4 \end{bmatrix} \begin{bmatrix} m \\ n \end{bmatrix} = \begin{bmatrix} \frac{4}{3} & \frac{4}{3} \end{bmatrix} \begin{bmatrix} m \\ n \end{bmatrix} = \left(\frac{4}{3}\right)(m + n)$$

$$= \frac{4}{3}(1) = \frac{4}{3} \text{ since } m + n \text{ always equals } 1.$$

Player B cannot reduce player A's expected value of $\frac{4}{3}$.

NOTE: If we had let player B find the strategy matrix $\begin{bmatrix} \frac{1}{3} \\ \frac{2}{3} \end{bmatrix}$ first and allowed player A to try to change the resulting expected value, we would find that player B would have a fix on the game.

$$V = [r \quad s]\begin{bmatrix} 8 & -2 \\ -4 & 4 \end{bmatrix}\begin{bmatrix} \frac{1}{3} \\ \frac{2}{3} \end{bmatrix} = [r \quad s]\begin{bmatrix} \frac{4}{3} \\ \frac{4}{3} \end{bmatrix} = \frac{4}{3}(r + s)$$

$$= \frac{4}{3} \text{ because } r + s = 1.$$

Restated simply, we claim that the value $V = \frac{4}{3}$ can be depended upon. If players A and B select their optimum strategy, then B can prevent A from winning more than $\frac{4}{3}$ dollars on the average and, conversely, A cannot be forced into a weaker position.

When this happens, the expected value V is called the value of the game. The basic assumption in game theory is that for any two-person zero-sum matrix game, a value for V can always be found. The following theorem is offered without proof.

Minimax Theorem Every two-person zero-sum game has a value.

We will now find, using matrix algebra and linear programming, the fixed value of the game just discussed. From player A's point of view, the hope is that his expected value $e \geq v$. From player B's point of view, it is $e \leq v$.

In our problem $A = [r \quad s]$, $B = \begin{bmatrix} m \\ n \end{bmatrix}$ and $G = \begin{bmatrix} 8 & -2 \\ -4 & 4 \end{bmatrix}$.

Remember $AGB = e$ and

From A's point of view	*From B's point of view*
$(AG)B = e$	$A(GB) = e$
$[r \quad s]\begin{bmatrix} 8 & -2 \\ -4 & 4 \end{bmatrix}\begin{bmatrix} m \\ n \end{bmatrix} = e$	$[r \quad s]\begin{bmatrix} 8 & -2 \\ -4 & 4 \end{bmatrix}\begin{bmatrix} m \\ n \end{bmatrix} = e$
$[(8r - 4s)(-2r + 4s)]\begin{bmatrix} m \\ n \end{bmatrix} = e$	$[r \quad s]\begin{bmatrix} (8m - 2n) \\ (-4m + 4n) \end{bmatrix} = e$
$m(8r - 4s) + n(-2r + 4s) = e$	$r(8m - 2n) + s(-4m + 4n) = e$

Hence, when player A must select $[r \quad s]$, player B will select $m = 0$, $n = 1$, if
$(8r - 4s) > (-2r + 4s)$,
or $n = 0$, $m = 1$ if
$(-2r + 4s) > (8r - 4s)$.
Therefore, player A will try to maximize e subject to
$8r - 4s = e \geq v$
and $-2r + 4s = e \geq v$.
Since A's strategy matrix is a probability matrix, $r + s = 1$, $r \geq 0$, $s \geq 0$.

Hence, when player B selects $\begin{bmatrix} m \\ n \end{bmatrix}$, player A will select $r = 0$, $s = 1$ if
$(8m - 2n) > (-4m + 4n)$
or $s = 0$, $r = 1$ if
$(-4m + 4n) > (8m - 2n)$.
Therefore, player B will try to minimize e subject to
$8m - 2n = e \leq v$
and $-4m + 4n = e \leq v$.
Since B's strategy matrix is a probability matrix, $m + n = 1$, $m \geq 0$, $n \geq 0$.

To simplify the algebra, we divide these constraints by v and we have

$$8\frac{r}{v} - 4\frac{s}{v} \geq 1$$
$$-2\frac{r}{v} + 4\frac{s}{v} \geq 1$$
$$\frac{r}{v} + \frac{s}{v} = \frac{1}{v}.$$

We define new **working variables** $x = \frac{r}{v}$, $y = \frac{s}{v}$ and we have

$$x + y = \frac{1}{v}$$

and $\begin{cases} 8x - 4y \geq 1 \\ -2x + 4y \geq 1. \end{cases}$

Player A wishes to minimize $x + y$ in order to maximize v.

To simplify the algebra, we divide these constraints by v and we have

$$8\frac{m}{v} - 2\frac{n}{v} \leq 1$$
$$-4\frac{m}{v} + 4\frac{n}{v} \leq 1$$
$$\frac{m}{v} + \frac{n}{v} = \frac{1}{v}$$

We define new **working variables,** $x = \frac{m}{v}$, $y = \frac{n}{v}$ and we have

$$x + y = \frac{1}{v}$$

and $\begin{cases} 8x - 2y \leq 1 \\ -4x + 4y \leq 1. \end{cases}$

Player B wishes to maximize $x + y$ in order to minimize v.

NOTE: Minimizing $x + y$ makes $\frac{1}{v}$ small, and, hence, makes v large.

Maximizing $x + y$ maximizes $\frac{1}{v}$ and minimizes v.

The graph of the feasible region defined by these constraints is

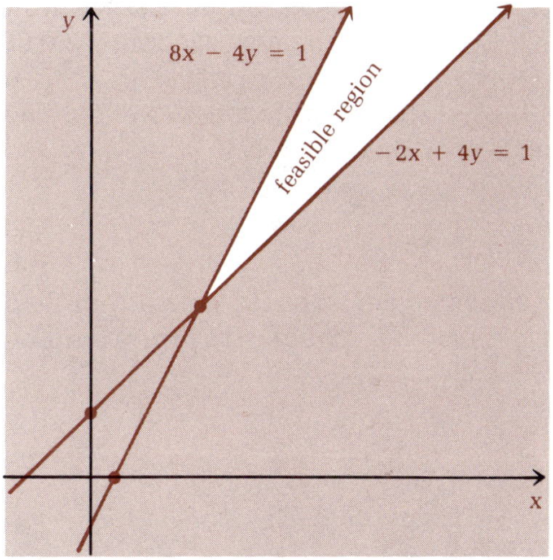

The graph has one corner point $\left(\frac{1}{3}, \frac{5}{12}\right)$ which is a minimum for $x + y$.

$x + y = \frac{1}{3} + \frac{5}{12} = \frac{9}{12} = \frac{3}{4}$ and

$v = \frac{1}{x + y} = \frac{4}{3}$. Since $x = \frac{r}{v}$ and

$y = \frac{s}{v}$ then $r = xv = \frac{1}{3} \cdot \frac{4}{3} = \frac{4}{9}$

and $s = yv = \frac{5}{12} \cdot \frac{4}{3} = \frac{5}{9}$.

Player *A*'s optimum strategy is $[\frac{4}{9} \quad \frac{5}{9}]$ and the value of the game is $\frac{4}{3}$. In summary, *A*'s optimum strategy is found by minimizing $x + y$ subject to

$$\begin{cases} x \geq 0,\ y \geq 0 \\ 8x - 4y \geq 1 \\ -2x + 4y \geq 1 \end{cases}$$

The graph of the feasible region defined by these constraints is

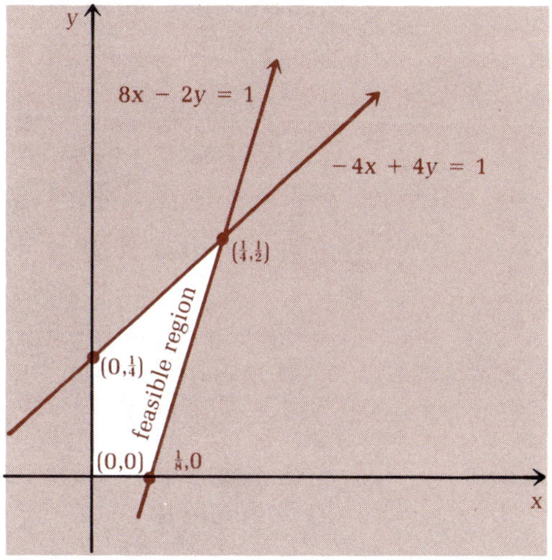

The corner points are (0,0), $\left(\frac{1}{8}, 0\right)$ $\left(0, \frac{1}{4}\right)$ and $\left(\frac{1}{4}, \frac{1}{2}\right)$ which has a maximum for $x + y$ at $\left(\frac{1}{4}, \frac{1}{2}\right)$.

$x + y = \frac{1}{4} + \frac{1}{2} = \frac{3}{4}$ and

$v = \frac{1}{x + y} = \frac{4}{3}$.

Since $x = \frac{m}{v}$ and $y = \frac{n}{v}$ then

$m = xv = \frac{1}{4} \cdot \frac{4}{3} = \frac{1}{3}$ and

$n = yv = \frac{1}{2} \cdot \frac{4}{3} = \frac{2}{3}$.

Player *B*'s optimum strategy is $\begin{bmatrix} \frac{1}{3} \\ \frac{2}{3} \end{bmatrix}$ and the value of the game is $\frac{4}{3}$.

In summary, *B*'s optimum strategy is found by maximizing $x + y$ subject to

$$\begin{cases} x \geq 0,\ y \geq 0 \\ 8x - 2y \leq 1 \\ -4x + 4y \leq 1 \end{cases}$$

The minimax approach was used to express the problem of finding a player's best mixed strategy as a linear programming problem. As these problems increase in size (more and unequal numbers of pure strategies), the solution will require more powerful tools. In the next section, we will employ the simplex technique to solve more complex games.

We will finish this section by suggesting a simple method of solving games where each player has exactly two choices. Recall

$$(AG)B \rightarrow m(8r - 4s) + n(-2r + 4s) = e$$

and $$A(GB) \rightarrow r(8m - 2n) + s(-4m + 4n) = e$$

where the opponent in each respective equation could assign 0 or 1 to the coefficient to void the higher number. The counterstrategy would be to make the terms equal.

Case 1.
Compute (AG)

$$\begin{cases} 8r - 4s = -2r + 4s \\ r + s = 1 \end{cases}$$

Solving:

$$\begin{cases} 10r - 8s = 0 \\ r + s = 1 \end{cases}$$

$$\begin{cases} r = \frac{4}{9} \\ s = \frac{5}{9} \end{cases}$$

$$A = \begin{bmatrix} \frac{4}{9} & \frac{5}{9} \end{bmatrix}$$

Case 2.
Compute (GB)

$$\begin{cases} 8m - 2n = -4m + 4n \\ m + n = 1 \end{cases}$$

Solving:

$$\begin{cases} 12m - 6n = 0 \\ m + n = 1 \end{cases}$$

$$\begin{cases} m = \frac{1}{3} \\ n = \frac{2}{3} \end{cases}$$

$$B = \begin{bmatrix} \frac{1}{3} \\ \frac{2}{3} \end{bmatrix}$$

■ Example 5. Given the matrix game $\begin{bmatrix} 4 & 3 \\ 2 & 5 \end{bmatrix}$

a. Find the value of the game and determine player A's optimal strategy.

b. Find player B's optimal counterstrategy without using linear programming.

Solution: We minimize the sum of working variables $x + y$ subject to

$$\begin{cases} x \geq 0,\ y \geq 0 \\ 4x + 2y \geq 1 \\ 3x + 5y \geq 1 \end{cases}$$

We graph the feasible region and identify the corner points.

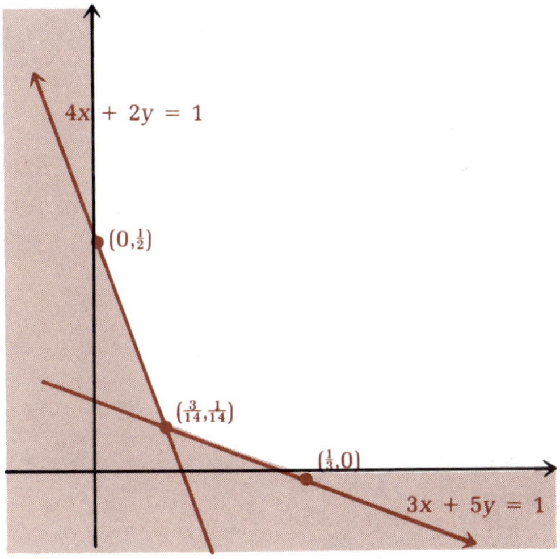

a. $\begin{cases} 3x + 5y = 1 \\ 4x + 2y = 1 \end{cases} \rightarrow \begin{cases} x = \frac{3}{14} \\ y = \frac{1}{14} \end{cases}$

$$x + y = \frac{3}{14} + \frac{1}{14} = \frac{2}{7}$$

The value of the game is

$$v = \frac{1}{x + y} = \frac{7}{2}$$

$$r = xv = \frac{3}{14} \cdot \frac{7}{2} = \frac{3}{4}$$

$$s = yv = \frac{1}{14} \cdot \frac{7}{2} = \frac{1}{4}$$

$$[r \quad s] = [\tfrac{3}{4} \quad \tfrac{1}{4}],\ v = \frac{7}{2}$$

b. Since $AGB = v$, $AGB = \frac{7}{2}$ for any A, if an optimal counterstrategy B is selected. So we let $A = [1 \quad 0]$. Then

$$[1 \quad 0]\begin{bmatrix} 4 & 3 \\ 2 & 5 \end{bmatrix}\begin{bmatrix} m \\ n \end{bmatrix} = \frac{7}{2}, \text{ where } m + n = 1.$$

Solve the system $\begin{cases} 4m + 3n = \frac{7}{2} \\ m + n = 1 \end{cases}$ to obtain $m = n = \frac{1}{2}$.

$$\begin{bmatrix} m \\ n \end{bmatrix} = \begin{bmatrix} \frac{1}{2} \\ \frac{1}{2} \end{bmatrix}.$$

■

Exercise Set 10.2

1. What is the average result of the game $\begin{bmatrix} -2 & 3 \\ 4 & -3 \end{bmatrix}$ if

   a. A elects strategy $[\frac{1}{2} \quad \frac{1}{2}]$ and B counters with $\begin{bmatrix} \frac{1}{4} \\ \frac{3}{4} \end{bmatrix}$?

   b. A elects strategy $[\frac{5}{6} \quad \frac{1}{6}]$ and B counters with $\begin{bmatrix} \frac{1}{3} \\ \frac{2}{3} \end{bmatrix}$?

   c. A elects strategy $[.2 \quad .8]$ and B counters with $\begin{bmatrix} .4 \\ .6 \end{bmatrix}$?

   d. A elects strategy $[.2 \quad .8]$ and B counters with $\begin{bmatrix} .3 \\ .7 \end{bmatrix}$?

2. What is the long range result of the game $\begin{bmatrix} 1 & -1 \\ -2 & 1 \end{bmatrix}$ if
   a. A elects to use the mixed strategy $[\frac{1}{3} \quad \frac{2}{3}]$ and B elects to use the mixed strategy $\begin{bmatrix} \frac{1}{4} \\ \frac{3}{4} \end{bmatrix}$?
   b. A elects to use the mixed strategy $[\frac{1}{2} \quad \frac{1}{2}]$ and B elects to use the mixed strategy $\begin{bmatrix} \frac{1}{5} \\ \frac{4}{5} \end{bmatrix}$?
3. Given the matrix game $\begin{bmatrix} 4 & 2 \\ 2 & 5 \end{bmatrix}$, find the value of the game and determine the optimal strategy for both players.
4. Given the matrix game $\begin{bmatrix} 3 & 2 \\ 0 & 1 \end{bmatrix}$, find the value of the game and determine the optimal strategy for both players.
5. Given the matrix game $\begin{bmatrix} -2 & 3 \\ 4 & -2 \end{bmatrix}$, find the best mixed strategy for both players and determine the value of the game.
6. Suggest why the optimal strategies for both players A and B are the same in each of the following games:

   a. $\begin{bmatrix} -1 & 3 \\ 0 & 2 \end{bmatrix}$ b. $\begin{bmatrix} 1 & 5 \\ 2 & 4 \end{bmatrix}$ c. $\begin{bmatrix} -3 & 1 \\ -2 & 0 \end{bmatrix}$

10.3 Simplex Solutions to Matrix Games

When a player has to choose between two pure strategies, and if he elects strategy 1, he will win \$100,000 half the time and win nothing the other half, his expectation is to win, on an average, \$50,000. If that player elects strategy 2 and strategy 2 will always result in a \$50,000 gain, then the expected values of both options are equal. Mathematically equal that is. If the player needs \$8,000 tomorrow as a matter of life or death, these two choices would be very different. Reasoning in the same way, the two games $G_1 = \begin{bmatrix} -2 & 3 \\ 1 & -1 \end{bmatrix}$ and $G_2 = \begin{bmatrix} 98 & 103 \\ 101 & 99 \end{bmatrix}$ are very different games, yet they are alike in one sense. If player A selects row 1 as his pure strategy, the amount he wins (-2 or 3 in G_1 and 98 and 103 in G_2) both differ by the same amount, 5. In fact, G_2 is G_1 with 100 added to each entry. The best mixed strategies for the participant in both of these games would be exactly the same. The value of game G_2 would be exactly 100 more than the value of game G_1.

Rule for Changing Game Matrix: We can always change the payoff matrix in any game by adding or subtracting a constant to each entry in the payoff matrix and **1.** the optimal strategies of both players will remain the same while **2.** the value of the game is changed by the amount added or subtracted.

■ Example 6. Solve the matrix game $R = \begin{bmatrix} 12 & 2 \\ 0 & 8 \end{bmatrix}$.

Solution: If we subtract 4 from each entry of this payoff matrix, we have $G = \begin{bmatrix} 8 & -2 \\ -4 & 4 \end{bmatrix}$, the payoff matrix of the game discussed in detail in section 10.2. The optimal strategies for the players then were $[\frac{4}{9} \quad \frac{5}{9}]$ for player A and $\begin{bmatrix} \frac{1}{3} \\ \frac{2}{3} \end{bmatrix}$ for player B. The value of the game was $\frac{4}{3}$. It follows, then, that the optimal strategies will not change and the value of the game is $\frac{4}{3} + 4 = \frac{16}{3}$. ■

Payoff matrices have always been read directly for one player (rows) and indirectly for his opponent (columns), and the entries in the payoff matrix took on the sign associated with a gain or a loss for player A. If we wished to look at a game from player B's point of view, we could rewrite the payoff matrix by changing all the signs and interchanging rows and columns. For example, $\begin{array}{cc} & B \\ A & \begin{bmatrix} 2 & -3 \\ 1 & 5 \end{bmatrix} \end{array}$ might become $\begin{array}{cc} & A \\ B & \begin{bmatrix} -2 & -1 \\ 3 & -5 \end{bmatrix} \end{array}$. In fact, if a player A and his opponent player were to play a matrix game G, then the payoff matrix from A's point of view is G. But from B's point of view, it is $[-G]^T$.

■ Example 7. Rewrite the payoff matrix for the opponent, player B, if the payoff matrix for player A is $G = \begin{bmatrix} 2 & 0 & -2 \\ 1 & -1 & 3 \end{bmatrix}$.

Solution: The payoff matrix for player $B = [-G]^T = \begin{bmatrix} -2 & 0 & 2 \\ -1 & 1 & -3 \end{bmatrix}^T = \begin{bmatrix} -2 & -1 \\ 0 & 1 \\ 2 & -3 \end{bmatrix}$. ■

When we investigated the game $G = \begin{bmatrix} 8 & -2 \\ -4 & 4 \end{bmatrix}$, we found that player A's optimal strategy was found by minimizing the sum of the working variables, $x + y$, subject to

$$\textbf{1.} \quad \begin{cases} x \geq 0,\ y \geq 0 \\ 8x - 4y \geq 1 \\ -2x + 4y \geq 1 \end{cases}$$

Also we found player B's optimal strategy by maximizing the sum of the working variables, $x + y$, subject to

$$\textbf{2.} \quad \begin{cases} x \geq 0,\ y \geq 0 \\ 8x - 2y \leq 1 \\ -4x + 4y \leq 1 \end{cases}$$

You will observe that minimizing **1** and maximizing **2** are dual problems (see section 3.7). It follows that both the player's and his opponent's optimal strategy can be obtained directly from the final **simplex tableau** if the initial set-up matrix is of the following form.

$$\left[\begin{array}{c|c|c|c} G & I_n & \begin{matrix} 0 \\ 0 \\ \vdots \\ 0 \end{matrix} & \begin{matrix} 1 \\ 1 \\ \vdots \\ 1 \end{matrix} \\ \hline 1 \cdots 1 & 0 \cdots 0 & -1 & 0 \end{array}\right]$$

CAUTION: All games have a value, hence a solution, but not all simplex tableaus have solutions.

Whenever a properly set-up simplex matrix has no solution, we must use an equivalent game with a sufficiently large amount added to each entry. It is common for solvers to adjust the game matrix in order to simplify the arithmetic by introducing as many zeros and small positive numbers as possible.

CAUTION: If values are added to the entries of G, the simplex tableau will give the value of the new game and an adjustment will be required to find the value of the game G.

CAUTION: The *direct* solutions read from the simplex are the working variables discussed in section 10.2. The fraction of time a specific pure strategy is to be used is given by $(x_i)v$, the product of the respective working variable x_i and the value of the game v.

The mixed strategy for the player of the game can be read directly from the bottom row of the final simplex tableau by dividing the entries in the last row by the last entry of that last row. You will discover that the next to the last entry will become the value of the game and the entries in the corresponding slack variable columns will form the optimal strategy for the player of the game.

■ **Example 8.** Determine the best mixed strategy for the game whose payoff matrix is $\begin{bmatrix} 3 & 2 & 0 \\ 1 & 4 & 5 \end{bmatrix}$.

Solution: Set up the simplex tableau as per instructions and solve to obtain

1. player *A*'s optimal strategy and the value of the game by reading the adjusted bottom row.
2. the working variables x, y, and z.
3. the value of the game a second way as an arithmetic check.
4. the optimal strategy for opponent *B*.

$$\left[\begin{array}{ccccccc} 3 & 2 & 0 & 1 & 0 & 0 & 1 \\ 1 & 4 & \textcircled{5} & 0 & 1 & 0 & 1 \\ 1 & 1 & 1 & 0 & 0 & -1 & 0 \end{array}\right]$$

Pick a pivot number and clear a column.

$$\left[\begin{array}{ccccccc} \textcircled{3} & 2 & 0 & 1 & 0 & 0 & 1 \\ 1 & 4 & 5 & 0 & 1 & 0 & 1 \\ 4 & 1 & 0 & 0 & -1 & -5 & -1 \end{array}\right]$$

Continue the simplex.

final tableau *adjusted bottom row*

$$\left[\begin{array}{ccccccc} 3 & 2 & 0 & 1 & 0 & 0 & 1 \\ 0 & 10 & 15 & -1 & 3 & 0 & 2 \\ 0 & -5 & 0 & -4 & -3 & -15 & -7 \end{array}\right] \longrightarrow \left[\begin{array}{ccccccc} & & & & & & \\ & & & & & & \\ - & - & - & \frac{4}{7} & \frac{3}{7} & \frac{15}{7} & 1 \end{array}\right]$$

1. Reading the adjusted bottom row, *A*'s optimal strategy is $[\frac{4}{7} \;\; \frac{3}{7}]$ and the value of the game is $\dfrac{15}{7}$.
2. Read the final tableau to find working variables x, y, and z. $x = \dfrac{1}{3}$, $y = 0$, $z = \dfrac{2}{15}$.
3. $v = \dfrac{1}{x + y + z} = \dfrac{1}{\frac{1}{3} + 0 + \frac{2}{15}} = \dfrac{1}{\frac{7}{15}} = \dfrac{15}{7}$

4. $a = xv = \left(\frac{1}{3}\right)\left(\frac{15}{7}\right) = \frac{5}{7}$

$b = yv = (0)\left(\frac{15}{7}\right) = 0$

$c = zv = \left(\frac{2}{15}\right)\left(\frac{15}{7}\right) = \frac{2}{7}$

Hence, B's optimal strategy is $\begin{bmatrix} \frac{5}{7} \\ 0 \\ \frac{2}{7} \end{bmatrix}$. ■

■ **Example 9.** Determine the best mixed strategy for the game whose payoff matrix is

$$G = \begin{bmatrix} 2 & 3 & -1 \\ 4 & 3 & 2 \\ -2 & 5 & 4 \end{bmatrix}.$$

Solution: Before attempting to solve using the simplex, check for dominant pure strategies.

$$\begin{bmatrix} \cancel{2} & \cancel{3} & \cancel{-1} \\ 4 & 3 & 2 \\ -2 & 5 & 4 \end{bmatrix} \rightarrow \begin{bmatrix} 4 & \cancel{3} & 2 \\ -2 & \cancel{5} & 4 \end{bmatrix} \rightarrow \begin{bmatrix} 4 & 2 \\ -2 & 4 \end{bmatrix}$$

Solve the following simplex matrix.

$$\begin{bmatrix} 4 & 2 & 1 & 0 & 0 & 1 \\ -2 & 4 & 0 & 1 & 0 & 1 \\ 1 & 1 & 0 & 0 & -1 & 0 \end{bmatrix}$$

$$\begin{bmatrix} 4 & 2 & 1 & 0 & 0 & 1 \\ 0 & 10 & 1 & 2 & 0 & 3 \\ 0 & 2 & -1 & 0 & -4 & -1 \end{bmatrix}$$

$$\begin{bmatrix} 20 & 0 & 4 & -2 & 0 & 2 \\ 0 & 10 & 1 & 2 & 0 & 3 \\ 0 & 0 & -6 & -2 & -20 & -8 \end{bmatrix} \rightarrow \begin{bmatrix} & & & & & \\ & & & & & \\ - & - & \frac{3}{4} & \frac{1}{4} & \frac{5}{2} & 1 \end{bmatrix}$$

Player A's optimal strategy is, then, $[\frac{3}{4} \quad \frac{1}{4}]$ and $v = \frac{5}{2}$. For Player B,

$x = \frac{2}{20} = \frac{1}{10}$, $y = \frac{3}{10}$, $v = \frac{1}{x + y} = \frac{1}{\frac{1}{10} + \frac{3}{10}} = \frac{10}{4} = \frac{5}{2}$. Then

$$m = \left(\frac{1}{10}\right)\left(\frac{5}{2}\right) = \frac{1}{4}$$

$$n = \left(\frac{3}{10}\right)\left(\frac{5}{2}\right) = \frac{3}{4}.$$

Therefore $\begin{bmatrix} m \\ n \end{bmatrix} = \begin{bmatrix} \frac{1}{4} \\ \frac{3}{4} \end{bmatrix}$. As a check, use the shortcut discussed in section 10.2 to find player A's optimal strategy.

$$[r \quad s]\begin{bmatrix} 4 & 2 \\ -2 & 4 \end{bmatrix} = [(4r - 2s)(2r + 4s)].$$

Solving $\begin{cases} 4r - 2s = 2r + 4s \\ r + s = 1 \end{cases}$ we have $[\frac{3}{4} \quad \frac{1}{4}]$.

Based on the original matrix, before dominant strategies were used to reduce, player A's optimal strategy is $[0 \quad \frac{3}{4} \quad \frac{1}{4}]$ and player B's optimal strategy is $\begin{bmatrix} \frac{1}{4} \\ 0 \\ \frac{3}{4} \end{bmatrix}$. Therefore, player A will use his strategy 2 three-fourths of the time and his strategy 3 the other fourth. Player B will counter using his strategy 1 one-fourth of the time and his strategy 3 three-fourths of the time. Player A will expect an average win of $\frac{5}{2}$ units per game played. ■

Exercise Set 10.3

Determine the optimal strategies for the players and find the value for each of the games defined by the following payoff matrices:

1. $\begin{bmatrix} 2 & -3 & 1 \\ 0 & -2 & -1 \end{bmatrix}$

2. $\begin{bmatrix} 3 & 2 & 1 \\ 1 & 2 & 1 \\ 1 & 0 & 1 \end{bmatrix}$

3. $\begin{bmatrix} 3 & 1 & 2 \\ 1 & 2 & 3 \\ 4 & 1 & -2 \end{bmatrix}$

4. $\begin{bmatrix} 4 & 2 & 1 \\ 1 & 5 & -2 \\ 2 & -1 & 2 \end{bmatrix}$

5. $\begin{bmatrix} 5 & 0 & 2 & 1 \\ 1 & 3 & 1 & 1 \\ 4 & 2 & 0 & 2 \end{bmatrix}$

6. $\begin{bmatrix} 2 & 1 \\ 1 & 3 \\ 0 & 4 \end{bmatrix}$

7. $\begin{bmatrix} 0 & 1 & 0 & 0 \\ 1 & 0 & 1 & 0 \\ -1 & 0 & 0 & 2 \\ 2 & -1 & 0 & 0 \end{bmatrix}$

8. $\begin{bmatrix} \frac{1}{2} & \frac{1}{3} & \frac{1}{6} \\ \frac{1}{4} & 0 & \frac{3}{4} \\ \frac{2}{3} & \frac{1}{3} & 0 \end{bmatrix}$

10.4 Chapter Review

Important Terms and Symbols in Chapter 10.

game theory (chapter introduction)
two-person zero-sum games (section 10.1)
strategy (section 10.1)
pure strategy (section 10.1)
mixed strategy (section 10.1)
payoff matrix (section 10.1)
matrix game (section 10.1)
dominant strategy (section 10.1)
value of the game (section 10.1)
fair game (section 10.1)
minimax (section 10.1)
saddle point (section 10.1)
strictly determined game (section 10.1)
maximin (section 10.1)
optimal strategy (section 10.2)
strategy matrix (section 10.2)
expected value of the game (section 10.2)
$AGB = v$ (section 10.2)
minimax theorem (section 10.2)
working variables (section 10.2)
Rule for Changing Game Matrix (section 10.3)
$[-G]^T$ (section 10.3)
simplex tableau set-up (section 10.3)

Review Exercises

1. Use the concept of dominant strategies to reduce each of the following payoff matrices as far as possible.

a. $\begin{bmatrix} 4 & 3 \\ -1 & -3 \end{bmatrix}$

b. $\begin{bmatrix} 3 & -1 & 2 & 1 \\ -1 & -2 & 3 & 2 \end{bmatrix}$

c. $\begin{bmatrix} 1 & 2 & 1 \\ 3 & -1 & 2 \\ 1 & -1 & 2 \end{bmatrix}$

d. $\begin{bmatrix} 1 & 2 & 2 \\ 3 & -1 & 5 \\ 4 & -2 & 4 \end{bmatrix}$

e. $\begin{bmatrix} 1 & 2 & 2 \\ 3 & -1 & 5 \\ -1 & 1 & 2 \end{bmatrix}$

f. $\begin{bmatrix} 3 & -1 & 3 & 4 \\ 3 & -2 & 1 & 2 \\ -2 & 1 & -1 & 3 \end{bmatrix}$

g. $\begin{bmatrix} 3 & 1 & 4 & -2 \\ -1 & 0 & 2 & -3 \\ 1 & 1 & 0 & -1 \\ 2 & -2 & 2 & 1 \end{bmatrix}$

h. $\begin{bmatrix} 3 & 1 & 4 & 2 \\ -1 & 0 & -2 & -1 \\ -2 & 4 & 4 & 3 \\ -2 & 1 & 3 & 1 \end{bmatrix}$

i. $\begin{bmatrix} 3 & 1 & 0 & -1 \\ 1 & 0 & -1 & -2 \\ 4 & 2 & 3 & 3 \\ -2 & 0 & 3 & 1 \end{bmatrix}$

j. $\begin{bmatrix} -1 & -3 & -4 & 1 & -1 & 0 \\ 2 & 3 & -3 & 2 & -1 & 2 \\ 4 & 3 & -1 & -3 & -2 & 3 \\ 3 & 1 & 0 & 5 & 4 & 5 \\ 3 & 1 & 4 & 4 & 3 & 5 \\ 5 & 4 & 0 & 4 & 3 & 4 \end{bmatrix}$

2. Determine which of the payoff matrices in exercise 1 have saddle points and find the saddle point.

3. Find the saddle point, if it exists, for each of the following:

a. $\begin{bmatrix} 2 & 1 \\ 0 & -1 \\ 3 & 2 \end{bmatrix}$

b. $\begin{bmatrix} 1 & 2 & 2 \\ -1 & 1 & 3 \end{bmatrix}$

c. $\begin{bmatrix} 1 & 0 & 2 \\ -1 & -1 & 0 \\ 0 & -2 & 0 \end{bmatrix}$

d. $\begin{bmatrix} 3 & 3 & 2 & 2 \\ 1 & 2 & 1 & 2 \\ 0 & 4 & 0 & 1 \\ 2 & 1 & 0 & 3 \end{bmatrix}$

$\begin{bmatrix} 2 & 1 & 3 & -2 \\ 3 & 0 & 2 & 1 \\ 1 & 1 & 2 & 3 \end{bmatrix}$

f. $\begin{bmatrix} 3 & -4 & 3 & -1 \\ 2 & 1 & 2 & 0 \\ -3 & 1 & 4 & 0 \\ 0 & 2 & -2 & -1 \end{bmatrix}$

4. Two people agree to play a game with cards. Player *A* has three cards showing −2, 1, and 2. Player *B* has two cards showing "Reverse" and "Double." Each player selects a card and the cards are played simultaneously. The payoff is determined by the two cards. If *A* selects the card showing 2 and *B* selects the card showing reverse, *A* pays *B* \$2; if *A* selects −2 and *B* selects reverse, *A* receives \$2 from *B*; if *A* selects −2 and *B* selects double, *A* pays *B* \$4; and so on. Write a payoff matrix for this game where *A* is the player and *B* is the opponent.

5. John and Jim agree to play a game with cards. John has four cards showing 5, 7, 9, and 10. Jim has two cards showing 6 and 8. The game is played as follows: John selects a card and lays it face down on the table. Jim selects a card and lays it down while at the same time calling "high" or "low." If Jim's card is lower than John's and Jim shouts "low," he receives the difference in value of the cards. If Jim's card is higher than John's and he shouts "high," Jim receives the difference in face value. If Jim's shout is wrong, John receives the difference in face value. Write the payoff matrix for this game where John is the player and Jim is the opponent.

6. Find the value of the game and the optimal strategies for each player for each part of exercise 1.
7. For exercise 4, find the value of the game and the optimal strategies for each player.
8. For exercise 5, find the value of the game and the optimal strategies for each player.

11 Math of Finance

Introduction

In this chapter, we will look at some of the institutional forms of mathematics frequently encountered in the planning and control of personal finances. The chapter is written with the consumer in mind. The theoretical insights necessary to make corporate financial decisions are, for the most part, not considered.

11.1 Percent (A Review)

Percent has a common reference of 100 parts. Six percent, written 6%, means 6 out of 100. Every number expressed as a percent can be written as a common fraction by dividing by 100. Hence, 6% is equivalent to $\frac{6}{100}$, which, when divided out, becomes the **decimal fraction** 0.06. Every percent can be changed to a decimal fraction by moving the decimal two places to the left, and, conversely, any decimal fraction can be changed to a percent by moving the decimal two places to the right and adding the symbol "%."

While percent is usually thought of as a part of the whole, it is not uncommon to think of percentages greater than 100. For example, when gasoline rose from $.60 to $1.20 a gallon, it doubled, or increased by 100%. When it rose from $.60 to $1.50 per gallon, it increased by 150%. It is also interesting to note that the increase from $1.20 a gallon to $1.50 a gallon is only a 25% increase. Hence, a 100% increase followed by a 25% increase is equivalent to a single 150% increase over the initial value.

In all arithmetic calculations, percent is used in its equivalent (decimal or common) fractional form. The basic relationship is

1. Part = (whole)(percent).

From this basic relationship, it is easy to show that

2. $\text{Percent} = \dfrac{\text{part}}{\text{whole}}$ or

3. $\text{Whole} = \dfrac{\text{part}}{\text{percent}}$.

It should be remembered that the "part" may be larger than the whole, in which case the percent exceeds 100%.

■ Example 1. An item originally marked at $37.00 is reduced to $29.60. What was the discount rate?

Solution: $\text{Percent} = \dfrac{29.60}{37.00} = 0.80$, which is 80%. Since $29.60 is 80% of $37.00, there has been a 20% discount. ■

■ Example 2. In 1940 a small bag of popcorn cost 5¢ at the local movie house. In 1980 that same size bag cost 45¢. What has been the percent of increase?

Solution: The amount of increase is 40¢. Therefore, $\text{Percent} = \dfrac{40}{5} = 8.0$ which is 800%. ■

■ Example 3. After a sale in a small retail store, the gross receipts were $4,095. What was the gross profit if the gross receipts include a 5% state sales tax and the items sold were sold at cost plus 30%?

Solution: Let s = selling price.
Gross receipts = selling price + tax

$$\$4{,}095 = s + .05s = s(1 + .05) = 1.05s$$

$$s = \frac{\$4{,}095}{1.05} = \$3{,}900$$

Let c = cost of goods sold.

Therefore s = (cost of goods) + (gross profit)

$$\$3{,}900 = c + .3c = 1.3(c)$$

$$c = \frac{\$3{,}900}{1.3} = \$3{,}000$$

$$\text{Gross profit} = \text{selling price} - \text{cost} = \$3{,}900 - \$3{,}000 = \$900$$

■

From the above example, we can generalize as follows. When the gross receipts (g) equal the net receipts (n) plus the tax collected (t), where the tax rate is r,

then $g = n + t$	(1)
and $t = nr$	(2)
hence $g = n + nr$	Combining (1) and (2)
$g = n(1 + r)$	

Solving for n we have the useful equation

$$n = \frac{g}{1 + r}.$$

■ **Example 4.** The gross receipts at the end of the day at May's Hamburger Heaven are \$8,990.92 which include May's selling price plus the state and city taxes collected at a rate of 6% of the selling price. What are the net receipts?

Solution:

$$g = \$8{,}990.92$$

$$r = .06$$

$$\text{Then } n = \frac{\$8{,}990.92}{1 + .06} = \frac{\$8{,}990.92}{1.06} = \$8{,}482.00.$$

■

At your garage sale you agree to include some items for sale which belong to your mother-in-law. When she drops them off she informs you that they are already priced and the price includes a 20% commission for your trouble. You keep her sales receipts separate and at the end of the day you have collected \$158.88 on the sale of several of her items. How much of that is yours?

Solution: The gross receipts are $g = \$158.88$
The commission rate is $r = 20\%$

$$\text{Then } n = \frac{158.88}{1 + .20} = \frac{158.88}{1.20} = \$132.40.$$

The \$132.40 belongs to your mother-in-law. The remainder, \$158.88 − \$132.40 = \$26.48 belongs to you.

Checking:
20% of \$132.40 = \$26.48. ■

Exercise Set 11.1

When dollars and cents are involved, give answers to the nearest cent. If only an integral number makes sense as an answer, give the answer to the nearest whole number. In other cases give the answer to the nearest tenth.

1. 100.8 is what percent of 42?
2. 13.2 is 22% of what number?
3. 33.25% of 50 is what number?
4. A certain quarterback completes on the average 55% of his attempted passes. If during the season he completed 150 passes, how many passes did he attempt?
5. A club has 35 members. At the second meeting seven new members join. What is the increase in membership expressed as a percent?
6. A pair of slacks on sale cost \$24.76. If they were marked down 20%, what was the regular, pre-sale price?
7. Western's enrollment this year is 10,253. If this is 108% of last year's enrollment, what was the enrollment last year?
8. During the first six months of year *X* the CPI (Consumer Price Index) rose from 230 to 250. What was the rate of increase expressed as a percent?
9. Joe Blow batted .280 last year and had 340 hits. How many times was he officially at bat?
10. The regular mark-up in business *X* is 30% of the selling price. If the wholesale cost of an item is \$45, what will the selling price be?
11. If your regular mark-up is $33\frac{1}{3}\%$ of your cost, what discount in percent could you give and still break even?
12. Express $\frac{1}{2}\%$ as a decimal fraction.
13. The bus drivers' union asks for a 30% increase in the base salary of \$1000 a month for the duration of their contract, which is 4 years. The city, short on cash, agrees but pleads for the 30% to be spread over the next 3 years, 10% per year. How much total money will the drivers lose if they accept the city's counter offer?

14. At the end of three years, what would be the base salaries of the two plans under discussion in problem 13?
15. If the regular markup is 80%, what is the maximum discount, expressed as a percent, which will bring the selling price down to cost?
16. If the cash registers check out the total receipts of $8,342.44 for the week, which includes 4% sales tax, what are the net receipts?
17. On gross receipts of $7,924.58 a sales tax of 5% has been collected, 4% for the state and 1% for the city. How much should the check to the city be written for?
18. A girl lost 10 pounds and reduced her weight to 118 pounds. Her friend lost 14 pounds to reduce his weight to 152 pounds. Who lost the greatest percent of weight?
19. Mary's weight has increased 15 pounds in the last year. She now weighs 142 pounds. Sue has gained 18 pounds to increase her weight to 160 pounds. Which of these women has had the greatest percent of increase?
20. After a hard scrimmage, football players may lose up to 5.2% of their total weight without any adverse effects. If a 220-pound tackle loses 12 pounds during practice, is he within the safe range?

11.2 Interest

> You shall not lend upon interest to your brother, interest on money, interest on victuals, interest on anything that is lent for interest.
> To a foreigner you may lend upon interest, but to your brother you shall not lend upon interest.
>
> (*Deuteronomy 23:19,20*)

As long as the free exchange of commodities has existed there have been debtors and creditors. Before the common use of currency a man's worth was sometimes best measured by the number of cattle he owned (the word *chattels* has its root there). When we talk of *pecuniary* affairs we are again referring back to cattle-currency (*pecunia* is Latin for "cattle").

When the factor of time is included with percent, the result is **interest.** Interest is the fee charged by the lender and has been studied in detail as early as the thirteenth century.

Our recorded history has well documented debt-slavery as an institution. Borrowing for constructive reasons rather than from absolute needs is only a recent trend. There are few, if any, businesses or governments which do not freely engage in finance in at least some of its many forms.

Equation 1
Simple interest is computed by the formula

$$I = Prt$$

where I = amount of interest
P = principal
r = rate of interest
and t = time in years

An understanding of **compound interest** and **present value** techniques is absolutely essential to the understanding of many business finance decisions. Interest is, simply, the fee that is paid for the use of someone else's money. If one lends his money he will generally be paid interest for its use. If one borrows someone else's money, he will be required to pay interest for its use.

In most cases, interest will be *compounded*. That simply means that at certain agreed upon time intervals, the accumulated interest will be added to the investment. Then in subsequent intervals the total amount of savings, plus accumulated interest, will grow at the agreed upon interest rate. Interest bearing bank accounts, certificates of deposit, stock, commodities, and real estate are examples of long-term investments which can earn **compound interest.**

Often, business finance decisions involve the current value of some future series of cash payments. This requires the determination of the present value of future cash payments. Present value concepts are employed in capital budgeting and financial structure decisions, lease versus purchase decisions, refunding operations, and certain security valuation techniques.

Compound value (V) is defined as the sum, principal plus interest, to which a single beginning amount (P_o) will grow if it is left to earn interest (I) for (n) time intervals.

If k is the number of times per year the principal is compounded, r is the interest rate, and t is the number of years, then $i = \frac{r}{k}$ and $n = kt$. The compound value may be computed by the formula

Equation 2
$$V_n = P_0(1 + i)^n$$

■ Example 6. Find the compound value after 3 years if $1000 is invested at 6% compounded quarterly.

Solution:

$$t = 3$$
$$k = 4$$
$$n = kt = 4(3) = 12$$
$$i = \frac{r}{k} = \frac{.06}{4} = .015$$
$$V_{12} = 1{,}000(1 + .015)^{12} = 1{,}000(1.015)^{12} = \$1{,}195.62$$ ■

The formula for simple interest is stated without proof. The compound interest formula is developed in section 11.4, example 19. The exercise sets are designed for computational drill where a pocket calculator with a $\boxed{y^x}$ button will be a great help.

In the appendix we have included an abbreviated set of tables. These tables, because of their size, cannot be used on some problems. Pocket calculators are more versatile and are used almost universally today. Because of the round-off errors built into calculators, there may be slight differences, depending on the brand and the capability of the instrument. Calculators are almost always more accurate than tables because the number of significant digits in a calculator is greater than that in most tables.

■ Example 7. Find the compound value after 3 years if \$5000 is invested at 8% compounded quarterly.

Solution:

$$t = 3$$
$$k = 4$$
$$n = kt = 4(3) = 12$$
$$i = \frac{r}{k} = \frac{.08}{4} = .02$$
$$V_{12} = 5{,}000(1 + .02)^{12} = 5{,}000(1.02)^{12} = 5{,}000(1.268) = \$6{,}340.$$

✓

NOTE: Using table I the entry under 2% for $n = 12$ is the number 1.268. Checking with a pocket calculator,
$V_{12} = 5{,}000(1.02)^{12} = 6{,}341.209 = \$6{,}341.20.$

The calculator value, \$6,341.20 is the correct solution. ■

The language used at most financial institutions is to quote a low interest rate, usually simple interest. This rate is called the **nominal interest rate.** This nominal rate, together with some qualifying adjective, is the quoted rate. Phrases such as 8% add-on or 9% bankers interest are misleading as 8% add-on is much larger than 8% and 9% bankers interest is larger than 9%.

The **United States Rule** is the most common method for computing interest on large installment loans and mortgage payments. The United

States Rule computes interest only on the unpaid balance. The method of computation is to divide the interest rate by the number of installments due per year and compute the interest due. The payment made is then divided into principal and interest for this installment period at this rate.

■ Example 8. The annual interest rate of 12% is being charged on a mortgage whose balance due is $26,842.38. If a monthly payment of $300 is entered in the mortgage payment book, the 12% (=.12) is divided by the number of installments, 12, to produce a 1% per month rate and 1% of the balance is the interest due. 1% of $26,842.38 = $268.42. So the payment will be divided into $268.42 to pay interest and $31.58 to reduce the principal. ■

■ Example 9. Using the United States Rule,

a. What is the monthly installment interest rate charged on the unpaid balance if the annual rate is 8%?
b. What is the quarterly installment interest rate charged on the unpaid balance if an annual interest rate of 12% is used?
c. What is the monthly interest rate which is equivalent to an annual interest rate of 18%?

Solution: a. $\dfrac{8\%}{12} = \dfrac{3}{4}\% = .0075 \quad \left(\dfrac{3}{4} \text{ of } 1\% \text{ per month}\right)$

b. $\dfrac{12\%}{4} = 3\% = .03 \quad (3\% \text{ per quarter})$

c. $\dfrac{18\%}{12} = 1.5\% = .015 \quad \left(1\dfrac{1}{2}\% \text{ per month}\right)$ ■

Under current law interest charges must be stated in terms of simple interest. This simple interest is called the **annual percentage rate (A.P.R.)** for short. The annual percentage rate is the rate to use when comparing finance problems. It has been common practice for financial institutions to take liberties with the language to better their position in financial transactions.

The language sometimes works the other way. For example, to attract more depositors a banking institution may compound your savings several times a year, thereby increasing the amount of interest earned on the deposit. By law, 6% may be the maximum rate payable. However, since the law does not say anything about how often the institution can compound this interest, 6% may be compounded monthly. An advertisement might read "we pay an effective rate of 6.17%." The following is the derivation of a formula used to compute effective rates. The effective rate is the true, or actual, rate being used. When this rate is expressed in simple interest terms, it is usually referred to as annual percentage rate or nominal interest rate.

If we let r = annual percentage rate, or **effective rate**
R = nominal interest rate to be compounded,
n = number of periods,

then the value of one dollar after one year is $(1 + r)$ when simple interest is computed, but is $\left(1 + \frac{R}{n}\right)^n$ when compounded n times. We arrive at a formula for computing the annual percentage rate by equating these two.

$$1 + r = \left(1 + \frac{R}{n}\right)^n \quad \text{so}$$

Equation 3

$$r = \left(1 + \frac{R}{n}\right)^n - 1$$

■ Example 10. A local bank has nominal interest rate of 6% paid on passbook savings. What is the annual percentage rate if

1. the bank uses simple interest?
2. the bank compounds 4 times a year?
3. the bank compounds 12 times a year?
4. the bank compounds daily (365 times a year)?

Solutions:

1. Nominal interest is simple interest. Therefore 6% simple is 6% nominal.
2. $r = \left(1 + \frac{.06}{4}\right)^4 - 1 = .06136 \simeq 6.14\%$ A.P.R.
3. $r = \left(1 + \frac{.06}{12}\right)^{12} - 1 \simeq 6.17\%$ A.P.R.
4. $r = \left(1 + \frac{.06}{365}\right)^{365} - 1 \simeq 6.18\%$ A.P.R. ■

NOTE: The symbol $\simeq$ means "is approximately equal to."

■ Example 11. In an effort to increase the A.P.R. of 6.18% produced by compounding daily the nominal interest rate of 6% set by law in the last example, the bank decides to divide by 360 (the number of days in an ordinary interest year) but use 365 (the number of days in an exact interest year) to compute a higher A.P.R. This is called *super daily* interest. Find this new A.P.R.

Solution:

$$r = \left(1 + \frac{.06}{360}\right)^{365} - 1 \cong 6.27\%.$$

On the other side of the coin, it has been common practice for installment loans to be quoted in terms of add-on interest rates. This method of calculating installment payments to pay out a loan is usually computed by taking the simple interest and adding that amount to the original principal and having the borrower repay this total amount in n equal monthly payments. These are not computed uniformly by all financial institutions. It suffices to note that 9% add-on over two years is equivalent to approximately 16.4% A.P.R.

Present value is simply the reverse of compounding. That is, the present value is the beginning amount (P_o) one would have to invest to ensure an ending value (V) which has earned (i) interest for (n) time intervals. Equation 2 can quite readily be transformed into a present value formula by solving for P_o. Solving $V = P_o(1 + i)^n$ for P_o, we obtain

Equation 4

$$P_o = \frac{V}{(1 + i)^n} \text{ or, equivalently } P_0 = V(1 + i)^{-n}.$$

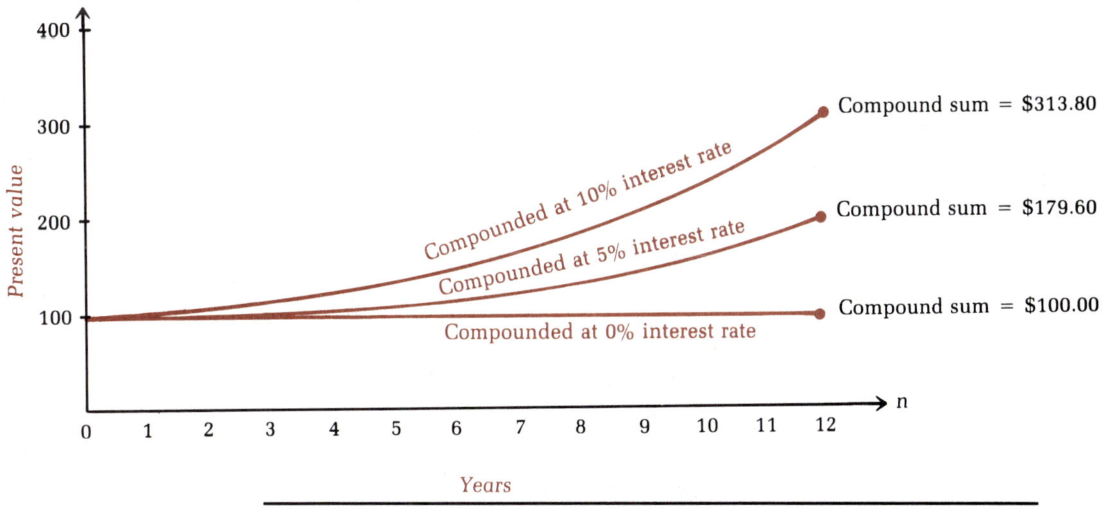

A Graphic View of the Present Value—Compound Sum Concept

■ Example 12. What amount must be invested in the money market if you expect a return of 15% compounded monthly which will produce a lump sum of $10,000 in 5 years?

Solution:

$$P_0 = 10{,}000\left(1 + \frac{.15}{12}\right)^{-5(12)} = 10{,}000\,(1.0125)^{-60}$$

$$= \$4{,}745.68.$$

Checking:
If $4,745.68 is invested at 15% compounded monthly for 5 years, what is the compound value?

$$V = P_0\,(1 + i)^n = (4{,}745.68)\left(1 + \frac{.15}{12}\right)^{60} \cong \$10{,}000.$$ ■

Exercise Set 11.2

Problems involving $I = Prt$

1. What is the simple interest due on a loan of $2,500 at 7% annually after $2\frac{1}{2}$ years?
2. If the simple interest on a loan of $10,000 after 3 years amounted to $2,550, what was the rate of interest charged?
3. If the simple interest at the rate of 12% annually on a loan amounted to $2,880 over a period of 3 years, how much was borrowed initially?
4. Consider the same situation as in problem 3, except that at the end of 3 years, the borrower in paying off the loan *and* the interest due at the end of 3 years wrote a check for $10,200. How much did he borrow initially?
5. What would be the monthly payments on an auto loan of $5,000 for 3 years at 7% simple interest calculated by the add on method? What, approximately, is the "true" interest rate (A.P.R.) being paid on this loan?

Problems involving compound interest. $V = P_0(1 + i)^n$

6. Find the compound amount if $2,500 is invested at 6% compounded quarterly for 3 years and 6 months.
7. Find the compound amount if $1,500 is invested at 6% compounded monthly for 25 years.
8. Mr. Jensen borrowed $1,000 at 12% compounded monthly three years ago today. What will it cost him to repay the loan today?

9. If I can invest $\$P$ now at 8% compounded quarterly, how much, to the nearest dollar, should I invest in order to have $10,000 at the end of 6 years?
10. If a man discovers an old bank book showing a deposit made 40 years ago to a savings account bearing 6% interest compounded semi-annually, how much was the deposit, to the nearest dollar, if there is $53,204 in the account now?

Problems involving present value. $P_0 = V(1 + i)^{-n}$

11. Find P_0, given $V = \$5{,}000$, $i = 1\%$, and $n = 24$.
12. Find P_0, given $V = \$442{,}873$ and $r = 9\%$ compounded 4 times a year for 10 years.
13. Find P_0, given $V = \$3{,}800$, $r = 8\%$, and $n = 24$ years compounded monthly.
14. How much should you be willing to pay for a $1,000 bond which will mature in three years if you expect a 12% annual return on your investment?
15. Rich Uncle George wishes to make an investment which will yield $10,000 in nine years. An insurance company offers to sell Uncle George an annuity for $4,200 which will mature in nine years for $10,000. Is Uncle George better off to buy savings deposits which have an 8.5% rate compounded quarterly?

Problems involving annual percentage rate.

16. Compute the annual percentage rate if the nominal rate is 15% and
    a. the principal is compounded quarterly.
    b. the principal is compounded monthly.
    c. the principal is compounded daily (using an ordinary interest year of 360 days).
    d. the principal is compounded daily using an exact interest year of 365 days.
    e. using super daily interest.
17. If two banks offer interest of 12% and 12.2% respectively but the first bank compounds daily and the second quarterly, which should you choose if
    a. you plan to keep your money on deposit for exactly one year?
    b. you plan to keep your money on deposit for five years?
18. A common practice is called *bank discount* where the interest is subtracted from the amount of the loan. For example, using bank discount, if a rate of 12% is charged on a $100 loan for one year the borrower is given $100 − $12 = $88 and must repay $100 in one year. What was the true interest rate?
19. Under the law, what is the annual percentage rate that must be quoted if bank discount of 8% is used?

11.3 Annuities—An Extension of Compound Interest and Present Value

An **annuity** is defined as a series of equal payments at regular, equal intervals. If one would make a series of payments (R) at the end of each time interval (n) and earn interest on these payments at the rate of (i), one would accumulate the **value amount** (VA). Algebraically this is expressed as

$$VA = R(1 + i)^{n-1} + R(1 + i)^{n-2} + \ldots + R(1 + i)^1 + R(1 + i)^0$$

$$VA = R\,[(1 + i)^{n-1} + (1 + i)^{n-2} + \ldots + (1 + i)^1 + 1]$$

VA is the series, each term of which is the value of a different payment.

$R(1 + i)^{n-1}$ is the final value of the first payment after it has been compounded $n - 1$ times.

$R(1 + i)^{n-2}$ is the final value of the second payment after it has been compounded $n - 2$ times.

⋮

$R(1 + i)^1$ is the final value of the next to last payment.

$R(1 + i)^0 = R$ is value of last payment. It has earned no interest.

✓ **NOTE:** n equals the number of payments and is easily misinterpreted. If you were to start an annuity, making a payment each month over the next four months, you would make *five*, not four, payments—one to start and one at the end of each of the next four months. A common deceptive practice is to hold the annuity an extra pay period on the pretext that the final payment has not had time to accumulate interest.

The following property reduces this series to a more convenient form. We have included a proof since we will be using the formula for a **geometric series** several times in this chapter. The sum of a geometric series of the form $S = 1 + a + a^2 + \ldots + a^{n-1}$ is

$$S = \frac{1 - a^n}{1 - a},\ a \neq 1.$$

Proof:

$$S - aS = (1 + a + a^2 + \ldots + a^{n-1}) - a(1 + a + a^2 + \ldots + a^{n-1})$$

$$S - aS = 1 + a + a^2 + \ldots + a^{n-1} - a - a^2 - a^3 - \ldots - a^{n-1} - a^n$$

$$S(1 - a) = 1 - a^n$$

$$S = \frac{1 - a^n}{1 - a}.$$

If we choose $a = 1 + i$ in the above formula, then

$VA = R[a^{n-1} + a^{n-2} + \ldots + a^2 + a + 1]$. By the above property 1,

$$VA = R\frac{1 - a^n}{1 - a} = R\left[\frac{1 - (1 + i)^n}{1 - (1 + i)}\right] = R\left[\frac{1 - (1 + i)^n}{-i}\right]$$

Equation 5

$$VA = R\left[\frac{(1 + i)^n - 1}{i}\right]$$

The value of an annuity then is the lump sum accumulated over a period of time and is the sum of several small (usually equal) payments into that annuity. Some common uses of an annuity are Christmas Funds, savings for college, or any cause for which the size of the annuity is predetermined. Insurance companies sell annuities, usually with an insurance package attached. A common arrangement is to buy an annuity on a child which will pay a lump sum, say $10,000, when the child is 19 and ready for college. The insurance is on the life of the buyer, not the child, and if that person dies the insurance company will still pay $10,000 on the due date. (Same annuity as on the death of the insuree.)

Looking at an annuity the other way around, someone might wish to start by depositing (a one-time investment) a large sum of money and receive in return a fixed periodic payment. In a rich and wise uncle's will, a young man might be left $50,000 per year for ten years; a widow may buy an annuity for herself with her husband's insurance; a ball player might prefer $200,000 per year for 5 years. The amount of money necessary to ensure these future payments is called *present value* of an annuity or present amount.

If one wanted to determine the value today, the present amount (PA), of a series of payments (R) at the end of each time interval (n) which earned interest at the rate of (i), the following equation would be used:

$$PA = R\left(\frac{1}{(1 + i)}\right) + R\left(\frac{1}{(1 + i)}\right)^2 + \ldots + R\left(\frac{1}{(1 + i)}\right)^n$$

$$PA = R\left(\frac{1}{(1 + i)} + \frac{1}{(1 + i)^2} + \ldots + \frac{1}{(1 + i)^n}\right)$$

Then,

$$PA = \frac{R}{1 + i}(1 + (1 + i)^{-1} + (1 + i)^{-2} + \ldots + (1 + i)^{-(n-1)}).$$

If we choose $a = (1 + i)^{-1}$, then

$$PA = \frac{R}{1 + i}(1 + a + a^2 + \ldots + a^{n-1}).$$

$$PA = \frac{R}{1 + i} \cdot \frac{1 - a^n}{1 - a}$$

$$PA = \frac{R}{1 + i} \cdot \frac{1 - [(1 + i)^{-1}]^n}{1 - (1 + i)^{-1}} = R\left[\frac{1 - (1 + i)^{-n}}{1 + i - 1}\right]$$

Equation 6

$$PA = R\left[\frac{1 - (1 + i)^{-n}}{i}\right]$$

CAUTION: This is often shortened to just present value, but present value of an annuity, $(PA) = R\left[\frac{1 - (1 + i)^{-n}}{i}\right]$, and present value, $P_0 = V(1 + i)^{-n}$, are not the same. PA makes a series of equal payouts of an amount R while P_0 makes exactly one payout of amount V.

■ Example 13. Below is a graphic illustration of an annuity, both present annuity (PA) and value annuity (VA), for 5 years at an interest rate of 10%.

Present annuity *is usually called* **present value.**

Value annuity *is usually called* **compound sum.**

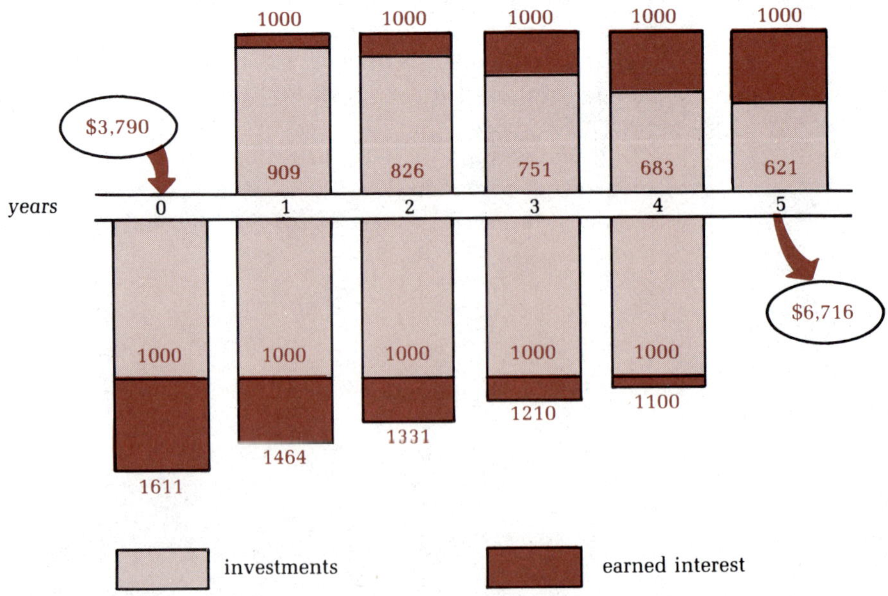

It is common practice in finance to round to the nearest dollar. The top half of the chart is present value and shows that

909 for 1 year at 10% returns 1,000
and 826 for 2 years at 10% returns 1,000
and 751 for 3 years at 10% returns 1,000
and 683 for 4 years at 10% returns 1,000
and 621 for 5 years at 10% returns 1,000
$3,790 – *total initial investment*

Hence an initial investment of $3,790 will induce 5 equal $1,000 payments. The present value of these future payments is $3,790.

Checking:

$$(PA) = R\left[\frac{1 - (1 + i)^{-n}}{i}\right]$$

$$= 1{,}000\left[\frac{1 - (1.1)^{-5}}{.1}\right] \approx \$3{,}791$$

The bottom half of the chart describes the compound sum (value annuity) at the end of a 5-year period. Note that 5 payments have been made over the 4-year period. The value will accumulate for five years as follows. The compound sum is the arithmetic sum of

an investment of $1000 at 10% for 5 years = $1,611
an investment of $1000 at 10% for 4 years = $1,464
an investment of $1000 at 10% for 3 years = $1,331
an investment of $1000 at 10% for 2 years = $1,210
an investment of $1000 at 10% for 1 year = $1,100
total $6,716

Checking:

$$VA = R\left[\frac{(1 + i)^n - 1}{i}\right] \quad \text{where } i = .1 \text{ and } n = 5, \; R = 1{,}000$$

$$VA = 1{,}000\left[\frac{(1.1)^5 - 1}{.1}\right] = 1{,}000\left[\frac{1.61051 - 1}{.1}\right]$$

$$= 1{,}000(6.1051) = 6{,}105$$

Since the value is to accumulate over 5 years, the money will draw one more year of interest: $6,105 at 10% yields $611 and the final sum is $6,105 + $611 = $6,716.

Checking again, if we let $n = 6$, the number of payments over 5 years,

$$VA = 1{,}000\left[\frac{(1.1)^6 - 1}{.1}\right] = 1{,}000\left[\frac{1.77156 - 1}{.1}\right]$$

$$= 7{,}716$$

and $7,716 − (1000)* = 6,716. ■

■ **Example 14.** Use present value of an annuity to determine the A.P.R. when an add-on interest rate is quoted. (By law, a lending institution must quote the A.P.R.). You have recently had a major wreck with your car. Repairs will cost $3,000. An auto repairman has agreed to finance the work for you at 8% add-on interest for two years (24 months). How much would your payments be and what is the equivalent annual interest rate compounded monthly?

Solution:

$3{,}000 \times .08 \times 2 =$	\$480	Amount you will pay in interest
$3{,}000 + 480 =$	\$3,480	Amount of loan plus interest
$3{,}480 \div 24 =$	\$145	Amount of monthly payment
$\dfrac{1 - (1 + i)^{-24}}{i} =$	$\dfrac{3{,}000}{145}$	i.e. $145\left[\dfrac{1 - (1 + i)^{-24}}{i}\right] \approx 3{,}000$ when $i = 1.223\% = .01223$

By programmable calculator $i = 1.223\%$ per month

$1.223 \times 12 = 14.676$ annual rate $= 14.68\%$ ■

Without a programmable calculator, you might not be able to solve

$$145\left[\frac{1 - (1 + i)^{-24}}{i}\right] = 3{,}000$$

You could, however, check their quoted rate. If your repairman gave you an offhand figure of 14%, you could check

$$145\left[\frac{1 - \left(1 + \frac{.14}{12}\right)^{-24}}{\frac{.14}{12}}\right].$$

Remember $\frac{.14}{12}$ is the monthly rate. This equals 3,020.35 which means his quote was low since you would have overpaid.

*This 1,000 was deposited and withdrawn at the same time, earning no interest.

CAUTION: In the problem above, 8% add-on over 2 years was really a rate in excess of 14%, but there is more bad news since 8% add-on over three years has an even higher A.P.R. Don't do business with people who use add-on interest rates.

The concepts described previously can be used to determine the following:

Personal Savings

1. How much one's savings will be worth in the future.
2. How much to deposit today to have a certain amount in the future.
3. The interest rate required for money to grow to a certain amount.
4. How long it will take a sum of money to grow to the desired amount.
5. How long it will take to save a certain amount with periodic deposits.
6. How much periodic deposits will be worth in a certain time period.

Consumer Loans

1. The total cost of the interest paid.
2. The payments required on a loan.
3. The total loan amount given the payment rate and the time interval.
4. The time needed to pay off a loan.

Home Mortgages and Real Estate

1. Buy versus rent decisions.
2. Monthly payments for home mortgages.
3. How much of a given payment on a home is principal and how much is interest.
4. Balance owed and principal paid at the end of a specified mortgage period.

Investment Decisions

1. Purchase price of a bond.
2. Intrinsic value of preferred or common stock.
3. Yield and rate of return on an investment.
4. Cost of whole life versus term insurance.
5. Value of annuities.
6. Benefits of tax shelter investments.

Financial Management Decisions

1. Investment decisions (capital budgeting) involving replacements and expansion.
2. Determining the value of advertising or research and development programs.
3. Bond refunding problems.
4. Cost of capital calculations.
5. Lease versus purchase decisions.
6. Financial structure decisions.

Exercise Set 11.3

Ordinary annuities

1. Find the amount and present value of an annuity of 25 quarterly payments of $200 if the interest rate is 4% compounded quarterly.
2. If a $100 birthday check is deposited in the bank every year for 10 years, how much will be in the bank, in your account, at the time of the tenth deposit if the bank pays 6% interest compounded annually?
3. What is the cash value of a house if you must pay $15,000 down and $450 per month for 10 years? The interest charged is 12% per year.
4. You need $25,000 in a sinking fund at the end of 10 years. How much will you need to set aside at the end of each year, including the end of the 10th year, if you can invest the money in a fund bearing 8% compounded annually?
5. You bought a $6,500 car and got a $1,200 trade-in allowance. How much down payment will you have to provide if you want to keep your monthly payments over $2\frac{1}{2}$ years down to $100? Assume the actual (not add-on) interest rate being charged is 12% per year. The first monthly payment is due one month after sale date and succeeding payments, 29 in number, at one-month intervals thereafter.
6. You need to borrow $20,000 for 10 years. The First National Bank will lend you the money at 10% per year and will allow you to amortize the loan by making equal size payments at the end of each year. The payments will include both principal and interest, interest being deducted first and then the remainder of the payment will reduce the principal. How large are your payments?

Compound value

7. If you deposit $500 in a 9% Savings and Loan account and you make no further deposits or withdrawals, how much would be in your account at the end of five years if interest is compounded annually?
8. You deposit $1,500 in a savings account that pays 6% annual interest compounded monthly. How much money would you have if you leave the money in the bank for 3 years?
9. You have accepted your first position after leaving college. Your new employer has agreed to contribute $1,500 each year toward your retirement plan. If you plan to retire in forty years and if the company invests these annual contributions at 10% for you, how much will you have on your retirement day?
10. A loan shark agrees to lend you $5,000 today to start a business. You must promise to return $9,535 to him three years from today. What rate of interest is the loan shark charging you?

Present value

11. You are considering the purchase of a bond as an investment. The bond has a $1000 face value and it pays $40 in annual interest (4% coupon rate). The bond will mature in five years. You have determined that this bond should pay its owner a minimum of 12% annually. How much would you pay for the bond to insure yourself a 12% annual return on your investment?

12. You plan to make a round-the-world trip five years from now. How much do you need to invest today at 8% interest, compounded annually, to insure that you will have the $15,000 needed for the trip?

13. You have been told that someone with a college degree will earn approximately $3,000 more per year than someone without a degree. If you estimate your working life at 35 years and if money will earn on the average 8% annually over the next 35 years, how much can you afford to spend on your education and still consider it a profitable investment?

14. You are considering borrowing $52,000 to finance the purchase of a new house at 12% annual interest compounded monthly. The mortgage will run for 30 years. How much will your monthly house payment be (excluding taxes and insurance)?

15. Suppose that next year you are to receive a $10,000 scholarship to be spent on your education over the next three academic years. If you place your money in a bank account yielding 6% annual interest and you make three equal withdrawals, one at the end of each year for the three years, how large a withdrawal can you make and leave yourself exactly a zero balance after the last withdrawal?

16. The XYZ Corporation is considering the purchase of a new machine to replace one that is less efficient. The new machine will cost $50,000 and will be financed at an 8% annual interest rate. According to engineering estimates the new machine will reduce labor, materials, and direct costs by $4,500 per year. The estimated life of the new machine is fifteen years. Is this a wise investment for the corporation (ignoring the effect of taxes, salvage value, etc.)?

17. It has been estimated that if you add insulation to your home you will save approximately $120 per year in heating costs. If insulation costs $1,500 to install and if it has a life of 20 years, what is the return on your investment you will realize from insulating your home? *HINT*: This problem requires a calculator or table which can solve the equation for the value of i.

18. You borrow $2,000 to buy a used car at 12% annual interest compounded monthly. With a monthly loan payment of $94.15, how long will it take you to pay off the loan? *HINT*: Use logarithms to solve

$$\frac{1 - (1 + i)^{-n}}{i} = \frac{P}{R} \text{ for } n.$$

11.4 Recursive Formula, Difference Equation

An equation which relates P_n to P_{n+1} coupled with a specific value for P_0 is called a **recursive equation.** The subscripts on P are just indexes.

P_n refers to the nth value of P.

$P_{(n+1)}$ refers to the $(n + 1)^{th}$ value of P.

P_0 refers to the initial value of P.

If P represents a payment, P_4 is the 4th payment.

If P were the remaining principal on a mortgage whose monthly rate is 1% (12% A.P.R.), using the United States Rule,

$$P_{(n+1)} = P_n - [R - (.01)P_n]$$

$P_{(n+1)} = (1.01)P_n - R$, where R is the amount of the fixed payment.

For example, suppose the balance due on a home mortgage is \$32,485.43 and a regular monthly payment of \$400 is made where an effective rate of 12% is being charged. Then the \$400 is broken down as 324.85 to defray interest—(.01)(32,485.43)—and 75.15 to reduce the principal. The balance next month is \$32,485.43 − \$75.15 which equals \$32,410.28. Using the recursive equation above,

$$\begin{aligned} P_{n+1} &= (1.01)(32{,}485.43) - 400 \\ &= 32{,}410.28. \end{aligned}$$

■ Example 15. Use the recursive equation $P_{n+1} = 0.1P_n + 3$ to find P_4 given $P_0 = 100$.

Solution:

$$\begin{aligned} P_1 &= 0.1P_0 + 3 = 0.1(100) + 3 = 13, \\ P_2 &= (0.1)(P_1) + 3 = 1.3 + 3 = 4.3, \\ P_3 &= (0.1)(P_2) + 3 = 0.43 + 3 = 3.43, \\ P_4 &= (0.1)(P_3) + 3 = 0.343 + 3 = 3.343. \end{aligned}$$ ■

■ Example 16. Find P_{301} given $P_{n+1} = (1.02)P_n - 318.13$ and $P_{300} = \$14,286.24$.

Solution: $P_{301} = (1.02)(14{,}286.24) - 318.13 = \$14{,}253.84$. ■

The recursive formula $P_{n+1} = aP_n + b$ is used in almost all installment transactions and is observed as the growth relationship in infectious diseases, human populations, radioactive decay, and a host of other natural occurrences which affect our everyday lives. If we expand using this formula, we obtain

$$\begin{aligned} P_1 &= aP_0 + b \\ P_2 &= aP_1 + b = a(aP_0 + b) + b = a^2P_0 + ab + b \\ &= a^2P_0 + b(a + 1) \end{aligned}$$

$$P_3 = aP_2 + b = a[a^2P_0 + b(a + 1)] + b$$
$$= a^3P_0 + ab(a + 1) + b = a^3P_0 + b(a^2 + a + 1)$$
$$P_4 = a^4P_0 + b(a^3 + a^2 + a + 1)$$

and in general

1. $P_n = a^nP_0 + b(a^{n-1} + a^{n-2} + \ldots + a^3 + a^2 + a + 1).$

If $a \neq 1$, the geometric series

$$a^{n-1} + a^{n-2} + \ldots + a^3 + a^2 + a + 1 = \frac{1 - a^n}{1 - a}.$$

Hence,

$$P_n = a^nP_0 + b\left[\frac{1 - a^n}{1 - a}\right]$$

$$P_n = a^nP_0 + \frac{b}{1 - a} - \frac{ba^n}{1 - a}$$ and finally

2. The difference equation.

$$P_n = \left[P_0 - \frac{b}{1 - a}\right]a^n + \frac{b}{1 - a}$$

When $a = 1$, equation **1** above becomes

$$P_n = a^nP_0 + b(n) = a^nP_0 + bn.$$

Equation **2** is called a difference equation and, when the recursive equation $P_{n+1} = aP_n + b$ is known, will calculate P_n directly.

Example 17. Rework example 15 where you were asked to compute P_4 given $P_0 = 100$ and $P_{n+1} = (0.1)P_n + 3$.

Solution: From the recursive formula, we observe $a = 0.1$ and $b = 3$. Using these values in the difference equation **2** above, we have

$$P_4 = \left(100 - \frac{3}{1 - 0.1}\right)(0.1)^4 + \frac{3}{1 - 0.1}$$
$$= \left(100 - \frac{3}{0.9}\right)(0.1)^4 + \frac{3}{0.9}$$
$$= 3.343 \text{ using a calculator to compute.}$$ ■

■ **Example 18.** A young family can borrow money from their local savings and loan and repay it at \$300 per month for the next 30 years. The interest is charged monthly at 1% of the unpaid balance. How much can they afford to borrow?

Solution: The recursive formula for the payback is $P_{n+1} = P_n + (.01)P_n - 300$; i.e., the new entry in the payment book each month, P_{n+1}, is computed by adding the incurred interest, $(.01)P_n$, and subtracting the payment. A little algebra produces $P_{n+1} = (1.01)P_n - 300$ where $a = 1.01$ and $b = -300$. Using the difference equations we wish to find P_0 when $P_{360} = 0$. (After 360 payments, the remaining balance is 0)

$$0 = \left(P_0 - \frac{-300}{1 - 1.01}\right)(1.01)^{360} + \frac{-300}{1 - 1.01}$$

$$0 = \left(P_0 - \frac{-300}{-(.01)}\right)(1.01)^{360} + \frac{-300}{-(.01)}$$

$$0 = (P_0 - 30{,}000)(1.01)^{360} + 30{,}000$$

$$0 = (1.01)^{360}P_0 - (1.01)^{360}(30{,}000) + 30{,}000$$

$$\text{so } P_0 = 30{,}000 - 30{,}000(1.01)^{-360}$$

$$= 30{,}000[1 - (1.01)^{-360}] = \$29{,}165.50.$$ ■

In examples 19 and 20, the difference equation will be used to develop some of the formulas seen earlier.

■ **Example 19.** If we look at compound value, then $V_{n+1} = V_n + i \cdot V_n = (1 + i)V_n$.

Solution: By applying the difference equation where $a = 1 + i$ and $b = 0$, we have

$$V_n = P_n = \left[P_0 - \frac{0}{1 - (1 + i)}\right](1 + i)^n + \frac{0}{1 - (1 + i)}, \text{ or}$$

$$V_n = P_0(1 + i)^n.$$ ■

■ **Example 20.** If we observe how each successive compound sum (VA) is generated, i.e., (new amount) = (old amount) + (interest earned) + one payment, then

$$(VA)_{n+1} = (VA)_n + i(VA)_n + R$$

$$(VA)_{n+1} = (1 + i)(VA)_n + R.$$

Solution: Substituting into the difference equation where $VA = P_n$, $a = 1 + i$, $b = R$, and $P_0 = 0$ we have

$$VA = P_n = \left(0 - \frac{R}{-i}\right)(1 + i)^n + \frac{R}{-i}$$

$$VA = \frac{R}{i}(1 + i)^n - \frac{R}{i} = R\left[\frac{(1 + i)^n - 1}{i}\right].$$

Also the present value of an annuity (*PA*) is obtained if we let $a = 1 + i$, $P_n = 0$, and $b = -R$. Then the difference equation becomes

$$0 = \left(P_0 - \frac{-R}{-i}\right) \cdot (1 + i)^n + \frac{-R}{-i}$$

$$\text{and } \left(P_0 - \frac{R}{i}\right) \cdot (1 + i)^n = -\frac{R}{i}$$

$$P_0 - \frac{R}{i} = -\frac{R}{i}(1 + i)^{-n}$$

$$P_0 = \frac{R}{i} - \frac{R}{i}(1 + i)^{-n}$$

$$P_0 = R\left[\frac{1 - (1 + i)^{-n}}{i}\right].$$

■

■ **Example 21.** The population of the United States is now about 220 million. If the population, due to a higher birth rate than death rate, is increasing at a rate of 0.6% per year and the country is accepting a quarter of a million new immigrants per year, what will be our population size in 10 years?

Solution:

$$P_{n+1} = P_n + .006P_n + 250{,}000$$

$$P_{n+1} = (1.006)P_n + 250{,}000$$

Then $P_0 = 220$, $a = 1.006$, and $b = \frac{1}{4}$, in millions. By the difference equation

$$P_n = \left(220 - \frac{.25}{-.006}\right)(1.006)^{10} - \frac{.25}{.006}$$

$$= \left(220 + \frac{.25}{.006}\right)(1.006)^{10} - \frac{.25}{.006} = 236.13.$$

So in ten years our population should be about 236 million if these trends continue. ■

■ **Example 22.** A new product is to be put on the market. Initial promotion will put the product into the hands of 10,000 people. Market analysis estimates that each month 12% of the people who have used the product that month will convince someone else to try it. Hence, the market will increase by 12% while additional advertising will add another 1,000 per month to the users. What is the number of sales in the 24th month? (Assume the product is a magazine and everyone buys a new one each month.)

Solution: $y_{n+1} = y_n + .12y_n + 1{,}000$

The number of people who buy the magazine this month = the number who bought last month + 12% of last month's sales + 1,000 or

$$y_{n+1} = 1.12y_n + 1{,}000$$

Using the difference equation where $a = 1.12$, $b = 1{,}000$, $n = 24$, and $P_0 = 10{,}000$,

$$y_{24} = \left(10{,}000 - \frac{1{,}000}{-0.12}\right)(1.12)^{24} + \frac{1{,}000}{-0.12}$$

$$y_{24} = 269{,}941$$

■ Example 23. Two old men were discussing the folly of their youth. Looking at some expensive real estate, the first old man said, "I could have bought this in '34 for \$1,500, and today it's worth a million." The second old man thought, "In '34 I invested \$1,500 at 7% compounded quarterly. I believe I might have enough to buy this land." What amount does he have if all this took place in '82, 48 years later?

Solution: $P = (1{,}500)\left[1 + \frac{.07}{4}\right]^{4(48)}$ = \$41,946.86 which is not a million yet.

Exercise Set 11.4

Given the following recursive formulas, use the difference equation to find P_n.

1. $P_{n+1} = 3P_n + 4$ when $n = 20$ and $P_0 = 10$
2. $P_{n+1} = 2P_n - 4$ when $n = 100$ and $P_0 = 80$
3. $P_{n+1} = \frac{1}{2}P_n$ when $P_0 = 10{,}000$ and $n = 24$
4. $P_{n+1} = (1.01)P_n - 285$ when $P_n = 24{,}000$ and $n = 720$
5. Given $P_{n+1} = (1.015)P_n - 360$ compute $P_1, P_2, P_3, P_4, P_{120}$ given $P_0 = 22{,}000$.
6. A couple borrows \$22,000 to buy a house at a nominal rate of 18%. They make monthly payments of \$360 where the interest incurred each month is subtracted from this payment and the balance is used to reduce the principal. How much will the couple owe in 10 years?
7. If the number of students in a college is decreasing at a rate of 3% per year, how many students will be in school five years from now if 12,000 are presently enrolled?
8. Use a recursive equation and the difference equation to compute the value after 3 years of an initial deposit of \$5,000 which is to be increased by a monthly deposit of \$60 if the bank compounds monthly at a rate of 6.27%.
9. You are left an estate which, after taxes, is worth \$750,000. If you could invest this sum at 7% in tax-free bonds, would you prefer to

have the \$750,000 to invest or have \$70,000 per year, again tax-free, for 20 years?

10. When a certain hot object is placed in a cooling room, it retains, at the end of each hour, 80% of its original heat less 5 degrees until it reaches room temperature. If, initially, the object is 314°F, how hot will it be in five hours?

11.5 Chapter Review

Important Terms and Formulas in Chapter 11.

percent (section 11.1)
decimal fraction (section 11.1)
interest (section 11.2)
simple interest (section 11.2)
compound interest (section 11.2)
present value (section 11.2)
compound value (section 11.2)
nominal interest rate (section 11.2)
United States Rule (section 11.2)
$P_0 = \frac{V}{[1 + i]^n}$ (section 11.2)
annual percentage rate (A.P.R.) (section 11.2)
simple interest formula (section 11.2)
$I = Prt$ (section 11.2)
$V_n = P_0(1 + i)^n$ (section 11.2)
$r = \left(1 + \frac{R}{n}\right)^n -$ (section 11.2)
$P_0 = V(1 + i)^{-n}$ (section 11.2)
effective rate (section 11.2)
annuity (section 11.3)
value amount, VA (section 11.3)
geometric series (section 11.3)
$S = \frac{1 - a^n}{1 - a}$ (section 11.3)
$VA = R\left[\frac{(1 + i)^n - 1}{i}\right]$ (section 11.3)
$PA = R\left[\frac{1 - (1 + i)^{-n}}{i}\right]$ (section 11.3)
present annuity, present value of (PA) (section 11.3)
value annuity, compound sum (section 11.3)
recursive equation (section 11.4)
difference equation (section 11.4)
$P_n = \left[P_0 - \frac{b}{1 - a}\right]a^n + \frac{b}{1 - a}$ (section 11.4)

Review Exercises

1. \$107.52 is 84% of what amount?
2. When 240 is increased to 276, what is the percent of increase?
3. If an amount P is increased by 10% and then this new amount is decreased by 10%, will the final result equal P?
4. A store manager knows that he may reduce his sales price by as much as 50% without going below his cost. What is his "mark-up"?
5. After 30 days on the "pound a day diet," a man reduced his weight from 248 pounds to 218 pounds. What percent of his original weight did he lose?

6. After 30 days on the "pound a day diet," a woman reduced her weight from 182 pounds to 152 pounds. What percent of her original weight did she lose?
7. A contract calls for a 40% increase of the base pay. The union members are asked to indicate which of two plans they like best.
Plan A: A one-time 40% increase to be in effect for 3 years.
Plan B: An initial 20% increase with two 10% increases, one for each of the last two years.
   a. Are these plans equal?
   b. What would be the amount paid on an initial base of $P over the 3 years using these plans?
   c. Is Plan B in any way better than Plan A?
   d. Which plan is better?
8. The gross receipts are $2,961.09. A sales tax of 4% was added to the retail price of the merchandise sold. What were the net receipts?
9. You buy several items at the local department store. Would you rather the clerk added the sales tax to each item before adding the grand total or would you prefer the clerk compute the grand total and then add the proper tax to this amount?
10. Prove the results of problem 9 for two items.
11. What is the interest due on a $2,000 loan at 8% annual interest rate for 3 years if
   a. simple interest is used?
   b. compounded yearly?
   c. compounded monthly?
12. On an 8% add-on rate for a $6,000 loan over 36 months
   a. what is the total interest paid to the lender?
   b. how much are the monthly payments?
13. Since the amount of interest paid is the same using simple interest and add-on interest, what is the major difference?
14. What amount, to the nearest dollar, must you invest at 15% compounded yearly in order to have $20,000 at the end of 10 years?
15. If you invest $4,000 at 8% interest compounded daily, how much would you have in the bank at the end of 2 years assuming your bank computed its interest on a 360-day year?
*16. If your bank compounds your investment continuously, they will use the formula $V = Pe^{rt}$ where

P = your deposit

r = interest rate

t = number of years

Use the formula to compute the amount your original deposit of $4,000 has become if the rate was 8% compounded continuously for 2 years. How does this compare with problem 15?

*$e \approx 2.71828...$, e is the base of the natural logarithm and is defined as $\lim_{n\to\infty}\left(1 + \frac{1}{n}\right)^n$

17. Compute the A.P.R. if the nominal rate is 12% and
    a. the principal is compounded quarterly.
    b. the principal is compounded monthly.
    c. the principal is compounded daily (using an ordinary interest year).
18. Find the sum of the geometric series

$$1 + \frac{1}{2} + \frac{1}{4} + \frac{1}{8} + \ldots + \frac{1}{1024}.$$

19. Find the compound sum of an annuity of 40 quarterly payments of \$300 at an annual rate of 12%
    a. on the day you make the 40th payment.
    b. forty months after the first payment is made.
20. Your child-support payments are \$200 per month and will remain in effect for the next 180 months. What will be the cost of an annuity which will make these 180 payments on time if your money earns 12% interest?
21. Under a sheltered income plan you can deposit \$100 a month for the next 20 years at an annual rate of 6% compounded monthly. While 6% isn't much, by law you will not have to pay federal income tax on this money until you withdraw it after you retire. Since you are now in a 32% tax bracket, you are buying a \$100 investment for \$68. How much will the account be worth in 20 years?
22. According to your best advice, you can afford to make house payments not to exceed 25% of your present take-home pay. In fact, this seems to be the limit you can borrow from your preferred lender. You and your spouse have a combined take-home pay of \$1,732 a month. What is the maximum amount you can expect if your loan cannot exceed 20 years and the current interest rate is 14%?
23. Given the following recursive formulas, use the difference equation to find P_n.
    a. $P_{n+1} = 1.03P_n - 200$ where $n = 10$ and $P_0 = 20{,}000$
    b. $P_{n+1} = (1.1)P_n + 12$ where $n = 6$ and $P_0 = 0$
24. A school band director assumes that 90% of the people who supported band fund drives one year will support them the next. Because of new band members, there seems to be a fixed number of new supporters, 30 each year. If we start with 100 loyal supporters, how many can we expect in 5 years?
25. Will the number of supporters in problem 24 continue to increase at the same rate? If the fact that your band was awarded the honor of marching in the state capital 4th of July parade increased your supporters to exactly 300, what would the future bring?

Appendix: Tables

I. Compound value of \$1 $(1 + i)^n$

II. Present value of \$1 $(1 + i)^{-n}$

III. Compound sum of an annuity of \$1 $\dfrac{(1 + i)^n - 1}{i}$

IV. Present value of an annuity of \$1 $\dfrac{1 - (1 + i)^{-n}}{i}$

V. Binomial probabilities $P(X = x)$

VI. Normal probabilities $P(0 < Z < z)$

Table I. Compound value of $1

n	1%	2%	3%	4%	5%	6%	7%	8%	9%	10%
1	1.010	1.020	1.030	1.040	1.050	1.060	1.070	1.080	1.090	1.100
2	1.020	1.040	1.061	1.082	1.102	1.124	1.145	1.166	1.188	1.210
3	1.030	1.061	1.093	1.125	1.158	1.191	1.225	1.260	1.295	1.331
4	1.041	1.082	1.126	1.170	1.216	1.262	1.311	1.360	1.412	1.464
5	1.051	1.104	1.159	1.217	1.276	1.338	1.403	1.469	1.539	1.611
6	1.062	1.126	1.194	1.265	1.340	1.419	1.501	1.587	1.677	1.772
7	1.072	1.149	1.230	1.316	1.407	1.504	1.606	1.714	1.828	1.949
8	1.083	1.172	1.267	1.369	1.477	1.594	1.718	1.851	1.993	2.144
9	1.094	1.195	1.305	1.423	1.551	1.689	1.838	1.999	2.172	2.358
10	1.105	1.219	1.344	1.480	1.629	1.791	1.967	2.159	2.367	2.594
11	1.116	1.243	1.384	1.539	1.710	1.898	2.105	2.332	2.580	2.853
12	1.127	1.268	1.426	1.601	1.796	2.012	2.252	2.518	2.813	3.138
13	1.138	1.294	1.469	1.665	1.886	2.133	2.410	2.720	3.066	3.452
14	1.149	1.319	1.513	1.732	1.980	2.261	2.579	2.937	3.342	3.797
15	1.161	1.346	1.558	1.801	2.079	2.397	2.759	3.172	3.642	4.177

$(1 + i)^n$

Table II. Present value of $1

n	1%	2%	3%	4%	5%	6%	7%	8%	9%	10%	12%	14%	15%
1	.990	.980	.971	.962	.952	.943	.935	.926	.917	.909	.893	.877	.870
2	.980	.961	.943	.925	.907	.890	.873	.857	.842	.826	.797	.769	.756
3	.971	.942	.915	.889	.864	.840	.816	.794	.772	.751	.712	.675	.658
4	.961	.924	.889	.855	.823	.792	.764	.835	.708	.683	.636	.592	.572
5	.951	.906	.863	.822	.784	.747	.713	.681	.650	.621	.567	.519	.497
6	.942	.888	.838	.790	.746	.705	.666	.630	.596	.564	.507	.456	.432
7	.933	.871	.813	.760	.711	.665	.623	.583	.547	.513	.452	.400	.376
8	.923	.853	.789	.731	.677	.627	.582	.540	.502	.467	.404	.351	.327
9	.914	.837	.766	.703	.645	.592	.544	.500	.460	.424	.361	.308	.284
10	.905	.820	.744	.676	.614	.558	.508	.463	.422	.386	.322	.270	.247

$(1 + i)^{-n}$

Table III. Compound sum of an annuity of $1

n	1%	2%	3%	4%	5%	6%	7%	8%
1	1.000	1.000	1.000	1.000	1.000	1.000	1.000	1.000
2	2.010	2.020	2.030	2.040	2.050	2.060	2.070	2.080
3	3.030	3.060	3.091	3.122	3.152	3.184	3.215	3.246
4	4.060	4.122	4.184	4.246	4.310	4.375	4.440	4.506
5	5.101	5.204	5.309	5.416	5.526	5.637	5.751	5.867
6	6.152	6.308	6.468	6.633	6.802	6.975	7.153	7.336
7	7.214	7.434	7.662	7.898	8.142	8.394	8.654	8.923
8	8.286	8.583	8.892	9.214	9.549	9.897	10.260	10.637
9	9.369	9.755	10.159	10.583	11.027	11.491	11.978	12.488
10	10.462	10.950	11.464	12.006	12.578	13.181	13.816	14.487

$$\frac{(1 + i)^n - 1}{i}$$

Table IV. Present value of an annuity of $1

n	1%	2%	3%	4%	5%	6%	7%	8%	9%	10%
1	0.990	0.980	0.971	0.962	0.952	0.943	0.935	0.926	0.917	0.909
2	1.970	1.942	1.913	1.886	1.859	1.833	1.808	1.783	1.759	1.736
3	2.941	2.884	2.829	2.775	2.723	2.673	2.624	2.577	2.531	2.487
4	3.902	3.808	3.717	3.630	3.546	3.465	3.387	3.312	3.240	3.170
5	4.853	4.713	4.580	4.452	4.329	4.212	4.100	3.993	3.890	3.791
6	5.795	5.601	5.417	5.242	5.076	4.917	4.766	4.623	4.486	4.355
7	6.728	6.472	6.230	6.002	5.786	5.582	5.389	5.206	5.033	4.868
8	7.652	7.325	7.020	6.733	6.463	6.210	6.971	5.747	5.535	5.335
9	8.566	8.162	7.786	7.435	7.108	6.802	6.515	6.247	5.985	5.759
10	9.471	8.983	8.530	8.111	7.722	7.360	7.024	6.710	6.418	6.145

$$\frac{1 - (1 + i)^{-n}}{i}$$

Table V. Binomial Probabilities $P(X = x)$

x \ p	.01	.05	.10	.20	.25	.30	.40	.50	.60	.70	.75	.80	.90	.95	.99
0	.895	.569	.314	.086	.042	.020	.004								
1	.099	.329	.384	.236	.155	.093	.027	.005	.001						
2	.005	.087	.213	.295	.258	.200	.089	.027	.005	.001					
3		.014	.071	.221	.258	.257	.177	.081	.023	.004					
4		.001	.016	.111	.172	.220	.236	.161	.070	.017	.006	.002			
5			.002	.039	.080	.132	.221	.226	.147	.057	.027	.010			
6				.010	.027	.057	.147	.226	.221	.132	.080	.039	.002		
7				.002	.006	.017	.070	.161	.236	.220	.172	.111	.016	.001	
8					.001	.004	.023	.081	.177	.257	.258	.221	.071	.014	
9						.001	.005	.027	.089	.200	.258	.295	.213	.087	.005
10							.001	.005	.027	.093	.155	.236	.384	.329	.099
11									.004	.020	.042	.086	.314	.569	.895

$n = 11$

x \ p	.01	.05	.10	.20	.25	.30	.40	.50	.60	.70	.75	.80	.90	.95	.99
0	.886	.540	.282	.069	.032	.014	.002								
1	.107	.341	.377	.206	.127	.071	.017	.003							
2	.006	.099	.230	.283	.232	.168	.064	.016	.002						
3		.017	.085	.236	.258	.240	.142	.054	.012	.001					
4		.002	.021	.133	.194	.231	.213	.121	.042	.008	.002	.001			
5			.004	.053	.103	.158	.227	.193	.101	.029	.012	.003			
6				.016	.040	.079	.177	.226	.177	.079	.040	.016			
7				.003	.012	.029	.101	.193	.227	.158	.103	.053	.004		
8				.001	.002	.008	.042	.121	.213	.231	.194	.133	.021	.002	
9						.001	.012	.054	.142	.240	.258	.236	.085	.017	
10							.002	.016	.064	.168	.232	.283	.230	.099	.006
11								.003	.017	.071	.127	.206	.377	.341	.107
12									.002	.014	.032	.069	.282	.540	.886

$n = 12$

x \ p	.01	.05	.10	.20	.25	.30	.40	.50	.60	.70	.75	.80	.90	.95	.99
0	.878	.513	.254	.055	.024	.010	.001								
1	.115	.351	.367	.179	.103	.054	.011	.002							
2	.007	.111	.245	.268	.206	.139	.045	.010	.001						
3		.021	.100	.246	.252	.218	.111	.035	.006	.001					
4		.003	.028	.154	.210	.234	.184	.087	.024	.003	.001				
5			.006	.069	.126	.180	.221	.157	.066	.014	.005	.001			
6			.001	.023	.056	.103	.197	.209	.131	.044	.019	.006			
7				.006	.019	.044	.131	.209	.197	.103	.056	.023	.001		
8				.001	.005	.014	.066	.157	.221	.180	.126	.069	.006		
9					.001	.003	.024	.087	.184	.234	.210	.154	.028	.003	
10						.001	.006	.035	.111	.218	.252	.246	.100	.021	
11							.001	.010	.045	.139	.206	.268	.245	.111	.007
12								.002	.011	.054	.103	.179	.367	.351	.115
13									.001	.010	.024	.055	.254	.513	.878

$n = 13$

x \ p	.01	.05	.10	.20	.25	.30	.40	.50	.60	.70	.75	.80	.90	.95	.99
0	.869	.488	.229	.044	.019	.007	.001								
1	.123	.359	.356	.154	.083	.041	.007	.001							
2	.008	.123	.257	.250	.180	.113	.032	.006	.001						
3		.026	.114	.250	.240	.194	.085	.022	.003						
4		.004	.035	.172	.220	.229	.155	.061	.014	.001					
5			.008	.086	.149	.196	.207	.122	.041	.007	.002				
6			.001	.032	.073	.126	.207	.183	.092	.023	.008	.002			
7				.009	.028	.062	.157	.209	.157	.062	.028	.009			
8				.002	.008	.023	.092	.183	.207	.126	.073	.032	.001		
9					.002	.007	.041	.122	.207	.196	.149	.086	.008		
10						.001	.014	.061	.155	.229	.220	.172	.035	.004	
11							.003	.022	.085	.194	.240	.250	.114	.026	
12							.001	.006	.032	.113	.180	.250	.257	.123	.008
13								.001	.007	.041	.083	.154	.356	.359	.123
14									.001	.007	.019	.044	.229	.488	.869

$n = 14$

Table V. *(Continued)*

x \ p	.01	.05	.10	.20	.25	.30	.40	.50	.60	.70	.75	.80	.90	.95	.99
0	.860	.463	.206	.035	.013	.005									
1	.130	.366	.343	.132	.069	.031	.005								
2	.009	.135	.267	.231	.156	.092	.022	.003							
3		.031	.129	.250	.225	.170	.063	.014	.002						
4		.005	.043	.188	.225	.219	.127	.042	.007	.001					
5		.001	.010	.103	.165	.206	.186	.092	.024	.003	.001				
6			.002	.043	.092	.147	.207	.153	.061	.012	.003	.001			
7				.014	.039	.081	.177	.196	.118	.035	.013	.003			
8				.003	.013	.035	.118	.196	.177	.081	.039	.014			
9				.001	.003	.012	.061	.153	.207	.147	.092	.043	.002		
10					.001	.003	.024	.092	.186	.206	.165	.103	.010	.001	
11						.001	.007	.042	.127	.219	.225	.188	.043	.005	
12							.002	.014	.063	.170	.225	.250	.129	.031	
13								.003	.022	.092	.156	.231	.267	.135	.009
14									.005	.031	.069	.132	.343	.366	.130
15										.005	.013	.035	.206	.463	.860

$n = 15$

x \ p	.01	.05	.10	.20	.25	.30	.40	.50	.60	.70	.75	.80	.90	.95	.99
0	.851	.440	.185	.028	.010	.003									
1	.138	.371	.329	.113	.054	.023	.003								
2	.010	.146	.274	.211	.134	.073	.015	.002							
3		.036	.142	.246	.208	.146	.047	.008	.001						
4		.006	.051	.200	.225	.204	.101	.028	.004						
5		.001	.014	.120	.180	.210	.162	.067	.014	.001					
6			.003	.055	.110	.165	.198	.122	.039	.006	.001				
7				.020	.052	.101	.189	.175	.084	.018	.006	.001			
8				.006	.020	.049	.142	.196	.142	.049	.020	.006			
9				.001	.006	.018	.084	.175	.189	.101	.052	.020			
10					.001	.006	.039	.122	.198	.165	.110	.055	.003		
11						.001	.014	.067	.162	.210	.180	.120	.014	.001	
12							.004	.028	.101	.204	.225	.200	.051	.006	
13							.001	.008	.047	.146	.208	.246	.142	.036	
14								.002	.015	.073	.134	.211	.274	.146	.010
15									.003	.023	.054	.113	.329	.371	.138
16										.003	.010	.028	.185	.440	.851

$n = 16$

x \ p	.01	.05	.10	.20	.25	.30	.40	.50	.60	.70	.75	.80	.90	.95	.99
0	.843	.418	.167	.022	.008	.002									
1	.145	.374	.315	.096	.043	.017	.002								
2	.012	.158	.280	.191	.114	.058	.010	.001							
3	.001	.042	.156	.239	.189	.124	.034	.005							
4		.008	.060	.209	.221	.187	.080	.018	.002						
5		.001	.018	.136	191	.208	.138	.047	.008	.001					
6			.004	.068	.128	.178	.184	.094	.024	.003	.001				
7			.001	.027	.069	.120	.193	.148	.057	.010	.003				
8				.008	.028	.064	.161	.186	.107	.028	.009	.002			
9				.002	.009	.028	.107	.186	.161	.064	.028	.008			
10					.003	.010	.057	.148	.193	.120	.069	.027	.001		
11					.001	.003	.024	.094	.184	.178	.128	.068	.004		
12						.001	.008	.047	.138	.208	.191	.136	.018	.001	
13							.002	.018	.080	.187	.221	.209	.060	.008	
14								.005	.034	.124	.189	.239	.156	.042	.001
15								.001	.010	.058	.114	.191	.280	.158	.012
16									.002	.017	.043	.096	.315	.374	.145
17										.002	.008	.022	.167	.418	.843

$n = 17$

x \ p	.01	.05	.10	.20	.25	.30	.40	.50	.60	.70	.75	.80	.90	.95	.99
0	.835	.397	.150	.018	.006	.002									
1	.152	.376	.300	.081	.034	.013	.001								
2	.013	.168	.284	.172	.096	.046	.007	.001							
3	.001	.047	.168	.230	.170	.105	.025	.003							
4		.009	.070	.215	.213	.168	.061	.012	.001						
5		.001	.022	.151	.199	.202	.115	.033	.004						
6			.005	.082	.144	.187	.166	.071	.014	.001					
7			.001	.035	.082	.138	.189	.121	.037	.005	.001				
8				.012	.038	.081	.173	.167	.077	.015	.004	.001			
9				.003	.014	.038	.128	.186	.128	.038	.014	.003			
10				.001	.004	.015	.077	.167	.173	.081	.038	.012			
11					.001	.005	.037	.121	.189	.138	.082	.035	.001		
12						.001	.014	.071	.166	.187	.144	.082	.005		
13							.004	.033	.115	.202	.199	.151	.022	.001	
14							.001	.012	.061	.168	.213	.215	.070	.009	
15								.003	.025	.105	.170	.230	.168	.047	.001
16								.001	.007	.046	.096	.172	.284	.168	.013
17									.001	.013	.034	.081	.300	.376	.152
18										.002	.006	.018	.150	.397	.835

$n = 18$

Table V. *(Continued)*

$n = 19$

x \ p	.01	.05	.10	.20	.25	.30	.40	.50	.60	.70	.75	.80	.90	.95	.99
0	.826	.377	.135	.014	.004	.001									
1	.159	.377	.285	.069	.029	.009	.001								
2	.014	.179	.285	.154	.080	.036	.005								
3	.001	.053	.180	.218	.152	.087	.018	.002							
4		.011	.080	.218	.202	.149	.047	.007							
5		.002	.027	.164	.202	.192	.093	.022	.002						
6			.007	.096	.157	.192	.145	.052	.008						
7			.001	.044	.097	.152	.180	.096	.024	.002					
8				.017	.049	.098	.180	.144	.053	.008	.002				
9				.005	.020	.051	.146	.176	.098	.022	.007	.001			
10				.001	.007	.022	.098	.176	.146	.051	.020	.005			
11					.002	.008	.053	.144	.180	.098	.049	.017			
12						.002	.024	.096	.180	.152	.097	.044	.001		
13							.008	.052	.145	.192	.157	.096	.007		
14							.002	.022	.093	.192	.202	.164	.027	.002	
15								.007	.047	.149	.202	.218	.080	.011	
16								.002	.018	.087	.152	.218	.180	.053	.001
17									.005	.036	.080	.154	.285	.179	.014
18									.001	.009	.029	.069	.285	.377	.159
19										.001	.004	.014	.135	.377	.826

x \ p	.01	.05	.10	.20	.25	.30	.40	.50	.60	.70	.75	.80	.90	.95	.99
0	.818	.358	.122	.012	.003	.001									
1	.165	.377	.270	.058	.021	.007									
2	.016	.189	.285	.137	.067	.028	.003								
3	.001	.060	.190	.205	.134	.072	.012	.001							
4		.013	.090	.218	.190	.130	.035	.005							
5		.002	.032	.175	.202	.179	.075	.015	.001						
6			.009	.109	.169	.192	.124	.037	.005						
7			.002	.054	.112	.164	.166	.074	.015	.001					
8				.022	.061	.114	.180	.120	.036	.004	.001				
9				.007	.027	.065	.160	.160	.071	.012	.003				
10				.002	.010	.031	.117	.176	.117	.031	.010	.002			
11					.003	.012	.071	.160	.160	.065	.027	.007			
12					.001	.004	.036	.120	.180	.114	.061	.022			
13						.001	.015	.074	.166	.164	.112	.054	.002		
14							.005	.037	.124	.192	.169	.109	.009		
15							.001	.015	.075	.179	.202	.175	.032	.002	
16								.005	.035	.130	.190	.218	.090	.013	
17								.001	.012	.072	.134	.205	.190	.060	.001
18									.003	.028	.067	.137	.285	.189	.016
19										.007	.021	.058	.270	.377	.165
20										.001	.003	.012	.122	.358	.818

$n = 20$

Table VI. Normal Probabilities

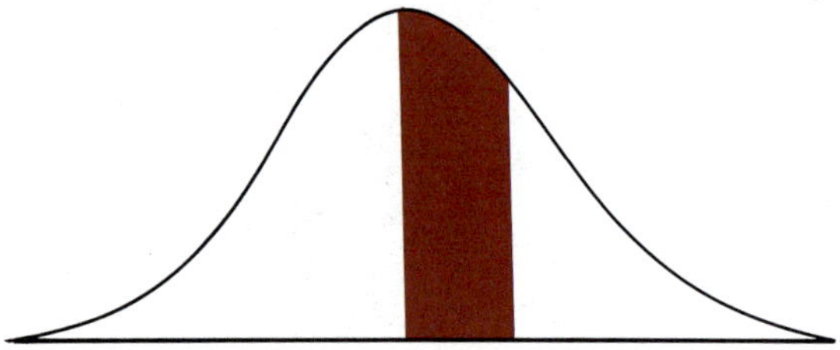

$P(0 \leq Z < z)$

Examples of Normal Probabilities

$P(0 \leq Z < 1.73) = 0.4582$

To find $P(Z < z)$ where $z > 0$, add 0.5000 to the table value.

$P(Z < 2.04) = 0.5000 + 0.4793 = 0.9793$

If $Z < 0$, use the symmetry of the normal distribution.

$P(Z > -1.35) = P(Z < 1.35) = 0.5000 + 0.4115 = 0.9115$

$$\begin{aligned} P(Z < -0.72) = P(Z > 0.72) &= 0.5 - P(0 \leq Z < 0.72) \\ &= 0.5000 - 0.2642 \\ &= 0.2358 \end{aligned}$$

$$\begin{aligned} P(-2.34 \leq Z \leq -1.05) = P(1.05 \leq Z \leq 2.34) &= P(0 \leq Z < 2.34) - P(0 \leq Z < 1.05) \\ &= 0.4904 - 0.3531 = 0.1373 \end{aligned}$$

$$\begin{aligned} P(-1.27 \leq Z \leq 0.96) &= P(-1.27 \leq Z \leq 0) + P(0 \leq Z \leq 0.96) \\ &= P(0 \leq Z \leq 1.27) + P(0 \leq Z \leq 0.96) \\ &= 0.3980 + 0.3315 \\ &= 0.7295 \end{aligned}$$

z	.00	.01	.02	.03	.04	.05	.06	.07	.08	.09
0.0	.0000	.0040	.0080	.0120	.0160	.0199	.0239	.0279	.0319	.0359
0.1	.0398	.0438	.0478	.0517	.0557	.0596	.0636	.0675	.0714	.0753
0.2	.0793	.0832	.0871	.0910	.0948	.0987	.1026	.1064	.1103	.1141
0.3	.1179	.1217	.1255	.1293	.1331	.1368	.1406	.1443	.1480	.1517
0.4	.1554	.1591	.1628	.1664	.1700	.1736	.1772	.1808	.1844	.1879
0.5	.1915	.1950	.1985	.2019	.2054	.2088	.2123	.2157	.2190	.2224
0.6	.2257	.2291	.2324	.2357	.2389	.2422	.2454	.2486	.2518	.2549
0.7	.2580	.2611	.2642	.2673	.2704	.2734	.2764	.2794	.2823	.2852
0.8	.2881	.2910	.2939	.2967	.2995	.3023	.3051	.3078	.3106	.3133
0.9	3159	.3186	.3212	.3238	.3264	.3289	.3315	.3340	.3365	.3389

z	.00	.01	.02	.03	.04	.05	.06	.07	.08	.09
1.0	.3413	.3438	.3461	.3485	.3508	.3531	.3554	.3577	.3599	.3621
1.1	.3643	.3665	.3686	.3708	.3729	.3749	.3770	.3790	.3810	.3830
1.2	.3849	.3869	.3888	.3907	.3925	.3944	.3962	.3980	.3997	.4015
1.3	.4032	.4049	.4066	.4082	.4099	.4115	.4131	.4147	.4162	.4177
1.4	.4192	.4207	.4222	.4236	.4251	.4265	.4279	.4292	.4306	.4319
1.5	.4332	.4345	.4357	.4370	.4382	.4394	.4406	.4418	.4429	.4441
1.6	.4452	.4463	.4474	.4484	.4495	.4505	.4515	.4525	.4535	.4545
1.7	.4554	.4564	.4573	.4582	.4591	.4599	.4608	.4616	.4625	.4633
1.8	.4641	.4649	.4656	.4664	.4671	.4678	.4686	.4693	.4699	.4706
1.9	.4713	.4719	.4726	.4732	.4738	.4744	.4750	.4756	.4761	.4767
z	.00	.01	.02	.03	.04	.05	.06	.07	.08	.09
2.0	.4772	.4778	.4783	.4788	.4793	.4798	.4803	.4808	.4812	.4817
2.1	.4821	.4826	.4830	.4834	.4838	.4842	.4846	.4850	.4854	.4857
2.2	.4861	.4864	.4868	.4871	.4875	.4878	.4881	.4884	.4887	.4890
2.3	.4893	.4896	.4898	.4901	.4904	.4906	.4909	.4911	.4913	.4916
2.4	.4918	.4920	.4922	.4925	.4927	.4929	.4931	.4932	.4934	.4936
2.5	.4938	.4940	.4941	.4943	.4945	.4946	.4948	.4949	.4951	.4952
2.6	.4953	.4955	.4956	.4957	.4959	.4960	.4961	.4962	.4963	.4964
2.7	.4965	.4966	.4967	.4968	.4969	.4970	.4971	.4972	.4973	.4974
2.8	.4974	.4975	.4976	.4977	.4977	.4978	.4979	.4979	.4980	.4981
2.9	.4981	.4982	.4982	.4982	.4984	.4984	.4985	.4985	.4986	.4986
z	.00	.01	.02	.03	.04	.05	.06	.07	.08	.09
3.0	.4987	.4987	.4987	.4988	.4988	.4989	.4989	.4989	.4990	.4990
3.1	.4990	.4991	.4991	.4991	.4992	.4992	.4992	.4992	.4993	.4993
3.2	.4993	.4993	.4994	.4994	.4994	.4994	.4994	.4995	.4995	.4995
3.3	.4995	.4995	.4996	.4996	.4996	.4996	.4996	.4996	.4996	.4997
3.4	.4997	.4997	.4997	.4997	.4997	.4997	.4997	.4997	.4998	.4998
3.5	.4998	.4998	.4998	.4998	.4998	.4998	.4998	.4998	.4998	.4998
3.6	.4998	.4999	.4999	.4999	.4999	.4999	.4999	.4999	.4999	.4999
3.7	.4999	.4999	.4999	.4999	.4999	.4999	.4999	.4999	.4999	.4999
3.8	.4999	.4999	.4999	.4999	.4999	.4999	.4999	.5000	.5000	.5000
3.9	.5000	.5000	.5000	.5000	.5000	.5000	.5000	.5000	.5000	.5000
z	.00	.01	.02	.03	.04	.05	.06	.07	.08	.09

$P(0 \le Z < z)$

Answers to Selected Exercises

Exercise Set 0.1 Page 16

1. 1 **3.** 5 **5.** -1 **7.** 1 **9.** $\frac{5}{2}$ **11.** 2 **13.** 2 **15.** $-\frac{2}{11}$ **17.** -5 **19.** -10 **21.** $\frac{9}{5}$

23. $-\frac{12}{5}$ -5 $\frac{11}{2}$ No solution

31. $x < 1$

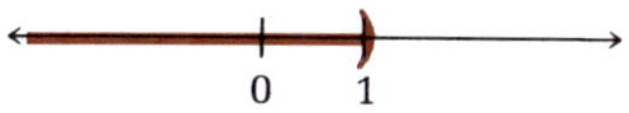

33. $x > 2$

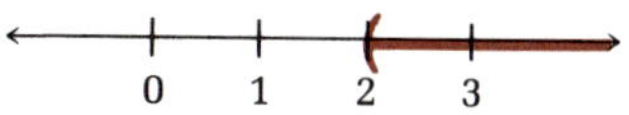

35. $x \le \frac{7}{3}$

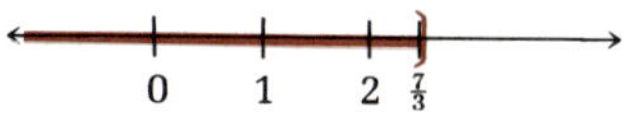

37. $x \ge 3$

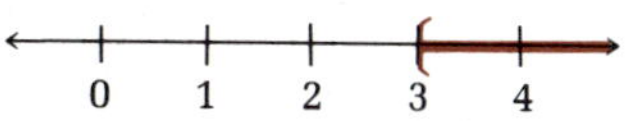

39. $x > -1$

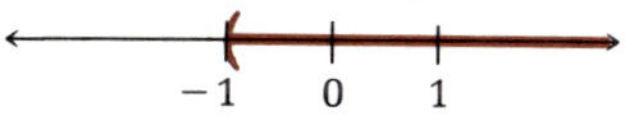

41. $x > -\frac{7}{3}$

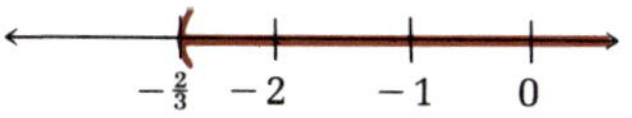

43. $x > -\frac{3}{2}$

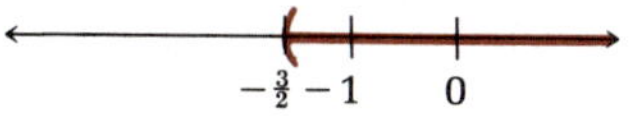

45. $x \le \frac{7}{3}$

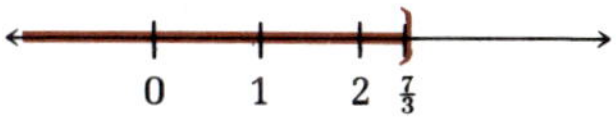

47. $x \ge \frac{2}{7}$

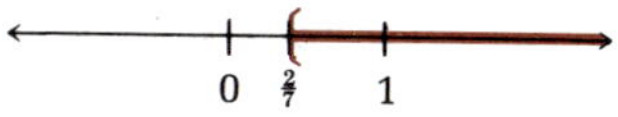

49. $x > -\frac{3}{2}$

51. $x < 2$ or $x > \frac{7}{2}$

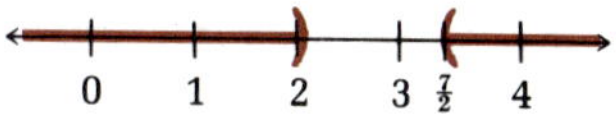

53. $-2 < x < -\frac{19}{12}$

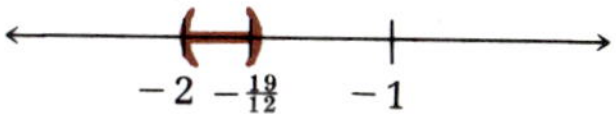

55. $2 < x < \frac{17}{3}$

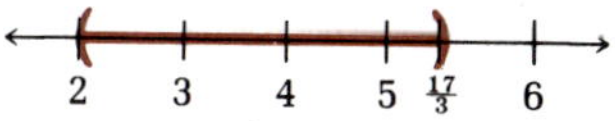

57. width = 25 meters
length = 30 meters

59. width = 100 yards, length = 300 yards

Exercise Set 0.2 Page 30

1. a. $x + 5y + 18 = 0$ **c.** $8x - 4y - 5 = 0$ **e.** $x - 3y + 3 = 0$ **g.** $x + 2y + 4 = 0$
$x + 3y + 6 = 0$ **k.** $2x - y + 8 = 0$ **2. a.** perpendicular **c.** neither **e.** perpendicular

3. a. $m = -\frac{2}{3}$; $b = 2$ **c.** $m = -1$; $b = \frac{1}{2}$ **e.** $m = 0$; $b = 4$ **g.** $m = -1$; $b = 0$ **5. a.** $y = -2x + 8$
$y = -2x + 3$ **7.** $x = 2$ **9.** $y = 6.5x$; \$162.50 **11.** $y = 0.05x + 100$; \$150; \$375

13. a. $y = \frac{9}{5}x + 40$ **b.** 148; 162.4; 175

Exercise Set 0.3 Page 38

1.

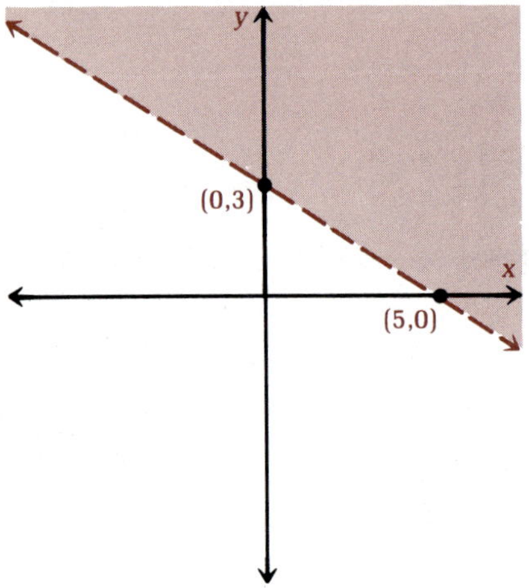

3.

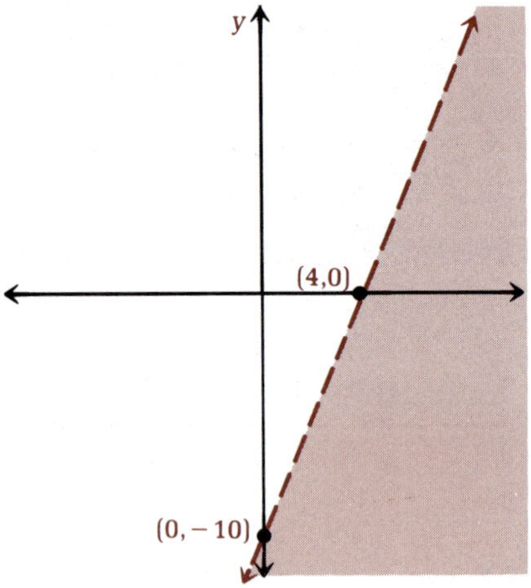

5.

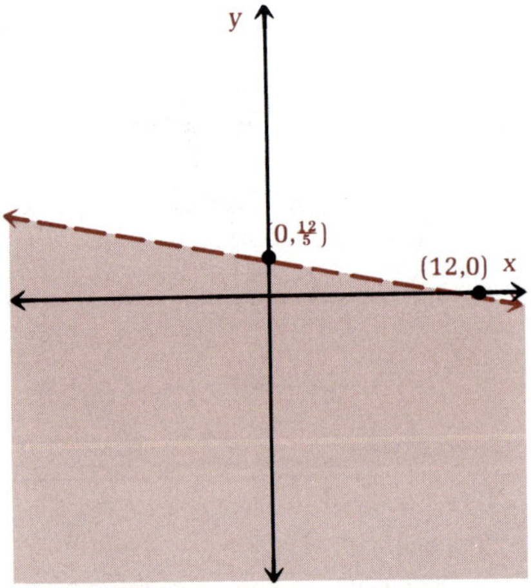

7.

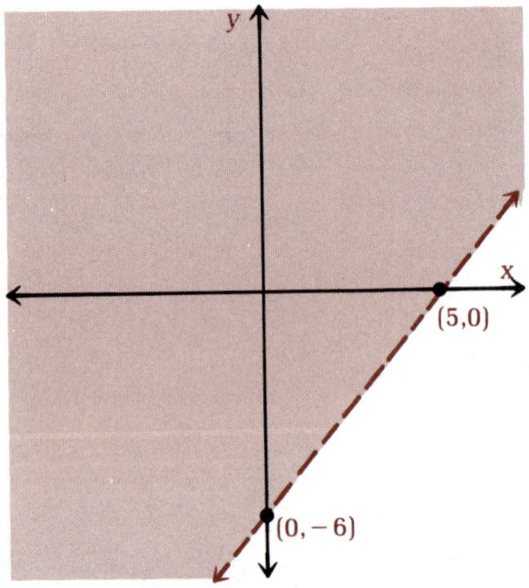

9.

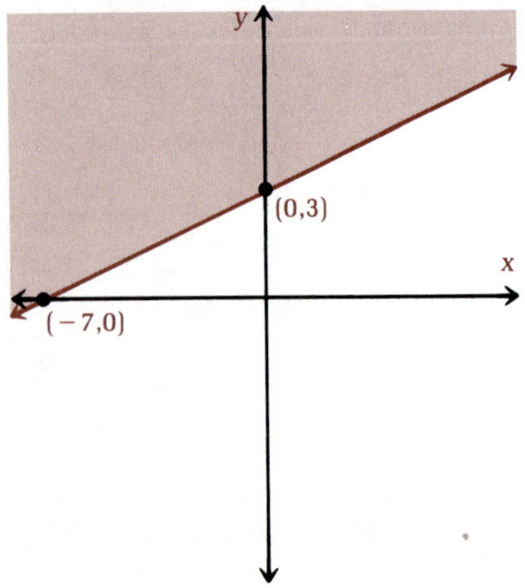

11.

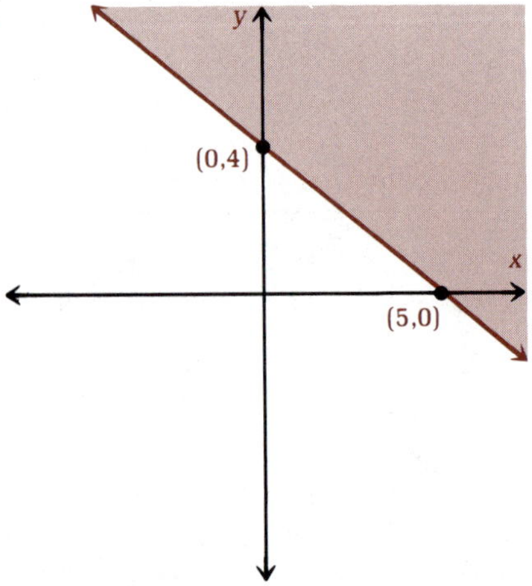

13.

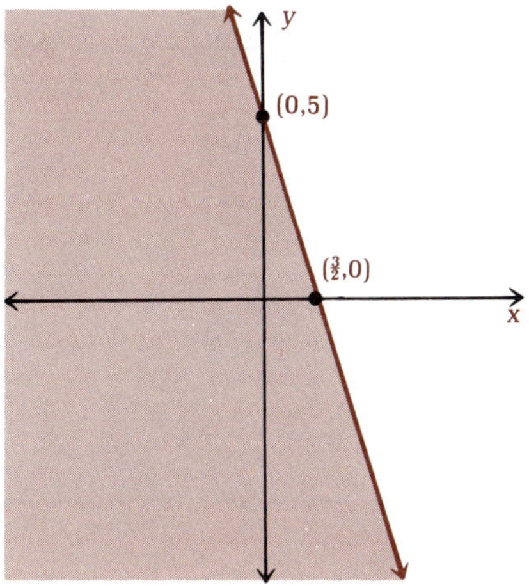

15.

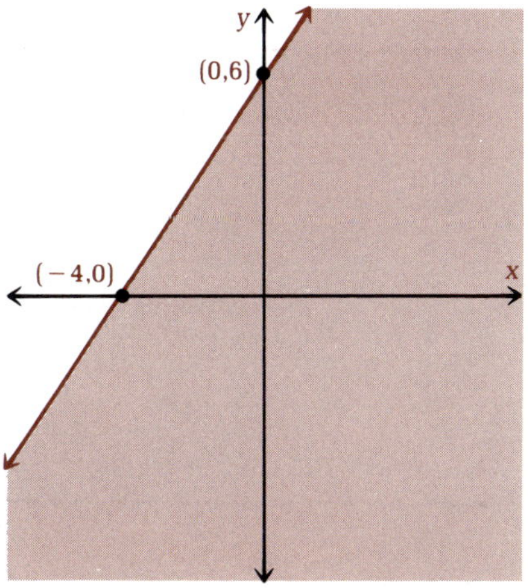

17. $7x + 8y < 56$ **19.** $15x + 12y \geq 90$ **21.** $2x - 2y \leq 4$ **23.** $5x - 2y > 10$ **25.** $0.5x + 0.4y \leq 10$ or $5x + 4y \leq 100$ **27.** $2x + 3y \geq 75$ **29.** $5x + 4y \leq 30$

Exercise Set 1.1 Page 51

1. a. 3×4 **c.** 3×2 **e.** 4×1 **3.** $\begin{bmatrix} 1 & 0 & 0 & 0 & 0 \\ 0 & 1 & 0 & 0 & 0 \\ 0 & 0 & 1 & 0 & 0 \\ 0 & 0 & 0 & 1 & 0 \\ 0 & 0 & 0 & 0 & 1 \end{bmatrix}$ **5. a.** $\begin{bmatrix} 3 & 5 & 3 \\ -1 & 2 & 1 \\ 4 & 3 & 4 \\ 2 & -5 & 1 \end{bmatrix}$

c. $\begin{bmatrix} 3 & 2 \\ 1 & -2 \\ -1 & 2 \end{bmatrix}$ **e.** I_3

Exercise Set 1.2 Page 59

1. a. $\begin{bmatrix} 5 & 2 & 2 & 2 \\ -2 & 6 & -1 & 3 \\ -6 & 2 & 7 & 2 \end{bmatrix}$; $\begin{bmatrix} -1 & 0 & 4 & 6 \\ 0 & 0 & -3 & -1 \\ -2 & 4 & 1 & 0 \end{bmatrix}$; $\begin{bmatrix} 1 & 0 & -4 & -6 \\ 0 & 0 & 3 & 1 \\ 2 & -4 & -1 & 0 \end{bmatrix}$;

$\begin{bmatrix} 4 & 2 & 6 & 8 \\ -2 & 6 & -4 & 2 \\ -8 & 6 & 8 & 2 \end{bmatrix}$; $\begin{bmatrix} -9 & -3 & 3 & 6 \\ 3 & -9 & -3 & -6 \\ 6 & 3 & -9 & -3 \end{bmatrix}$

5.

	Shift		
Dep't.	1	2	3
A	60	40	20
B	92	80	48
C	76	76	8
D	108	124	20

7. Multiply old salary schedule by 1.1.

Exercise Set 1.3 Page 70

1. **a.** $\begin{bmatrix} 3 & -1 \\ 4 & 12 \end{bmatrix}$ **c.** $\begin{bmatrix} -11 & 31 \\ 4 & 12 \\ -19 & 17 \end{bmatrix}$ **e.** $\begin{bmatrix} 17 & -5 \\ 11 & 7 \end{bmatrix}$ **g.** $\begin{bmatrix} -8 & 9 & -12 \\ -5 & 15 & 18 \\ 9 & 15 & 7 \end{bmatrix}$ **3.** 3A

5. **a.** $\begin{bmatrix} 13 & 1 \\ 6 & 0 \end{bmatrix}$ **c.** Not defined

Exercise Set 1.4 Page 91

1. **a.** $\begin{bmatrix} 2 & -3 \\ -1 & 2 \end{bmatrix}$ **c.** $\frac{1}{2}\begin{bmatrix} 1 & -1 \\ -1 & 3 \end{bmatrix}$ **e.** $\begin{bmatrix} 1 & -1 \\ 2 & 2 \end{bmatrix}$ **g.** $\frac{1}{11}\begin{bmatrix} -2 & 9 & -5 \\ -1 & -1 & 3 \\ 5 & -6 & 7 \end{bmatrix}$

i. $\frac{1}{10}\begin{bmatrix} 5 & 0 & -5 \\ 3 & 4 & -5 \\ -1 & 2 & 5 \end{bmatrix}$ **k.** Not invertible **m.** $\frac{1}{2}\begin{bmatrix} -1 & 1 & -1 & -1 \\ 3 & 3 & 1 & -7 \\ 5 & 3 & 1 & -9 \\ -2 & -2 & 0 & 6 \end{bmatrix}$

2. **a.** $\begin{bmatrix} 1 & 2 & -1 \\ -1 & -1 & 2 \\ 1 & -1 & -3 \end{bmatrix}$ **c.** $\frac{1}{2}\begin{bmatrix} 2 & -4 & 1 \\ 0 & -1 & 1 \\ -1 & 3 & -1 \end{bmatrix}$ **e.** $\begin{bmatrix} 16 & 25 & 11 \\ 3 & 5 & 2 \\ -10 & -16 & -7 \end{bmatrix}$ **3.** **a.** T **c.** F

Exercise Set 1.5 Page 96

1. **a.** $\begin{bmatrix} 3 & 2 & 1 & 4 \\ 0 & 5 & 10 & -5 \\ 0 & 2 & 1 & 3 \\ 0 & 1 & 8 & 2 \end{bmatrix}$ **c.** $\begin{bmatrix} 4 & 3 & 4 & 1 & 2 & 3 \\ 0 & -7 & -8 & 3 & 2 & 5 \end{bmatrix}$

Exercise Set 2.1 Page 113

1. (3,1) **3.** No solution **5.** $\left(\frac{1+2y}{3}, y\right)$ **7.** $(-1,-2)$ **9.** (4,3) **11.** (2,1,1) **13.** No solution

15. $(-1,-1,2)$ **17.** $\left(\frac{1}{4},\frac{3}{4},\frac{1}{2}\right)$ **19.** **a.** surplus **b.** 18.25

21. $q - 20p = 80$ supply
$p + q = 248$ demand
$p = 8, q = 240$ **23.** $q = 100, r = c = 500$ **25.** 30 pair

Exercise Set 2.2 Page 128

1. (3,1) **3.** No solution **5.** $\left(\frac{1}{4},\frac{1}{2},\frac{3}{4}\right)$ **7.** $\left(4-2x_3,\frac{5x_3-3}{2},x_3\right)$ **9.** $\left(\frac{23}{60},\frac{1}{5},-\frac{4}{15}\right)$ **11.** (1,1,2)

13. $\left(\frac{4+z}{5},\frac{-3+8z}{5},z\right)$ **15.** (2,1,0,1) **17.** (4,2,−1) **19.** $\left(\frac{3}{4},\frac{3}{4},\frac{1}{4},\frac{1}{4}\right)$ **21.** (2,2 + z,z)

23. (43 + 2y,y,18) **25.** (2,1,2) **27.** (2 + z,3 − 3z,z) **29.** $\left(\frac{108+13z}{49},\frac{25-6z}{7},\frac{251-51z}{49},z\right)$

Exercise Set 2.3 Page 138

1. a. $\begin{cases} 3x + y + 4z = 6 \\ x + 2y + 3z = -2 \\ 2x + y + 4z = 3 \\ x + 3y - z = 4 \end{cases}$ **c.** $\begin{cases} 2w - x + 3y + 4z = 5 \\ w + x - y + 3z = 1 \\ 4w + 2x + y + z = 2 \end{cases}$ **3. a.** $x = \begin{bmatrix} 1 \\ 2 \\ 1 \end{bmatrix}$ **c.** $X = \begin{bmatrix} 1 \\ 3 \\ 1 \end{bmatrix}$ **e.** $X = \begin{bmatrix} 2 \\ 1 \\ 2 \end{bmatrix}$

4. a. $X = \begin{bmatrix} 3 & 2 & 1 \\ 1 & 1 & 1 \end{bmatrix}$ **c.** $X = \begin{bmatrix} \frac{2}{9} & \frac{5}{9} \\ \frac{5}{9} & \frac{17}{9} \\ \frac{2}{3} & -\frac{1}{3} \end{bmatrix}$ **5.** $X = \begin{bmatrix} 1 & 1 & 2 & 3 & 3 \\ 1 & 2 & 1 & 1 & 2 \\ 1 & 1 & 2 & 2 & 1 \end{bmatrix}$

Exercise Set 2.4 Page 154

1. $\frac{200}{7}$ gallons of $\frac{1}{2}$% and $\frac{150}{7}$ gallons of 4%.

3.

Type	options 1	2	3
A	7	4	1
B	17	19	21
C	0	1	2

5. Operate factory I for 100 hours, factory II for 50 hours, and factory III for 50 hours.
7. Solution is given by (7 − z,−4 + z,z) where z = 4,5,6,7.
9. Solution is (15 + c,15 − 2c,c) where c = 0,1,2,3,4,5,6,7.
11. Operate I 6 hours, II 5 hours, III 3 hours, and IV 8 hours.
13. a. $\left(\frac{d}{3},\frac{c-d}{4},\frac{a-c}{12},\frac{b-c}{12}\right)$ **15. a.** 2B, 2C, and 6D; 2B, 7D, and 1E; or 5B and 5E.
c. Yes, under certain conditions. The total number of boxes needed must be divisible by 6 for A, by 3 for B, and by 2 for C to avoid using part boxes.

17.

Type	Farmer 1	3	7
I	10	10	$\frac{17}{2}$
II	8	10	11
III	12	10	$\frac{21}{2}$

Exercise Set 2.5 Page 162

1. b. $\begin{bmatrix} 0.9 & -0.2 & -0.3 \\ -0.4 & 0.7 & -0.2 \\ -0.2 & -0.2 & 1.0 \end{bmatrix}$ **3. b.** $\begin{bmatrix} 63{,}781 \\ 52{,}407 \\ 42{,}754 \\ 45{,}370 \end{bmatrix}$ **5.** $\begin{bmatrix} 942.11 \\ 523.68 \end{bmatrix}$

Exercise Set 3.1 Page 175

1.

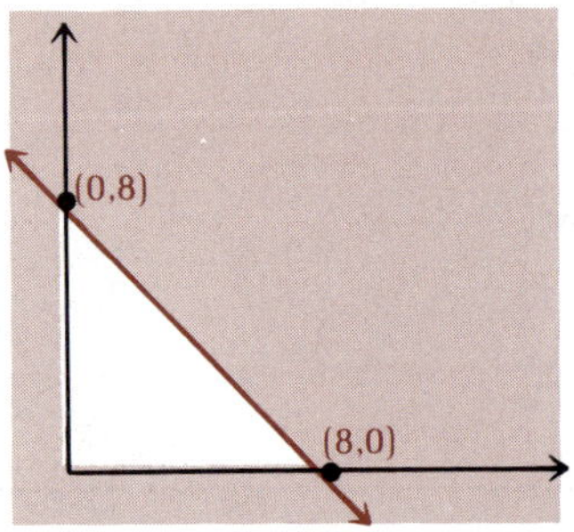

3.

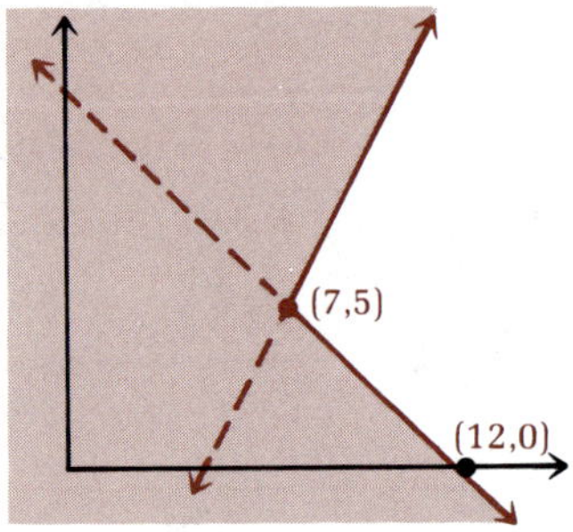

5.

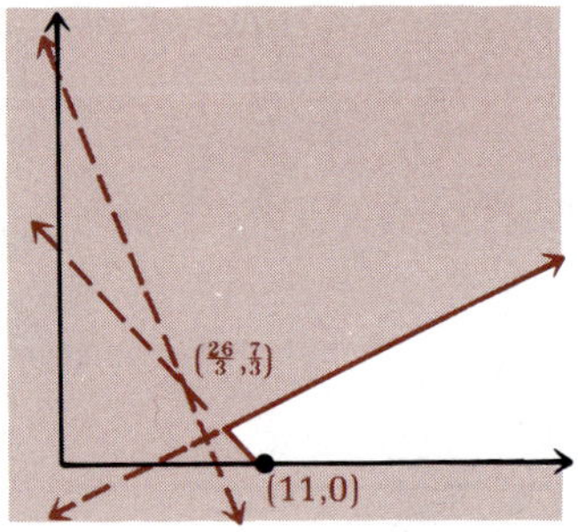

7. $\begin{cases} x \geq 0,\ y \geq 0 \\ -4x + 3y \leq 12 \\ x + y \leq 11 \\ 3x + y \leq 21 \end{cases}$ **9.** $\begin{cases} x \geq 0,\ y \geq 0 \\ x + y \geq 5 \\ 3x + 4y \geq 18 \end{cases}$

Exercise Set 3.2 Page 181

1. 26 at (4,5) **3.** $Z = \dfrac{122}{3}$, $T = -4$ **5.** No minimum **7.** No maximum **9.** $R(\max) = 159.13$, $R(\min) = 119.03$

Exercise Set 3.3 Page 185

1. a.

	Time	A	B
B.C.	36	2	3
A	48	3	4
T.&F.	14	1	1
Profit	—	120	150

b. $P = 120x + 150y$

$\begin{cases} x \geq 0,\ y \geq 0 \\ 2x + 3y \leq 36 \\ 3x + 4y \leq 48 \\ x + y \leq 14 \end{cases}$

c.

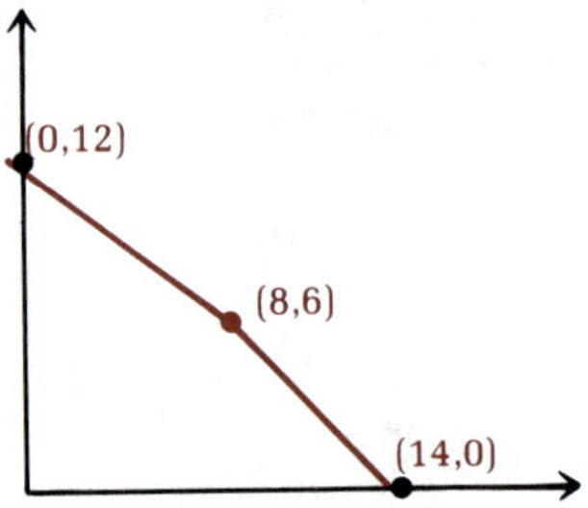

d. \$1860 for 8 model *A* and 6 model *B*. **3. a.** \$9600 **c.** \$22,400
5. Use 12 RH5 and 2 SS3 for 440 flavor units. **7.** \$181.67

Exercise Set 3.4 Page 201

1. a. $\begin{bmatrix} 3 & -2 & 1 & 0 & 0 & 14 \\ 1 & 1 & 0 & 1 & 0 & 8 \\ 2 & 1 & 0 & 0 & -1 & 0 \end{bmatrix}$ **c.** $\begin{bmatrix} 1 & 1 & 1 & 0 & 0 & 0 & 4 \\ 2 & -1 & 0 & 1 & 0 & 0 & 8 \\ 3 & 4 & 0 & 0 & 1 & 0 & 15 \\ 2 & 1 & 0 & 0 & 0 & -1 & 0 \end{bmatrix}$ **3. a.** 3

c. 3 **5. a.** $x = 2,\ y = 4$; $s_1 = s_2 = 0$; $P = 16$ **7.** $x = 200,\ y = 0,\ z = 1600$; $s_1 = s_2 = 0$; $P = \$144.00$ **9.** $x = 8,\ y = z = 0$; $s_1 = s_3 = 0$; $s_2 = 28,\ s_4 = 7$; $P = 64$

Exercise Set 3.5 Page 212

1. $x = 8, y = 0$, $s_1 = 0, s_2 = 10$, $R = 32$ **3.** $x = 5, y = 0$, $s_1 = 2, s_2 = 6, s_3 = 0$, $B = 20$ **5.** No solution **7. a.** 6 oz. of A, 10 oz. of B **b.** 68¢

11. $x = 3, y = 0, s_1 = 0, s_2 = 5, R = 12$ **13.** $x = 10, y = 10, s_1 = s_2 = 0, P = 150$

Exercise Set 3.6 Page 217

1. Max. $x = 4, y = 3, P = 11$; Min. $x = 2, y = 6, C = 26$ **3.** $x = 4, y = 1$, $R = 11$ **5.** $x = 4.2, y = 0.8$, $M = 19.2$ **7.** $x = \frac{5}{2}, y = \frac{1}{2}$, $T = 26$

9. Ship $\frac{5}{2}$ pounds of 3RG2. Ship $\frac{1}{2}$ pounds of 6BR7. **11.** Vice-President I, 6 hours, II and III no hours, overtime cost $180

Exercise Set 3.7 Page 228

1. $\begin{bmatrix} 2 & 3 & 1 & 0 & 0 & 1 \\ 3 & 2 & 0 & 1 & 0 & 2 \\ 18 & 12 & 0 & 0 & -1 & 0 \end{bmatrix}$ **3.** $\begin{bmatrix} 3 & 2 & 3 & 1 & 0 & 0 & 0 & 3 \\ 2 & 3 & 2 & 0 & 1 & 0 & 0 & 2 \\ -1 & 1 & 2 & 0 & 0 & 1 & 0 & 1 \\ 11 & 15 & 18 & 0 & 0 & 0 & -1 & 0 \end{bmatrix}$

5. $\begin{bmatrix} 2 & 3 & -1 & 1 & 0 & 0 & 2 \\ 3 & 2 & -1 & 0 & 1 & 0 & 3 \\ 15 & 11 & -12 & 0 & 0 & -1 & 0 \end{bmatrix}$

7. $\begin{bmatrix} -2 & -3 & 1 & 1 & 0 & 0 & 0 & 3 \\ -3 & -1 & 2 & 0 & 1 & 0 & 0 & 2 \\ -3 & -1 & 4 & 0 & 0 & 1 & 0 & -1 \\ -30 & -28 & 12 & 0 & 0 & 0 & -1 & 0 \end{bmatrix}$ **9.** $x = \frac{-1}{-1} = 1, y = \frac{-1}{-1} = 1, C = \frac{-3}{-1} = 3$

11. $x = \frac{-8}{-4} = 2, y = \frac{-12}{-4} = 3, C = \frac{-52}{-4} = 13$ **13.** $x = \frac{-10}{-5} = 2, y = \frac{0}{-5} = 0, z = \frac{-25}{-5} = 5,$ $C = \frac{-60}{-5} = 12$ **15.** $w = \frac{-15}{-5} = 3, x = \frac{-10}{-5} = 2, y = \frac{-25}{-5} = 5, z = \frac{-5}{-5} = 1, C = \frac{-55}{-5} = 11$ **17.** $x = \frac{2}{3}$, $y = \frac{4}{3}$, $C = 8$ **19.** $x = 2, y = 2, C = 40$ **21.** $x = \frac{1}{3}, y = \frac{4}{3}, C = 6$ **23.** $x = 0, y = \frac{1}{3}, z = \frac{7}{3}, C = \frac{134}{3}$

25. $x = 2, y = 10, C = 6$ **27.** $x = \frac{25}{4}, y = \frac{35}{4}, z = 0, C = \frac{15}{4}$

Exercise Set 4.1 Page 242

1.

100				100
50	50			100
	200			200
		100	200	300
150	250	100	200	

120						120
	130					130
	50	100				150
		100	100			200
			90	160	50	300
					100	100
120	180	200	190	160	150	

Exercise Set 4.2 Page 247

1. 11 **3.** for 1 – (1)(2)(26)(29)(17)(18)(24)(19)
for 6 – (6)(2)(26)(29)(17)(18)
for 36 – (36)(24)(19)(37)(40)(34)

Exercise Set 4.3 Page 253

1. Several shipping schedules at a total cost of $1910. **3.** Several shipping schedules at a total cost of $360.

Exercise Set 4.4 Page 257

1. a. T.S.C. = 260

—	50	—	—
—	—	—	40
30	—	10	—
10	10	—	—

—	50	—	—
—	—	—	40
20	10	10	—
20	—	—	—

2. a. T.S.C. = 335

—	5	—	10
—	15	10	—
—	—	—	10
20	—	—	—
20	—	10	—

Many alternate optimal solutions.

3. $440 **5.** There are several solutions, each producing a total of 838 hours. This is far short of the 960 that would be produced by the regular workers.

Exercise Set 4.5 Page 262

1.

	A	B	C	D	E	
I	[1] $\textcircled{3000}$	[2]	[3] $\textcircled{2000}$	[4]	[3]	5000
II	[4]	[2] $\textcircled{2000}$	[5]	[4]	[2] $\textcircled{3000}$	5000
III	[4]	[2] $\textcircled{2000}$	[4] $\textcircled{2000}$	[3] $\textcircled{2000}$	[3]	6000
OT	[9]	[9]	[11]	[9] $\textcircled{1000}$	[9]	1000
	3000	4000	4000	3000	3000	

3. \$500

Exercise Set 4.6 Page 270

1. $\begin{bmatrix} \circled{3} & 3 & 4 & 4 \\ 4 & \circled{3} & 3 & 4 \\ 5 & 4 & 6 & \circled{4} \\ 6 & 5 & \circled{4} & 5 \end{bmatrix}$

3. $\begin{bmatrix} 3 & 3 & 4 & \circled{4} \\ \circled{4} & 3 & 3 & 4 \\ 5 & 4 & \circled{6} & 4 \\ 6 & \circled{5} & 4 & 5 \end{bmatrix} \begin{bmatrix} 3 & 3 & 4 & \circled{4} \\ 4 & \circled{3} & 3 & 4 \\ 5 & 4 & \circled{6} & 4 \\ \circled{6} & 5 & 4 & 5 \end{bmatrix} \begin{bmatrix} 3 & \circled{3} & 4 & 4 \\ 4 & 3 & 3 & \circled{4} \\ 5 & 4 & \circled{6} & 4 \\ \circled{6} & 5 & 4 & 5 \end{bmatrix}$

5. 8.4

7. a. $\begin{bmatrix} 3 & 3 & 4 & 5 & 6 & \circled{3} \\ 5 & 9 & 5 & \circled{5} & 5 & 5 \\ 7 & 5 & \circled{5} & 7 & 7 & 7 \\ \circled{3} & 7 & 3 & 4 & 7 & 9 \\ 8 & 9 & 7 & 8 & \circled{4} & 9 \\ 10 & \circled{5} & 6 & 7 & 9 & 9 \end{bmatrix}$ **b.** no

Exercise Set 5.1 Page 284

1. a. T **c.** F **e.** T **g.** T **i.** T **k.** F **m.** T **o.** T **q.** F **s.** F **3. a.** {0,2,4,6} **e.** {4,5,6} **5. a.** T **c.** T **e.** F **g.** T **i.** F **7. a.** {0,2,3,4,6,8,9} **c.** {1,3,4,5,7,9} **e.** A **g.** B **i.** $\emptyset$ **k.** B **m.** B **o.** A **q.** {1,7} **s.** {4} **u.** C **w.** {0,1,2,4,5,6,7,8} **9.** $B \subset A$ **11.** $A = B = \emptyset$

Exercise Set 5.2 Page 292

3. a.

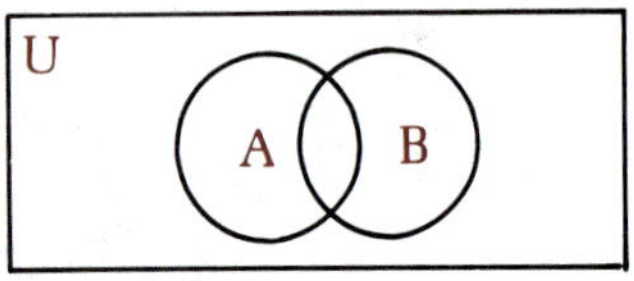

c.

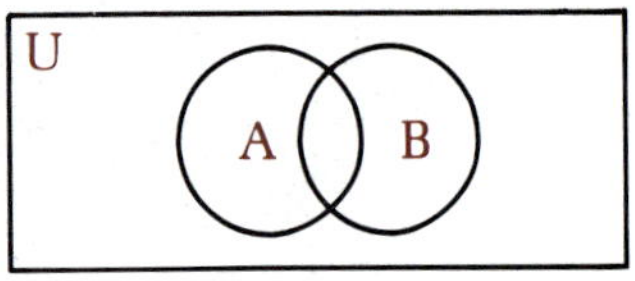

e.

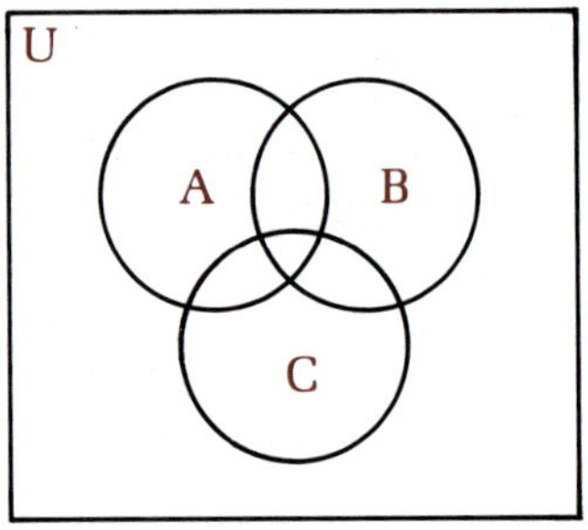

g.

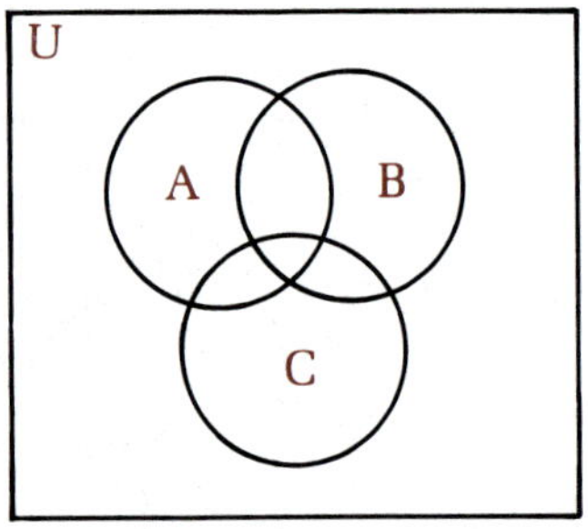

5. 45 **7.** 20 **9.** 85 **11.** 160 **13.** 40 **15.** 20 **17.** 1885

Exercise Set 5.3 Page 300

1. 8 **3.** 245 **5.** 5 **7. a.** 165 **c.** 65 **e.** 95 **9.** 400 **11.** 45,55 **13.** 30, driving package
15. b. 14% **d.** A+, AB−

Exercise Set 5.4 Page 305

1. 160 **3.** 720 **5.** 162 **7.** 6,760,000 **9.** $26^2 \cdot 10^4 + 26^3 \cdot 10^3 + 10^5$ **11.** 2^{20} **13. a.** 24 **b.** 60

Exercise Set 5.5 Page 312

1. a. 3,628,800 **c.** 336 **e.** 495 **g.** $\frac{7}{66}$ **3. a.** $P(26,3) = 15{,}600$ **b.** $C(26,3) = 2{,}600$ **7. a.** $\binom{20}{2}$
c. $\binom{8}{2}$ **9. a.** $\binom{5}{3}$ **b.** $\binom{9}{2}$ **c.** Because $\binom{5}{3} = \binom{5}{2}$ **11.** $C(13,5)$ **15.** 150 **17.** $\binom{12}{4}\binom{8}{4}\binom{4}{4}$
19. $C(100,93)$ **21.** $\binom{10}{4}\binom{6}{3}\binom{3}{3}$ **23.** 3 **25.** 4 **27.** $4\binom{13}{5}$ **29. a.** 336 **b.** 56

Exercise Set 5.6 Page 319

1. a. 90 **c.** 1260 **e.** 60 **g.** 1,247,400 **3.** $\binom{8}{3}$ **5.** $\binom{8}{2,2,2,2}$ **7.** $\binom{30}{22,6,2}$ **9. a.** $\binom{7}{5}$
c. $\binom{7}{1}$ **11. a.** 12 **c.** 240 **13.** p^4

Exercise Set 6.1 Page 337

1. $\frac{1}{36},\frac{2}{36},\frac{3}{36},\frac{4}{36},\frac{5}{36},\frac{6}{36},\frac{5}{36},\frac{4}{36},\frac{3}{36},\frac{2}{36},\frac{1}{36}$ 3. $\frac{10}{32}$ 5. a. $\frac{C(13,5)}{C(52,5)}$ c. $\frac{4\binom{13}{4}\binom{39}{1}}{\binom{52}{5}}$ 7. .006 9. $\frac{4}{15}$ 11. $\frac{5}{19}$

13. $\frac{1}{2}$ 15. 0.7183 17. Not equally likely 19. $\frac{1}{6},\frac{1}{12}$

Exercise Set 6.2 Page 348

1. a. 0.3 c. 0.29 3. a. 0.6 c. 0.15 5. Yes 7. 0.10 9. a. 0.64 c. 0.01 e. 0.15

11. 0.10 13. a. $\frac{8}{36}$ c. $\frac{10}{36}$ e. $\frac{21}{36}$ 15. 0.3

Exercise Set 6.3 Page 353

1. 2.9% 3. a. 0.39666 . . . c. 0.11 5. 0.38 7. 0.1984

Exercise Set 6.4 Page 366

1. a. $\frac{6}{11}$ c. $\frac{9}{11}$ 3. a. $\frac{130}{300}$ c. $\frac{62}{118}$ e. $\frac{12}{130}$ 5. a. $\frac{1}{2}$ c. $\frac{35}{58}$ 7. a. $\frac{27}{45}$ c. $\frac{18}{45}$ e. $\frac{2}{11}$

9. a. $\frac{24}{60}$ c. They are independent 11. 32% 13. 35.7% 15. a. 0.25 c. 0.4

Exercise Set 6.5 Page 380

1. b. $\frac{6}{23}$ c. $\frac{1}{3}$ 3. a. $\frac{4}{55}$ b. $\frac{7}{8}$ 5. a. 0.47 c. 0.0709 7. a. 15% b. $\frac{6}{15}$ 9. a. 0.023

b. $\frac{8}{23}$ 11. a. 0.389 c. $\frac{138}{389}$

Exercise Set 6.6 Page 390

1. a. $\frac{1}{8},\frac{7}{8}$ c. $\frac{5}{11},\frac{6}{11}$ f. $\frac{3}{5},\frac{2}{5}$ h. $\frac{7}{10},\frac{3}{10}$ j. $\frac{11}{14},\frac{3}{14}$ 3. 7 to 3 5. a. 5 to 1 c. 13 to 5 7. \$8.00

9. −50¢ 11. \$8.00

Exercise Set 7.1 Page 408

1.

x_i	f_i	cf_i
0	2	2
1	4	6
2	6	12
3	5	17
4	5	22
5	3	25

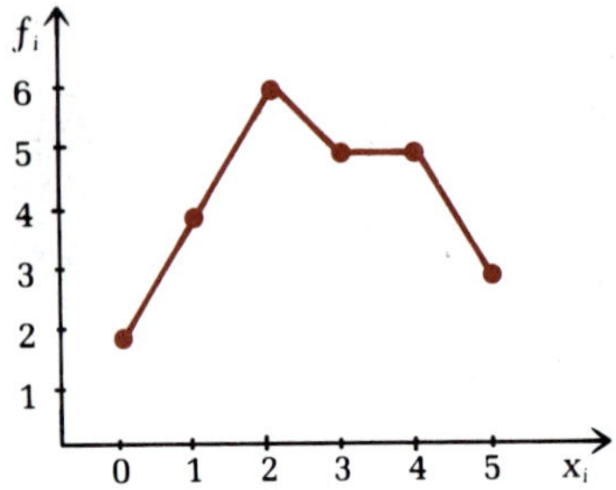

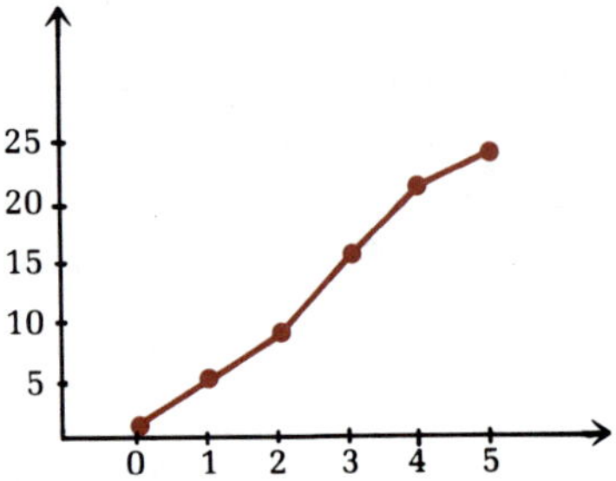

3.

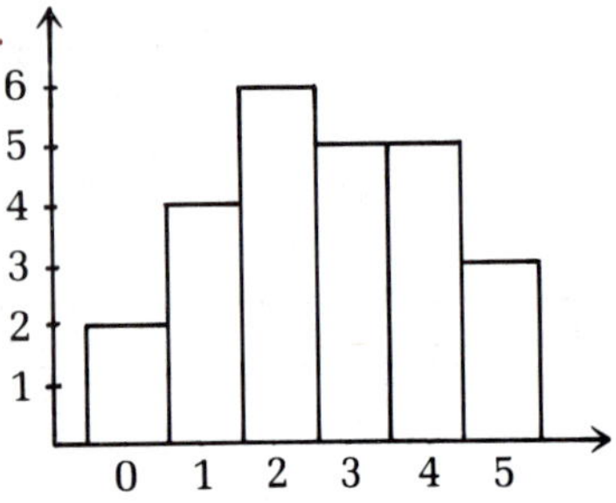

5.

x_i	rf_i
0	.08
1	.16
2	.24
3	.20
4	.20
5	.12

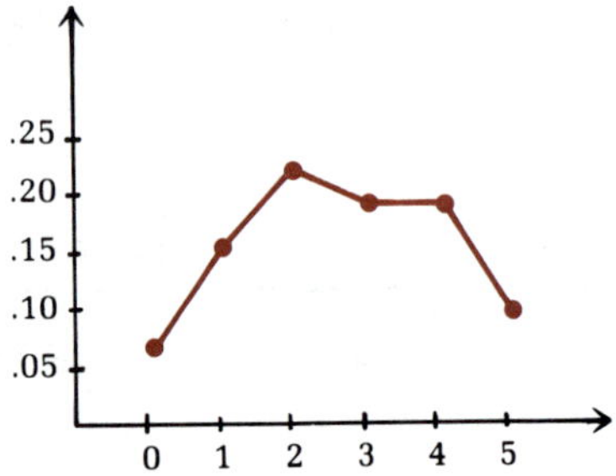

7.

Class	f_i	cf_i
21–30	1	1
31–40	1	2
41–50	2	4
51–60	3	7
61–70	6	13
71–80	13	26
81–90	9	35
91–100	5	40

9.

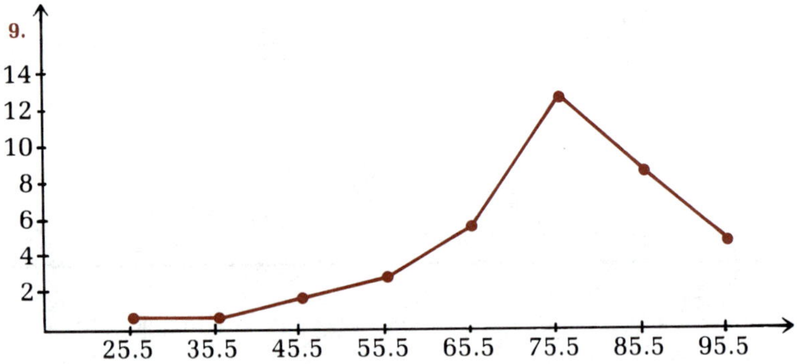

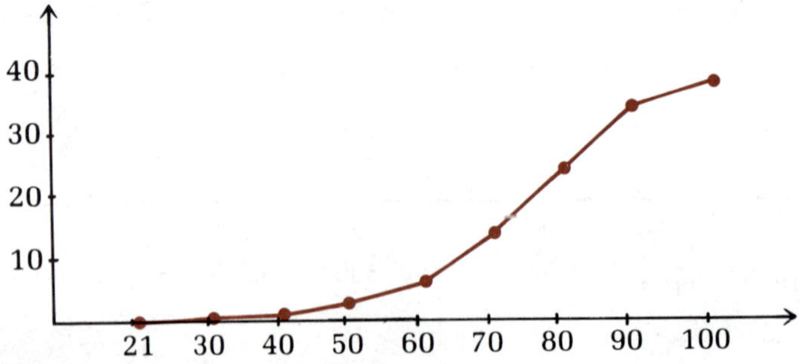

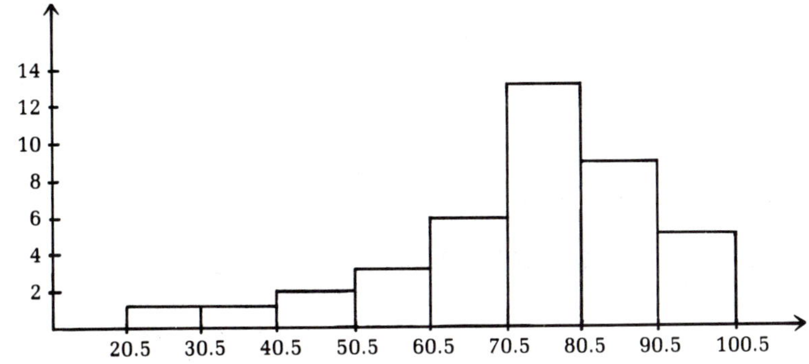
14
12
10
8
6
4
2
20.5
30.5
40.5
50.5
60.5
70.5
80.5
90.5
100.5

11.

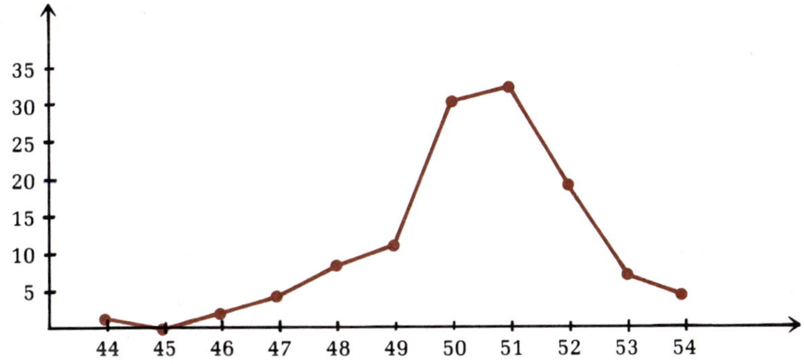
35
30
25
20
15
10
5
44
45
46
47
48
49
50
51
52
53
54

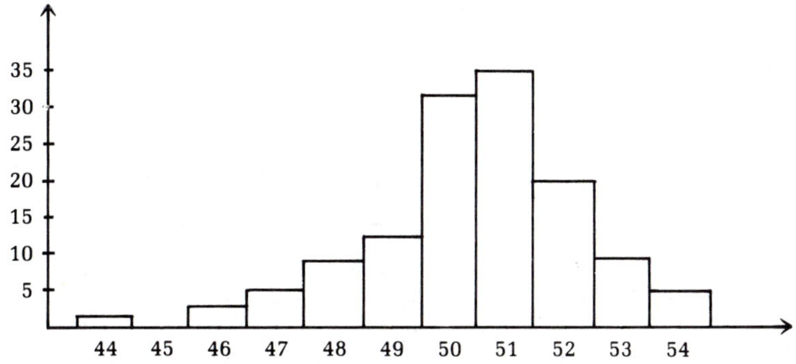
35
30
25
20
15
10
5
44
45
46
47
48
49
50
51
52
53
54

13.

Class	rf
5.00–5.49	.01
5.50–5.99	.01
6.00–6.49	.02
6.50–6.99	.05
7.00–7.49	.13
7.50–7.99	.36
8.00–8.49	.31
8.50–8.99	.10
9.00–9.49	.01

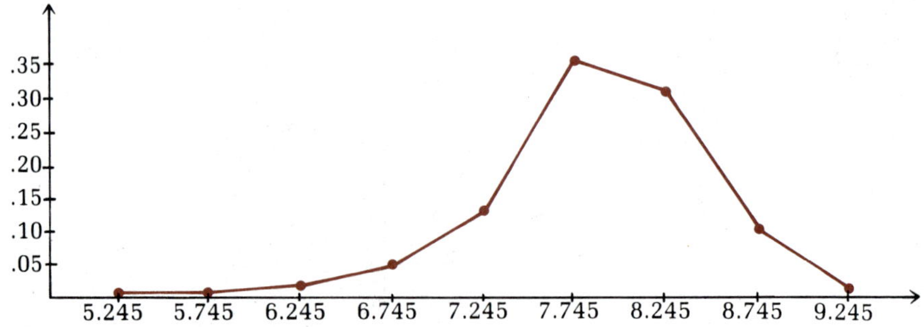

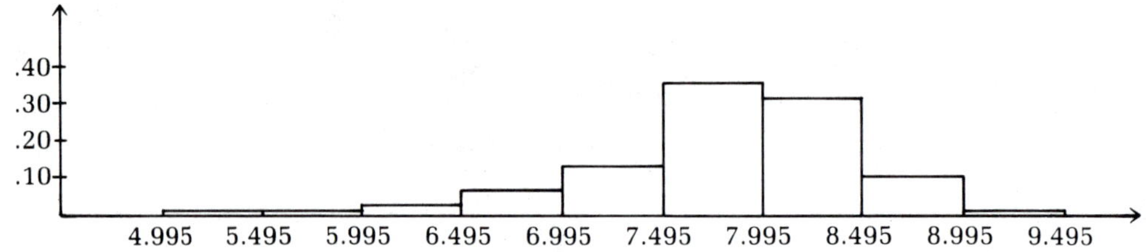

Exercise Set 7.2 Page 413

1. 5050 **3.** 5370 **5.** 173,880 **7.** 85 **9.** 121 **11.** 28.5 **13.** 85,850 **15.** 100 **17.** 120 **19.** 100

Exercise Set 7.3 Page 424

1. mean 73.25, median 72, mode 75,69 **3.** mean 1160, median 1100, mode 900 **5.** mean 2675, median 2550, mode no information **7.** 49.32 **9.** 6.55 **11.** 105.79 **13.** 7.825

Exercise Set 7.4 Page 435

1. 46, 10, $050, 48, $2000, 24 **3.** **a.** $\sigma^2 = 129.5875$, $\sigma = 11.3837$ **b.** $\sigma^2 = 89{,}733.33$, $\sigma = 299.56$ **c.** $\sigma^2 = 293{,}541.67$, $\sigma = 541.79$ **5.** **a.** $\sigma^2 = 1.5776$, $\sigma = 1.256$ **b.** $\sigma^2 = 5.0142$, $\sigma = 2.2392$ **c.** $\sigma^2 = 242.6327$, $\sigma = 15.5767$ **d.** $s^2 = 0.4279$, $s = 0.6541$ **7.** $\mu = 77.5$, $\sigma^2 = 79.6484$, $\sigma = 8.92$ **9.** $\mu = 18.55$, $\sigma^2 = 0.7622$, $\sigma = 0.8730$ **11.** $\mu = 49.715$, $\sigma^2 = 35.121$, $\sigma = 5.926$ **13.** $\bar{x} = 1292$, $s^2 = 17{,}983$, $s = 134.10$

Exercise Set 7.5 Page 442

1. **a.** 4th **b.** 78th **c.** 90th **d.** 50 **e.** 48 **f.** 51 3. **a.** 32nd **b.** 85th **c.** 96th **d.** 7 **e.** 8 **f.** 15 5. **a.** 1.67 **b.** 0.69 **c.** −0.49 **d.** −1.48 **e.** 1.28 **f.** −0.89 7. best: high jump, worst: long jump

Exercise Set 8.1 Page 453

1. **a.**

x	P(X = x)
0	0.1
1	0.6
2	0.3

b. $\mu = 1.2$
c. $\sigma^2 = 0.36, \sigma = 0.6$

3. **a.**

x	P(X = x)
0	$\frac{27}{64} = 0.421875$
1	$\frac{27}{64} = 0.421875$
2	$\frac{9}{64} = 0.140625$
3	$\frac{1}{64} = 0.015625$

b. $\mu = \frac{3}{4} = 0.75$
c. $\sigma^2 = \frac{9}{16} = 0.5625$
$\sigma = \frac{3}{4} = 0.75$

5. **a.**

x	P(X = x)
0	$\frac{1}{64} = 0.015625$
1	$\frac{6}{64} = 0.09375$
2	$\frac{15}{64} = 0.234375$
3	$\frac{20}{64} = 0.3125$
4	$\frac{15}{64} = 0.234375$
5	$\frac{6}{64} = 0.09375$
6	$\frac{1}{64} = 0.015625$

b. $\mu = 3$
c. $\sigma^2 = 1.5, \sigma = 1.225$

7. **a.**

x	P(X = x)
10%	$\frac{3}{5} = 0.6$
25%	$\frac{3}{10} = 0.3$
50%	$\frac{1}{10} = 0.1$

b. $\mu = 18.5\%$
c. $\sigma^2 = 155.25, \sigma = 12.46$

9. **a.** $E[y] = 3.75$ **b.** $E[x^2] = 1.125$ **c.** $E[x^2] - \mu^2 = 0.5625$ which is σ^2.

Exercise Set 8.2 Page 463

1. **a.** 0.2734 **c.** 0.1445 3. 0.9298 5. About 0.0006 7. 0.1562 9. 6; 0.2508 13. **a.** 0.021 **b.** 0.252 **c.** 0.862 **d.** 0.406 15. **a.** 18 **b.** 0.957 **c.** 0.002 17. **a.** 0.517 **b.** 0.051 **c.** 0.595 19. **a.** 0.985 **b.** 0.174 **c.** 0.001 21. **a.** $\mu = 4.5, \sigma^2 = 3.15, \sigma = 1.775$ **b.** 0.723 **c.** 0.704 **d.** 0.263 23. **a.** $\mu = 6.4, \sigma^2 = 3.84, \sigma = 1.96$ **b.** 0.526 **c.** 0.833 **d.** 0.792 25. **a.** $\mu = 5.7, \sigma^2 = 3.99, \sigma = 1.997$ **b.** 0.083 **c.** 0.474 **d.** 0.181 27. $\mu = 150, \sigma^2 = 37.5, \sigma = 6.124$

Exercise Set 8.3 Page 472

1. 0.9904 **3.** 0.1469 **5.** 0.1562 **7.** 0.7852 **9.** 0.30000 **11.** 0.7682 **13.** 0.2731 **15.** 0.9426 **17. a.** about 38% **b.** about 24% **c.** about 24% **d.** about 7% **e.** about 7% **19. a.** 0.119 **b.** 0.0094 **c.** 0.3616 **d.** 0.0392 **e.** 0.9906 **21. a.** 0.1335 **b.** 0.0132 **c.** 0.8533 **d.** 0.1331 **e.** 0.0132

Exercise Set 8.4 Page 477

1. 0.9429 **3.** 0.9922 **5.** 0.1788 **7.** 0.9550 **9. a.** 0.1093 **b.** 0.2810 **c.** 0.6742 **d.** 0.2850 **e.** 0.2771 **f.** 0.2771 **11. a.** 0.025 **b.** 0.0016 **c.** 0.4538 **d.** 0.3438 **e.** 0.2158

Exercise Set 9.1 Page 490

1. (2), (4), (5) **3. a.** $\begin{bmatrix} .7 & .2 & .1 \\ .3 & .5 & .2 \\ .4 & .4 & .2 \end{bmatrix}$ **b.** $\begin{matrix} [40{,}000 & 30{,}000 & 20{,}000] \\ & \text{next weeks sales} & \\ [45{,}000 & 31{,}000 & 14{,}000] \end{matrix}$ **c.** $[\frac{4}{9} \quad \frac{3}{9} \quad \frac{2}{9}]$

This week, the probability that a customer orders (1) a big-burger is $\frac{4}{9}$, (2) a chicken-wich is $\frac{3}{9}$, and (3) a fish-wich is $\frac{2}{9}$. **5. a.** $[\frac{7}{36} \quad \frac{13}{18} \quad \frac{1}{12}]$ **b.** $[\frac{1}{9} \quad \frac{61}{72} \quad \frac{1}{24}]$ **c.** $[\frac{5}{9} \quad \frac{1}{18} \quad \frac{7}{18}]$

d. $[\frac{163}{216} \quad \frac{6}{216} \quad \frac{47}{216}]$ **e.** $[\frac{5}{9} \quad \frac{1}{18} \quad \frac{7}{18}] \begin{bmatrix} \frac{1}{2} & \frac{1}{2} & 0 \\ 0 & 1 & 0 \\ \frac{1}{6} & \frac{1}{3} & \frac{1}{2} \end{bmatrix} = [\frac{37}{108} \quad \frac{50}{108} \quad \frac{21}{108}]$ **7.** $T = \begin{bmatrix} \frac{3}{4} & \frac{1}{4} \\ \frac{1}{2} & \frac{1}{2} \end{bmatrix}$

$D = [\frac{1}{2} \quad \frac{1}{2}]$ $DT^3 = [\frac{85}{128} \quad \frac{43}{128}]$

Exercise Set 9.2 Page 497

1. a. $[\frac{1}{2} \quad \frac{1}{4} \quad \frac{1}{4}]$ For parts b, c, d, and e check to see that $DX = D$.
3. [60% 10% 25% 5%] ⇒ $[\frac{49}{84} \quad \frac{9}{84} \quad \frac{21}{84} \quad \frac{5}{84}]$ or about [58% 11% 25% 6%]. *A* will lose about 2%, *B* will gain about 1%, *C* will remain the same, and *D* will gain about 1%. **5.** First state will get everything. Matrix is not regular. **7.** [70 140 10]

Exercise Set 9.3 Page 502

1. For *A*; $R = [0 \quad \frac{1}{2}]$, $Q = [\frac{1}{2}]$. For *C*; $R = [\frac{1}{4} \quad \frac{1}{3}]$, $Q = [\frac{5}{12}]$ For *E*; $R = [0 \quad \frac{1}{4} \quad \frac{1}{4}]$, $Q = [\frac{1}{2}]$. For *F*; $R = [.3 \quad .4 \quad .1]$, $Q = [.2]$. **3.** 2; $\frac{12}{7}$; 2; 1.25. **5. a.** $[\frac{1}{2} \quad 0 \quad \frac{1}{2}]$ **b.** $[\frac{11}{18} \quad 0 \quad \frac{7}{18}]$ **c.** $[\frac{4}{7} \quad \frac{3}{7} \quad 0]$ **d.** $[0 \quad \frac{1}{3} \quad \frac{2}{3}]$ **7. a.** [400 600 0] **b.** [0 200 800] **9.** [.8 0 .05 .15] **b.** [.8375 .15 .0125 0] **c.** [.8696 0 0 .1304]

11. Same as $\begin{bmatrix} \frac{1}{2} & \frac{1}{4} & \frac{1}{4} \\ \frac{1}{2} & \frac{1}{2} & 0 \\ 0 & \frac{1}{3} & \frac{2}{3} \end{bmatrix}$

Exercise Set 10.1 Page 513

1. a. $[2]$ **b.** $\begin{bmatrix} 3 & 1 \\ 1 & 2 \end{bmatrix}$ **c.** $[-1]$ **d.** will not reduce **e.** $\begin{bmatrix} -2 & 3 \\ 3 & 2 \end{bmatrix}$ **3. a.** $\begin{bmatrix} 2 & \textcircled{1} \\ 1 & 0 \end{bmatrix}$

b. $\begin{bmatrix} 5 & 6 & \textcircled{4} \\ -2 & 3 & 4 \\ 5 & 2 & 1 \end{bmatrix}$ **c.** $\begin{bmatrix} 0 & -1 & -1 & 2 \\ 1 & 3 & 0 & 1 \\ 1 & 2 & \textcircled{1} & 2 \\ 2 & 1 & -2 & 3 \end{bmatrix}$ **d.** $\begin{bmatrix} 8 & 4 & 5 & 6 \\ 6 & 4 & 4 & 5 \\ 5 & 6 & \textcircled{5} & 8 \\ 5 & 3 & 3 & 9 \end{bmatrix}$

5.

	e0 s0	e0 s1	e0 s2	e1 s1	e1 s2	e1 s3	e2 s2	e2 s3	e2 s4
extend 0, shout 0	0	1	1	−1	0	0	−1	0	0
extend 0, shout 1	−1	0	0	0	1	1	−1	0	0
extend 0, shout 2	−1	0	0	−1	0	0	0	1	1
extend 1, shout 1	1	0	1	0	−1	0	0	−1	0
extend 1, shout 2	0	−1	0	1	0	1	0	−1	0
extend 1, shout 3	0	−1	0	0	−1	0	1	0	1
extend 2, shout 2	1	1	0	0	0	−1	0	0	−1
extend 2, shout 3	0	0	−1	1	1	0	0	0	−1
extend 2, shout 4	0	0	−1	0	0	−1	1	1	0

7. Yes. When played with the proper strategy, the game always ends in a tie.

9. a.

	Bid by labor 5.56	5.60	5.64	5.68	5.72	5.76	5.80
5.56	5.56	5.60	5.64	5.68	5.72	5.76	5.68
5.60	5.56	5.60	5.64	5.68	5.72	5.68	5.60
5.64	5.56	5.60	5.64	5.68	(5.68)	5.64	5.64
Offer 5.68	5.56	5.60	5.64	5.68	5.68	5.68	5.68
5.72	5.56	5.60	5.68	5.72	5.72	5.72	5.72
5.76	5.56	5.68	5.76	5.76	5.76	5.76	5.76
5.80	5.68	5.80	5.80	5.80	5.80	5.80	5.80

b. management offers \$5.64, labor bids \$5.72 **c.** Yes, \$5.68.

Exercise Set 10.2 Page 521

1. a. $\frac{1}{4}$ **b.** 1 **c.** 0.4 **d.** −0.42 **3.** $A = [\frac{3}{5} \quad \frac{2}{5}]$, $B = \begin{bmatrix} \frac{3}{5} \\ \frac{2}{5} \end{bmatrix}$, $v = \frac{16}{5}$. **5.** $A = [\frac{5}{11} \quad \frac{6}{11}]$, $B = \begin{bmatrix} \frac{6}{11} \\ \frac{5}{11} \end{bmatrix}$, $v = \frac{8}{11}$

Exercise Set 10.3 Page 527

1. $A = [0 \quad 1]$, $B = \begin{bmatrix} 0 \\ 1 \\ 0 \end{bmatrix}$, $v = -2$. **3.** $A = [0 \quad \frac{3}{4} \quad \frac{1}{4}]$, $B = \begin{bmatrix} \frac{1}{3} \\ \frac{7}{12} \\ \frac{1}{12} \end{bmatrix}$, $v = \frac{7}{4}$. **5.** By dominant strategy, reduce to $\begin{bmatrix} 0 & 2 & 1 \\ 3 & 1 & 1 \\ 2 & 0 & 2 \end{bmatrix}$. $A = [\frac{1}{2} \quad \frac{1}{4} \quad \frac{1}{4}]$, $B = \begin{bmatrix} 0 \\ \frac{1}{8} \\ \frac{3}{8} \\ \frac{1}{2} \end{bmatrix}$, $v = \frac{5}{4}$. **7.** $A = [\frac{5}{12} \quad \frac{1}{3} \quad \frac{1}{6} \quad \frac{1}{12}]$, $B = \begin{bmatrix} \frac{1}{3} \\ \frac{1}{3} \\ 0 \\ \frac{1}{3} \end{bmatrix}$, $v = \frac{1}{3}$.

Exercise Set 11.1 Page 535

1. 240% **3.** 16.625 **5.** 20% **7.** 9,494 **9.** 1,214 **11.** 25% **13.** $2,736 **15.** 44.4%
17. $75.47 **19.** Sue

Exercise Set 11.2 Page 542

$I = Prt$
1. $437.50 **3.** $8000
$V = P(1 + i)^n$
6. $3079.39 **8.** $1,430.77
$P_0 = V(1 + i)^{-n}$
11. $3937.83 **13.** $560.67

Exercise Set 11.3 Page 550

9. $663,888.83 **11.** $711.62 **13.** $34,963.71 **15.** $3,741.10 **17.** 4.96%

Exercise Set 11.3 Page 556

See *Instructor's Manual*

Answers to Review Exercises

Chapter 0

1. $\frac{4}{3}$ **2.** $\frac{4}{3}$ **3.** $\frac{3}{2}$ **4.** 2 **5.** -3 **6.** 2 **7.** 2 **8.** $-\frac{2}{3}$ **9.** 2 **10.** $-\frac{4}{3}$ **11.** 7 **12.** -5
13. $-\frac{1}{3}$ **14.** 1 **15.** All real numbers **16.** 2 **17.** No solution **18.** -1 **19.** -2 **20.** $-\frac{5}{6}$
21. -13 **22.** 0 **23.** 6 **24.** No solution **25.** $y < 1$ **26.** $y \le 5$ **27.** $x \le -1$ **28.** $y < 2$
29. $x > \frac{3}{4}$ **30.** $y \ge \frac{7}{3}$ **31.** $t < 4$ **32.** $y > -1$ **33.** $x > -\frac{5}{3}$ **34.** $y \le -\frac{1}{3}$ **35.** $x < -\frac{1}{7}$
36. $x \ge -\frac{3}{2}$ **37.** $x \ge 5$ **38.** $x \le -4$ **39.** $x > \frac{7}{3}$ **40.** $x < -5$ **41.** $x > \frac{9}{5}$ **42.** $t < \frac{9}{5}$
43. $x \le -\frac{1}{2}$ **44.** $x \le 2$ **45.** $x \le 1$ **46.** $x \le \frac{3}{7}$ **47.** $2 < x \le \frac{9}{4}$ **48.** $x < \frac{26}{15}$ or $x > 2$ **49.** $x < -3$ or $x \ge \frac{23}{4}$ **50.** $-\frac{11}{3} < x < -\frac{3}{2}$ **51.** $y = \frac{4}{7}x + \frac{1}{7}$ **52.** $y = -x - 1$ **53.** $x = 3$ **54.** $y = -\frac{1}{2}x + 4$
55. $y = -4$ **56.** $y = -\frac{7}{4}x + 7$ **57.** $y = -3x + 15$ **58.** $y = \frac{1}{2}x + \frac{3}{2}$ **59.** $y = -\frac{2}{3}x + \frac{14}{3}$
60. $y = \frac{4}{5}x - \frac{8}{5}$ **61.** $y = -\frac{3}{2}x + \frac{5}{2}$ **62.** $y = \frac{5}{2}x + \frac{1}{2}$ **63.** $y = -2x + 3$ **64.** $y = \frac{1}{2}x - 2$
65. $y = -\frac{3}{2}x - 1$ **66.** $y = \frac{2}{3}x + 6$ **67.** $y = 2x - 7$ **68.** $y = \frac{1}{2}x + 4$ **69.** $y = -\frac{3}{2}x + \frac{13}{2}$
70. $y = -\frac{4}{3}x - 5$ **71.** $x = 3$ **72.** $y = -4$ **73.** $y = \frac{1}{2}x + \frac{5}{2}$ **74.** $y = -\frac{1}{2}x - \frac{5}{2}$ **75.** $y = \frac{4}{3}x - \frac{4}{3}$
76. $y = \frac{3}{5}x - \frac{19}{5}$ **77.** $y = \frac{3}{2}x + \frac{3}{4}$ **78.** $y = -\frac{2}{3}x + 7$ **79.** $y = -5$ **80.** $x = -4$

81.

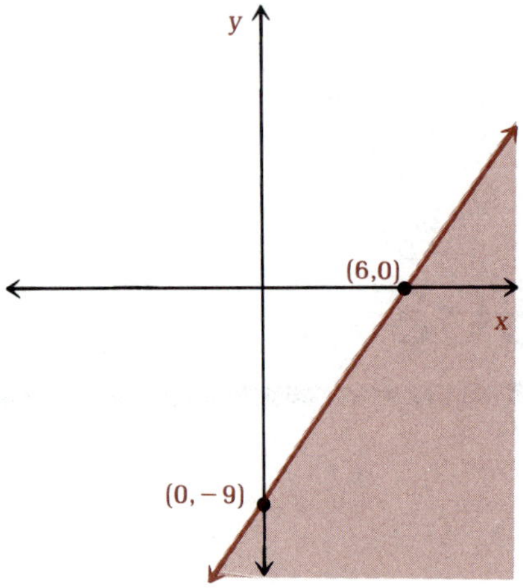

82.

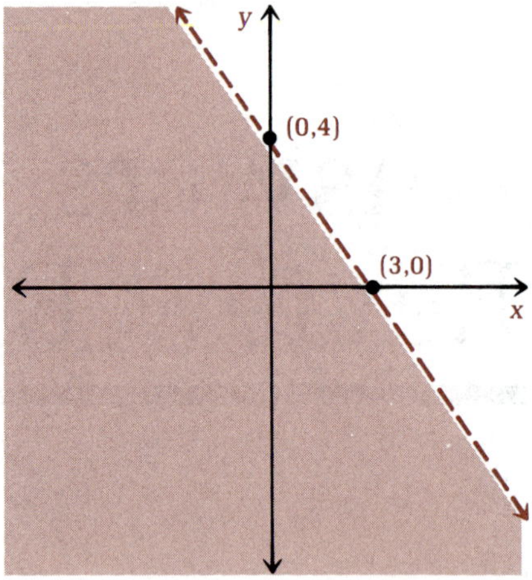

83.

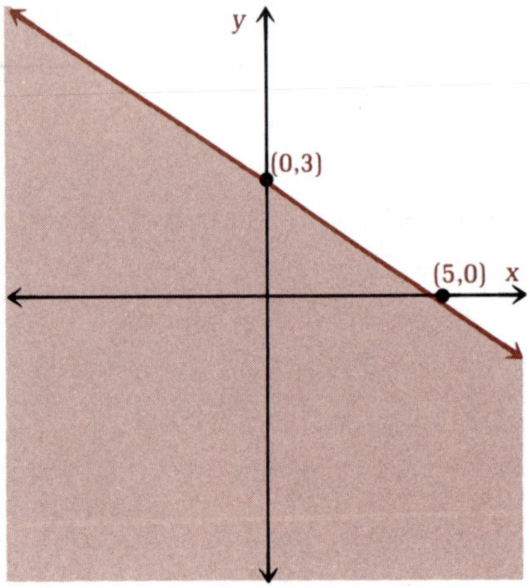

84.

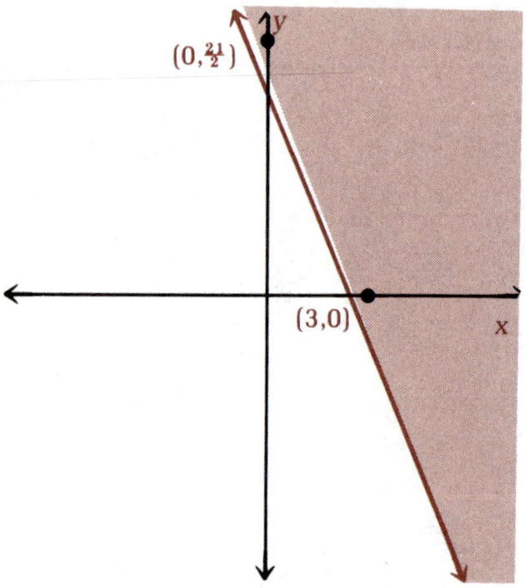

85.

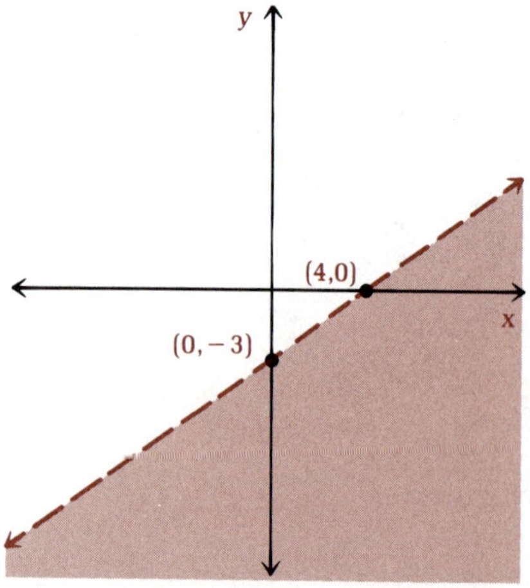

86.

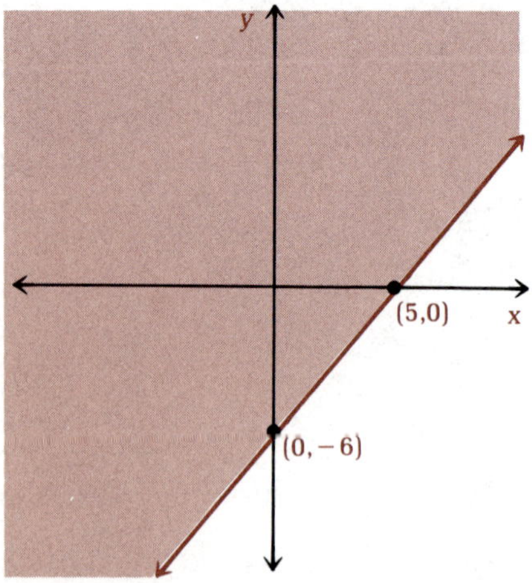

87.

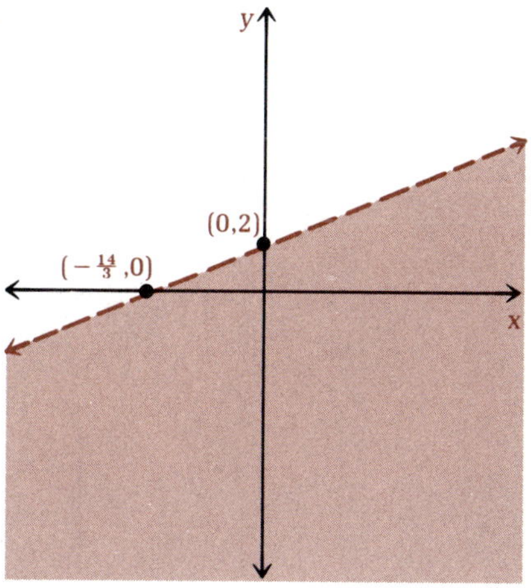

88.

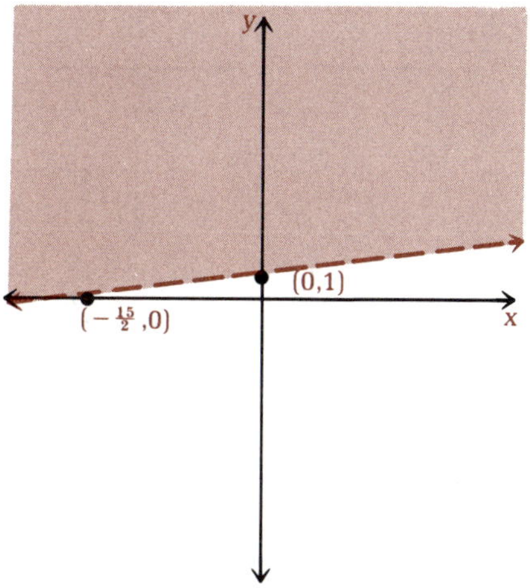

89.

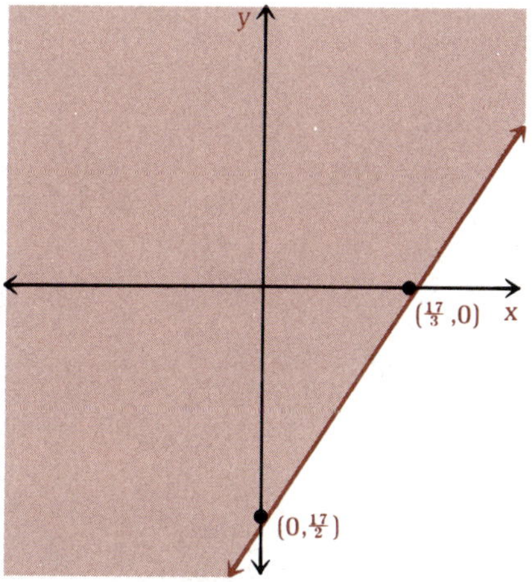

90.

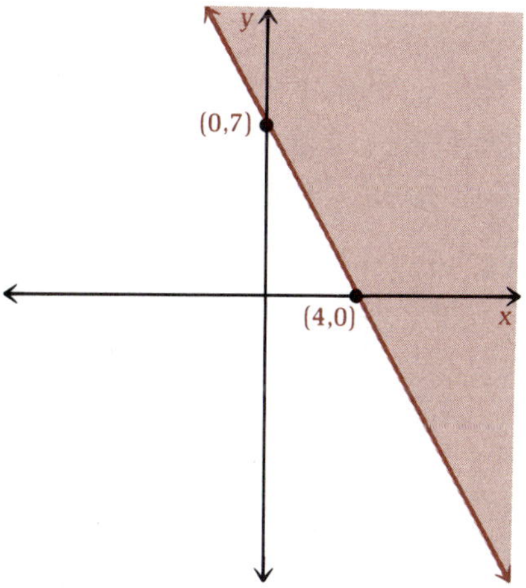

91. $x + y \le 4$ **92.** $x - y \ge 3$ **93.** $-3x + 4y < 12$ **94.** $2x + 3y > 12$ **95.** $-2x + y > 10$
96. $3x + 2y \ge 12$ **97.** $3x + 4y < 24$ **98.** $-3x + 5y \le 15$ **99.** $-2x + 6y \ge 15$
100. $34x + 22y \le 187$ **101.** $w = 20, l = 60$ **102.** $w = 20, l = 65$ **103.** 150 **104.** $w = 50, l = 240$
105. a. $y = 4.20x$ **b.** \$84 **c.** \$105 **106. a.** $y = 6x$ **b.** \$12 **c.** \$9 **d.** \$27
107. a. $y = 200 + 0.05x$ **b.** \$450 **c.** \$875 **108. a.** \$27.50 **b.** \$65 **c.** \$96.25
109. a. $F = \frac{9}{5}C + 32$ **b.** 68 **c.** 86 **110. a.** $C = \frac{5}{9}(F - 32)$ **b.** 25 **c.** 37 **111.** $-40°$
112. a. $15x + 12y \ge 1000$ **113. a.** $2x + 3y \ge 40$ **114. a.** $4x + 5y \le 128$ **115. a.** $10x + 30y \le 250$
116. a. $q = -200p + 1100$ **b.** 100 **c.** 350 **d.** \$3.50 **117. a.** $y = 0.075x - 3.25$ **b.** 2.375, 3.125, 3.5, 3.875 **118. a.** $y = 400x - 200$ **b.** 760; 960; 1100

Chapter 1

1. **a.** $\begin{bmatrix} 5 & 0 & -3 \\ 4 & 5 & 5 \end{bmatrix}$ **b.** $\begin{bmatrix} 0 & 0 & 2 \\ -2 & -1 & 2 \\ 2 & 0 & -4 \end{bmatrix}$ **c.** $\begin{bmatrix} -7 & 7 & -8 \\ 13 & -2 & 7 \\ 2 & -3 & 3 \end{bmatrix}$ **d.** Not defined

e. $\begin{bmatrix} 1 & 6 & -5 \\ -2 & 12 & -14 \end{bmatrix}$ **f.** $\begin{bmatrix} 33 & 21 \\ -2 & 11 \end{bmatrix}$ 2. **a.** $\frac{1}{10}\begin{bmatrix} 2 & 1 \\ -4 & 3 \end{bmatrix}$ **b.** $\frac{1}{2}\begin{bmatrix} 6 & -4 \\ 2 & -1 \end{bmatrix}$

c. Not invertible **d.** $\frac{1}{7}\begin{bmatrix} 2 & -5 \\ -1 & 6 \end{bmatrix}$ **e.** $\frac{1}{22}\begin{bmatrix} 8 & -6 & 1 \\ -10 & 2 & 7 \\ -6 & 10 & 2 \end{bmatrix}$ **f.** $\frac{1}{28}\begin{bmatrix} 1 & 10 & -7 \\ 3 & 2 & 7 \\ -7 & 14 & -7 \end{bmatrix}$

g. Not invertible **h.** Not invertible **i.** $\frac{1}{110}\begin{bmatrix} -20 & 30 & -20 & 10 \\ -23 & -4 & 21 & 28 \\ 29 & -38 & 7 & 46 \\ 5 & 20 & 5 & -30 \end{bmatrix}$

j. $\frac{1}{32}\begin{bmatrix} -29 & -4 & 28 & -11 \\ 11 & -4 & -4 & 13 \\ -21 & -4 & 28 & -19 \\ 26 & 8 & -24 & 22 \end{bmatrix}$

3. **a.**

Job class	Shift 1	2	3
I	9.66	9.89	10.35
II	9.89	10.12	10.81
III	10.12	10.58	11.27
IV	10.58	11.04	11.96

b. $\begin{bmatrix} 1.26 & 1.29 & 1.35 \\ 1.29 & 1.32 & 1.41 \\ 1.32 & 1.38 & 1.47 \\ 1.38 & 1.44 & 1.56 \end{bmatrix}$

4. **a.** $[400 \quad 300 \quad 200 \quad 100]\begin{bmatrix} 8.40 & 8.60 & 9.00 \\ 8.60 & 8.80 & 9.40 \\ 8.80 & 9.20 & 9.80 \\ 9.20 & 9.60 & 10.40 \end{bmatrix} = [8620 \quad 8880 \quad 9420]$

b. [9913 10212 10833] **c.** 40[8620 8880 9420] = [344,800 355,200 376,800]
d. [396,520 408,480 433,320] **e.** 161,520 5. Overtime salary schedules

a.

Class	Shift 1	2	3
I	12.60	12.90	13.50
II	12.90	13.20	14.10
III	13.20	13.80	14.70
IV	13.80	14.40	15.60

b.

Class	Shift 1	2	3
I	14.49	14.835	15.525
II	14.835	15.18	16.215
III	15.18	15.87	16.905
IV	15.87	16.56	17.94

6. a. $\begin{matrix} T & C & S.C. \\ [6 & 30 & 4] \end{matrix}$ **b.** $\begin{matrix} & T.C. & F & C.T. & V \\ T & 7 & 2 & 6 & 0 \\ C & 2 & 0 & 1 & 2 \\ S.C. & 3 & 1 & 2 & 0 \end{matrix}$

c. $[6 \quad 30 \quad 4]\begin{bmatrix} 7 & 2 & 6 & 0 \\ 2 & 0 & 1 & 2 \\ 3 & 1 & 2 & 0 \end{bmatrix} = [114 \quad 16 \quad 74 \quad 60]$ **d.** Write the cost per unit of material as a column matrix and multiply by the row matrix obtained as the solution to part c.

7. a. $\begin{bmatrix} -10 & 32 & -3 \\ -1 & 0 & 0 \\ -3 & 11 & -1 \end{bmatrix}$ **b.** $\begin{bmatrix} 210 & 76 & -115 \\ -86 & -31 & 47 \\ -113 & -41 & 62 \end{bmatrix}$ **c.** $\frac{1}{2}B^{-1}$ **d.** $3C^{-1}$ **e.** $\begin{bmatrix} 1 & 0 & 0 \\ 0 & 1 & 0 \\ 0 & 0 & 1 \end{bmatrix}$

f. $\begin{bmatrix} -76 & 244 & -29 \\ 63 & -202 & 24 \\ 29 & -93 & 11 \end{bmatrix}$ **g.** $\begin{bmatrix} -35 & -79 & -4 \\ 30 & 67 & 4 \\ 13 & 30 & 1 \end{bmatrix}$ **h.** $\frac{1}{3}A$ **i.** $4C$ **j.** $3A$

Chapter 2

1. $x = 2, y = 5$ **2.** $x = 15, y = 12$ **3.** $x = 2, y = 2 + z, z = z$ **4.** $x = 3, y = -5, z = 4$ **5.** $x = 2, y = -3 + z, z = z, w = 3$ **6.** $x = 8, y = 7, z = 6, w = 5$ **7.** $x = 4, y = 1$ **8.** $x = \frac{1}{4}, y = \frac{5}{4}$ **9.** No solution **10.** $x = 1 - z, y = z, z = z$ **11.** $x = \frac{3}{4}, y = \frac{1}{4}, z = -\frac{1}{2}$ **12.** $x = 3 + 2z, y = -2, z = z$ **13.** No solution **14.** $x = -2, y = \frac{-5 + z}{2}, z = z$ **15.** $x = 4, y = 2, z = 3, w = 5$ **16.** No solution **17.** $x = 8 - 5z, y = -5 + 4z, z = z$ **18.** $x = -3 - 2z, y = 1, z = z$ **19.** $x = \frac{5 - 9w}{7}, y = \frac{12 + 5w}{7}, z = \frac{-8 + 6w}{7}, w = w$ **20.** $v = \frac{13 + 6z}{10}, w = \frac{4 - 2z}{5}, x = \frac{9 - 2z}{10}, y = \frac{9 - 7z}{5}, z = z$

21. $p = -\frac{1}{20}q + 100$ demand
$p = \frac{1}{200}q + 1$ supply
$p = 10, q = 1800$

22. $p = -\frac{1}{25}q + 14$ demand
$p = \frac{1}{25}q - 10$ supply
$p = 2, q = 300$

23. $p = \frac{1}{24}q - 600$ supply; $p = \frac{-4}{400}q + 62.5$ demand; $p = 25,\ q = 15{,}000$

24. $p = \frac{1}{500}q + 0.1$ supply; $p = -\frac{1}{25}q + 208$ demand; $p = 10,\ q = 4950$

25. $p = \frac{1}{200}q - \frac{1}{40}$ supply; $p = -\frac{1}{20}q + \frac{39}{2}$ demand; $p = \frac{7}{4},\ q = 355$

26. $p = \frac{1}{40}q - 150$ supply; $p = -\frac{1}{60}q + \frac{1265}{12}$ demand; $p = 3.25,\ q = 6130$

27. $q = 20,\ r = c = 500$ 28. $q = 10,\ r = c = 6500$ 29. $q = 200,\ r = c = 3600$ 30. $q = 30,\ r = c = 450$ 31. $q = 1000,\ r = c = 9750$ 32. $q = 20,\ r = c = 249$ 33. $X = \begin{bmatrix} -1 & -2 & -5 \\ -1 & -2 & -3 \end{bmatrix}$

34. $X = \begin{bmatrix} 13 & -3 & 11 & 8 \\ -5 & 2 & -4 & -3 \end{bmatrix}$ 35. $X = \frac{1}{3}\begin{bmatrix} 12 & 7 & 7 & 4 & 6 \\ 9 & 7 & 3 & 1 & 4 \\ -6 & -5 & -4 & -2 & 2 \end{bmatrix}$

36. $X = \begin{bmatrix} 3 & 1 & 2 & 1 & 1 & 2 \\ 1 & 3 & 2 & 1 & 2 & 1 \\ 2 & 1 & 2 & 1 & 3 & 2 \end{bmatrix}$ 37. Each man should work 3 hours.

38.

	Week 1	2	3	4	5	6
Smith	2	3	1	2	3	4
Brown	1	1	2	3	2	3
Allen	2	1	3	2	3	4

39. Multiple solutions of the form $(7 - 2y, y, 3)$. Since $y \geq 0$ and $x \geq 0$, $7 - 2y \geq 0$ or $y \leq \frac{7}{2}$. Hence $y = 0, 1, 2, 3$. Practical solutions (7,0,3), (5,1,3), (3,2,3), (1,3,3).

40.

Carton type	N.E.	M.A.	S.E.	N.C.	M.W.	R.M.	S.W.	W.C.
A	300	200	200	300	250	100	200	300
B	100	200	100	200	−25	200	200	400
C	200	200	300	100	425	300	100	200

*For the midwest, the company cannot exactly fill the order since they can't ship −25 type *B* cartons. Some kind of decision would have to be made.

41. Business $7,000,000
Industry $7,500,000

42.

	76	77	78	79	80	81
B	13	15	18.5	25	35	43
I	13	13.75	17.5	24	34.25	43.75

43. Agriculture $556 million, industry $394 million, and business $548 million.

44.

	1976	1977	1978	1979	1980	1981
A	27.8	39.5	49.4	52	64.6	85.3
B	19.7	27.5	34.1	37	45.4	59.2
C	27.4	41.5	50.2	54.5	71.3	91.4

Chapter 3

1.

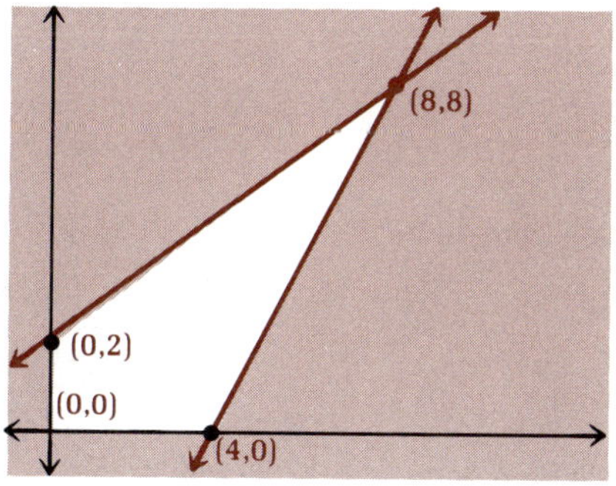

2. $2x - y \geq 4$
$2x + 3y \geq 20$
$2x + y \leq 20$

3.

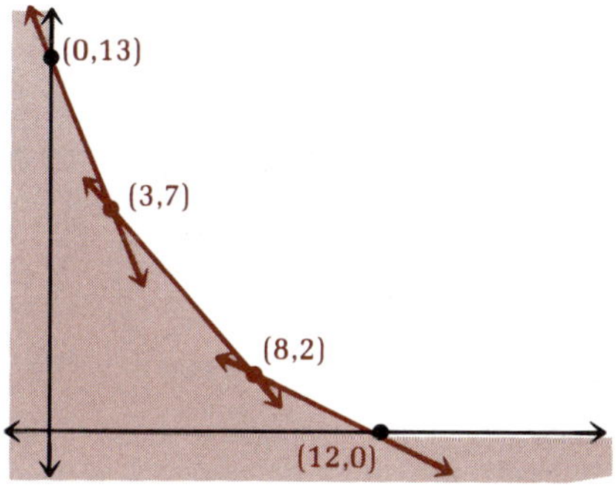

4. a, b, d, and e are in the half-plane $2x + 5y \leq 38$; c and f are not.

5. $P = 9$ when $x = 5$, $y = 3$ **6.** $C = 64$ at $x = 0$, $y = 8$

7. $P = 9$
$x = 5$
$y = 3$
$s_1 = s_2 = 0$

8. $C = 64$
$x = 0$
$y = 8$
$s_1 = 0$, $s_2 = 25$

9. $A = 9$
$x = 5$
$y = 3$
$s_1 = 0$
$s_2 = 0$
$s_3 = 20$

10. $A = -11$
$x = 1$
$y = 7$
$s_1 = 0$
$s_2 = 20$
$s_3 = 0$

11.
$$\begin{bmatrix} -1 & -1 & -2 & 1 & 0 & 0 & 0 & -6 \\ -1 & 1 & -2 & 0 & 1 & 0 & 0 & -4 \\ 1 & 2 & 3 & 0 & 0 & 1 & 0 & 9 \\ 1 & 1 & 1 & 0 & 0 & 0 & 0 & 5 \\ 2 & -3 & 4 & 0 & 0 & 0 & -1 & 0 \end{bmatrix}$$

12.
$$\begin{bmatrix} -1 & -1 & -2 & 1 & 0 & 0 & 0 & -6 \\ -1 & 1 & -2 & 0 & 1 & 0 & 0 & -4 \\ 1 & 2 & 3 & 0 & 0 & 1 & 0 & 9 \\ 1 & 1 & 1 & 0 & 0 & 0 & 0 & 5 \\ -3 & 2 & 1 & 0 & 0 & 0 & 1 & 0 \end{bmatrix}$$

13. $P = 27$
$x = 0$
$y = 12$
$z = 13$
$s_1 = s_3 = 0$
$s_2 = 5$

14. $B = 25$, $x = 0$, $y = 12$, $z = 13$, $s_1 = s_3 = 0$, $s_2 = 5$

15. No minimum exists.

16. $C = 19$, $x = \frac{5}{2}$, $y = \frac{1}{2}$, $z = \frac{9}{2}$, $s_1 = s_2 = s_3 = 0$

17. a. Run Factory I for 10 hours. Run Factory II for 15 hours. Run Factory III for 10 hours. **b.** Total cost = 10 (210) + 15 (210) + 10 (280) = 8050
c. $s_1 = 0$ Exactly our order for 70 Product *A*.
$s_2 = 0$ Exactly our order for 65 Product *B*.
$s_3 = 0$ Exactly our order for 50 Product C.
Hence, no excess.

18. a. Process I, 2 days; Process II, 6 days. **b.** Excess of 180 ounces of gold. **c.** 2,600 gallons

Chapter 4

1.

120					120
164	166	10			340
		280			280
		52	108	100	260
284	166	342	108	100	

2. (1) *cihb* (2) *eqpjhb* (3) *sabhjpqw* (4) *mabhjp* **3.** (1) *acig* (2) *dpnhic* (3) *qnhice* (4) *lxvpnh* (5) *wvpnhice* **4.** Add 10 from (2,4) to 20 in (4,4) and add 10 from (4,1) to 20 in (2,1).
5. a. 2, 3, −1, −1, −2, 0, 1 **b.** 10, 20, 10, 20, 10, 20, 20 **c.** +20, +60, −10, −20, −20, 0, +20
6. 1, 2, 6, 4, 3, 0, −3, −3, −4, −2 **7.** 0, 50, 10, 10, 10, 20, 20, 20, 20, 20 **8.** 0, +100, +60, +40, +30, 0, −60, −60, −80, −40

9.

From \ To	1	2	3	4
A	20			10
B				30
C		30		10
D			50	

or

From \ To	1	2	3	4
A				30
B	20			10
C		30		10
D			50	

Minimum shipping cost $480.

10.

From \ To	1	2	3	4	5
A	50				
B					60
C	10			90	40
D		60	80	10	

Other shipping schedules will also produce a minimum T.S.C. of $1060.

11.

To / From	A	B	C
1	20		
2		30	
3		10	15
4	10		35

T.S.C. = $760

or

To / From	A	B	C
1			20
2		30	
3		10	15
4	30		15

12.

To / From	1	2	3	4
I				150
II		125	75	
III	150	25		25

T.S.C. = $2175

13.

To / From	1	2	3
A	5		25
B	15		
C		15	

T.S.C. = $110

14.

To / From	A	B	C	D	W
1		30		70	
2	50		50		
3	20			10	70
4	50				

T.S.C. = $430

15.

To / From	1	2	3
A	200		
B	100		250**
C		250*	

*100 from overtime
**50 from overtime
T.S.C. = $2700

or

To / From	1	2	3
A	200		
B	100*		250
C		250**	

*50 from overtime
**100 from overtime
T.S.C. = $2700

16.

To / From	Phil.	Atl.	Don.	L.A.	Donation	Cannery	W.H.
Jackson	1000	1000					2000
Little Rock	500		2000	400	100		
Shreveport				2100		900	

T.C. = $10,540

17. Two Solutions

	No. 1	No. 2
A	3	3
B	5	1
C	4	4
D	1	2
E	2	5

18.

	Job
A	4
B	3
C	2
D	5
E	1

19. *A*—2
B—4
C—1
D—3

20. *A*—3
B—2
C—4
D—1

21. Doris—Sales
Joy—Advertising
Judy—Personnel
Sue—Public Relations

22. Two Solutions

	No. 1	No. 2
James	2	3
Jerry	4	4
Terry	3	2
Vern	1	1

23. Three of the Solutions

	No. 1	No. 2	No. 3
Tom	Set-up	Tickets	Clean-up
Ron	Ref.	Ref.	Ref.
J.D.	Donation	Don.	Don.
Fred	Tickets	Set-up	Set-up
Joe	Clean-up	Clean-up	Tickets
Dick	Adv.	Adv.	Adv.

24. Mary—Colchester
Judi—Industry
Pat—Blandinsville
Connie—Good Hope
Vicki—Adair
Marilyn—Bushnell
Candy—Tennessee

25. Two Solutions

	No. 1	No. 2
Brent	Const.	Const.
Erik	Pharm.	Teacher
Gina	Nurse	Nurse
Kirk	Merch.	Merch.
Nickie	Teach.	Pharm.
Ronnie	Dent.	Dent.
Timmy	Doct.	Doct.
Wendy	Law	Law

Chapter 5

1. a. {p,e,n,s,y,l,v,a,i} **b.** {12,14,16,18} **c.** {1,2,3,4,5,6,7} **d.** { } **2. a.** {6} **b.** {1,2,3,4,5,6,7} **c.** {0,6} **d.** {1,7} **e.** {0,1,2,3,4,5,6,7} = U **f.** {1,3,5,7} = C **g.** {3,5} **h.** {2,4,6} = A **i.** {0,1,3,5,7} **j.** {0,3,6} = B **k.** {2,4} **l.** {0,1,2,3,5,6,7} **m.** {1,3,4,5,7} **n.** {1,3,5,7} **o.** { } or ϕ **p.** {0,1,3,5,7}

3. a.

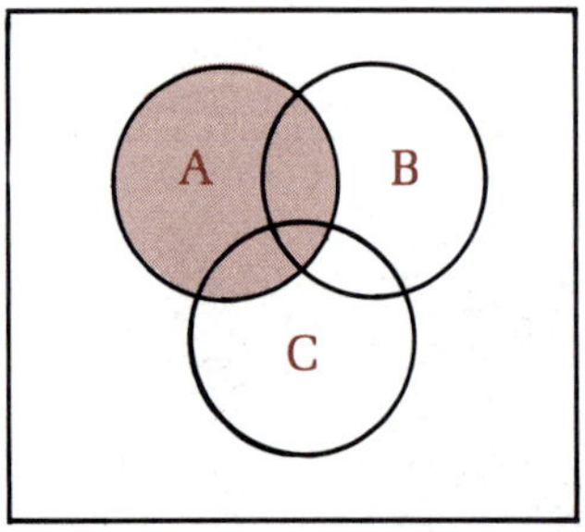

b.

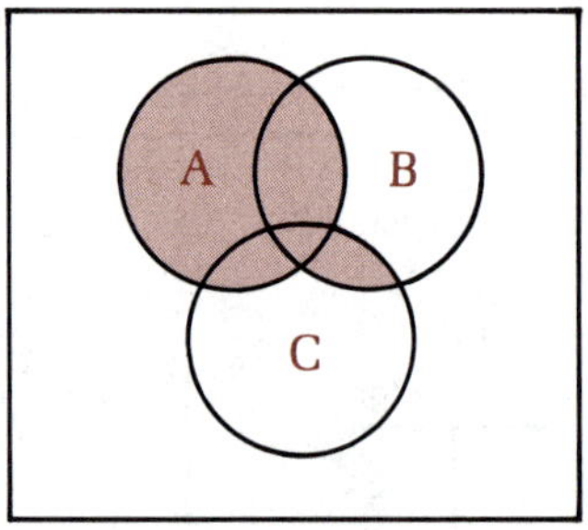

c.

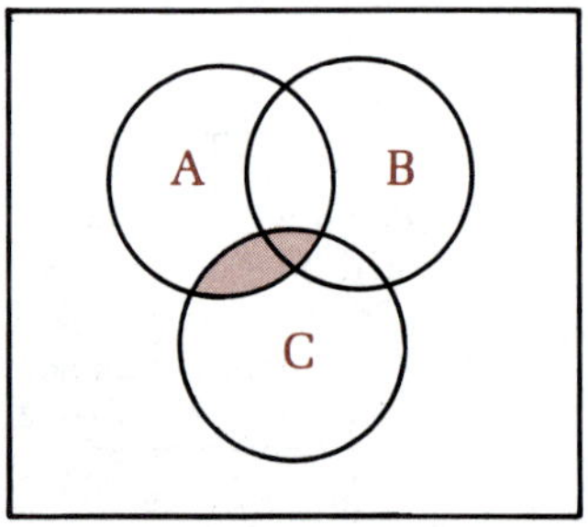

d.

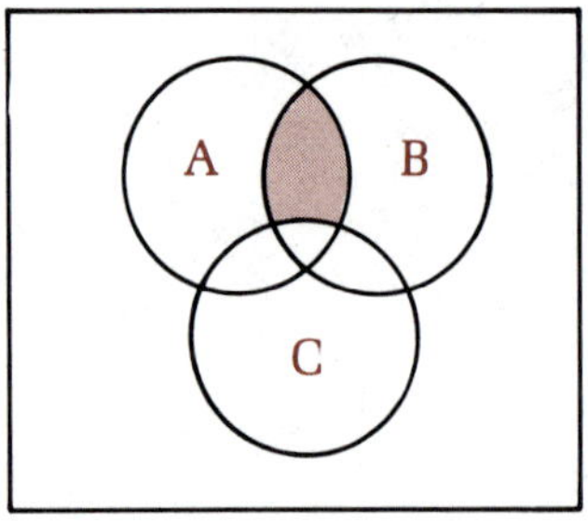

e.

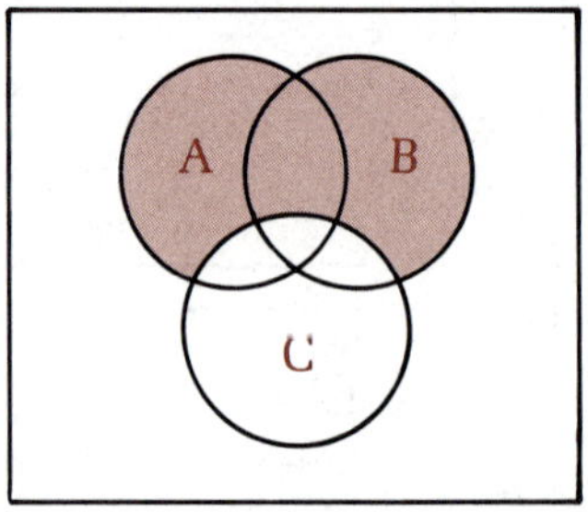

f.

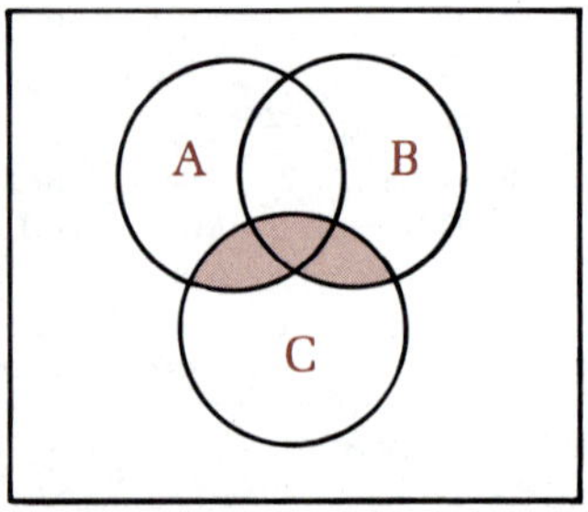

4. a. $(A \cap C') \cup (B \cap C') = (A \cup B) \cap C'$ **b.** $(C \cap A') \cup (A \cap B \cap C)$ **5.** 105 **6.** 10 **7.** 7000 **8.** 0 **9.** 400 **10.** 5000 **11.** 40 **12.** 390 **13.** 40 **14.** 145 **15. a.** 6720 **b.** 1,816,214,400 **c.** 60,480 **d.** 11,880 **e.** 7 **f.** 39,916,800 **g.** 56 **h.** 5005 **i.** 84 **j.** 495 **k.** 7 **l.** 1 **16. a.** 277,200 **b.** 420 **c.** 12,600 **d.** 1,663,200 **e.** 1680 **f.** 720,720 **17. a.** 138,600 **b.** 210 **c.** 12,600 **d.** 138,600 **e.** 280 **f.** 720,720 **18. a.** 126 **b.** 36 **c.** 84 **d.** 9 **19. a.** 160 **b.** 60 **c.** 240 **d.** 192 **20. a.** $60a^4$ **b.** $160a^3$ **c.** $240a^2$ **d.** $12a^5$ **21.** (26)(26)(10)(10)(10)(26)(26) = 456,976,000 **22. a.** (26)(25)(10)(9)(8)(24)(23) = 258,336,000 **b.** (26)(25)(10)(9)(9)(26)(25) = 342,225,000 **23. a.** 35 **b.** 5 **24. a.** 1302 **b.** 3929 **25. a.** 25 **b.** 225 **c.** 250 **d.** 175 **26.** 36 **27. a.** 60 **b.** 40 **c.** 0 **d.** 5 **e.** 25 **f.** 25 **28. a.** 940 **b.** 60 **c.** 600 **d.** 340 **29. a.** 480 **b.** 20 **c.** 200 **d.** bingo **30. a.** 12! = 479,001,600 **b.** 1320 **c.** 220 **31. a.** 132 **b.** 144 **32. a.** 720 **b.** 120 **33.** P(10,3) = 720 **34.** P(50,5) = 254,251,200 **35.** C(50,5) = 2,118,760 **36.** $\binom{20}{6,5,4,5} = 9{,}777{,}287{,}520$

37. $\binom{20}{5,5,5,5} \div 4! = 488{,}864{,}376$ **38. a.** 17,153,136 **b.** 2,858,856 **39. a.** 0 **b.** 2 **c.** 6

40. Won by 1,042 votes **b.** 9851 **c.** Wife—2 votes less than son, 5,485 less than Brown.

Chapter 6

1. a. $\frac{1}{6} = 0.17$ **b.** $\frac{9}{29} = 0.31$ **c.** $\frac{88}{203} = 0.43$ **d.** $\frac{21167}{23751} = 0.89$ **2. a.** $\frac{4}{47}$ **b.** $\frac{8}{47}$ **c.** $\frac{1}{47}$ **d.** $\frac{2}{47}$ **e.** $\frac{8}{47}$ **f.** $\frac{7}{47}$ **3. a.** .58 **b.** .75 **c.** .79 **d.** .37 **4. a.** $\frac{27}{216} = \frac{1}{9}$ **b.** $1 - \frac{3}{8} = \frac{5}{8}$ **c.** $\frac{1}{2}$

5. a. $\frac{1}{12}$ **b.** $\frac{5}{12}$ **c.** $\frac{1}{2}$ **d.** $\frac{1}{4}$ **6. a.** $\frac{7}{15}$ **b.** $\frac{11}{30}$ **c.** $\frac{1}{6}$ **7. a.** .6 **b.** .18 **c.** .26 **d.** .3 **e.** .2 **f.** $\frac{9}{13}$ **g.** $\frac{4}{7}$ **8.** 0.49 **9. a.** $\frac{126}{1820} = \frac{9}{130} = .069$ **b.** $1 - \frac{9}{130} = \frac{121}{130} = .931$ **c.** $\frac{756}{1820} = \frac{27}{65} = .415$ **10.** $P(W) = \frac{1}{6} \cdot \frac{4}{13} + \frac{5}{6} \cdot \frac{1}{13} = \frac{9}{78} = \frac{3}{26} = .115$ **11. a.** $\frac{64}{625}$ **b.** $\frac{7}{75}$ **12.** $\frac{2}{5}$

13. a. .4 **b.** .8 **c.** $\frac{4}{7}$ **d.** .3 **e.** .2 **f.** .6 **g.** .1 **h.** .2 **14. a.** 0 **b.** .6 **c.** .3 **d.** 0 **e.** 1 **f.** $\frac{6}{7}$ **15.** Mutually exclusive ($P(E \cap F) = 0$). **16. a.** .27 **b.** .43 **c.** .6 **d.** .45 **e.** .33 **f.** .73 **g.** .6 **h.** .55 **17.** Independent ($P(E) \cdot P(F) = P(E \cap F)$) **18.** 40% **19. a.** 2 to 3 in favor of *A* **b.** 3 to 2 against *A* **20.** $\frac{2}{9}$ **21.** $\left(\frac{12}{52}\right)(-10) + \left(\frac{4}{52}\right)(1) + \left(\frac{4}{52}\right)(2) + \ldots + \left(\frac{4}{52}\right)(10) = \frac{-120}{52} + \frac{4}{52}(1 + 2 + \ldots + 10) = \frac{-120}{52} + \left(\frac{4}{52}\right)\frac{10 \cdot 11}{2} = \frac{-120 + 220}{52} = \frac{100}{52} = \frac{50}{26} = \frac{25}{13}$ **22. a.** $\frac{1}{560}$ **b.** $\frac{137}{280}$

23. 3 to Kellie **24.** $\binom{6}{3}\left(\frac{1}{2}\right)^6 = \frac{6!}{3!3!} \cdot \frac{1}{64} = \frac{20}{64} = \frac{5}{16}$ **25. a.** .054 **b.** $\frac{4}{9} = .444$ **c.** $\frac{5}{27} = 185$

26. a. .017 **b.** $\frac{2}{17} = .118$ **c.** $\frac{3}{17} = .176$ **d.** $\frac{4}{17} = .235$ **e.** 1,700

Chapter 7

1. a.

x_i	f_i
0	3
1	7
2	10
3	9
4	7
5	4

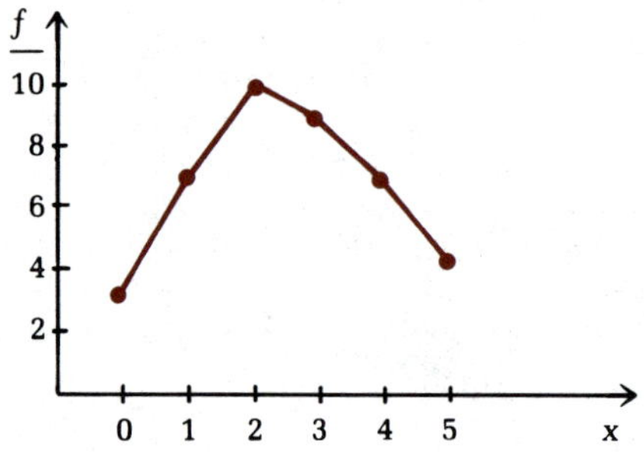

b.

x_i	f_i
2	2
3	2
4	1
5	3
6	4
7	9
8	5
9	1
10	0
11	2
12	1

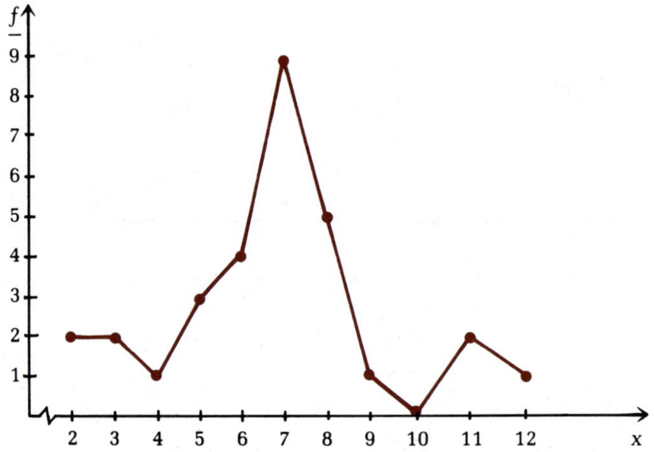

2. a.

x_i	cf_i
0	3
1	10
2	20
3	29
4	36
5	40

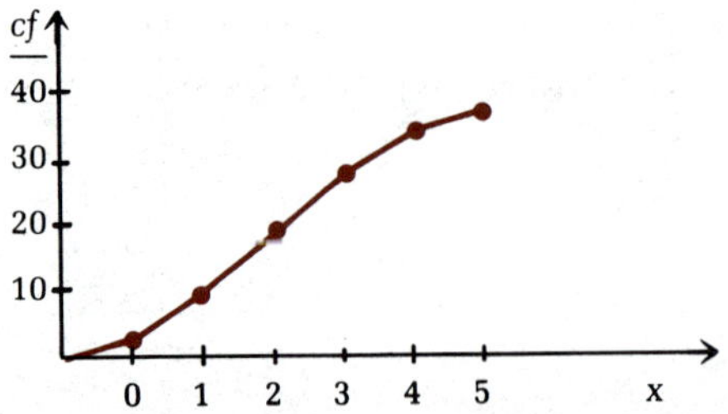

b.

x_i	cf_i
2	2
3	4
4	5
5	8
6	12
7	21
8	26
9	27
10	27
11	29
12	30

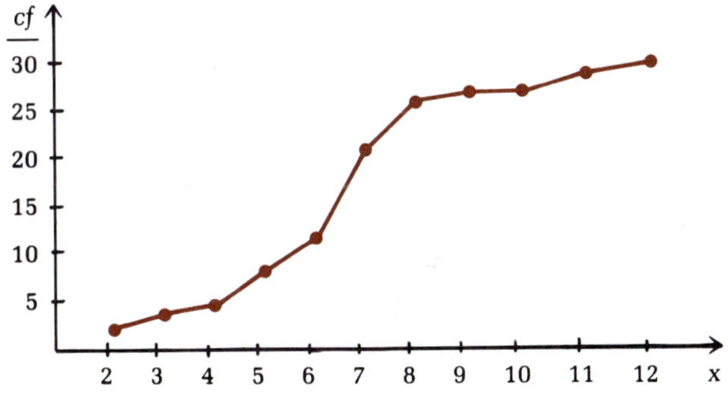

3.

a.

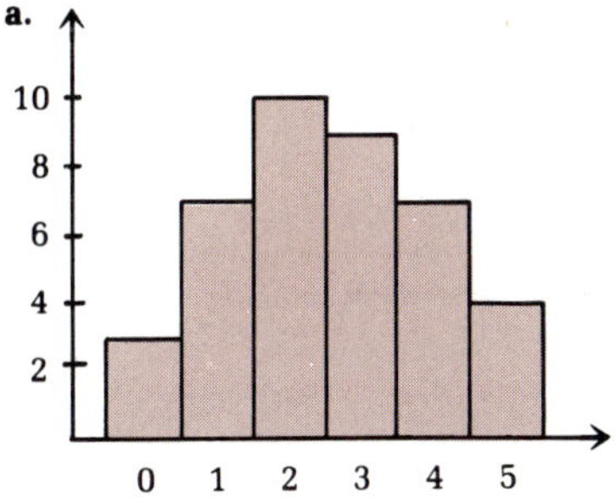

b.

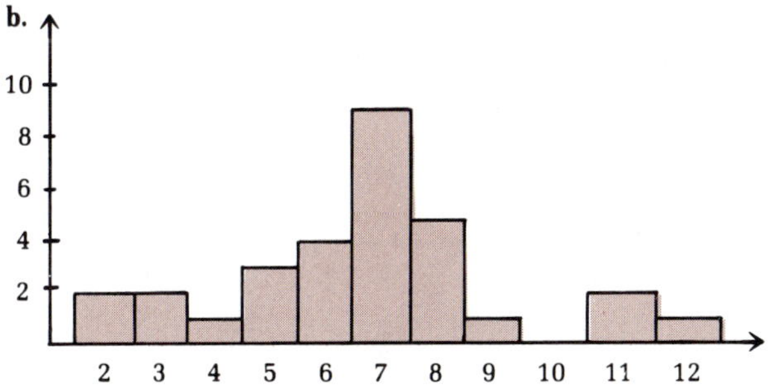

4. Answers may vary.

a.

Class	f_i
4.5–4.9	7
5.0–5.4	5
5.5–5.9	11
6.0–6.4	4
6.5–6.9	4
7.0–7.4	5
7.5–7.9	2
8.0–8.4	2

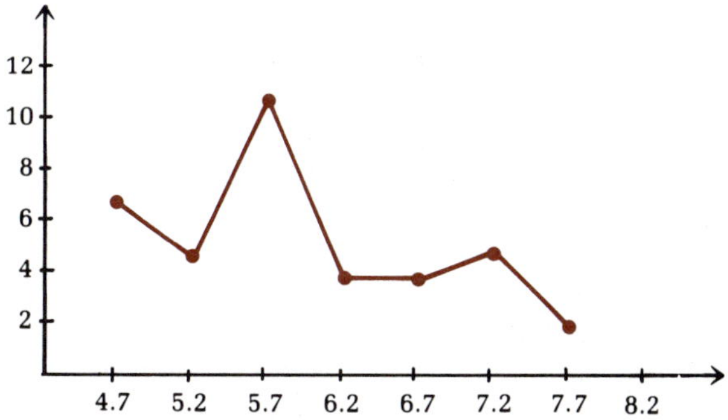

b.

Class	f_i
1.01–1.50	1
1.51–2.00	3
2.01–2.50	4
2.51–3.00	15
3.01–3.50	9
3.51–4.00	8

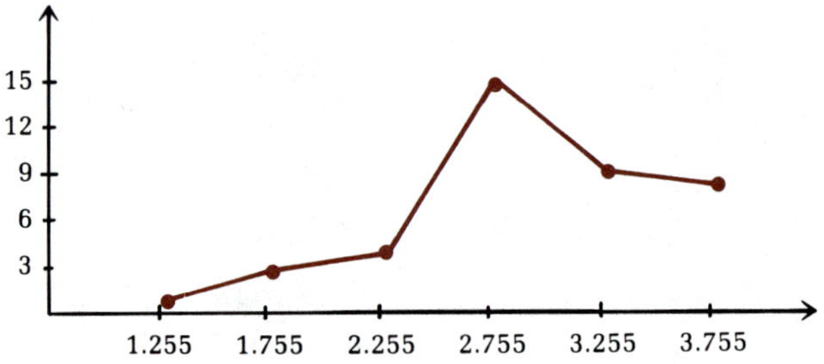

5.

Class	cf_i
4.5–4.9	7
5.0–5.4	12
5.5–5.9	23
6.0–6.4	27
6.5–6.9	31
7.0–7.4	36
7.5–7.9	38
8.0–8.4	40

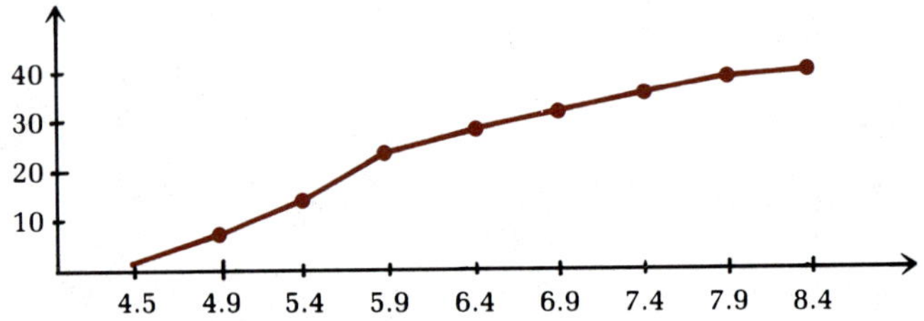

6.

Class	cf_i
1.01–1.50	1
1.51–2.00	4
2.01–2.50	8
2.51–3.00	23
3.01–3.50	32
3.51–4.00	40

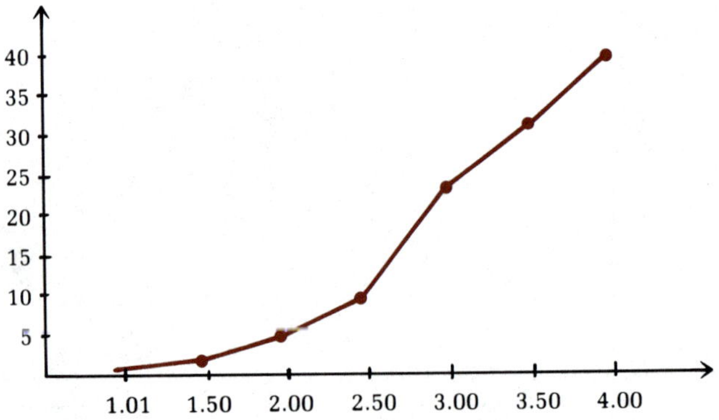

7. a.

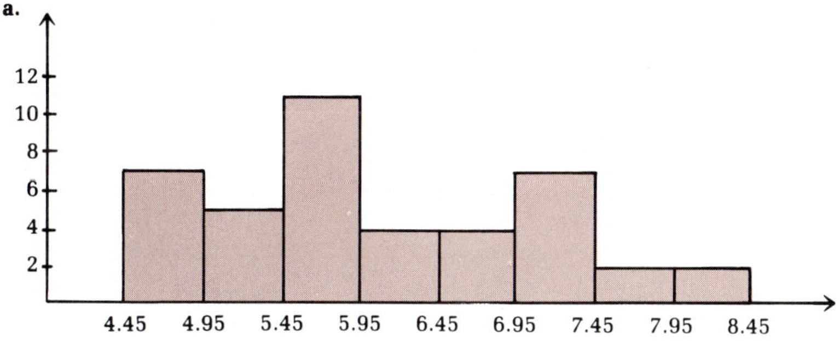

b.

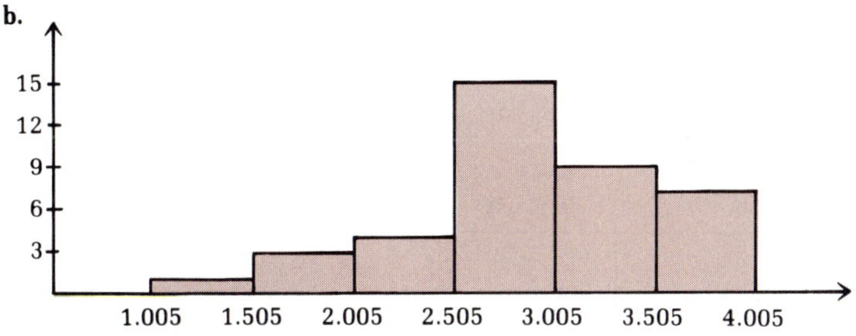

8.

x_i	rf_i	crf_i
0	.075	.075
1	.175	.250
2	.250	.500
3	.225	.725
4	.175	.900
5	.100	1.000

9. a. 89.2 or 89 **b.** 53.6 or 54 **c.** 97.6 or 98 **d.** 12 **e.** 10 **f.** 8
10. for 2(a) – 2 for 4(a) – modal class 5.5–5.9 (actual 5.8)
for 2(b) – 7 for 4(b) – modal class 2.51–3.00 (actual 2.96)
11. a. 5.85 **b.** 2.96 **12. a.** $\bar{x} = 6.07$, $s^2 = 0.9375$, $s = 0.9683$ **b.** $\bar{x} = 6.025$, $s^2 = 1.0455$, $s = 1.0225$
13. a. $\mu = 2.91$, $\sigma^2 = 0.3895$, $\sigma = 0.6241$ **b.** $\mu = 2.91$, $\sigma^2 = 0.3900$, $\sigma = 0.6245$ **14.** $\mu = 9.8$, $\sigma^2 = 8.5360$, $\sigma = 2.9216$ **15. a.** mode 84, median 76 **b.** $\mu = 73.4$, $\sigma^2 = 218.1775$, $\sigma = 14.7708$
16. a. trimodal—93, 90, 83, median 83 **b.** $\mu = 79.0$, $\sigma^2 = 237.3627$, $\sigma = 15.4066$ **17. a.** 0.217 **b.** 1.194 **c.** -1.520 **d.** 2.172 **e.** -2.172 **f.** -0.651 **18. a.** 0.967 **b.** -2.072 **c.** 1.796 **d.** 3.453 **e.** -0.967 **f.** 2.348 **19. a.** highest—math **b.** lowest—accounting **20. a.** hopping **b.** carrying water

Chapter 8

1.

x_i	$P(X = x_i)$
10	$\frac{25}{50} = \frac{1}{2} = 0.5$
20	$\frac{15}{50} = \frac{3}{10} = 0.3$
30	$\frac{6}{50} = \frac{3}{25} = 0.12$
40	$\frac{3}{50} = 0.06$
50	$\frac{1}{50} = 0.02$

2.

x_i	$P(X = x_i)$
0	$\frac{\binom{9}{3}}{\binom{20}{3}} = \frac{7}{95} = .074$
1	$\frac{\binom{11}{1}\binom{9}{2}}{\binom{20}{3}} = \frac{33}{95} = .347$
2	$\frac{\binom{11}{2}\binom{9}{1}}{\binom{20}{3}} = \frac{33}{76} = .434$
3	$\frac{\binom{11}{3}}{\binom{20}{3}} = \frac{11}{76} = .145$

3.

x_i	$P(X = x_i)$
0	$\binom{3}{0}\left(\frac{11}{20}\right)^0\left(\frac{9}{20}\right)^3 = .099$
1	$\binom{3}{1}\left(\frac{11}{20}\right)\left(\frac{9}{20}\right)^2 = .334$
2	$\binom{3}{2}\left(\frac{11}{20}\right)^2\left(\frac{9}{20}\right) = .408$
3	$\binom{3}{3}\left(\frac{11}{20}\right)^3\left(\frac{9}{20}\right)^0 = .166$

4. a.

x	$P(X = x)$
0	$\frac{1}{32} = 0.03125$
1	$\frac{5}{32} = 0.15625$
2	$\frac{10}{32} = 0.31250$
3	$\frac{10}{32} = 0.31250$
4	$\frac{5}{32} = 0.15625$
5	$\frac{1}{32} = 0.03125$

b.

$y = x + 5$	$P(Y = y)$
5	$\frac{1}{32}$
6	$\frac{5}{32}$
7	$\frac{10}{32}$
8	$\frac{10}{32}$
9	$\frac{5}{32}$
10	$\frac{1}{32}$

c.

$z = x^2$	$P(Z = z)$
0	$\frac{1}{32}$
1	$\frac{5}{32}$
4	$\frac{10}{32}$
9	$\frac{10}{32}$
16	$\frac{5}{32}$
25	$\frac{1}{32}$

5. a. 14 **b.** 4.2 **c.** 2.049 **d.** 0.417 **e.** 0.048 **f.** 0.112 **g.** 0.535 **h.** 0.964 **i.** 0.779 **j.** 0.381 **6. a.** 7.6 **b.** 4.56 **c.** 2.135 **d.** 0.180 **e.** 0.180 **f.** 0.657 **g.** 0.913 **h.** 0.034 **i.** 0.802 **j.** 0.809 **7.** P(all had shots) = $\binom{5}{5}(.8)^5(.2)^0 = .3277$ or .33, P(at least one has not) = 1 − P(all have) = 1 − .33 = .67 **8. a.** .346 **b.** .087 **c.** .663 **d.** .078 **9. a.** .552 **b.** 1 − .552 = .448 **c.** .884 **d.** .011 **10. a.** .166 **b.** .833 **c.** .633 **11. a.** 0.998 **b.** 0.055 **c.** 0.184 **12. a.** 0.043 **b.** 0.698 **c.** 0.758 **d.** 0.044 **13. a.** 0.056 **b.** 0.438 **c.** 0.760 **d.** 0.981 **14. a.** 0.570 **b.** 0.022 **c.** 0.570 **d.** 0 **15. a.** 0.0885 **b.** 0.8438 **c.** 0.9049 **d.** 0.2266 **e.** 0.1195 **f.** 0.1370 **g.** 0.6218 **h.** 0.8836 **i.** 0.8647 **j.** 0.3849 **k.** 0.4564 **l.** 0.4997 **16. a.** 0.9525 **b.** 0.2514 **c.** 0.9082 **d.** 0.1151 **e.** 0.6653 **f.** 0.3669 **g.** 0.6280 **h.** 0.2476 **17. a.** 0.3264 **b.** 0.6904 **c.** 0.2514 **18. a.** 0.0062 **b.** 0.0062 **c.** 0.7888 **d.** 0.9876 **19. a.** 0.0668 **b.** 0.6915 **c.** 0.0228 **d.** 0.3830 **20. a.** 0.0017 **b.** 0.0256 **c.** 0.9854 **d.** 0.6730 **21. a.** 0.0062 **b.** 0.1056 **c.** .0009 **d.** 0.9920 **22. a.** 0.8944 **b.** 0.0301 **c.** 0.1736 **d.** 0.7208 **23. a.** 320 **b.** 0.0052 **c.** 0.0073 **d.** 0.8997 **24. a.** 300 **b.** 0.9608 **c.** 0.0026 **d.** 0.7435

Chapter 9

1. a. $x = \frac{1}{2}, y = \frac{2}{3}$ **b.** $x = 0.1, y = 0.3, z = 0.4$ **c.** Not a transition matrix, since in row 1, $\left(\frac{1}{2}\right) + \left(\frac{1}{3}\right) + \left(\frac{1}{4}\right) \neq 1$. **2. a.** $[14 \quad 11 \quad 11]$ **b.** $[12.5 \quad 14.5 \quad 9]$ **3. a.** $\begin{bmatrix} \frac{1}{2} & \frac{1}{2} \\ \frac{1}{3} & \frac{2}{3} \end{bmatrix}$ **b.** April 320, May 313 **c.** $[312 \quad 468]$ **d.** 40% **4. a.** $[\frac{1}{8} \quad \frac{7}{8}]$ **b.** $[\frac{18}{43} \quad \frac{9}{43} \quad \frac{16}{43}]$ **5.** State 2 gets everything, i.e. $[0 \quad 1 \quad 0]$ **6.** No **7.** State 1 is inactive and neither gains nor loses. States 2 and 3 form a 2×2 regular transition matrix whose steady state is $[\frac{3}{5} \quad \frac{2}{5}]$. Hence the solution is $[\frac{1}{2} \quad \frac{3}{10} \quad \frac{2}{10}]$. That is, state 1 keeps its $\frac{1}{2}$, state 2 keeps $\frac{3}{5}$ of the remaining half $[(\frac{3}{5}) \cdot (\frac{1}{2}) = \frac{3}{10}]$, state 3 keeps $\frac{2}{5}$ of the remaining half $[(\frac{2}{5}) \cdot (\frac{1}{2}) = \frac{2}{10}]$. **8. a.**

$$\begin{array}{c|cc|cc} & B & D & A & C \\ \hline B & 1 & 0 & 0 & 0 \\ D & 0 & 1 & 0 & 0 \\ \hline A & .3 & .1 & .4 & .2 \\ C & .1 & 0 & .2 & .7 \end{array}$$

b. $R = \begin{bmatrix} .3 & .1 \\ .1 & 0 \end{bmatrix}, Q = \begin{bmatrix} .4 & .2 \\ .2 & .7 \end{bmatrix}$

c. $N = \left(\frac{5}{7}\right)\begin{bmatrix} 3 & 2 \\ 2 & 6 \end{bmatrix}$ **d.** $\begin{array}{cccc} A & B & C & D \\ [0 & 770 & 0 & 230] \end{array}$ **9.** $24[\frac{1}{4} \quad \frac{1}{6} \quad \frac{1}{4} \quad \frac{1}{3}] = [6 \quad 4 \quad 6 \quad 8]$ Tank 1 should hold 6 gallons, tank 2 should hold 4 gallons, tank 3 should hold 6 gallons, and tank 4 should hold 8 gallons. **10.** $36[\frac{1}{4} \quad \frac{1}{6} \quad \frac{1}{4} \quad \frac{1}{3}] = [9 \quad 6 \quad 9 \quad 12]$
11. $24[.1805 \quad .2569 \quad .3122 \quad .2504] \approx [4.3 \quad 6.2 \quad 7.5 \quad 6.0]$.

Chapter 10

1. a. [3] **b.** [−1] **c.** Cannot be reduced **d.** $\begin{bmatrix} 1 & 2 \\ 3 & -1 \\ 4 & -2 \end{bmatrix}$ **e.** $\begin{bmatrix} 1 & 2 \\ 3 & -1 \end{bmatrix}$ **f.** $\begin{bmatrix} 3 & -1 \\ -2 & 1 \end{bmatrix}$ **g.** $\begin{bmatrix} 1 & -1 \\ -2 & 1 \end{bmatrix}$ **h.** $\begin{bmatrix} 3 & 1 & 2 \\ -2 & 4 & 3 \end{bmatrix}$ **i.** [2] **j.** $\begin{bmatrix} 1 & 0 & 4 \\ 1 & 4 & 3 \\ 4 & 0 & 3 \end{bmatrix}$ **2. a.** Has saddle point 3 in (1,2) entry **b.** Has saddle point −1 in (1,2) entry **i.** Has saddle point 2 in (3,2) entry **3. a.** 2 in (3,2) entry **b.** 1 in (1,1) entry **c.** 0 in (1,2) entry **d.** 2 in (1,3) entry **e.** 1 in (3,2) entry **f.** No saddle point entry

4. $\begin{array}{ccc} & R & D \\ \end{array}$ $\begin{bmatrix} -2 & 2 & -4 \\ 1 & -1 & 2 \\ 2 & -2 & 4 \end{bmatrix}$ **5.** $\begin{array}{c} \quad 6L \quad 6H \quad 8L \quad 8H \end{array}$ $\begin{bmatrix} 5 & 1 & -1 & 3 & -3 \\ 7 & -1 & 1 & 1 & -1 \\ 9 & -3 & 3 & -1 & 1 \\ 10 & -4 & 4 & -2 & 2 \end{bmatrix}$ **6. a.** Player $[1 \quad 0]$, opponent $\begin{bmatrix} 0 \\ 1 \end{bmatrix}$, $v = 3$ **b.** Player $[1 \quad 0]$, opponent $\begin{bmatrix} 0 \\ 1 \\ 0 \\ 0 \end{bmatrix}$, $v = -1$ **c.** Player has 2 choices $[\frac{3}{4} \quad \frac{1}{4} \quad 0]$

or $[\frac{3}{4} \quad \frac{1}{8} \quad \frac{1}{8}]$, opponent $\begin{bmatrix} 0 \\ \frac{1}{4} \\ \frac{3}{4} \end{bmatrix}$, $v = \frac{5}{4}$ **d.** Player $[\frac{6}{7} \quad 0 \quad \frac{1}{7}]$, opponent $\begin{bmatrix} \frac{4}{7} \\ \frac{3}{7} \\ 0 \end{bmatrix}$, $v = \frac{10}{7}$ **e.** Player $[0 \quad \frac{1}{3} \quad \frac{2}{3}]$, opponent $\begin{bmatrix} \frac{1}{3} \\ \frac{2}{3} \\ 0 \end{bmatrix}$, $v = \frac{1}{3}$ **f.** Player $[\frac{3}{7} \quad 0 \quad \frac{4}{7}]$, opponent $\begin{bmatrix} \frac{2}{7} \\ \frac{5}{7} \\ 0 \end{bmatrix}$, $v = \frac{1}{7}$ **g.** Player $[0 \quad \frac{3}{5} \quad 0 \quad \frac{2}{5}]$, opponent $\begin{bmatrix} 0 \\ \frac{2}{5} \\ 0 \\ \frac{3}{5} \end{bmatrix}$, $v = -\frac{1}{5}$ **h.** Player $[\frac{3}{4} \quad 0 \quad \frac{1}{4} \quad 0]$, opponent $\begin{bmatrix} \frac{3}{8} \\ \frac{5}{8} \\ 0 \\ 0 \end{bmatrix}$, $v = \frac{7}{4}$

i. Player $[0 \quad 0 \quad 1 \quad 0]$, opponent $\begin{bmatrix} 0 \\ 1 \\ 0 \\ 0 \end{bmatrix}$, $v = 2$ **j.** Player $[0 \quad 0 \quad 0 \quad 0 \quad \frac{4}{7} \quad \frac{3}{7}]$, opponent $\begin{bmatrix} 0 \\ \frac{4}{7} \\ \frac{3}{7} \\ 0 \\ 0 \\ 0 \end{bmatrix}$, $v = \frac{16}{7}$ **7.** Player $A[\frac{1}{3} \quad \frac{2}{3} \quad 0]$, opponent $B\begin{bmatrix} \frac{2}{3} \\ \frac{1}{3} \end{bmatrix}$, $v = 0$. The game is fair. **8.** John $[\frac{3}{5} \quad 0 \quad 0 \quad \frac{2}{5}]$, Jim $= \begin{bmatrix} \frac{1}{2} \\ 0 \\ 0 \\ \frac{1}{2} \end{bmatrix}$, $v = -1$ (Jim wins \$1)

Chapter 11

1. \$128 **2.** 15% **3.** no **4.** His selling price is double his cost. This practice is very common and is referred to by the code word "keystone." 100% mark-up. **5.** Approximately 12% **6.** Approximately $16\frac{1}{2}$% **7. a.** no **b.** $4.2P$ (plan A), $3.972P$ (plan B) **c.** Yes, at the end of this contract the base pay for plan B is $1.452P$ while the base pay for plan A is only $1.4P$. **d.** The answer to this question is subjective. For example, on an initial base salary of \$20,000 Plan A produces over 3 years \$84,000 and a new base salary of \$28,000 per year while Plan B produces over 3 years \$79,440 and a new base salary of \$29,040 per year. **8.** \$2847.20 **9.** It doesn't matter. **10.** Hint: $(P_1 + rP_1) + (P_2 + rP_2) = (1 + r)(P_1 + P_2)$ **11. a.** \$480 **b.** \$519.42 **c.** \$540.47 **12. a.** \$1440 **b.** \$206.67 **13.** At simple interest you have the use of the money for the duration of the loan. Add-on interest charges you interest on money you have paid back. **14.** $P_0 = (20{,}000)(1.15)^{-10} = \$4944$ **15.** $V = 4000\left(1 + \frac{.08}{360}\right)^{720} = \$4693.96$ **16.** \$4694.04 **17. a.** 12.55%

b. 12.68% **c.** 12.75% **18.** $S = \frac{1 - a^n}{1 - a} = \frac{1 - \left(\frac{1}{2}\right)^{11}}{1 - \frac{1}{2}} = \frac{2047}{1024}(2)$ **19. a.** $300\left[\frac{(1 + .03)^{40} - 1}{.03}\right] =$ \$22,620.38 **b.** $(1.03)\left\{300\left[\frac{(1 + .03)^{40} - 1}{.03}\right]\right\} = \$23{,}298.99$ **20.** $PA = 200\left[\frac{1 - (1 + .01)^{-180}}{.01}\right] =$ \$16,664.33 **21.** $VA = 100\left[\frac{(1.005)^{240} - 1}{.005}\right] = \$46{,}204.09$ **22.** \$34,820.49 **23. a.** 42,504.44 **b.** 92.59 **24.** $P_5 = \left[100 - \frac{30}{.1}\right](.9)^5 + \frac{30}{.1} \approx 194$ supporters **25.** Equilibrium. The number of supporters would continue to equal 300.

Index